Sulfur-Containing Polymers

Sulfur-Containing Polymers

From Synthesis to Functional Materials

Edited by
Xing-Hong Zhang and Patrick Theato

WILEY-VCH

Editors

Prof. Xing-Hong Zhang
Zhejiang University
Department of Polymer Chemistry
Polymer Building 421
Yu-Quan Campus
310027 Hangzhou
China

Prof. Patrick Theato
Karlsruhe Institute of Technology (KIT)
Institute for Chemical Technology and Po
Engesser Str. 18, Building 11.
D-76131 Karlsruhe
Germany

Library of Congress Card No.:
applied for

British Library Cataloguing-in-Publication Data:
A catalogue record for this book is available from the British Library.

Bibliographic information published by the Deutsche Nationalbibliothek
The Deutsche Nationalbibliothek lists this publication in the Deutsche Nationalbibliografie; detailed bibliographic data are available on the Internet at <http://dnb.d-nb.de>.

Print ISBN: 978-3-527-34670-7
ePDF ISBN: 978-3-527-82379-6
ePub ISBN: 978-3-527-82380-2
oBook ISBN: 978-3-527-82381-9

Typesetting SPi Global, Chennai, India

Printing and Binding CPI Group (UK) Ltd,

Croydon, CR0 4YY

Printed on acid-free paper

C096834_260221

Contents

Introduction

Jia-Liang Yang[1], Xiao-Han Cao[1], Xing-Hong Zhang[1], and Patrick Theato[2,3]

[1] *MOE Key Laboratory of Macromolecular Synthesis and Functionalization, Department of Polymer Science and Engineering, Zhejiang University, Hangzhou 310027, China*
[2] *Institute for Chemical Technology and Polymer Chemistry, Karlsruhe Institute of Technology (KIT), Engesser Str. 18, D-76131 Karlsruhe, Germany*
[3] *Soft Matter Synthesis Laboratory, Institute for Biological Interfaces III (IBG3), Karlsruhe Institute of Technology (KIT), Hermann-von-Helmholtz-Platz 1, D-76344 Eggenstein-Leopoldshafen, Germany*

1 Scope of This Book

Various elements in nature are continuously cycled through a variety of chemical, biochemical, and physical processes. As the sixth abundant nonmetal element in the earth, sulfur and its natural cycle are closely related to the ecological balance and human life (Figure 1). The ocean is a major part of natural sulfur sources. Sulfur exists in the form of sulfates in the ocean and can also be converted to hydrogen sulfide (H_2S) or carbonyl sulfide (COS), which are released to the atmosphere. In addition to marine sources, natural releases from land including volcanic eruptions, and human activities also emit a variety of sulfur-containing gases into the atmosphere, including methyl sulfide ($(CH_3)_2S$), H_2S, COS, carbon disulfide (CS_2), methyl mercaptan (CH_3SH), and dimethyl disulfide (CH_3SSCH_3) [1, 2]. Sulfur-containing compounds in the atmosphere undergo a series of photochemical reactions and return to the ocean and land through rain. Herein we want to address another conversion route for sulfur-containing compounds: the synthesis of sulfur-containing polymers [3, 4] (Figure 1).

Polymer materials play an incredibly important role in modern life. The richness of polymer materials derives from the diversity of their constituent elements, such as carbon, nitrogen, oxygen, silicon, and sulfur. Unlike other chemical commodities, sulfur is mostly "involuntary" produced. It mainly arises from petroleum refineries and natural gas processing, thus creating a global surplus of sulfur. Production of sulfur reached more than 60 million tons in 2013 with a rising trend [5]. Global sulfur capacity tends to be in excess thus the use of sulfur needs to find a new direction. Actually, sulfur has been recognized as a valuable chemical agent since antiquity: employed as far back as 1600 B.C.E. by the Egyptians to bleach cotton fabric. Since the "chemical" invention of sulfur in explosives by the Chinese, the use of sulfur

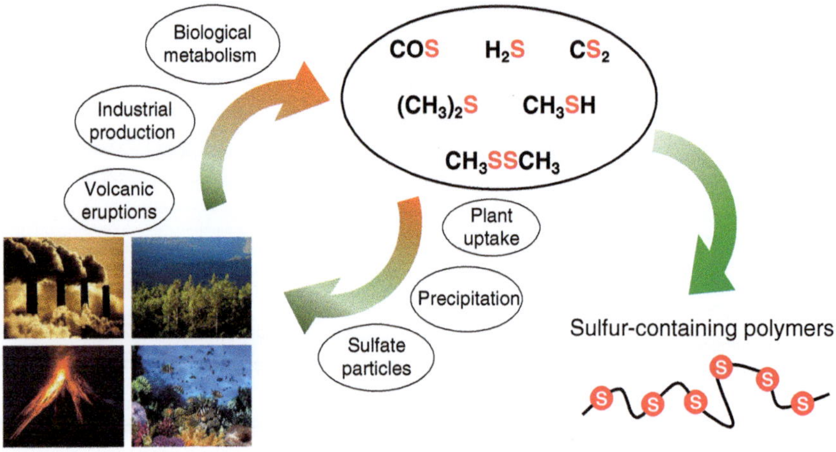

Figure 1 Sulfur cycle in nature and the formation of sulfur-containing polymers.

is closely linked with the development of chemical science. From the raw material of sulfuric acid to rubber industry auxiliaries, the mid-19th century witnessed the first applications of sulfur in polymer chemistry such as the vulcanizing agent of rubber. After the development of polymer chemistry in the last century, sulfur and sulfur-containing moieties have been increasingly involved in the construction of various polymers.

The purpose of this book is to summarize the latest development of sulfur-containing polymers, focusing on the utilization of sulfur-containing compounds and corresponding polymerization methods, analyzing the intrinsic relationship between polymer structure and its properties and applications. This book covers the state-of-the-art syntheses of a spectrum of sulfur-containing polymers from various sulfur resources including elemental sulfur, CS_2, COS, and mercaptan. In-depth mechanistic understanding relating to the chain polymerization is particularly presented for the readers. Of course, various types of sulfur-containing polymers, including poly(thioester)s, poly(thioether)s, poly(thioamide)s, poly(thiocarbonate)s, poly(thiourethane)s, and poly(thiourea)s with linear or hyperbranched (dendrimer) architectures, are discussed. High-tech applications of these sulfur-containing polymers in energy, optical and biomaterials are discussed in detail. This book will provide the latest knowledge in this area in a comprehensive way to the readers including students in the colleges, engineers in research organizations in the polymer community.

2 Sulfur-Containing Polymers

2.1 Building Feedstocks

The presence of sulfur atoms in polymer structure, depending on the kind of functional group, can bring some important properties, such as mechanical, electrical, optical, and then adhesion to metals, resistance to heat, chemicals, radiation, bacteria, biocompatibility, and so on. These are some of the functional groups: sulfide (–S–), polysulfide (–S_n–), sulfonyl (–SO_2–), sulfinyl (–SO–), sulfo (–SO_3H), as well as some organic groups containing sulfur atoms, such as thioester [–(C=O)–S–], monothio- or dithiocarbonate [–O–C(=S)–O–], [–O–(C=O)–S–], or [–S–(C=O)–S–], as well as thiourethane [–NH–(C=O)–S–] or [–NH–(C=S)–O–] and other sulfur-nitrogen groups. Sulfur-containing polymers can be used as high-performance engineering plastics, chemically stable ion-exchange membranes in electromembrane processes, proton-conducting electrolytes, as well as optical, optoelectronic and photochemical materials. Some polymers are employed in biomedical applications, for example, sulfopolymers as biomembranes and blood-compatible materials, polysulfates and polysulfonates as antithrombotic or antiviral agents. The presence of sulfur, particularly in the form of disulfide bridges, has played an important role in biopolymers and self-healing materials.

With a series of synthetic breakthroughs in recent years, new chemistries suitable for the synthesis of various polymers containing sulfur atoms have been developed, and thus many novel materials with promising properties have been reported. As a way to provide sustainable polymers, some sulfur-containing polymers are synthesized from *waste* of petrochemical resources, which is an advance in waste utilization. Over the past 20 years, researchers have synthesized and prepared a wide variety of sulfur-containing polymers, but there are few books on comprehensive introduction of the synthesis and applications of sulfur-containing polymers. The rapid development of this field can be visualized by the increase of research reports published in recent years (Figure 2). However, there are relatively few monographs on sulfur-containing polymers.

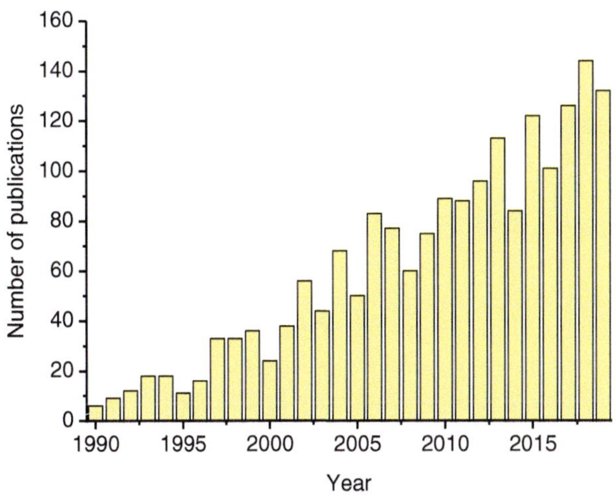

Figure 2 Scientific publications with the key word "sulfur-containing polymers" published from 1990 to 2019 (Web of Science).

Furthermore, books that focus specifically on the synthetic methods and the correlation between polymer microstructure and macroscopic properties are still lacking. This book is intended to meet the development needs in this field, introduce the synthesis methods of various sulfur-containing polymers in detail, and provide directions for the design of new methods and new polymerization systems.

Via the in-depth study of sulfur chemistry, diversified chemical transformation strategies have been developed to synthesize various sulfur-containing compounds, including thiourea, episulfide, and a large category of sulfur-containing medicinal compounds. To date, sulfur can also be used directly in the synthesis of sulfur-containing polymers by multicomponent polymerizations (MCPs). Except for sulfur, many sulfur-containing moieties and their comonomers have been developed for the synthesis of sulfur-containing polymers, which are summarized in Figure 3. Noteworthy, sulfur-containing one-carbon (C1) monomers (e.g. CS_2, COS) are a kind of important resource as two bulk and cheap compounds from which polythiocarbonates, polythioethers, polythioureas, and poly(thioester)s can be obtained by many polymerization methods. Meanwhile, sulfur-containing moieties with disulfide bond, thioester function groups, or basing on thiophene can also provide sulfur-containing polymers with particular properties. For each type of monomer, researchers have developed and advanced various polymerization methods to achieve an accurate and efficient synthesis of the target products.

2.2 Synthetic Methods

Roughly, Chapters 1, 4, 6, and 8 introduce step polymerization processes for sulfur-containing polymers. MCP is an emerging polymer synthetic method that combines three or more monomers in a one-pot reaction, resulting in a highly efficient, huge structural diversity, high atomic economy, and a well-defined structure polymer. By way of elemental sulfur based MCPs (Figure 4, (1)), polythioamides and polythioureas can be synthesized directly, as summarized by Prof. Rong Rong Hu in Chapter 1. While based on other sulfur containing moieties, many polymers with different microstructures can also be produced by MCPs. Sulfonyl group-containing polymers can be derived from the sulfonyl azide or sulfonyl hydrazide-based MCPs, and the sequence-controlled sulfur-containing polymers can be prepared from the multicomponent tandem polymerizations of thiol or sulfur-containing heterocycle monomers.

Except for MCPs, thiol-based click polymerization features remarkable advantages and has also been nurtured into a powerful technique for the preparation of sulfur-containing polymers (Figure 4, (2)), as introduced in Chapter 4 by Prof. An Jun Qin and Prof. Ben Zhong Tang. Owing to the high efficiency of click chemistry, a variety of click reactions can be utilized including thiol-ene, thiol-yne, thiol-epoxy, thiol-isocyanate, and thiol-halogen clicks. Polymerization based on these click reactions have yielded various sulfur-containing polymers. Thiol-ene click polymerization can produce poly(thioether)s via a radical process or thiol-Michael

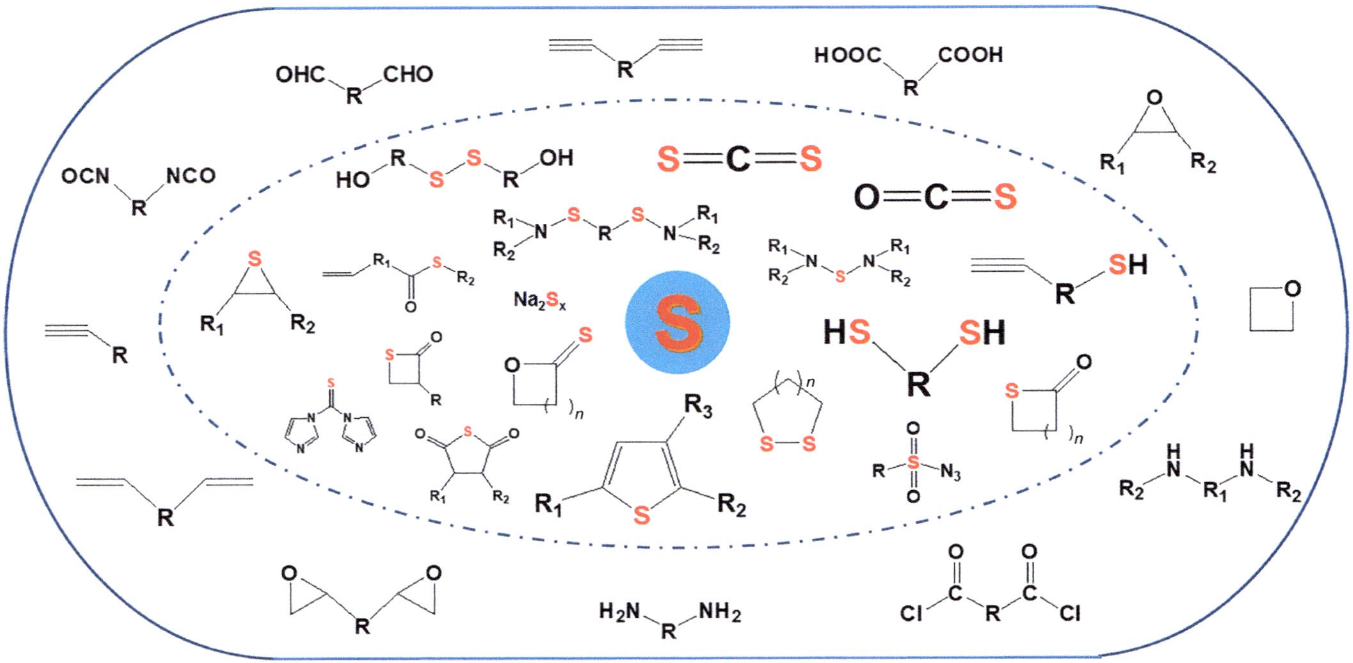

Figure 3 Sulfur-containing compounds and the comonomers.

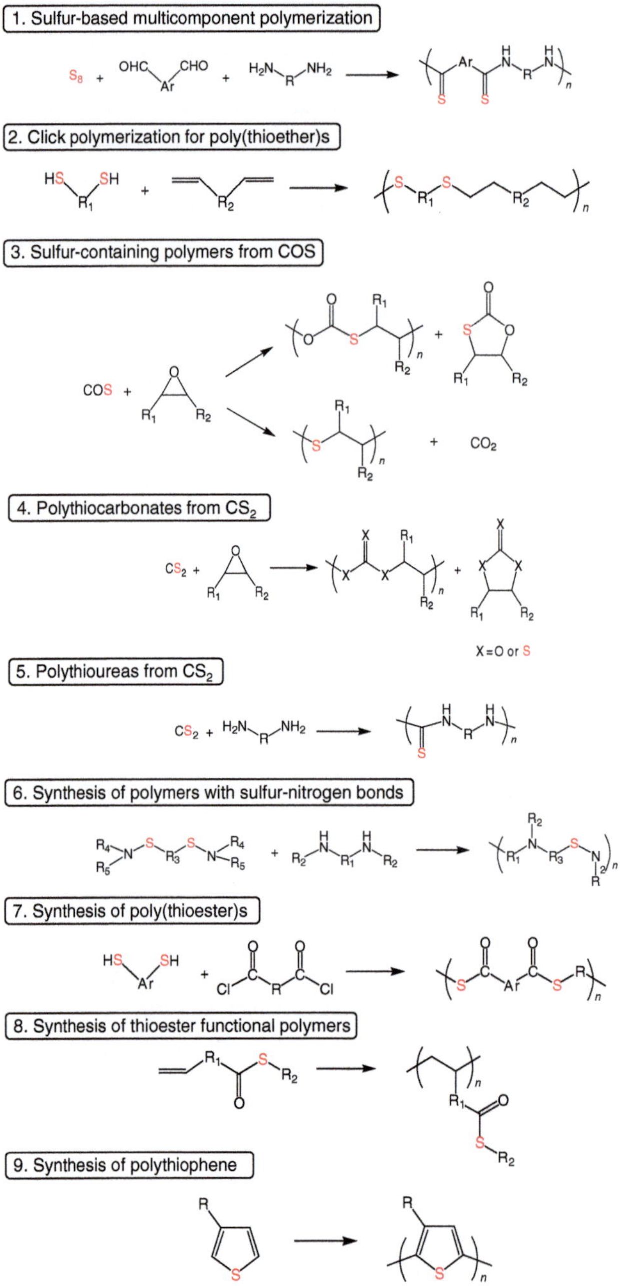

Figure 4* Selected synthetic methods for sulfur-containing polymers.

* IUPAC nomenclature suggests square brackets for repeating units, while this book retain round brackets in some places.

addition process. Thiol-yne click polymerization, a complement of the former, can proceed in a bis-addition or mono-addition manner to furnish poly(thioether)s or poly(vinyl sulfide)s with different regio-/stereoregularity. Other thiol-based click polymerizations, including thiol-epoxy, thiol-isocyanate, and thiol-halogen click polymerizations, can be applied to prepare poly(β-hydroxythioether)s, poly(thiourethane)s, and poly(thioether)s.

Next, polymers containing sulfur-nitrogen functional groups are unique functionalized polymers featuring the advantages of sulfur and nitrogen solely. In Chapter 6, Prof. Patrick Theato discussed the synthetic methods comprehensively for this kind of sulfur-containing polymers, reactions involving disulfenamide derivatives and diamines and other delicate designed are discussed thoroughly. On the other hand, poly(disulfide)s have their utility in biological applications owing to the dynamic nature of the disulfide bond, which offers attractive options for intra-cellular delivery of drugs or other materials specifically in cancer cells. A variety of synthetic methodologies for poly(disulfide)s are introduced in Chapter 11, contributed by Prof. Suhrit Ghosh.

Polythiophene and its derivatives have very promising prospects in organic electronic applications. In Chapter 8, Prof. Xiao Jie Zhang introduced the advanced methods for the synthesis of thiophene-based polymers (Figure 4, (9)). In this chapter, several classical or innovative synthetic methods are introduced, along with their results in the structural characteristics. Some recent methods, like Ni-catalyzed catalyst transfer homopolymerization allowing for a precise control of the regioregularity and the molecular weight, transmetalation copolymerization with remarkable control and coupling methods (e.g. solid-state polymerization) for highly crystallizable material are presented.

Chapters 2, 3, 5, and 7 mainly present the chain polymerization reaction for synthesizing sulfur-containing polymers. CS_2 and COS, an important class of sulfur-containing C1 compounds, have been utilized to copolymerize with epoxides via chain polymerization process, as described in chapters 2 and 3, wrote by Prof. Donald J. Darensbourg and his coworkers and Prof. Xing-Hong Zhang and his coworkers, respectively. Both monomers can be derived directly from sulfur. The inexpensive and abundant COS is industrially synthesized from natural gas and elemental sulfur, meanwhile COS can be derived from CS_2 or sulfur and carbon monoxide. Since 2013, many efforts have been made for developing a variety of new COS-based copolymers, including poly(monothiocarbonate)s and poly(thioether)s (Figure 4, (3)) [6–8]. These materials are excellent candidates for applications as structural or functional materials because of their unique thermal, mechanical, optical, and electrical properties. Utilization of CS_2 in forming polymer materials also attracted more and more attention (Figure 4, (4)) [9]. CS_2 can also be employed in the synthesis of versatile poly(thiourea)s at low temperature and pressure in the presence of amines and pyridine (Figure 4, (5)). Of special note, *oxygen-sulfur exchange reaction* (O/S ER) is a common phenomenon in the copolymerization of CS_2 (or COS) with epoxides (Figure 5). O/S ER has been well investigated for its complex intrinsic quality and great potential functionality. It can be simply defined as the exchange reaction of sulfur

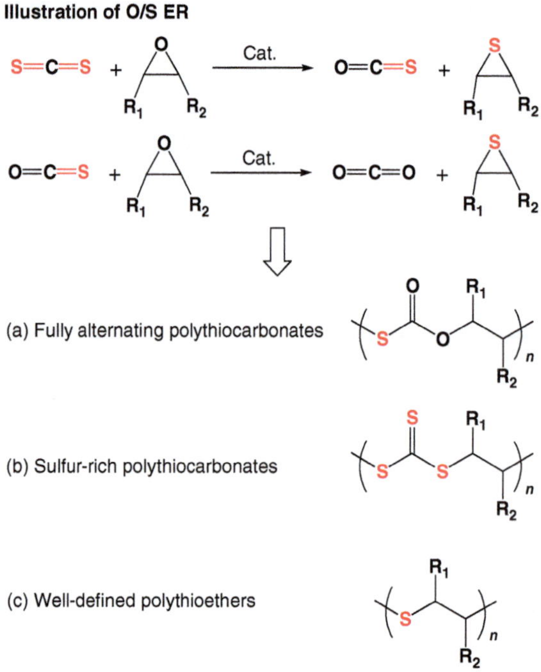

Figure 5 Definition of *oxygen-sulfur exchange reaction* (O/S ER) and application of O/S ER for synthesizing sulfur-containing polymers.

atom in CS_2 (COS) with the oxygen atom in epoxides generates COS (CO_2) and episulfides under the specific catalytic conditions. By inhibiting O/S ER, fully alternating poly(monothiocarbonate)s (Figure 5a) can be synthesized from the copolymerization of COS with epoxides. Importantly, O/S ER can also be utilized to synthesize well-defined poly(thioether)s from COS (or CS_2) as the sulfur source (Figure 5b). A deep understanding of O/S ER is conducive to enriching the preparation of sulfur-containing polymers; meanwhile it can also serve as an efficient way to produce high value-added, functional materials from renewable resources.

In synthetic methodology, *supramolecular anion* strategy has been proposed for controlling the polymerization of oxygen-containing monomers via organocatalysis. It has also been developed for polymerization of many types of monomers with heteroatoms (e.g. O, S, N). Supramolecular anions incorporate Lewis acids in the growing species, such as triethyl borane (TEB) or (thio)ureas [(T)Us]. In these systems, the basicity and nucleophilicity of these growing anions capped with TEB or (T)Us can also be mediated by the types, structures of these Lewis acids, as well as the related supramolecular interactions that are typically involved with hydrogen and organic coordination bonds. That is, the performance of the growing oxyanions can be regulated by altering these neutral Lewis acids.

Thioester-containing polymer is an emerging polymer having sulfur-substituted ester group. In Chapter 5, Prof. Wei Min Ren and his coworkers describe the recent research progress on synthesizing polythioesters, including aromatic, semi-aromatic, and aliphatic polymers. They focus on the copolymerization of cyclic thioanhydride with epoxides or episulfides to afford the polythioesters with well-defined structures. On the other hand, the synthesis of polymers having thioester groups in different position in the polymer backbone is well demonstrated by Prof. Remzi Becer, in Chapter 7. They introduced several types of thioester-containing polymers, such as thioester groups in the unit (via ring-opening polymerization of cyclic thioesters, condensation, or thia-Michael polymerization), side chain (using acrylate monomers), and the end of chain (using ATRP or RAFT initiators). The introduction of thioesters will change the flexibility of the backbone or the sidechain in contrast to its ester or amide analogues. The side or end thioesters groups can also be converted to the free thiol via post modification. These variations of thioester-containing polymers bring the opportunities towards fantastic applications.

Selenium is a semimetallic element lying in the group XVI (chalcogen element, with O and S) of the periodic table. Of course people want to incorporate selenium element into polymer backbone for achieving special polymers. In this book, Prof. Hua Ping Xu and his coworkers provide a special chapter demonstrating the synthesis and properties of selenium-containing polymers via step growth polymerization, radical polymerization, and ring-opening polymerization in recent years (see Chapter 10). Impressively, this chapter described a series of selenium-containing dynamic covalent bonds like Se—Se, Se—S, Se—Te, and Se—N and the applications in protein-based, intelligent, and adaptive materials.

2.3 Applications

Sulfur-containing polymers possess various classes and cover a broad property range, such as excellent thermal stability, high refractivity, and selective absorption to some heavy metal ions, self-healing property, enhanced dielectrical property, and unique electrochemical property. Sulfur-containing polymers have attracted increasing scientific attention and industrial interest in the past two decades. Sulfur-containing polymers generally possess large number of heteroatom-containing functional groups such as thioamides, thioureas, thiophene, thiazol, thiocarbonates, thioesters, thioethers, sulfonamide, and sulfonyl groups, which endow the polymer materials with the aforementioned properties. The authors in this book have discussed the applications of those sulfur-containing polymers.

Of special note, sulfur plays a crucial role in the area of high refractive index polymers (HRIPs) owing to the high polarizability and chemical versatility. New materials with improved optical properties, e.g. a high refractive index, low dispersion, and low birefringence, have been reported from sulfur containing HRIPs. Sulfur moieties can be incorporated into all relevant optical polymer classes such as polyimides and polyamides, polycarbonates and polyesters, poly(meth)acrylates, and new polymers introduced in this book to improve their optical properties. Details of HRIP materials are summarized in Chapter 9, which was contributed by Prof. Patrick Theato.

Owing to the sulfur-metal coordination, the copolymers of thioamides (—NH—(C=S)—NH—) show excellent metal-ion removal ability and can extract Pd and Au salts from their aqueous solution in an easy and efficient way (see Chapter 1). Polymers with disulfide groups have also been used as self-healing materials result from sulfur–sulfur interactions, which is a research hotspot in the field of polymers.

This book also emphasizes the application of sulfur-containing polymers in medical and biomaterials areas. Disulfide-containing polymers also had been exploited extensively for applications of targeted drug and gene carriers, bioimaging, drug conjugates, and self-healing materials, which are intensively discussed in Chapter 12 by Prof. De Cheng Wu. With proper design of these properties and other functionalities integrated in the monomer structures, a group of practical applications such as optical materials, optoelectronic devices, biological materials, self-healing materials, metal ion removal materials, engineering plastics, and rechargeable batteries could be realized (Figure 6). However, not all reports provide full insight into the properties of sulfur-containing polymers, which may reduce the attractiveness of new materials in terms of follow-up investigations and ultimately their real-life applications. In this book, we selected and highlighted

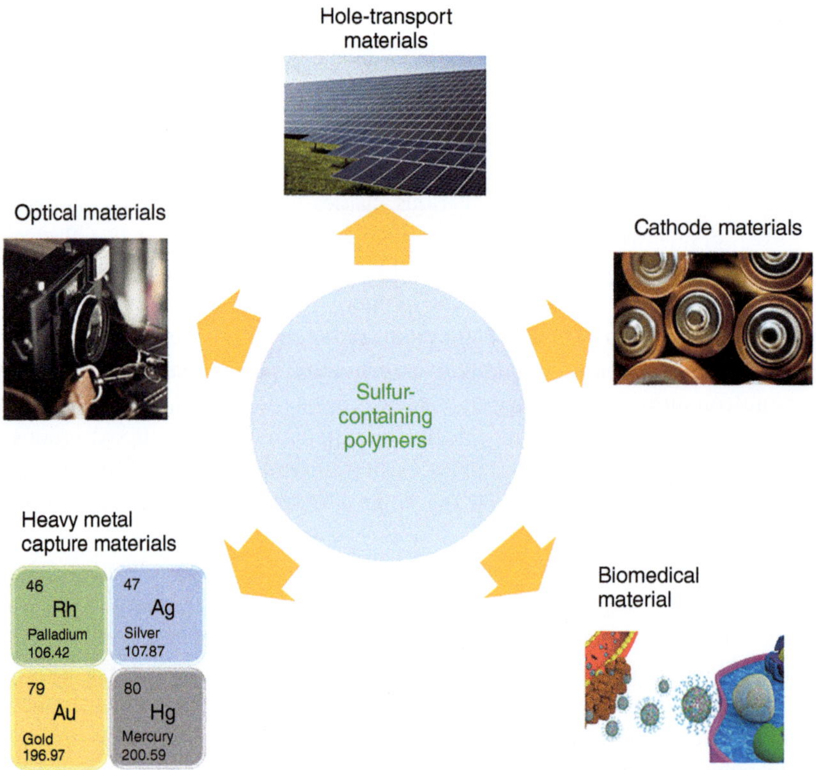

Figure 6 The application of sulfur-containing polymers in diverse areas.

some typical examples to raise the reader's interest in these newly emerged polymer structures.

3 Concluding Remarks

To provide readers with an in-depth and systematic understanding of sulfur-containing polymers, synthetic methods and applications of various sulfur-containing polymers are summarized in this book. Due to the extensive attention received by sustainable polymers in scientific research and industrial production, this book focuses on sulfur-containing monomers from natural sources, including sulfur, CS_2, and COS. In order to promote the rapid development of synthetic methods for sulfur-containing polymers and enrich the types and properties of sulfur-containing polymers, relationship between the synthetic methods and polymer structure as well as polymer structure and its property are emphasized in each chapter. To the best of our knowledge, this book might be the first monograph that comprehensively discusses sulfur-containing polymers, meeting the demands of the rapid development in this field. We are particular grateful to all contributing authors for their hard work to have this book completed as scheduled for the readers.

While it must be pointed out that the sulfur-containing polymers covered in this book are part of them [10, 11], for example, very recent research results in sulfur-containing chitin and chitosan derivatives field have not been included [12]. Also polyphenylene sulfide, which has been industrially produced on a large scale, is not discussed in detail [13]. This book also does not involve the direct application of elemental sulfur in free radical mechanisms [14]. Meanwhile, another well-known application of sulfur-containing polymers is not collected in this book, polymers with high sulfur contents are a class of electroactive cathode materials for Li–S batteries, which exhibit high specific capacity and energy density. Sulfonated polymers generally possess high proton conductivity and mechanical strength, which have been employed as ion-exchange membranes in fuel cell. Details can be found in recent review papers [15–19]. Nevertheless, we hope that this book could be useful to the readers in the entire field in sulfur chemistry.

References

1 Lliott, S. and Rowland, F.S. (1990). The effect of metal complexation on hydrogen sulfide transport across the sea-air interface. *Atmos. Chem.* 10: 315–327.

2 Bates, T.S., Cline, J.D., Gammon, R.H., and Kelly-Hansen, S.R. (1987). Regional and seasonal variations in the flux of oceanic dimethylsulfide to the atmosphere. *J. Geophys. Res.* 92 (C3): 2930.

3 Boyd, D.A. (2016). Sulfur and its role in modern materials science. *Angew. Chem.* 55 (50): 15486–15502.
4 Griebel, J.J., Glass, R.S., Char, K., and Pyun, J. (2016). Polymerizations with elemental sulfur: a novel route to high sulfur content polymers for sustainability, energy and defense. *Prog. Polym. Sci.* 58: 90–125.
5 Chung, W.J., Griebel, J.J., Kim, E.T. et al. (2013). The use of elemental sulfur as an alternative feedstock for polymeric materials. *Nat. Chem.* 5 (6): 518–524.
6 Luo, M., Zhang, X.-H., and Darensbourg, D.J. (2016). Poly(monothiocarbonate)s from the alternating and regioselective copolymerization of carbonyl sulfide with epoxides. *Acc. Chem. Res.* 49 (10): 2209–2219.
7 Luo, M., Zhang, X.-H., Du, B.-Y. et al. (2013). Regioselective and alternating copolymerization of carbonyl sulfide with racemic propylene oxide. *Macromolecules* 46 (15): 5899–5904.
8 Zhang, C.-J., Zhu, T.-C., Cao, X.-H. et al. (2019). Poly(thioether)s from closed-system one-pot reaction of carbonyl sulfide and epoxides by organic bases. *J. Am. Chem. Soc.* 141 (13): 5490–5496.
9 Zhang, X.-H., Liu, F., Sun, X.-K. et al. (2008). Atom-exchange coordination polymerization of carbon disulfide and propylene oxide by a highly effective double-metal cyanide complex. *Macromolecules* 41 (5): 1587–1590.
10 Kausar, A., Zulfiqar, S., and Sarwar, M.I. (2014). Recent developments in sulfur-containing polymers. *Polym. Rev.* 54 (2): 185–267.
11 Li, C., Cheng, J., Yang, F. et al. (2013). Synthesis and cationic photopolymerization of a difunctional episulfide monomer. *Prog. Org. Coat.* 76 (2–3): 471–476.
12 Yong, S.K., Bolan, N.S., Lombi, E. et al. (2013). Sulfur-containing chitin and chitosan derivatives as trace metal adsorbents: a review. *Crit. Rev. Environ. Sci. Technol.* 43 (16): 1741–1794.
13 Lenz, R.W., Handlovits, C.E., and Smith, H.A. (1962). Phenylene sulfide polymers. 3. Synthesis of linear polyphennylene sulfide. *J. Polym. Sci. Part A-2 Polym. Phys.* 58: 351–367.
14 Lim, J., Pyun, J., and Char, K. (2015). Recent approaches for the direct use of elemental sulfur in the synthesis and processing of advanced materials. *Angew. Chem. Int. Ed.* 54 (11): 3249–3258.
15 Kim, H., Lee, J., Ahn, H. et al. (2015). Synthesis of three-dimensionally interconnected sulfur-rich polymers for cathode materials of high-rate lithium-sulfur batteries. *Nat. Commun.* 6: 7278.
16 Pirayesh, A., Salami-Kalajahi, M., Roghani-Mamaqani, H., and Najafi, F. (2018). Polysulfide polymers: synthesis, blending, nanocomposites, and applications. *Polym. Rev.* 59 (1): 124–148.
17 Liu, J., Wang, M., Xu, N. et al. (2018). Progress and perspective of organosulfur polymers as cathode materials for advanced lithium-sulfur batteries. *Energy Storage Mater.* 15: 53–64.

18 Asensio, J.A., Sanchez, E.M., and Gomez-Romero, P. (2010). Proton-conducting membranes based on benzimidazole polymers for high-temperature PEM fuel cells. A chemical quest. *Chem. Soc. Rev.* 39 (8): 3210–3239.

19 Li, Q., He, R., Jensen, J.O., and Bjerrum, N.J. (2003). Approaches and recent development of polymer electrolyte membranes for fuel cells operating above 100 degrees. *Chem. Mater.* 15: 4896–4915.

1

Synthesis of Sulfur-Containing Polymers Through Multicomponent Polymerizations

Yuzhang Huang[1], Rongrong Hu[1], and Ben Zhong Tang[1,2]

[1] *South China University of Technology, State Key Laboratory of Luminescent Materials and Devices, Guangdong Provincial Key Laboratory of Luminescence from Molecular Aggregates, Center for Aggregation-Induced Emission, No. 381 Wushan Road, Tianhe District, Guangzhou 510640, China*
[2] *The Hong Kong University of Science & Technology, Hong Kong Branch of Chinese National Engineering Research Center for Tissue Restoration and Reconstruction, Department of Chemistry, Clear Water Bay, Kowloon, Hong Kong, China*

1.1 Introduction

Sulfur-containing polymers represent a group of popular advanced materials that have attracted much attention owing to their broad range of fascinating properties such as high refractivity [1], coordination with metal ion [2], self-healing property [3], dielectrical property [4], and unique electrochemical property [5]. The well-known sulfur-containing polymers include polythioethers [6], polythiophenes [7], polyoligosulfides [8], and vulcanized rubbers [9], which are generally prepared from classical thiol-ene/yne addition reactions [10], thiol-epoxy reaction [11], transition-metal catalyzed polycouplings [12], and other ring-opening reactions [13]. High sulfur-content materials with excellent optical properties were also reported to be prepared through inverse vulcanization with elemental sulfur [14, 15]. However, the widely studied sulfur-containing polymers are still quite limited. Sulfur-containing polymers with new and complex structures remain rare, and their design and synthesis still face many challenges such as complicated and unstable monomer structures [16], poor solubility of polymers [17], chaotic product structures [18], etc. One effective approach is to develop new sulfur-containing polymer structures based on the exploration of new synthetic methodologies.

Multicomponent polymerizations (MCPs) as emerging polymer synthetic approaches combine three or more monomers in a one-pot reaction to construct complex polymers with high efficiency, great structural diversity, high atom economy, and well-defined structures [19]. During the past decade, with the inspiration from the progress in organic multicomponent reactions (MCRs), great effort has been made by polymer chemists to solve the problems regarding polymer solubility, structural regularity, narrow monomer scope, low conversion, molecular weight

Sulfur-Containing Polymers: From Synthesis to Functional Materials,
First Edition. Edited by Xing-Hong Zhang and Patrick Theato.

control, inhibition of side reactions, sequence control and topological control of the structures, and so on, and a series of efficient MCPs have been developed [20]. For example, the three-component Passerini polymerization of dialdehyde, dicarboxylic acids, and isocyanides was reported to access a class of polyesters [21]. Different combinations of various monomers were also applied to obtain sequence-regulated poly(ester-amine)s and hyperbranched polymers [22, 23]. Many other MCPs such as Ugi four-component polymerizations [24], Biginelli polymerizations [25], Kabachnik–Fields polymerizations [26], alkyne, aldehyde, and amine A^3-polycouplings [27], and a group of alkyne-based MCPs were reported [28–32]. Moreover, through diversity-oriented synthesis from different combinations of multiple types of monomers [33, 34], MCPs can easily afford libraries of polymer materials with structural similarities.

Recently, MCPs prove to be powerful and popular tools for the construction of sulfur-containing polymers with unique structures and functionalities. The sulfur-related MCPs have been developed rapidly nowadays, and their recent progress will be summarized in this chapter. With their rich chemical properties and high reactivity, sulfur-containing reactive reagents including elemental sulfur, sulfonyl azides, sulfonyl hydrazides, thiols, and cyclic thiolactones were utilized as the sulfur sources of the MCPs; newly reported organic MCRs such as sulfur and amine-based MCRs, sulfonyl azide and alkyne-based MCRs, and thiol-related addition or ring-opening reactions were applied in polymer synthesis as new MCPs; sulfur-containing linear or hyperbranched polymers with unique structures such as polythioamides, polythioureas, polysulfonylamides, polysulfonylimines, and polythioethers have been accessed, which were proved to find potential applications in transition metal ion detection, removal, enrichment, cell imaging, and so on. This chapter will only focus on the synthesis, structures, and functionalities of sulfur-containing polymers prepared from MCPs, and other synthetic approaches or polymer structures will be introduced in other chapters.

1.2 Multicomponent Polymerizations of Elemental Sulfur

Elemental sulfur, as the abundantly existed and inexpensive nature source, is an ideal raw material for the preparation of sulfur-containing functional polymer materials. Large amount of sulfur was produced from the worldwide petroleum industry and the major applications of sulfur are limited to the production of few chemical commodities such as sulfuric acid, fertilizers, and vulcanized rubber [35]. Whereas there are several allotropes of elemental sulfur existed in nature, the well-known stable species of sulfur at room temperature is the crown-shaped eight-member ring S_8 with rich chemical properties. For example, sulfur ring could be easily opened with the nucleophilic attack of amine to form polysulfide anion (R-NH-S_x^-) with high reactivity [36], which can act as new nucleophile to attack other active centers to afford various product structures. Attracted by the unique reactivity of elemental sulfur and amine, much effort has been made to develop sulfur and amine-based MCPs for the preparation of sulfur-containing polymers.

Of all the sulfur-containing polymers, polythioamides are unique types that share similar chemical structures with the widely used traditional polyamide materials. Thioamides, as analogues of amides, also play important roles in organic chemistry [37]. Introducing thioamide group into the repeating unit of the polymer backbone to afford polythioamides is desired considering the fascinating properties such as selective coordination with transition metal ions and the potential hydrogen bonds those could be brought by the NH—C=S moieties [38, 39]. However, the synthesis of polythioamides remained to be a great challenge. To the best of our knowledge, polythioamides were only reported to be prepared from bis(thioester)s or bis(dithioester)s and diamines with harmful byproduct released [40, 41], until recent elemental sulfur-based MCP approaches emerged as to be discussed below.

1.2.1 Multicomponent Polymerization of Sulfur, Dialdehydes, and Diamines

Willgerodt–Kindler reaction of elemental sulfur, aldehyde, and amine reported in 1923 is a classical MCR for the synthesis of thioamides [42]. This reaction can perform smoothly upon heating without any metal catalyst, and the readily available commercial sulfur, aldehyde, and amine reactants could easily generate a large number of diverse thioamide structures from this MCR. Willgerodt–Kindler reaction was hence firstly introduced into polymer synthesis for the construction of polythioamides in 1999 (Scheme 1.1a) [43, 44]. In this reaction, polysulfide anion **A** was firstly formed from the nucleophilic attack of amine to the sulfur ring, which was then reacted with the Schiff base **B** formed from aldehyde and amine, to furnish thioamide.

The MCP of elemental sulfur **1**, isophthalaldehyde **2a**, and 4,4′-trimethylenedipiperidine **3a** was firstly carried out in *N,N*-dimethylformamide (DMF) at 115 °C for one hour to afford polythioamide with a molecular weight (M_w) of 15 300 g mol^{-1} in 85% yield (Scheme 1.1b). The polymerization temperature above the melting point of sulfur (112 °C) is crucial for the MCP, but no obvious change was observed when the temperature was further raised. Among the tested solvents, DMF and dimethylacetamide (DMAc) showed better performance compared with nonpolar solvents such as *o*-xylene, in which insoluble precipitate was formed. Slight excess of sulfur was beneficial to the MCP, and further prolonging the polymerization time increased the yield but decreased the M_w.

Under the optimal polymerization conditions, various aromatic dialdehydes and aliphatic/benzyl diamines were tested for the MCP. However, it was found that simultaneous addition of sulfur, terephthalaldehyde, and *p*-xylylenediamine into the polymerization solution could not afford desired well-defined polythioamide product. It is likely that Schiff base moiety was formed from aldehyde and amine as insoluble precipitate, which inhibited the formation of thioamide. To solve that problem, sulfur and diamine were firstly mixed to form polysulfide anion, and the freshly added dialdehyde was then reacted with amine, followed by the reaction

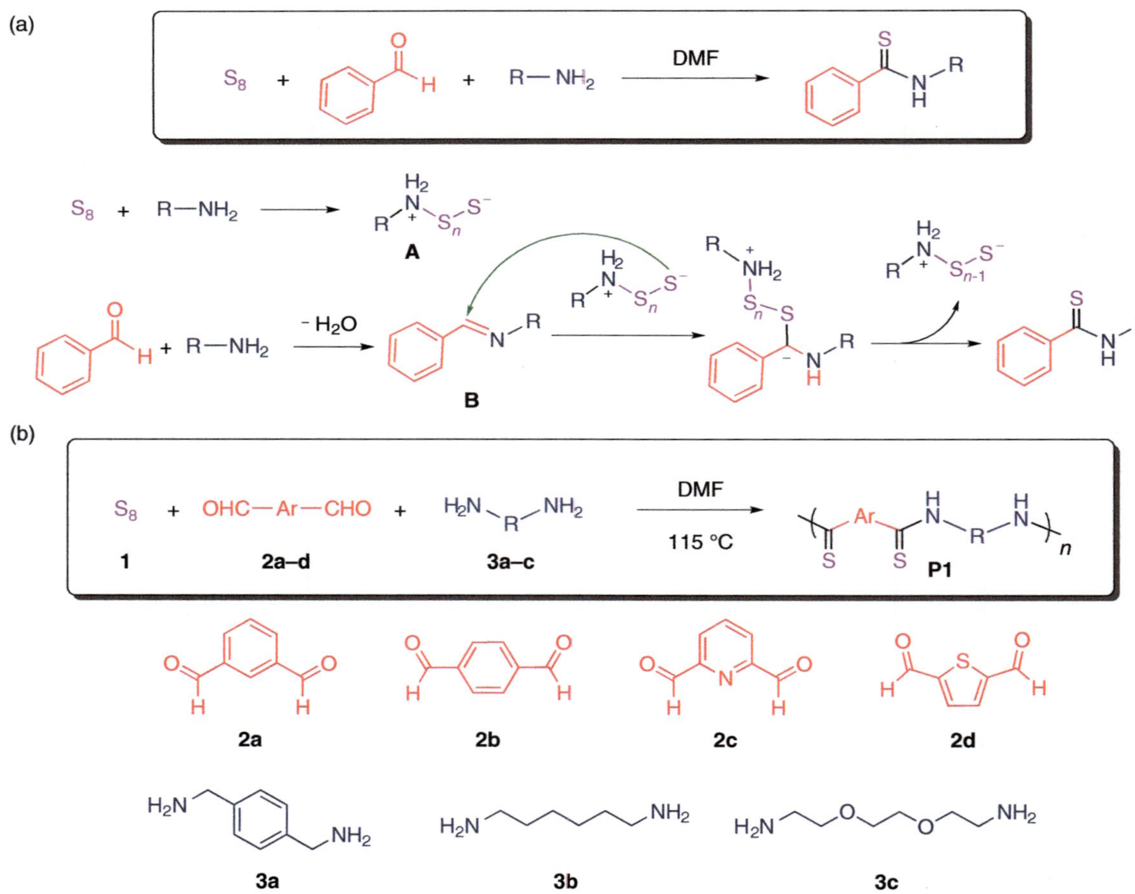

Scheme 1.1 (a) Willgerodt–Kindler reaction of sulfur, aldehyde, and amine. Source: Refs. [43, 44], and (b) the MCPs of elemental sulfur **1**, dialdehydes **2a–d**, and diamines **3a–c**.

with polysulfide anion to furnish well-defined polythioamide with a high M_w of 69 300 g mol^{-1}.

A series of aromatic dialdehydes and diamines were investigated for the MCP, suggesting that only aliphatic/benzyl amines were applicable for this MCP. Due to the low nucleophilicity of the aromatic amines, the key intermediate polysulfide anion could not be generated, and insoluble Schiff base precipitate was formed instead of thioamides.

1.2.2 Multicomponent Polymerization of Sulfur, Diynes, and Aliphatic Amines

The high reactivity and rich chemical property of polysulfide anion may bring many opportunities in MCRs with various reactants. For example, alkynes with carbon–carbon triple bonds are highly reactive species and their MCRs are extensively reported [45]. In fact, an efficient catalyst-free three-component reaction of sulfur, alkynes, and aliphatic amines was reported by Nguyen's group to afford thioamides [46]. Similarly with Willgerodt–Kindler reaction, the polysulfide anions **A** was first generated, which was then reacted with the terminal C≡C bonds of alkyne to yield vinyl polysulfide **C**. Another molecule of amine then attacked the vinyl group to form thioaminal intermediate **D**, and the thioamide product was produced following the cleavage of the S—S bond and the release of the polysulfide anion (Scheme 1.2a).

This promising MCR was developed to an efficient MCP for the synthesis of diverse polythioamides with well-defined structures [47]. A group of aromatic diynes and benzyl/aliphatic diamines were designed as monomers to polymerize with elemental sulfur through the MCP in pyridine, which generally proceed smoothly under the optimal condition to afford polythioamides with high M_ws of up to 127 900 g mol^{-1} in high yields of up to 98% (Scheme 1.2b). High temperature at 100 °C and high diamine monomer concentration of 1.0 M are important for this MCP to obtain satisfying result. Besides the improved reactivity at such high temperature, considering the strong intermolecular hydrogen bonds among thioamide moieties in the product, high temperature could help breaking down these intermolecular interactions and reducing the viscosity of the reaction solution to facilitate the polymerization. Temperature control is hence crucial to this polymerization because the solubility of the monomers/polymers as well as the viscosity of the solution is highly dependent on the temperature. Further increasing the monomer concentration to 1.5 M resulted in gelation and early termination of the polymerization. Despite that, neat polymerization could also proceed to afford polythioamide product. The stoichiometric ratio of alkyne and amine only showed slight effect on the M_w, while excess amount of alkyne could obviously increase the yield.

The obtained polythioamide P**2a** exhibited satisfying thermal stability with its decomposition temperature at 5 wt% weight loss at 284 °C and good solubility in polarized solvents such as DMF and dimethyl sulfoxide (DMSO). Most importantly, unexpected fluorescence was observed for these polythioamides and their small molecular model compounds, even though they did not contain any conventional

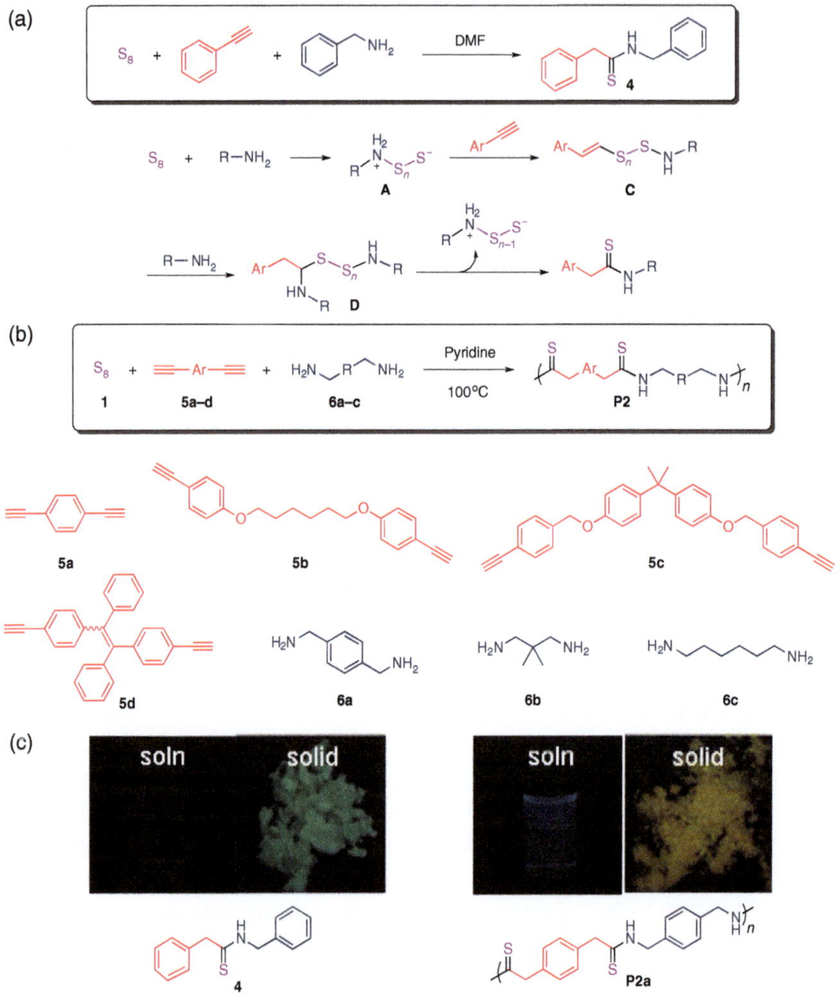

Scheme 1.2 (a) The MCR of sulfur, alkyne, and benzyl/aliphatic amine. (b) MCP of sulfur **1**, diynes **5a–d**, and benzyl/aliphatic diamines **6a–c**. (c) The photographs of model compound **4** and P**2a** under UV illumination. Source: Reproduced from Ref. [47]. Copyright 2015 American Chemical Society.

chromophore or large conjugation system. Both DMF solutions and solid powders of the polythioamides emitted blue light upon UV irradiation (Scheme 1.2c). Solvent and temperature influences were investigated to reveal the underlying effect of the intermolecular interactions among thioamide groups. When methanol was added into the DMF solution of P**2a** to break down the hydrogen bonding between thioamide groups, the PL intensity at 470 nm was decreased with hypsochromically shifted emission maximum. Upon decreasing temperature from 60 to −60 °C, the PL intensity at 470 nm was gradually increased because of the enhanced hydrogen bonding at low temperature. This unusual emission behavior of polythioamides was interpreted as the formation of "heterodox clusters" composed of a large number of lone-pair electron-containing electron-rich heteroatoms, owing to the interchain and intramolecular interactions such as hydrogen bonding and $n \rightarrow \pi^*$ interaction between thioamide moieties, which may serve as the chromophore similar to the reported fluorescent hyperbranched poly(amido amine)s (PAMAMs) [48].

1.2.3 Multicomponent Polymerization of Sulfur, Benzyl Diamines, and Aliphatic Diamines

In addition to nucleophilic attack, benzylamine and sulfur can also undergo oxidation reaction upon heating to form benzaldimine **E**, one of the azomethine derivative [49], which can undergo further reactions to form thioamide. Indeed, a catalyst-free multicomponent selective oxidative reaction of sulfur, benzylamines, and aliphatic amines was reported by Nguyen's group to prepare thioamides at high temperature [50]. According to the proposed mechanism of this reaction (Scheme 1.3a), the benzylamine group was first oxidized by sulfur to afford benzaldimine **E**, which may undergo two different pathways: (i) further oxidation by sulfur to form thiobenzamide **F**, followed by a transthioamidation with amine to afford thioamide, releasing ammonia; or (ii) transimination of benzaldimine and amine to form another azomethine **G**, followed by oxidation with sulfur to afford thioamide product.

Later, Chen's group reported a polymerization of sulfur and diamines based on this MCR (Scheme 1.3b) [51]. The polymerization of sulfur and *p*-xylylenediamine **7a** was performed in *N*-methyl-2-pyrrolidinone (NMP) at 110 °C for 48 hours to afford polymer with a M_w of 55 100 g mol^{-1}. Compared with the optimal temperature at 110 °C, lower temperature at 90 °C gave product with a relatively low M_w of 12 500 g mol^{-1} in 48% yield, while higher temperature at 130 °C provided insoluble products, probably due to the cross-linking formation (C-S-S-C) by the side reaction between thioamides at such temperature [40]. Moreover, similar to the aforementioned MCPs of sulfur, slight excess amount of sulfur was generally beneficial to the polymerization.

It should be noticed that the product structure of the MCP of **1** and **7a** is a mixture of three different repeating units in the polymer chain noted as "S-S", "N-N", and "S-N" structures shown in Scheme 1.3c, which referred to the repeating units whose benzene ring was substituted with two C=S groups, two CH_2NH groups, one C=S group and one CH_2NH group, respectively. According to the aforementioned mechanism, each CH_2NH_2 group of *p*-xylylenediamine could randomly undergo either oxidation reaction with sulfur, or transthioamidation/transimination reaction, to afford NH—C=S group or CH_2NH group, respectively, proved by the characteristic ^{1}H NMR peaks of the aromatic protons located at δ 7.79 ("S-S"), δ 7.36 ("N-N"), and δ 7.89/7.41 ("S-N").

When another diamine was copolymerized with *p*-xylylenediamine and sulfur through the MCP, the expected product structure was even more complicated. For example, the MCP of *p*-xylylenediamine **7a**, *m*-xylylenediamine **7b**, and sulfur produced a hybrid structure with six possible disubstituted benzene ring structures and even more kinds of repeating structures generated from homo- or cross-coupling of the diamines coexisted in the polymer chain, indicated by the ^{1}H NMR spectrum,

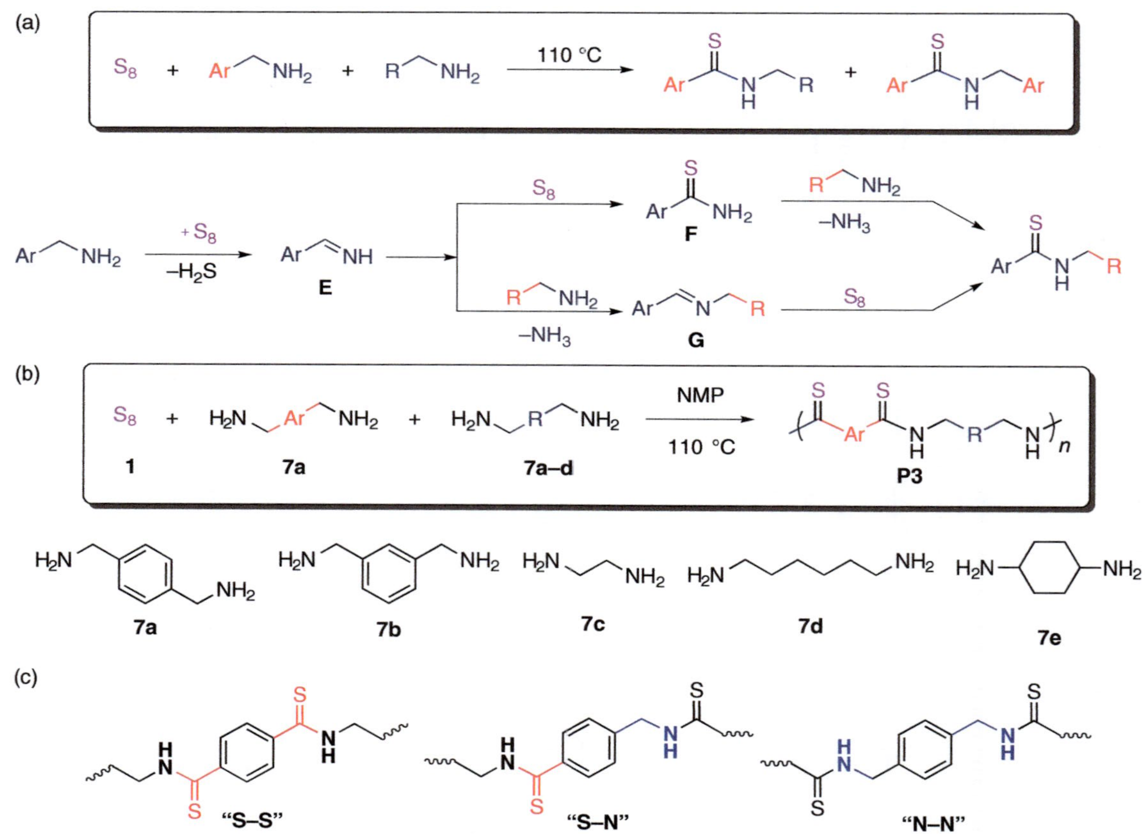

Scheme 1.3 (a) Proposed mechanism of MCR of sulfur, benzyl amine, and aliphatic primary amine. (b) MCP of sulfur **1**, benzyl diamines **7a**, and aliphatic primary diamines **7a–e**. (c) The structures of three possible units "S–S", "S–N", and "N–N."

suggesting that the oxidation reaction with sulfur, transthioamidation, and transimination reactions randomly occurred regardless of homo- or cross-coupling. This problem has brought great difficulty in obtaining well-defined regular polymer structures.

Nonetheless, effort has been made to control the polymer structures and other alkyl diamines such as ethylenediamine **7c**, 1,6-hexamethylenediamine **7d**, and 1,4-cyclohexanediamine **7e**, which can hardly be oxidized by sulfur, but can undergo transthioamidation/transimination reactions, were selected as the third component monomers of the MCP. As a result, in the product of the MCP of sulfur, *p*-xylylenediamine **7a**, and one of these alkyl diamines, besides the homocoupling structure of the *p*-xylylenediamine, there are mainly alternating structures with both two kinds of diamines existed, and the homocoupling structure of the alkyl diamines should not exist. In fact, the MCPs of sulfur, *p*-xylylenediamine **7a**, and 1,6-hexamethylenediamine **7d** or ethylenediamine **7c** afforded high ratios of alternating structures of 80% or 78% in the product, respectively, proving that alternating structure is preferred with alkyl amine monomers. Raising the loading ratio of alkyl diamines is found to be an effective approach to increase the proportion of alternating structure in the polymer product.

1.2.4 Multicomponent Polymerization of Sulfur, Diarylacetic Acids, and Aliphatic/Aromatic Diamines

The aforementioned sulfur-based MCPs generally utilized benzyl diamines or alkyl diamines with strong nucleophilicity as monomers, and less nucleophilic aromatic diamines were not reported in these polymerizations because they could not undergo the crucial ring-opening reaction with sulfur to afford the key intermediate, reactive polysulfide anions. Polythioamides with rigid backbone and conjugated structures from aromatic diamines were hence not accessed yet. The MCR of elemental sulfur, carboxylic acids, and amines was reported by Singh's group to yield thioamide derivatives with the assistance of K_2CO_3 [52], and it's worth mentioning that various amines including aromatic amines were applicable for this MCR with gratifying yields.

To transform this reaction into an efficient polymerization, a series of experimental evidences were obtained to investigate its reaction mechanism and the following steps were proposed. Under basic condition of aliphatic amine or additional inorganic base, cyclic S_8 ring was first reacted with aliphatic or aromatic amine to afford polysulfide anion **A**. The proton of benzyl acid was removed under such basic condition to form **H**, which was then reacted with **A** to release CO_2 and generate intermediate **I**, followed by further reaction with **A** again to afford intermediate **J**. **J** was then transformed to the thiocarbonyl group-containing intermediate **K**, followed by the attack with amine on the thiocarbonyl group to afford thioamide product (Scheme 1.4a).

Encouraged by the rich chemical structures of the polythioamide structures that could be accessed from this reaction, the MCPs of elemental sulfur, diarylacetic acids, and aliphatic/aromatic diamines were developed (Scheme 1.4b) [53]. The

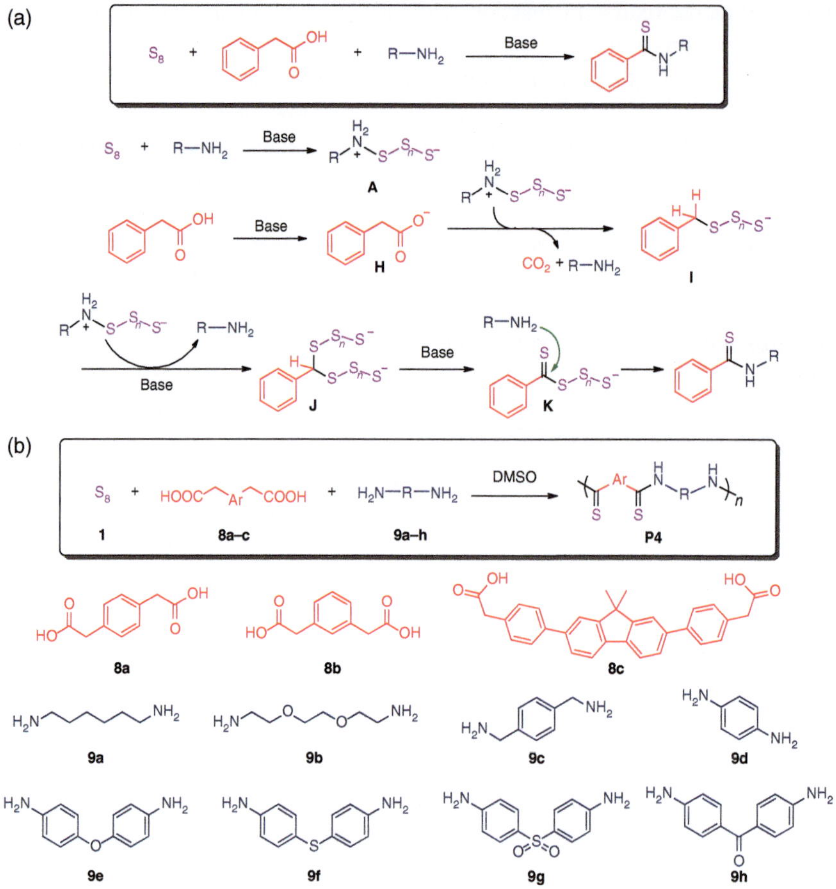

Scheme 1.4 (a) Proposed mechanism of the MCR of sulfur, diarylacetic acids, and amine. (b) MCPs of sulfur **1**, diarylacetic acids **8a–c**, and diamines **9a–h**.

MCP of 1,4-phenylenediacetic acid **8a**, aliphatic 1,6-hexanediamine **9a**, and sulfur was first carried out in DMSO at 100 °C for 10 hours under nitrogen with a high diamine concentration of 0.5 M to afford polythioamide with a M_w of 52 300 g mol^{-1} in 81% yield. Prolonging polymerization time to 15 hours could significantly improve the yield to 96% but show little influence on the M_w. The polymerization temperature at 100 °C is again proved to be critical to obtain satisfying results. Compared with the result obtained at 100 °C, both yield and M_w of the product obtained below 100 °C is decreased to 1/3 of the former one, and the M_w of the product obtained above 100 °C is decreased to the half of the former one, probably due to the poor solubility of insoluble salts formed from acid and amine at lower temperature, and the possible side reactions at higher temperature. While the excess amount of sulfur was generally beneficial to the sulfur-based MCPs, it was unexpected that the best polymerization results was obtained with excess amount of diacid and the monomer loading ratio of sulfur, diacid, and diamines of 3.0 : 1.5 : 1.0, unlike other conventional polycondensations that generally require strict stoichiometric balance among the monomers. This is because (i) the diacid and aliphatic diamine can form insoluble salt rapidly, which may inhibit further reactions, and (ii) the reaction between the deprotonated acid and the polysulfide anion is a crucial step for the whole polymerization. Under such optimal polymerization condition without any catalyst or additive, this MCP could proceed smoothly with a series of aliphatic diamines to afford polymers with the M_ws ranging from 13 600 to 54 600 g mol^{-1} in up to 96% yields.

Although aromatic diamines could not react with sulfur even upon heating, with the assistance of additional inorganic or organic base such as K_2CO_3, they could efficiently react with diacids and sulfur at 100 °C under nitrogen in DMSO to produce polythioamides. The best polymerization result was obtained with the monomer loading ratio of sulfur, diacid, and aromatic diamines of 3 : 1 : 1, considering that aromatic diamines with poor alkalinity could not form salt with diamine and hence no excess amount of acid is needed. Five different aromatic diamines were explored for the MCP, which generally afforded polythioamides with high M_ws ranging from 14 900 to 86 200 g mol^{-1} in 69–95% yields. Compared with the electron-poor aromatic diamine monomers, the electron-rich aromatic diamines generally produced polymers with higher M_ws and yields. Secondary diamines were not applicable for this MCP, probably due to their large steric hindrance.

Furthermore, it's noteworthy that the MCP of sulfur, diacids, and diamines could be conducted with a 200-fold scale under the same condition to afford similar high yields and slightly improved M_ws of the products, regardless of aliphatic or aromatic diamines. Considering the commercial availability of the inexpensive monomers and catalyst-free mild condition of the MCPs, this scalable polymerization might potentially be developed into a convenient and economic approach for the large-scale production of polythioamides.

1.2.5 Multicomponent Polymerization of Sulfur, Diisocyanides, and Aliphatic Diamines

Unlike polythioamides with limited reports, polythioureas with the carbonyl groups in the classical polyurea materials replaced by thiocarbonyl groups are a group of popular sulfur-containing polymers, which have been extensively studied, owing to their applications as dielectric materials [54], self-healing materials [55], and noble metal ion absorbents [56]. The current synthetic approaches for polythioureas generally use sulfur-containing reagents such as carbon disulfide, thiourea, thiophosgene, diisothiocyanate, or 1,1′-thiocarbonyldiimidazole, and the availability of reactants, poor atom economy, and harsh reaction condition have limited the exploration of diverse polythiourea structures.

Recently, a MCR of sulfur, isocyanides, and aliphatic amines was reported to synthesize asymmetric thiourea derivatives (Scheme 1.5a) [57]. In this reaction, the carbenoid carbon atom of isocyanide was first reacted with polysulfide anion **A** to form intermediate **L**, which then underwent S—S bond cleavage to furnish isothiocyanate **M**. The amine then rapidly attacked on the carbon atom of **M** to afford thiourea product. This MCR is more efficient than the abovementioned sulfur-based MCRs, considering that isocyanides are more reactive species compared with alkynes, aldehydes, or acids, which might realize efficient conversion at mild temperature in sterically demanding condition. Moreover, high functional group tolerance of this reaction enables the transformation of hydroxyl group, C=C bonds, C≡C bonds, or ester groups-bearing reactants. It is hence developed into an MCP of sulfur, diisocyanides, and diamines for the facile construction of diverse polythioureas (Scheme 1.5b) [58].

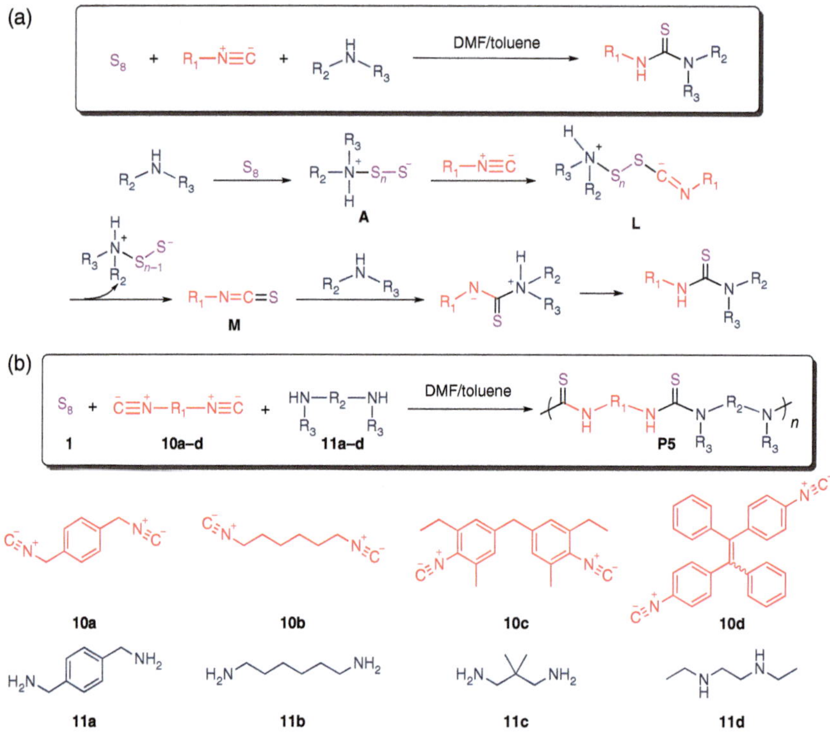

Scheme 1.5 (a) Possible mechanism of MCR of sulfur, isocyanide, and amine [57]. (b) The catalyst-free MCP of sulfur **1**, diisocyanides **10a–d**, and diamines **11a–d**. [58]

Compared with the synthesis of small molecular thioureas, one major challenge of the preparation of polythioureas is their poor solubility in common organic solvents resulted from the strong intramolecular or interchain hydrogen bonding among thiourea moieties embedded in the polymer main chain. DMF/toluene mixed solvent was hence found to be the optimal solvent with balanced reactivity and product solubility for the MCP of sulfur, diisocyanides, and diamines. As expected, with the high reactivity of diisocyanide, the MCP could proceed smoothly at room temperature to afford polythioureas with excellent yields and M_ws. The MCP was also robust, which could be conducted under both nitrogen atmosphere or in air, and there was only slight influence on the yields and M_ws of the polythioureas. The optimal monomer loading ratio of sulfur, diisocyanide, and diamine was found to be 4 : 1 : 1, and higher monomer concentration generally resulted in higher M_ws of the product, but did not show obvious effect on the yields. Neat polymerization of sulfur, diisocyanide **10a**, and diamine **11a** was also succeeded at 100 °C to afford polythiourea with an M_w of 22 100 g mol^{-1} after one hour.

Furthermore, *in situ* IR technique was applied to monitor the formation of polythiourea structures at different temperatures from the characteristic stretching vibration peaks of the newly formed C=S and C—N bonds at ~1540 and 1340 cm^{-1}, respectively, suggesting the rapid completion of the MCPs. The real-time kinetics of the MCP of sulfur, diisocyanide **10a**, and diamine **11a** suggested that polymerization could proceed rapidly and the reaction equilibrium could be reached in two hours at room temperature in air to afford polythiourea with a M_w of 24 400 g mol^{-1} in 82% yield. At elevated temperature of 100 °C, the MCP could even be finished in 10 minutes in air, affording polythiourea with a M_w of 47 600 g mol^{-1} in 81% yield.

Besides the high efficiency and rapid completion of the MCP, the general monomer applicability of the MCP was also an appealing feature. Four aliphatic and aromatic diisocyanides **10a–d** and four primary and secondary aliphatic diamines **11a–d** were selected as monomers to represent different structures, and the MCPs of all the 16 monomer combinations could afford satisfactory results with a great structural diversity of polythioureas obtained with high M_w of up to 242 500 g mol^{-1} in 64–95% yields, demonstrating the wide monomer scope of this MCP.

The success MCP of sulfur, diisocyanides, and diamines was also extended from sulfur to selenium [59], and we have recently reported the efficient MCP of selenium, diisocyanides, and diamines under similar catalyst-free mild condition [60]. Compared with the sulfur-based MCP, the selenium-based MCP is even more efficient, which can be completed in 1 minute at 100 °C or 45 minutes at 40 °C. Moreover, the monomer scope of the selenium-based MCP includes not only primary and secondary aliphatic diamines, but also aromatic diamines, enabling the construction of a large variety of polyselenoureas with unreported structures, good stability, and high M_ws.

The MCPs of elemental sulfur represent a series of economic, convenient, and efficient approaches to directly convert industrial waste sulfur into profitable functional sulfur-containing polymers.

1.3 Cu(I)-Catalyzed Multicomponent Polymerizations of Sulfonyl Azides/Hydrazides

Sulfonyl group-containing polymers, enjoying a series of advantages such as high dielectric ability [61, 62], and excellent mechanical properties [63], are another group of sulfur-containing polymers, which were less explored, due to the lack of efficient synthetic approach. The Cu(I)-catalyzed azide-alkyne cycloaddition as the well-known click reaction can afford stable 1,2,3-triazole product [64]. If a strong electron withdrawing sulfonyl group is installed on the triazole ring, it may undergo ring-opening reaction to release nitrogen gas and form a highly reactive ketenimine species, which can undergo a wide range of different reactions and afford a great diversity of sulfonyl-group containing heteroatom-rich compounds. Based on that, the recently emerged alkyne and sulfonyl azide-based MCPs have brought great opportunity to access sulfonyl group-containing polymers with unique and complex structures as well as functionalities.

1.3.1 Multicomponent Polymerization of Sulfonyl Azides, Alkynes, and Amines/Alcohols

For example, a Cu(I)-catalyzed MCR of terminal alkynes, sulfonyl azides, and amines was reported to synthesize sulfonyl amidine derivatives by Chang and coworkers [65]. The MCR underwent three steps including the cycloaddition of terminal alkyne and sulfonyl azide, the ring-opening reaction of the unstable triazole ring to generate the highly reactive ketenimine intermediate **O**, and its subsequent nucleophilic attack by amine (Scheme 1.6). Based on this mechanism, similar MCRs with the amine replaced by other nucleophiles such as alcohol and water were successively reported [69].

Choi and coworkers first developed the MCRs of alkynes, sulfonyl azides, and amines into efficient MCPs for the preparation of poly(*N*-sulfonylamidine)s P**6** (Scheme 1.6a) [66]. After a series of polymerization condition optimization on solvents and Cu(I) catalysts, the MCP could be conducted at room temperature in DMF with the catalysis of CuCl to afford improved results. Unlike the sulfur-based MCPs, strict stoichiometric balance between diynes and diamines was required to obtain satisfying polymerization results, while an excess amount of *p*-toluenesulfonyl azide is beneficial to the monomer conversion. On the other hand, organic amine additives such as diisopropylethylamine, 2,6-lutidine, tris(benzyltriazolylmethyl) amine, or triethylamine were found to increase the catalytic activity because they can inhibit the protonation of the diamine monomers and facilitate the formation of the key intermediate Cu-acetylide complex. A great diversity of alkyl groups, diene, aromatic groups, and ethylene glycol-containing diyne monomers, alkylsulfonyl azide and phenylsulfonyl azide monomers with electron-withdrawing groups, electron-donating groups, or sterically hindered groups substituted at different positions of the benzene ring, as well as primary and secondary dialkylamines, and aniline derivatives were selected as monomers for this Cu(I)-catalyzed MCP, affording 26 well-defined poly(*N*-sulfonylamidine)s with M_n values of up to 75 100 g mol^{-1}.

Later, this MCP was extended to a group of optically active amino ester monomers by Tang and coworkers [70]. The Cu(I)-catalyzed MCP of alkynes, sulfonyl azides, and chiral amino esters could be conducted in chloroform at room temperature, affording chiral poly(*N*-sulfonylamidine)s **P7** with high M_ws of up to 35 900 g mol^{-1} in up to 87% yields (Scheme 1.6b). With the chiral centers introduced from the amino

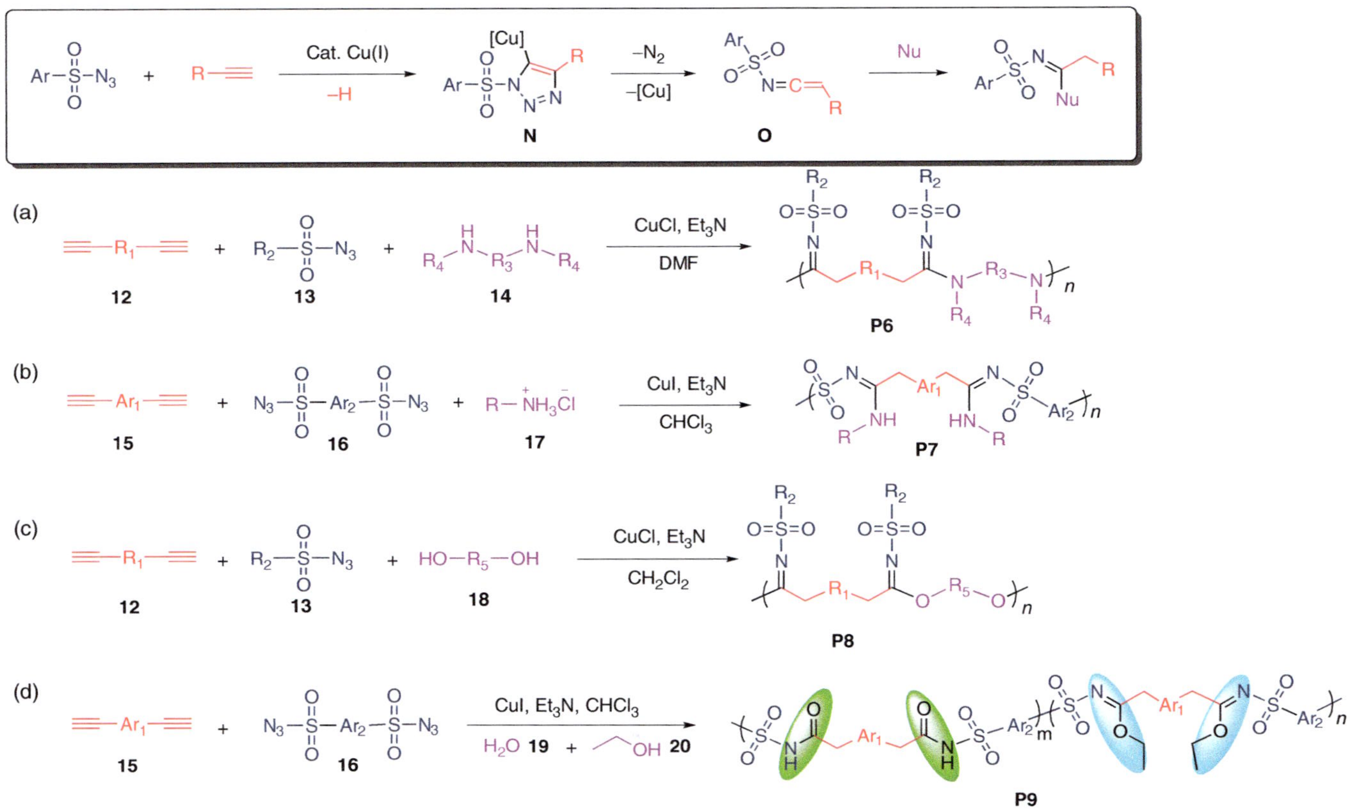

Scheme 1.6 Cu(I)-catalyzed MCPs of (a) diynes **12**, sulfonyl azides **13**, and diamines **14** [66]. (b) diynes **15**, disulfonyl azides **16**, and chiral amino esters **17**, (c) diynes **12**, sulfonyl azides **13**, and diols **18** [67]. and (d) diynes **15**, disulfonyl azides **16**, water **19**, and alcohol **20** [68].

ester monomers, the resultant polymer backbone could be induced to helically rotate and possess chiroptical properties.

Similarly, alcohols can be used as the third component monomer to replace amines as nucleophiles and react with the ketenimine intermediate to proceed the Cu(I)-catalyzed MCP and afford poly(*N*-sulfonylimidate)s **P8** (Scheme 1.6c) [67]. There are two differences of the alcohol-based MCPs compared with the amine-based MCPs. Firstly, the nucleophilicity of alcohols is not as strong as amines, hence the molecular weights of the poly(*N*-sulfonylimidate)s product are relatively low compared with the poly(*N*-sulfonylamidine)s. Secondly, this MCP is sensitive to water because even trace amount of water can compete with the diol monomers to form *N*-sulfonyl amide structures and terminate the polymerization. Moisture should hence be strictly avoided and dry solvents such as $CHCl_3$ and CH_2Cl_2 should be used instead of DMF. The MCPs of various diynes, sulfonyl azides, and a great variety of primary, secondary, aromatic, cyclic, and bicyclic diols generally occurred smoothly to produce 24 different poly(*N*-sulfonylimidate)s with high M_ws of up to 33 500 g mol^{-1}.

The side reaction with water could also be utilized in the MCPs if disulfonyl azide monomers were used. For instance, a four-component polymerization of diyne, disulfonyl azide, water, and alcohol was reported in $CHCl_3$ in the presence of CuI and triethylamine under nitrogen at room temperature (Scheme 1.6d) [68]. In the resultant polymer product **P9**, both amide and imidate structures co-existed, and their ratio in the polymer chain could be controlled by adjusting the loading ratio of water and ethanol. With the concentrations of water and ethanol were 0.1 and 0.3 M, respectively, a polymer with an amide/imidate ratio of 39/61 and a M_w of 83 900 g mol^{-1} was obtained. In the absence of water, a soluble poly(*N*-sulfonylimidate) product with a high M_w of 257 100 g mol^{-1} could be obtained in four hours. It should be noted that the MCP without ethanol may form poly(*N*-sulfonylamide) as a gel, due to the formation of intermolecular hydrogen bonding among the polyamide chains.

Besides the high efficiency of these MCPs, their functional group tolerance and broad monomer scope are quite unique. Satisfying polymerization results could be obtained regardless of the electronic nature, substitution position, or steric hindrance of the sulfonyl azides. The rigid diamines, diols, or diynes could generally produce polymers with higher M_w compared with flexible monomers, owing to the suppression of the intramolecular cyclizations.

1.3.2 Multicomponent Polymerization of Sulfonyl Azides, Alkynes, and Other Monomers

Besides simple nucleophiles such as amines and alcohols, the ketenimine intermediate formed from alkynes and sulfonyl azides could react with many other reactants in a more complex manner [71–75]. The successful introduction of these MCRs into polymer synthesis could provide access to various heteroatom, heterocycle, or fused heterocycle-containing polymers with unique structures, which could not be synthesized from other reported methods.

For example, Cu(I)-catalyzed MCPs of diynes, disulfonyl azides, salicylaldehyde, or *o*-hydroxyacetophenone were reported at room temperature under nitrogen in the presence of triethylamine in DMAc to produce poly(iminocoumarin)s P**10** with M_ws of up to 64 600 g mol^{-1} in satisfactory yields of up to 99% (Scheme 1.7) [76]. In this MCP, the ketenimine intermediate formed from alkyne and sulfonyl azide was quickly attacked by the OH group of *o*-hydroxyacetophenone to form an anionic intermediate, which then underwent an intramolecular nucleophilic addition on the carbonyl group. After protonation and dehydration, the poly(iminocoumarin) was formed.

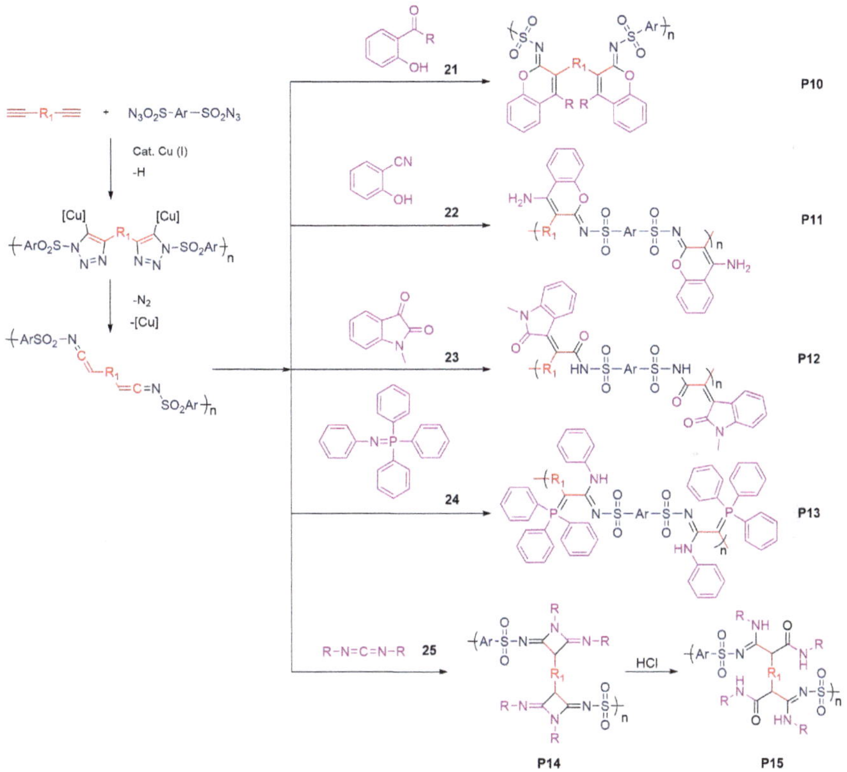

Scheme 1.7 The synthetic routes of Cu(I)-catalyzed MCPs of diynes, disulfonyl azides, and various monomers **21–25**.

Similar mechanism also applies to the MCR of diynes, disulfonyl azides, and 2-hydroxybenzonitrile or 2-aminobenzonitrile, and the OH or NH_2 groups from substituted benzonitriles first attack the ketenimine to produce an intermediate, which then undergoes subsequent intramolecular cyclization, protonation, and isomerization to afford amino-substituted iminocoumarin or quinoline structures [72]. These two MCPs were developed under the catalysis of CuCl and triethylamine in CH_2Cl_2 under nitrogen, generating iminocoumarin or quinoline-containing poly(*N*-sulfonylimine)s P**11** with M_ws of up to 37 700 g mol^{-1} in high yields up to 96% (Scheme 1.7) [77]. Moreover, the fused heterocycles iminocoumarin and aminoquinoline structures built *in situ* from the MCP have brought the polymers interesting fluorescence properties.

N-protected isatins with their unique chemical reactivity could also be developed as the third monomer of the alkyne and sulfonyl azide-based MCPs. In fact, the carbonyl group linked on the benzene ring of the *N*-protected isatin could undergo a [2+2] cycloaddition with metal alkynamide intermediate generated from alkyne and sulfonyl azide to form a four-membered ring oxetene structure. In the presence of LiOH, the oxetene underwent ring-opening reaction and acidification to afford 3-alkenyloxindole; in the presence of H_2O and Na_2CO_3, the oxetene underwent isomerization, nucleophilic attack with OH^- to open the four-member ring, and subsequent acidification to afford 3-hydroxyindole [73]. Following these routes, the MCPs of diynes, disulfonyl azides, and *N*-protected isatins were developed at room temperature in CH_2Cl_2/*t*-BuOH under the catalysis of CuI in the presence of LiOH, generating oxindole-containing poly(*N*-acylsulfonamide)s P**12** with M_ws of up to 30 600 g mol^{-1} in 92–98% yield (Scheme 1.7) [78]. Furthermore, a four-component polymerization of diynes, disulfonyl azides, *N*-protected isatins, and water was developed under similar condition in DMF with Na_2CO_3 as the base, affording random copolymers consisting 3-alkenyloxindole moieties and two chiral center-containing 3-hydroxyindole moieties. Interestingly, these poly(*N*-acylsulfonamide)s possess reversibly tunable solubility and hydrophilicity, which can be well-dissolved in alcohol or alcohol/water mixtures, but not in THF or CH_2Cl_2. Opposite solubility of these polymers is realized upon acidification of the polymers.

Iminophosphorane can also undergo [2+2] cycloaddition reaction with ketenimine intermediate, followed by ring-opening reaction to afford a phosphorus amidine product [74], which could be considered as the third monomer of the alkyne and sulfonyl azide-based MCPs to construct heteroatom-rich polymers. We have reported a facile MCP of diynes, disulfonyl azides, and iminophosphorane, which proceeds at room temperature in THF in the presence of CuI and triethylamine with the alkyne monomer concentration of 0.1 M and the monomer loading ratio of alkyne, sulfonyl azide, and iminophosphorane to be 1.0 : 1.0 : 2.5, furnishing N, O, S, and P-containing heteroatom-rich poly(phosphorus amidine)s P**13** with high M_ws of 85 600 g mol^{-1} in 91% yield (Scheme 1.7) [79]. The MCPs of a great diversity of diynes and disulfonyl azides could proceed smoothly and efficiently, including aliphatic and aromatic diynes.

These alkyne and sulfonyl azide-based MCPs were not only convenient for the preparation of polymers with five- or six-membered heterocycles embedded in the backbone as discussed above, but also proved to be effective for the construction of polymers with readily openable small ring heterocycles, which remained to be a great challenge due to their synthetic difficulties and poor stability. This was realized by a recently reported room temperature MCP of terminal diynes, disulfonyl azides, and carbodiimides under nitrogen with the catalysis of CuI, producing heteroatom-rich multisubstituted azetidine ring-bearing polymers P**14** with good stability in air and common organic solvents, and M_ws of up to 74 500 g mol^{-1}

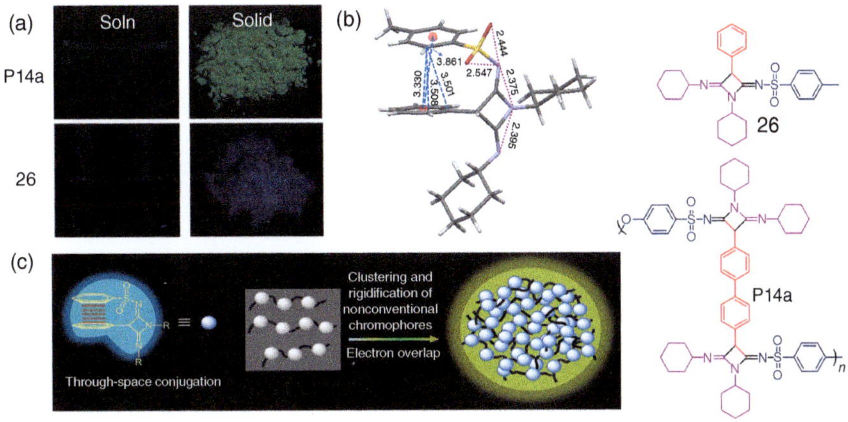

Figure 1.1 (a) The fluorescence photos of solutions and powders of **P14a** and **26** taken under 365 nm UV irradiation. (b) Single crystal structure of **26** with indicated distances (Å) between π–π stacked aromatic rings and distances between atoms with lone-pair electrons. (c) Schematic illustration of the unconventional luminescence of model compound **26** and polymers. Source: Reproduced from Ref. [80]. Copyright 2018 American Chemical Society.

in yields up to 89% (Scheme 1.7) [80]. With the large ring strain of the azetidine rings, these four-membered rings in polymer skeleton could facilely transformed into amide and amidine moieties through acid-mediated ring-opening reactions to afford polymers with new structures and tunable properties.

Interestingly, unexpected fluorescence at 502–536 nm was observed in the solid powders of the azetidine ring-containing polymers and model compounds, considering that there was no conventional chromophore existed in these structures (Figure 1.1). Single crystal analysis of model compound **26** suggested the existence of through-space π–π interaction, lone-pair electronic interactions among N and O atoms, and intermolecular C—H· · ·O and C—H· · ·π interactions, which rigidified the conformation of the molecules, maintained the stability of spatial configuration, and strengthened the through-space π–π conjugation. The unconventional luminescence of these polymers is hence attributed to the clusteroluminogens formed by through-space electronic interactions of heteroatoms and phenyl rings, evidenced by X-ray photoelectron spectroscopy (XPS) analysis. In the solid state, the chain entanglement and intra-/interchain interactions led to the rigidification of nonconventional chromophores and high density of clustering, resulting in bright emission; in the solution state, these through-space interactions and intra-/interchain interactions were broken and thus no emission could be observed.

These alkyne and sulfonyl azide-based Cu(I)-catalyzed MCPs generally enjoy high efficiency, mild condition, excellent functional group tolerance, and high atom economy, which can proceed smoothly at room temperature, release nitrogen gas as the only byproduct, and generate polymers with defect-free well-defined structures and high M_ws in high yields, demonstrating fascinating synthetic approaches for the construction of polymers with unique and complex heteroatom-rich structures.

1.3.3 Multicomponent Polymerization of Sulfonyl Hydrazide, Alkynes, and Diphenyl Dichalcogen

Sulfonyl hydrazides, which are usually odorless, stable, and easy to be synthesized, are another type of popular sulfonyl source in organic synthesis [81]. For example, an efficient Cu(I)-catalyzed regio- and stereospecific selenosulfonation of alkynes, arylsulfonohydrazides, and diphenyl diselenide was reported to generate selenovinyl sulfones by Liu et al. [82] A radical mechanism was proposed for this MCR: firstly, sulfonylhydrazide was reacted in the presence of copper salt and $K_2S_2O_8$ to release N_2 and generate a sulfonyl radical, which was then regiospecifically added on alkynes to form a relatively stable β-sulfonyl vinyl radical. The phenyl selenol radical generated from diphenyl diselenide was then coupled with the vinyl radical to afford thermodynamically stable (E)-β-selenovinyl sulfone product. This MCR was then developed into the MCPs of alkynes, arylsulfonohydrazides, and diphenyl dichalcogen for the construction of chalcogen-rich poly(vinyl sulfone)s (Scheme 1.8) [83]. The MCP of diyne **27a**, arylsulfonohydrazides **28**, and diphenyl diselenide **29a** was found to proceed smoothly in the presence of $Cu(MeCN)_4PF_6$ and $K_2S_2O_8$ in DMF/CH_3CN (v/v, 1/1) at room temperature for 24 hours. This MCP applies to various aromatic and aliphatic diynes **27a–d** and diphenyl dichalcogens **29a–c**, producing regio- and stereoselective poly(vinyl sulfone)s P**16** with M_ws of up to 20 800 g mol^{-1} in yields of up to 92%.

One unique feature of the selenium-containing poly(vinyl sulfone) is that it can be completely transformed to poly(ethynyl sulfone) P**17** with the phenyl selenol species cleaved from the polymer upon the post-modification by oxidation with H_2O_2, enabling facile modulation of the polymer properties. For example, the fluorescence quantum yield of a tetraphenylethene-containing poly(vinyl sulfone) is 2.7%, which increased about five-fold to 13.4% after it was oxidized to remove the phenyl selenol species, because the heavy atom effect caused from selenium atoms disappeared.

1.3.4 Topological Polymers Prepared from Sulfonyl Azides and Alkyne-Based MCPs

Topological polymers with different connection of repetitive units of the polymers have attracted much attention recently in fields regarding biomaterials, coatings, and photoelectric materials [84–86]. Encouraged by the high efficiency and mild condition of the aforementioned Cu(I)-catalyzed MCPs, sulfonyl group-containing polymers with unique topological structures could be designed and synthesized. For example, a series of mono-functional alkyne-containing polymer or dendronized macromonomers were designed for the Cu(I)-catalyzed MCPs of mono-functionalized alkynes, bis-sulfonyl azides, and diamines/diols [87]. Through diversity-oriented synthesis and "graft through" strategy, 54 samples of graft and dendronized polymers P**18** and P**19** with large bulky side chains and high M_n of up to 1 041 000 g mol^{-1} were prepared, despite of the severe steric hindrance from the macromonomers (Scheme 1.9a). The construction of topological polymers from

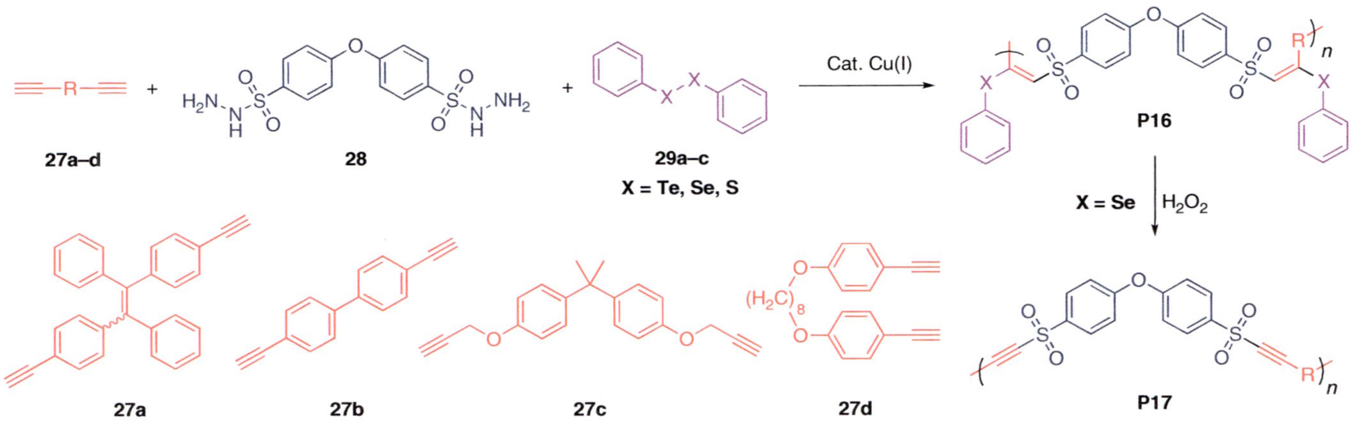

Scheme 1.8 Cu(I)-catalyzed MCPs of diynes **27a–d**, aryl sulfonyl hydrazide **28**, and diphenyl dichalcogen **29a–c** towards regio- and stereoselective chalcogen-rich poly(vinyl sulfone)s and the post-modification of the selenium-containing poly(vinyl sulfone) by oxidation. [83]

MCP approaches enables simultaneous manipulation of both main chain and side chain structures.

MCP is also a powerful tool for the construction of hyperbranched polymers from simple tuning of monomer combinations. For example, we have reported the preparation of N, O, S, P-containing heteroatom-rich functional hyperbranched polymers from the MCPs of alkyne monomers with three or four functional groups, disulfonyl azide, and iminophosphorane (Scheme 1.9b) [88]. Hyperbranched poly(phosphorus amidine)s **P20** with different topological structures and high M_ws of up to 441 700 g mol^{-1} were obtained in high yield at room temperature in the presence of CuI and triethylamine from the monomer combination strategies such as $A_3 + B_2 + C$ or $A_4 + B_2 + C$.

The high efficiency, mild condition, high atom economy, wide monomer scope, and versatile monomer combinations of these sulfonyl azide and alkyne-based MCPs have enabled the facile construction of functional polymers with various topological structures.

1.4 Multicomponent Polymerizations with Thiol-Related Monomers

Thiols are a group of simple, readily available, and reactive compounds for the synthesis of sulfur-containing materials from a series of versatile reactions, especially nucleophilic additions considering that thiols are generally strong nucleophiles. In MCPs where the simple nucleophilic addition of thiol was combined with other reactions in a one-pot procedure, sulfur-containing polymers with complex structures and advanced functionalities could be accessed.

1.4.1 One-Pot Multicomponent Tandem Polymerization of Alkyne, Carbonyl Chloride, and Thiol

For example, a one-pot, two-step, three-component tandem polymerization of alkyne, carbonyl chloride, and thiol was designed, combining the Sonogashira coupling reaction of alkyne and carbonyl chloride, as well as the subsequent thiol addition on the alkynone intermediate in a single polymerization. Both two reactions share high efficiency, common solvent THF, and compatible reaction condition, enabling the multicomponent tandem polymerization (MCTP) to proceed efficiently at room temperature in the presence of $Pd(PPh_3)_2Cl_2$, CuI, triethylamine, and n-Bu_3P to afford sulfur-rich alkylthiol-substituted polymers with high regioselectivity, good stereoselectivity, high M_ws (up to 59 100 g mol^{-1}), and high yields (up to 98%) [89]. The MCTP is applicable to both aliphatic and aromatic thiols, and polymers **P21–23** with tunable polymer backbones and optical/photophysical properties could be accessed from the same MCTP with different monomer combination strategies such as "$A_2 + B_2 + C_1$," "$A_2 + B_1 + C_2$," or "$A_1 + B_2 + C_2$" (Scheme 1.10a).

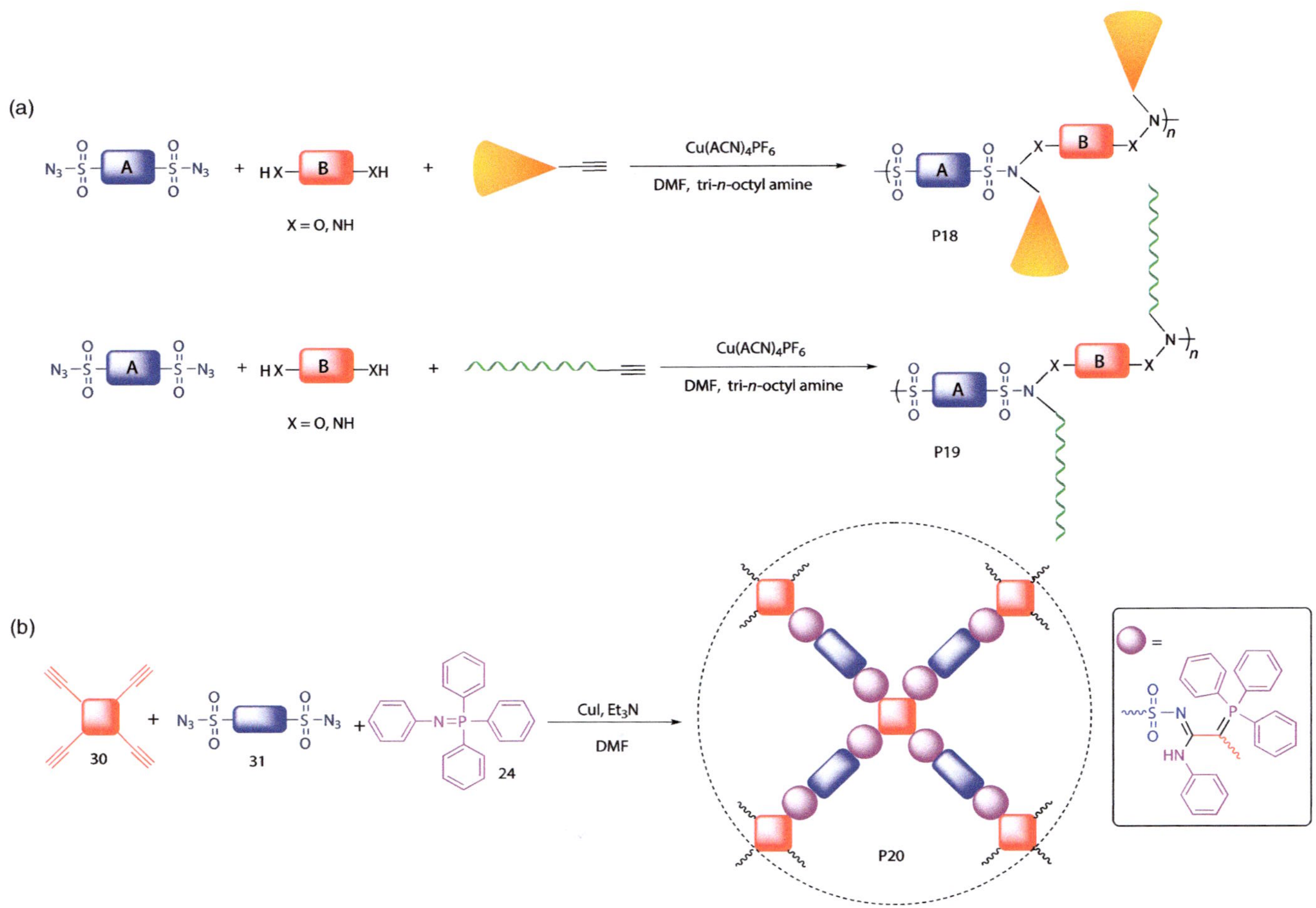

Scheme 1.9 One-pot synthesis of a library of (a) graft, dendronized, and (b) hyperbranched polymers by Cu(I)-catalyzed sulfonyl azide and alkyne-based MCPs. Source: Reproduced from Ref. [88]. Copyright 2020 Georg Thieme Verlag KG.

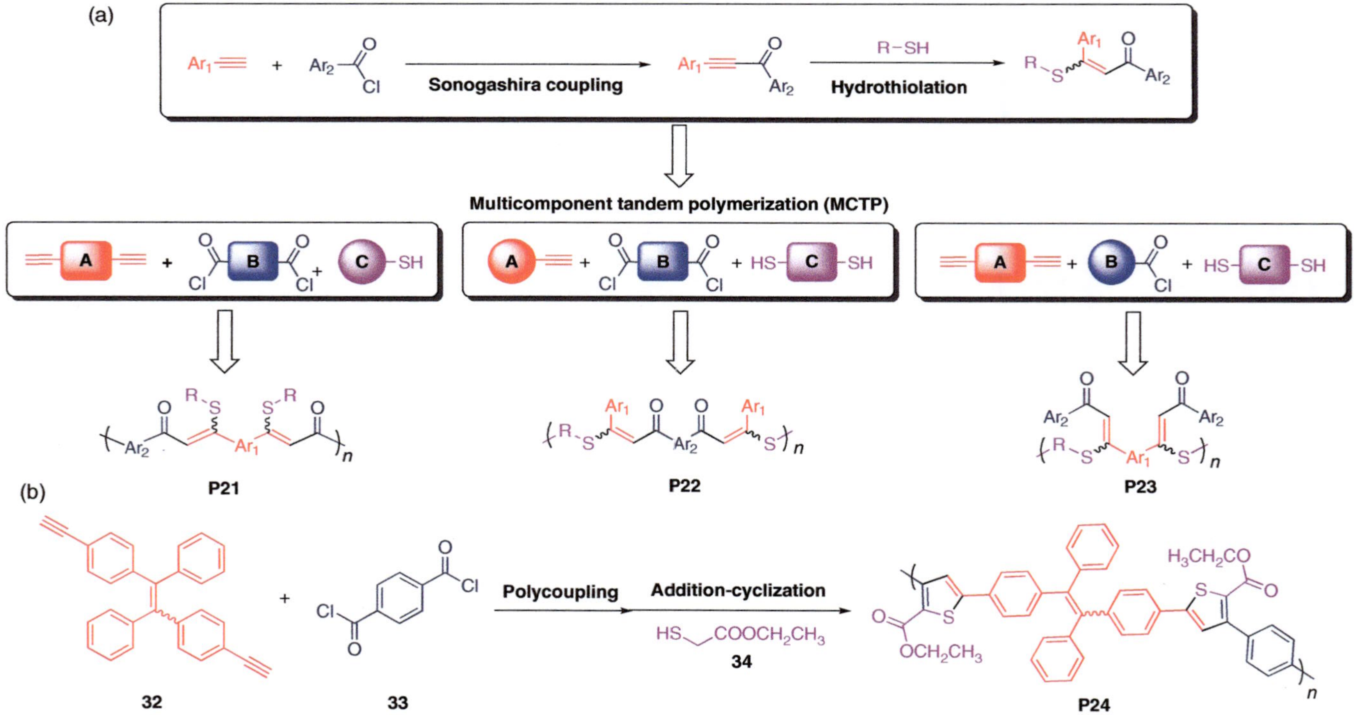

Scheme 1.10 Multicomponent tandem polymerizations of alkynes, carbonyl chlorides, and thiols/ethyl 2-mercaptoacetate for the synthesis of (a) P**21–23** and (b) poly(arylene thiophenylene) P**24**. [90]

More reactions could be integrated into this MCTP when normal thiol in the aforementioned MCTP of alkyne, carbonyl chloride, and thiol was replaced with ethyl 2-mercaptoacetate. After the Sonogashira coupling reaction and hydrothiolation, the intermediate may further undergo the Fiesselmann cyclocondensation to furnish a multisubstituted thiophene ring in a one-pot procedure [91]. The MCTP of tetraphenylethene-containing diyne **32**, the terephthaloyl dichloride **33**, and ethyl 2-mercaptoacetate **34** was hence reported to afford thiophene-containing polymer P**24** with well-defined structures and high M_w of 156 000 g mol^{-1} in 93% yield (Scheme 1.10b) [90]. In this MCTP, the polymerization not only linked the functional groups from monomers together, but also built new functional unit such as multisubstituted aromatic thiophene ring embedded in the polymer backbone, demonstrating its high synthetic efficiency.

1.4.2 Multicomponent Polymerizations with Cyclic Dithiocarbonate

To solve the problem that thiol compounds are generally smelly, too reactive to prepare and store limited the exploration of thiol monomers. Sulfur-containing heterocycle such as five-membered cyclic dithiocarbonate, which could generate thiol group in situ, would serve as thiol precursor to develop MCPs from the combination with highly efficient thiol-click reactions [92]. Through the ring-opening reaction of sulfur-containing heterocycle with nucleophile to generate thiol group, and the sequential thiol-based click reaction, sulfur-containing polymers with strict moiety sequence could be afforded.

For example, the MCP of cyclic dithiocarbonate **35**, diamine **36**, and diyne **38** was reported by Ochiai et al. in the presence of 2,2′-azobisisobutyronitrile (AIBN) to produce sequence-controlled polymer P**25** (Scheme 1.11) [93]. The nucleophilic ring-opening reaction of cyclic dithiocarbonate was induced by diamine to generate mercaptothiourethane intermediates **37**, and the resulting thiol groups of **37** then underwent thiol-yne radical addition with diyne monomer **38** to provide desirable polymers P**25** with well-defined sequence of the functional units. The polymer product could be fine-tuned by changing the amount of AIBN, which may influence the concentration of thiyl radicals. Excess amount of AIBN led to insoluble polymer due

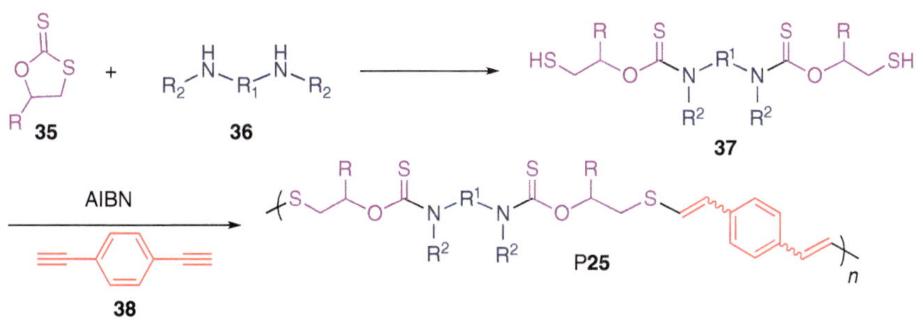

Scheme 1.11 MCPs of five-membered cyclic dithiocarbonates **35**, diamines **36**, and diyne **38**. [93]

to the crosslinking caused by further addition of thiyl radical to double bonds in the polymer product. The MCP of diyne, various diamines and dithiocarbonates proceeded smoothly to afford sulfur-containing polymers with M_ws ranging from 9 000 to 81 700 g mol^{-1} in excellent yields of up to 99%.

1.4.3 Multicomponent Polymerizations with Cyclic Thiolactone

Similarly, cyclic thiolactones can be converted to sulfhydryl species through ring-opening reaction with various nucleophiles such as amines, which are also popular monomers of MCPs for the preparation of sequence-controlled sulfur-containing polymers [94]. For example, Hong's group reported the catalyst-free MCP of thiolactone **39**, diamines **40**, and bis(methacrylate)s **42** through the sequential ring-opening reaction of thiolactone by amine, and thiol-Michael addition reaction on the $CH{=}CH_2$ group of methylacrylates (Scheme 1.12a), producing a series of ABAC-type sequence-controlled polymer P**26** with M_ns ranging from 10 700 to 48 000 g mol^{-1}. In this MCP, the primary diamine monomers serve two roles including nucleophile to open the thiolactone ring, and activating the $CH{=}CH_2$ bonds of dimethylacrylate. On the other hand, secondary amine could also activate the vinyl bonds but could not open the thiolactone ring, which is not applicable to this MCP.

Taking advantages of the different reactivity of primary and secondary amines, an asymmetric amine **44** with both a primary amine and a secondary amine group was designed for a four-component polymerization (Scheme 1.12b) [95]. Firstly, the primary amine group of **44** was reacted with thiolactone **43** to open the ring and generate thiol group, which then underwent thiol-ene addition with dimethylacrylates **45** to generate an BACAB-type compound **46** with two terminal secondary amine groups. Dipropiolate **47** was then added to proceed the amino-yne addition reaction to afford an BACABD-type sequence-controlled polymer P**27** with a M_n of 59 800 g mol^{-1} in 81% yield.

Based on this similar tandem strategy, a five-component polymerization of thiolactone **43**, diamine **48**, propargyl methacrylate **49**, sulfonyl azide **51**, and diol **52** was designed in a one-pot procedure to directly afford a sequence-controlled polymer P**28** with multiple building units (Scheme 1.12c) [96]. In this MCP, the propargyl methacrylate **49** with both an activated $CH{=}CH_2$ group and a terminal alkyne group is designed as the orthogonal monomer to combine two sequential MCRs. In the first step, monomers **43**, **48**, and **49** were reacted through the abovementioned amine-thiol-ene reaction to afford the intermediate **50** with two remaining terminal alkyne groups. The sulfonyl azide **51** and diol **52** were then added into the reaction solution together with copper catalyst and organic base to undergo the Cu(I)-catalyzed MCR of alkyne, sulfonyl azide, and alcohol. The DCABACDE-type sequence-controlled polymer P**28** was then obtained with a M_n of 72 600 g mol^{-1}.

It is worth mentioning that this tandem MCP strategy could also be applied for the preparation of sequence-controlled topological polymers by using monomers with multiple functional groups such as diethylenetriamine [97], demonstrating the

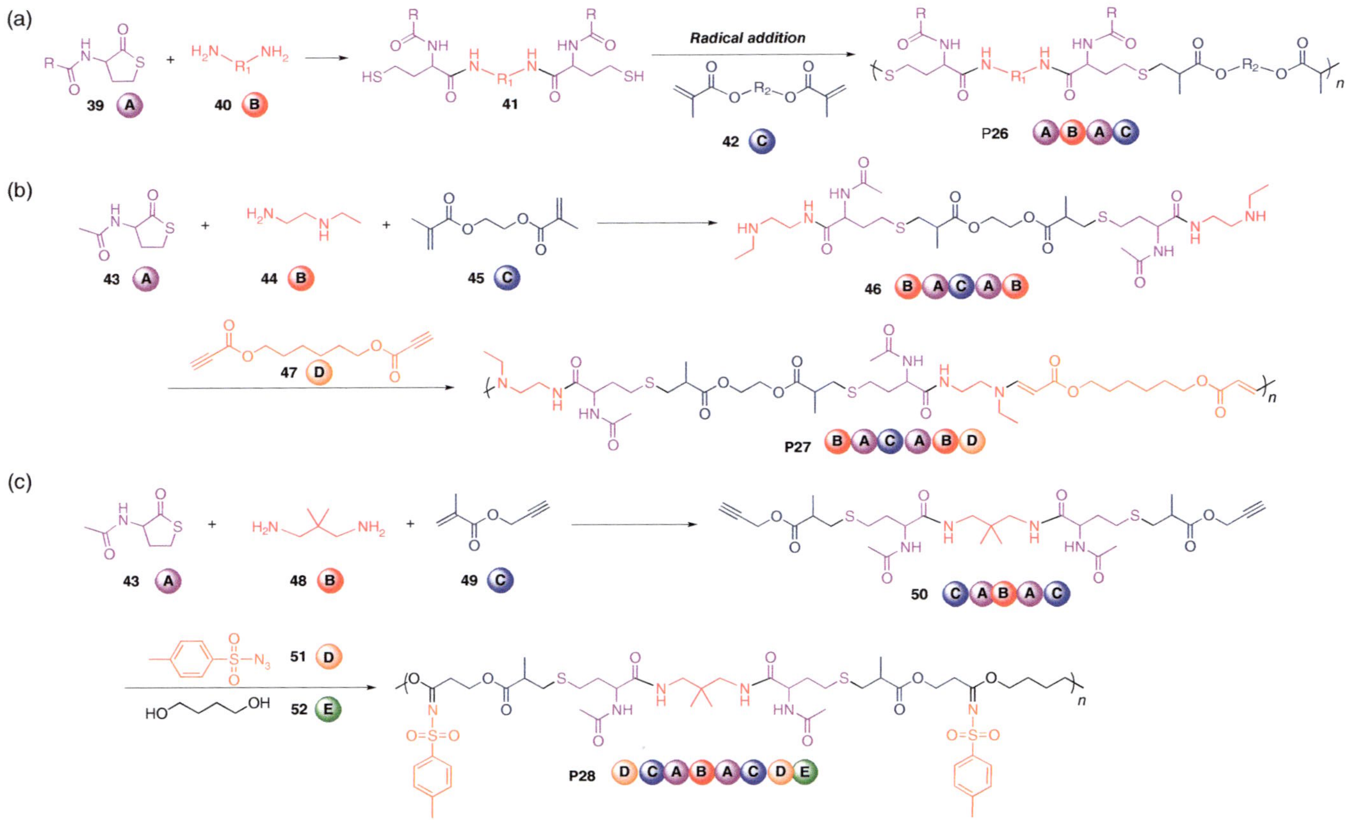

Scheme 1.12 Sequence-controlled sulfur-containing polymers prepared from (a) three-component polymerization of thiolactones **39**, primary diamines **40**, and bis(methacrylate)s **42**, (b) four-component polymerization of thiolactone **43**, asymmetric diamine **44**, dimethylacrylate **45**, and dipropiolate **47** [95], and (c) five-component polymerization of thioactone **43**, diamine **48**, propargyl methacrylate **49**, sulfonyl azide **51**, and diol **52**. [96]

great potential of these MCPs in the construction of diverse topological functional materials.

1.5 The Applications of Sulfur-Containing Polymers Prepared from MCP

The abovementioned MCP products generally possess large number of heteroatom-containing functional groups such as thioamides, thioureas, sulfonyl groups, and thioethers, which may endow the polymer materials with a series of functionalities such as the coordination with various metal ions and intermolecular interactions, considering that there are many strong and unique sulfur-involved interactions such as sulfur–metal coordination, sulfur–sulfur interaction, and hydrogen bonding between S=C groups and active protons. With proper design of these properties and other functionalities integrated in the monomer structures, there are a group of practical applications such as fluorescence chemosensors, metal ion removal or enrichment, and bioimaging could be realized.

1.5.1 Chemosensors

Detection of transition metal ion is of scientific significance and practical implication in environmental and industrial applications. With the promising selective interaction between the heteroatoms of these sulfur-containing MCP polymer products and particular transition metal ions, together with a fluorescence signal output that is dependent on the coordination of transition metal ion analyte, these polymers could serve as chemosensors for transition metal ions. For example, aggregation-induced emission (AIE) property could be facilely introduced to these MCP product by simply adopting AIEgen-containing monomers to produce AIE-active metal ion-responsive sulfur-containing polymers [98, 99].

The poly(phosphorus amidine)s **P13a** prepared from the Cu(I)-catalyzed MCP of tetraphenylethene-containing diyne, disulfonyl azide, and iminophosphine was found to be a fluorescence chemosensor of Pd^{2+} (Figure 1.2a) [79]. The fluorescence intensity of the nanoaggregates of **P13a** in DMSO/water mixture with 50 vol% water was gradually decreased with the progressive addition of $PdCl_2$ solution while the emission profile remained the same. The relative PL intensity (I_0/I) versus Pd^{2+} concentration fitted a linear relationship with a large quenching constant of 207 600 M^{-1}, suggesting high sensitivity towards Pd^{2+}. Moreover, high selectivity of Pd^{2+} compared with other metal ions was observed under same condition, attributing to the specific strong interaction between Pd^{2+} and the phosphorus atoms on the polymer chain. Similarly, the amino substituted iminocoumarin-containing poly(*N*-sulfonylimine) **P11a** was found to be a fluorescence chemosensor of Ru^{3+} with the quenching constant of 89 200 M^{-1} (Figure 1.2b) [77]. The PL intensity of the nanoaggregates of **P11a** in DMSO/water mixture with 50 vol% water was quenched upon the addition of Ru^{3+} selectively, while the other tested 15 metal ions did not show obvious interference. Several other AIE-active sulfur-containing

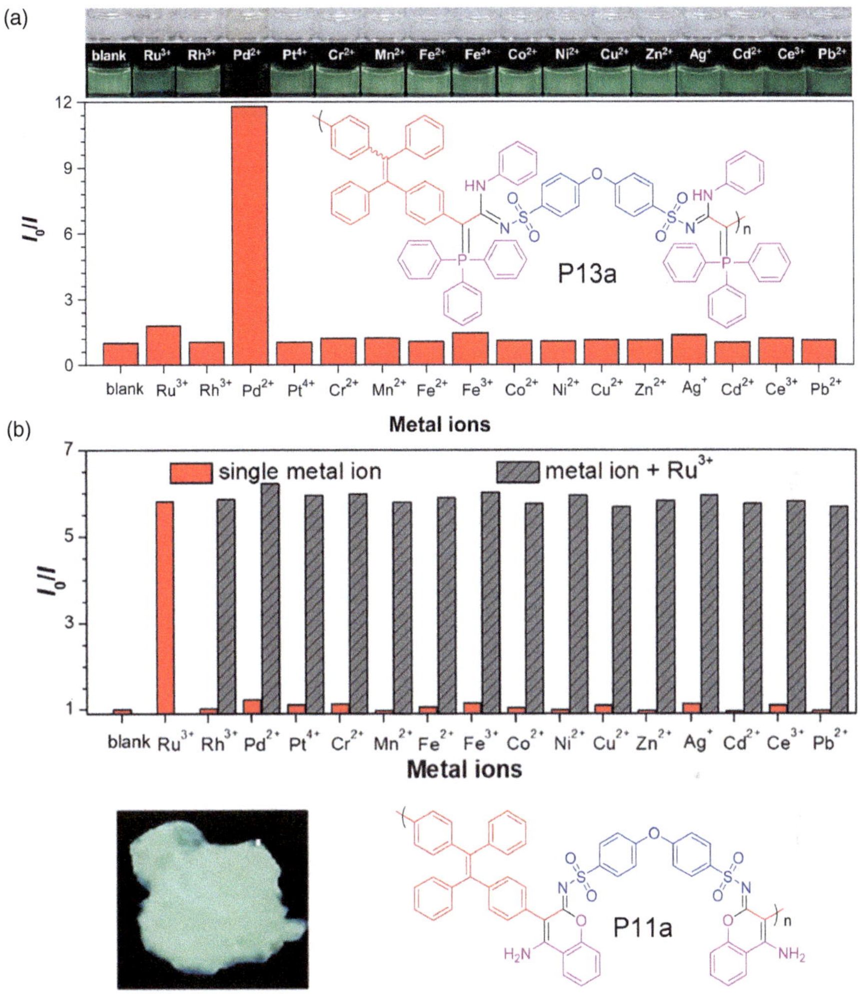

Figure 1.2 (a) Fluorescence Photos of P**13a** in DMSO/water mixtures with 50 vol% water containing different metal ions taken (upper) in daylight and (lower) under UV lamp. Relative intensity (I_0/I) of solution of (a) P**13a** and (b) P**11a** versus different metal ions. I_0 = PL intensity without metal ions. Source: Reproduced from Refs. [77, 79]. Copyright 2017, 2019 American Chemical Society.

MCP polymer products were found to be sensitive and selective fluorescence chemosensors for transition metal ions [88, 90], proving their potential application as efficient chemosensors.

1.5.2 Metal Ion Removal/Enrichment

Besides the detection of existence of metal ions, it is also crucial to remove low concentration toxic metal ions or recover trace amount of precious metal ions from aqueous system. Thanks to the high affinity of sulfur with transition metal ions, the strong coordination between trace amount of metal ions and the sulfur coordination sites in the repeating unit of the sulfur-rich polymers may crosslink the polymer and result in poor solubility of the polymer-metal complexes, which hence precipitated out from the solution to realize removal or enrichment of the trace amount of metal ions.

For example, polythiourea P**5a** prepared from the MCP of sulfur, isocyanide, and amine had shown great potential in mercury detection and removal [58]. Into the Hg^{2+}-containing aqueous solution, when the polythiourea was added, the thiourea groups rapidly coordinated with Hg^{2+} to crosslink the polymers and form insoluble polythiourea-Hg^{2+} complex precipitate, accompanying with the emission quenching of the fluorescent polythiourea (Figure 1.3). The polythioureas could hence be utilized to detect mercury pollution with high sensitivity (K_{sv} = 224 900 M^{-1}) and high selectivity, clean Hg^{2+} contaminant in the polluted water with high removal efficiency (>99.99%) to achieve drinking water standard from a single treatment, and monitor the real-time mercury removal process by fluorescence change. Both DMF solution and solid powder of P**5a** could efficiently absorb Hg^{2+} from aqueous solution and the latter possessed better performance.

Thioamide was also a well-known ligand for precious metal ions such as Au^{3+}. Similarly, the polythioamide P**4a** prepared from the MCP of sulfur, acid, and amine hence could be used to extract Au^{3+} with superior selectivity among 16 different metal ions, high efficiency (>99.99%), high sensitivity (10 ppb), and high extraction capacity (up to 0.60 g Au^{3+}/g polymer) rapidly (within one minute) (Figure 1.4) [53]. The sulfur atoms on the thioamide groups could coordinate with Au^{3+} in a 2 : 1 ratio to crosslink the polymers and form precipitates in the aqueous solution. Most importantly, the gold enrichment test was carried out directly from the acidic leaching solution of discarded electronic waste with high concentrations of interference metal ions such as Ni^{2+} and Cu^{2+}. The polythioamide could still selectively extract Au^{3+} with a high recovery efficiency of 97%, demonstrating great performance of the polythioamide under practical condition. Moreover, elemental gold with high purity could be facilely recovered by simple pyrolysis of the polythioamide-Au^{3+} complex precipitates from the aqueous solution.

1.5.3 Cell Imaging

The fluorescent sulfonylimine-containing polymer P**15a** could also be applied in cell imaging [80]. The cell staining capability of P**15a** and LysoTracker Red (LTR), a lysosome imaging dye, were investigated in Hela cell, which showed remarkable blue and red fluorescence, respectively, under UV irradiation with the excitation wavelength of 405 nm. By merging two images, the cell staining region were almost completely coincident with Pearson's correlation coefficient of 0.96, indicating good cell staining specificity to lysosomes (Figure 1.5), which may be attributed to the weak alkalinity from amine structure. Besides, P**15a** enjoyed excellent performance in both photobleaching and cell viability tests, suggesting its potential application as lysosome-specific dye in biological imaging.

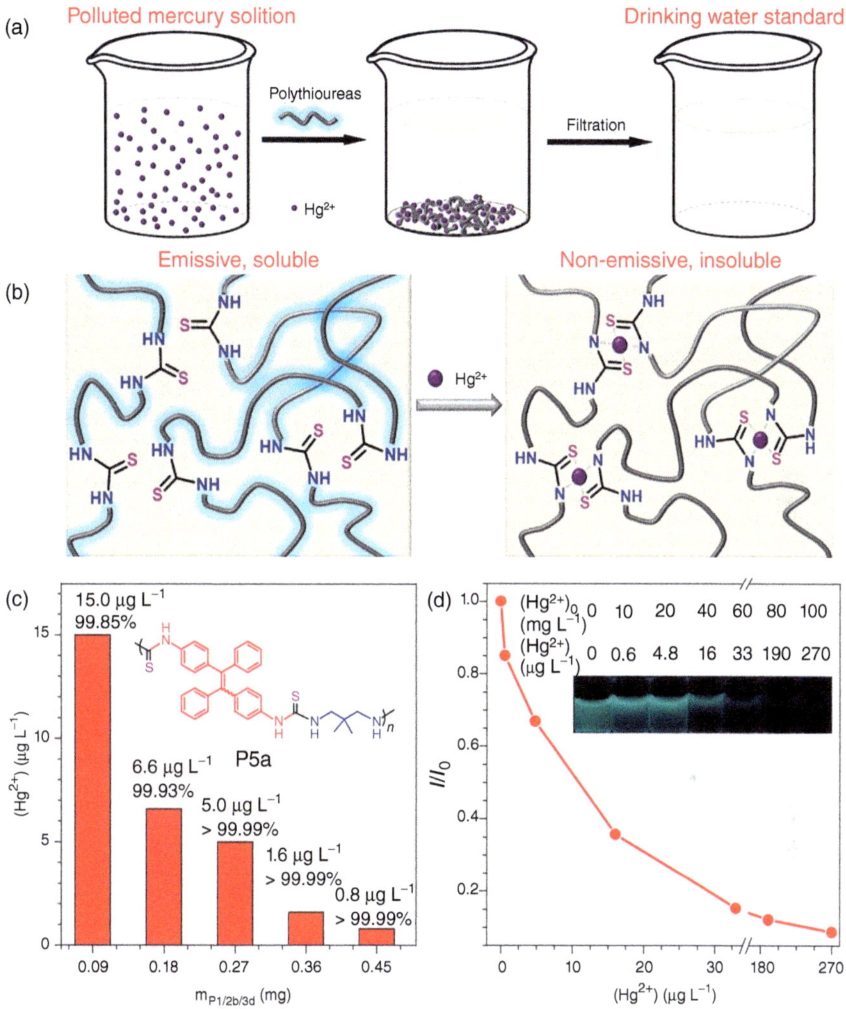

Figure 1.3 (a) Schematic diagram of facile mercury removal process. (b) The proposed mechanism for mercury removal with polythioureas. (c) The mercury removal efficiencies of different amount of P**5a** in DMF solution with 2 mL of Hg^{2+} solution ($[Hg^{2+}]_0 = 10\,mg\,L^{-1}$). (d) The plot of relative emission intensity (I/I_0) versus $[Hg^{2+}]$ when DMF solution of P**5a** ($0.27\,mg\,mL^{-1}$) was mixed with $HgCl_2$ solutions with $[Hg^{2+}]_0$ ranging from 10 to $100\,mg\,L^{-1}$. I_0 = PL intensity in the absence of Hg^{2+}. Inset: fluorescence photographs of the resultant solutions taken upon UV irradiation. Source: Reproduced from Ref. [58]. Copyright 2018 American Chemical Society.

1.6 Conclusion

The various sulfur-containing functional groups such as thioamide, thiourea, sulfonamide, sulfonimine, and thiyl groups have endowed new functionalities to the MCP products, in addition to the classical sulfur-containing polymer materials such

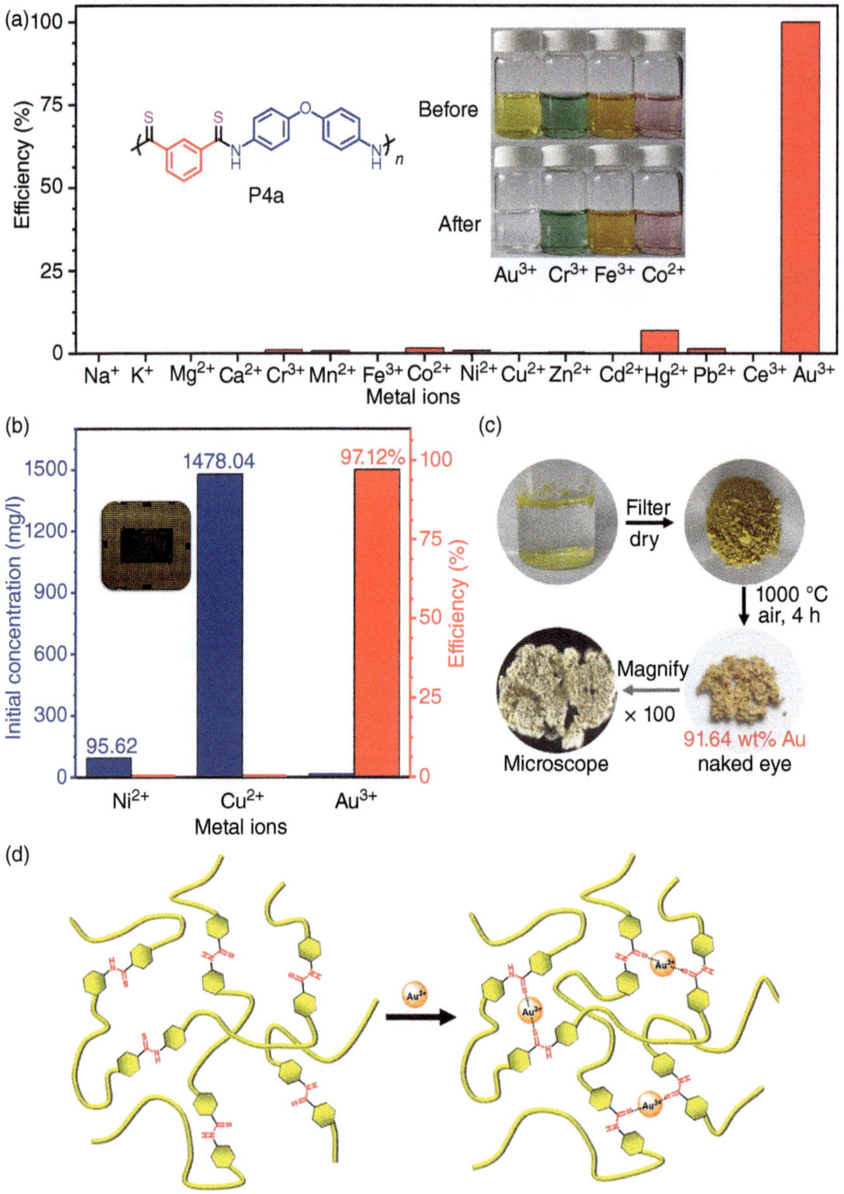

Figure 1.4 (a) Selective enrichment of Au^{3+} by P**4a** in 16 mixed metal ions. (b) The initial concentrations of Au^{3+}, Ni^{2+}, and Cu^{2+} and the selective extraction of Au^{3+} by P**4a** from the leaching solution of a discarded CPU. (c) The gold recovery process with P**4a**. (d) The proposed mechanism for the extraction of Au^{3+} with polythioamides. Source: Reproduced from Ref. [53]. Copyright 2019 American Chemical Society.

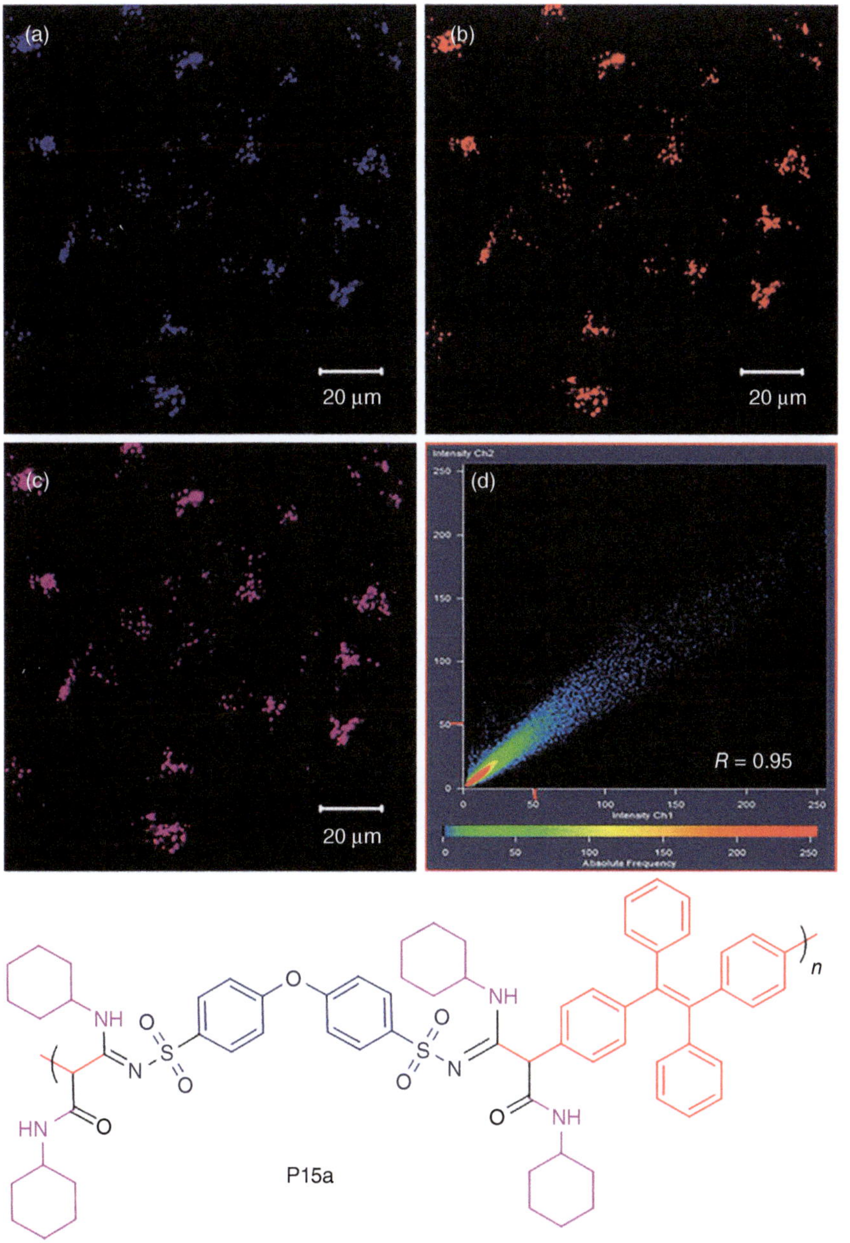

P15a

Figure 1.5 Confocal images of HeLa cells stained with (a) P**15a** and (b) LysoTracker Red (LTR). (c) Merged image of panels a and b. (d) Scatter plot indicating an overlap coefficient between the imaging result of P**15a** and LTR (inset: calculated Pearson's correlation coefficient). Source: Reproduced from Ref. [80]. Copyright 2018 American Chemical Society.

as polythiophenes, polythioethers, etc. MCPs prove to be economic, convenient, and efficient powerful synthetic approaches for the preparation of sulfur-containing polymers from economic and readily available monomers, which generally enjoy high efficiency, atom economy, operational simplicity, and wide monomer scope.

Elemental sulfur can be converted to C=S group-containing polythioamides or polythioureas from its catalyst-free MCPs with amines, together with aldehydes, alkynes, benzyl amines, carboxylic acids, or isocyanides; Sulfonyl-containing moieties can be easily introduced into polymer skeleton from the Cu(I)-catalyzed room temperature MCPs of sulfonyl azides, alkynes, and nucleophiles or other monomers, as well as the MCP of sulfonyl hydrazide, alkyne, and diphenyl dichalcogen; MCTP strategy could be used to combine two or more successive highly efficient reactions into a one-pot procedure, and from thiol, or five-membered cyclic dithiocarbonate or thiolactone, regio- and stereoselective or sequence-controlled sulfur-containing polymers could be obtained. It should be mentioned that with the unique feature of MCPs, through different monomer combinations, sulfur-containing polymers with different main chain structures and topological structures can be facilely constructed, and both mainchain and sidechain of the polymer product can be simultaneously tuned, to afford a large library of polymers. Preliminary investigation of their applications in fluorescent chemosensors, heavy metal ion removal or precious metal ion recovery, and bioimaging were conducted, suggesting the great potential of these sulfur-containing polymers. It is anticipated that from the fascinating MCP approaches, a wide range of sulfur-containing polymer materials could be explored to find their practical applications related to environment, health, and materials.

Acknowledgments

This research was partially supported by the National Science Foundation of China (21822102, 21774034, and 21788102), and the Natural Science Foundation of Guangdong Province (2016A030312002 and 2016A030306045).

References

1 Anderson, L.E., Kleine, T.S., Zhang, Y. et al. (2017). *ACS Macro Lett.* 6 (5): 500–504.
2 Hasell, T., Parker, D.J., Jones, H.A. et al. (2016). *Chem. Commun.* 52 (31): 5383–5386.
3 Arslan, M., Kiskan, B., and Yagci, Y. (2017). *Sci. Rep.* 7 (1): 1–11.
4 Burlingame, Q., Wu, S., Lin, M. et al. (2013). *Adv. Energy Mater.* 3 (8): 1051–1055.
5 Chung, W.J., Griebel, J.J., Kim, E.T. et al. (2013). *Nat. Chem.* 5 (6): 518–524.
6 Agar, S., Baysak, E., Hizal, G. et al. (2018). *J. Polym. Sci. Part Polym. Chem.* 56 (12): 1181–1198.

7 Tsuda, T. and Takeda, A. (1996). *Chem. Commun.* 11: 1317–1318.
8 Patrick, J.C. (1936). *Trans. Faraday Soc.* 32 (0): 347–357.
9 Hulst, R., Seyger, R.M., van Duynhoven, J.P.M. et al. (1999). *Macromolecules* 32 (22): 7521–7529.
10 Lowe, A.B. (2010). *Polym. Chem.* 1 (1): 17–36.
11 Stuparu, M.C. and Khan, A. (2016). *J. Polym. Sci. Part Polym. Chem.* 54 (19): 3057–3070.
12 Malenfant, P.R.L. and Fréchet, J.M.J. (2000). *Macromolecules* 33 (10): 3634–3640.
13 Mutlu, H., Ceper, E.B., Li, X. et al. (2019). *Macromol. Rapid Commun.* 40 (1): 1800650.
14 Griebel, J.J., Nguyen, N.A., Namnabat, S. et al. (2015). *ACS Macro Lett.* 4 (9): 862–866.
15 Kleine, T.S., Nguyen, N.A., Anderson, L.E. et al. (2016). *ACS Macro Lett.* 5 (10): 1152–1156.
16 Zhang, C.J. and Zhang, X.H. (2020). *Macromolecules* 53 (1): 233–239.
17 Park, S., Lee, D., Cho, H. et al. (2019). *ACS Macro Lett.* 8 (12): 1670–1675.
18 Wu, X., Smith, J.A., Petcher, S. et al. (2019). *Nat. Commun.* 10 (1): 647.
19 Hu, R., Li, W., and Tang, B.Z. (2016). *Macromol. Chem. Phys.* 217 (2): 213–224.
20 Hu, R. and Tang, B.Z. (2015). Multicomponent polymerization of alkynes. In: *Multi-Component and Sequential Reactions in Polymer Synthesis* (ed. T. Page), 17–42. Heidelberg: Springer.
21 Kreye, O., Tóth, T., and Meier, M.A.R. (2011). *J. Am. Chem. Soc.* 133 (6): 1790–1792.
22 Deng, X.X., Cui, Y., Du, F.S. et al. (2014). *Polym. Chem.* 5 (10): 3316–3320.
23 Deng, X.X., Li, L., Li, Z.L. et al. (2012). *ACS Macro Lett.* 1 (11): 1300–1303.
24 Sehlinger, A., Dannecker, P.K., Kreye, O. et al. (2014). *Macromolecules* 47 (9): 2774–2783.
25 Wu, H., Fu, C., Zhao, Y. et al. (2015). *ACS Macro Lett.* 4 (11): 1189–1193.
26 Moldenhauer, F., Kakuchi, R., and Theato, P. (2016). *ACS Macro Lett.* 5 (1): 10–13.
27 Chan, C.Y.K., Tseng, N.W., Lam, J.W.Y. et al. (2013). *Macromolecules* 46 (9): 3246–3256.
28 Wei, B., Li, W., Zhao, Z. et al. (2017). *J. Am. Chem. Soc.* 139 (14): 5075–5084.
29 Kayser, L.V., Vollmer, M., Welnhofer, M. et al. (2016). *J. Am. Chem. Soc.* 138 (33): 10516–10521.
30 Tang, X., Zheng, C., Chen, Y. et al. (2016). *Macromolecules* 49 (24): 9291–9300.
31 Fu, W., Kong, L., Shi, J. et al. (2019). *Macromolecules* 52 (2): 729–737.
32 Qi, C., Zheng, C., Hu, R. et al. (2019). *ACS Macro Lett.* 8 (5): 569–575.
33 Fu, W., Shi, J., Tong, B. et al. (2019). *Macromolecules* 52 (9): 3319–3326.
34 Xue, H., Zhao, Y., Wu, H. et al. (2016). *J. Am. Chem. Soc.* 138 (28): 8690–8693.
35 Worthington, M.J.H., Kucera, R.L., and Chalker, J.M. (2017). *Green Chem.* 19 (12): 2748–2761.
36 Rauchfuss, T. (2011). *Nat. Chem.* 3 (8): 648–648.
37 Murai, T. (ed.) (2019). *Chemistry of Thioamides*. Singapore: Springer.

38 Kagaya, S., Tanaka, E., Kawai, N. et al. (2009). *J. Inorg. Organomet. Polym. Mater.* 19 (1): 67–73.

39 Kagaya, S., Miyazaki, H., Ito, M. et al. (2010). *J. Hazard. Mater.* 175 (1–3): 1113–1115.

40 Sanui, K., Kishimoto, Y., and Ogata, N. (1971). *Polym. J.* 2 (3): 422–425.

41 Delfanne, I. and Levesque, G. (1989). *Macromolecules* 22 (6): 2589–2592.

42 Kindler, K. (1923). *Justus Liebigs Ann. Chem.* 431 (1): 187–230.

43 Kawai, Y., Kanbara, T., and Hasegawa, K. (1999). *J. Polym. Sci. Part Polym. Chem.* 37 (12): 1737–1740.

44 Kanbara, T., Kawai, Y., Hasegawa, K. et al. (2001). *J. Polym. Sci. Part Polym. Chem.* 39 (21): 3739–3750.

45 D'Souza, D.M. and Müller, T.J.J. (2007). *Chem. Soc. Rev.* 36 (7): 1095–1108.

46 Nguyen, T.B., Tran, M.Q., Ermolenko, L. et al. (2014). *Org. Lett.* 16 (1): 310–313.

47 Li, W., Wu, X., Zhao, Z. et al. (2015). *Macromolecules* 48 (21): 7747–7754.

48 Wang, R., Yuan, W., and Zhu, X. (2015). *Chin. J. Polym. Sci.* 33 (5): 680–687.

49 Sasaki, Y. and Olsen, F.P. (1971). *Can. J. Chem.* 49 (2): 283–293.

50 Nguyen, T.B., Ermolenko, L., and Al-Mourabit, A. (2012). *Org. Lett.* 14 (16): 4274–4277.

51 Sun, Z., Huang, H., Li, L. et al. (2017). *Macromolecules* 50 (21): 8505–8511.

52 Guntreddi, T., Vanjari, R., and Singh, K.N. (2014). *Org. Lett.* 16 (14): 3624–3627.

53 Cao, W., Dai, F., Hu, R. et al. (2020). *J. Am. Chem. Soc.* 142 (2): 978–986.

54 Wu, S., Li, W., Lin, M. et al. (2013). *Adv. Mater.* 25 (12): 1734–1738.

55 Yanagisawa, Y., Nan, Y., Okuro, K. et al. (2018). *Science* 359 (6371): 72–76.

56 Kausar, A., Zulfiqar, S., Ahmad, Z. et al. (2012). *J. Appl. Polym. Sci.* 124 (1): 373–385.

57 Nguyen, T.B., Ermolenko, L., and Al-Mourabit, A. (2014). *Synthesis* 46 (23): 3172–3179.

58 Tian, T., Hu, R., and Tang, B.Z. (2018). *J. Am. Chem. Soc.* 140 (19): 6156–6163.

59 Tuten, B.T., Bloesser, F.R., Marshall, D.L. et al. (2018). *ACS Macro Lett.* 7 (8): 898–903.

60 Wu, X., Lin, H., Dai, F. et al. (2020). *CCS Chem.* 2 (3): 191–202.

61 Zhu, Y.F., Zhang, Z., Litt, M.H. et al. (2018). *Macromolecules* 51 (16): 6257–6266.

62 Tong, H., Fu, J., Ahmad, A. et al. (2019). *Macromol. Mater. Eng.* 304 (4): 1800709.

63 Dong, J., Sharpless, K.B., Kwisnek, L. et al. (2014). *Angew. Chem. Int. Ed.* 53 (36): 9466–9470.

64 Rostovtsev, V.V., Green, L.G., Fokin, V.V. et al. (2002). *Angew. Chem. Int. Ed.* 41 (14): 2596–2599.

65 Bae, I., Han, H., and Chang, S. (2005). *J. Am. Chem. Soc.* 127 (7): 2038–2039.

66 Lee, I.H., Kim, H., and Choi, T.L. (2013). *J. Am. Chem. Soc.* 135 (10): 3760–3763.

67 Kim, H. and Choi, T.L. (2014). *ACS Macro Lett.* 3 (8): 791–794.

68 Deng, H., Han, T., Zhao, E. et al. (2016). *Polym. Chem.* 7 (36): 5646–5654.

69 Cho, S.H., Yoo, E.J., Bae, I. et al. (2005). *J. Am. Chem. Soc.* 127 (46): 16046–16047.

70 Deng, H., Zhao, E., Li, H. et al. (2015). *Macromolecules* 48 (10): 3180–3189.

71 Cui, S.L., Lin, X.F., and Wang, Y.G. (2006). *Org. Lett.* 8 (20): 4517–4520.
72 Yi, F., Zhang, S., Huang, Y. et al. (2017). *Eur. J. Org. Chem.* 2017 (1): 102–110.
73 Cheng, D., Ling, F., Zheng, C. et al. (2016). *Org. Lett.* 18 (10): 2435–2438.
74 Cui, S.L., Wang, J., and Wang, Y.G. (2008). *Org. Lett.* 10 (6): 1267–1269.
75 Xu, X., Cheng, D., Li, J. et al. (2007). *Org. Lett.* 9 (8): 1585–1587.
76 Deng, H., Han, T., Zhao, E. et al. (2016). *Macromolecules* 49 (15): 5475–5483.
77 Xu, L., Zhou, T., Liao, M. et al. (2019). *ACS Macro Lett.* 8 (2): 101–106.
78 Xu, L., Zhou, F., Liao, M. et al. (2018). *Polym. Chem.* 9 (13): 1674–1683.
79 Xu, L., Hu, R., and Tang, B.Z. (2017). *Macromolecules* 50 (16): 6043–6053.
80 Han, T., Deng, H., Qiu, Z. et al. (2018). *J. Am. Chem. Soc.* 140 (16): 5588–5598.
81 Yang, F.L. and Tian, S.K. (2017). *Tetrahedron Lett.* 58 (6): 487–504.
82 Liu, Y., Zheng, G., Zhang, Q. et al. (2017). *J. Org. Chem.* 82 (4): 2269–2275.
83 Gao, Q., Xiong, L.H., Han, T. et al. (2019). *J. Am. Chem. Soc.* 141 (37): 14712–14719.
84 Imbesi, P.M., Fidge, C., Raymond, J.E. et al. (2012). *ACS Macro Lett.* 1 (4): 473–477.
85 Wu, W., Tang, R., Li, Q. et al. (2015). *Chem. Soc. Rev.* 44 (12): 3997–4022.
86 Deng, J., Zhou, Y., Xu, B. et al. (2011). *Biomacromolecules* 12 (3): 642–649.
87 Kim, H., Bang, K.T., Choi, I. et al. (2016). *J. Am. Chem. Soc.* 138 (27): 8612–8622.
88 Xu, L., Yang, K., Hu, R. et al. (2018). *Synlett* 29 (19): 2523–2528.
89 Zheng, C., Deng, H., Zhao, Z. et al. (2015). *Macromolecules* 48 (7): 1941–1951.
90 Deng, H., Hu, R., Zhao, E. et al. (2014). *Macromolecules* 47 (15): 4920–4929.
91 Teiber, M. and Müller, T.J.J. (2012). *Chem. Commun.* 48 (15): 2080–2082.
92 Hoyle, C.E., Lowe, A.B., and Bowman, C.N. (2010). *Chem. Soc. Rev.* 39 (4): 1355–1387.
93 Ochiai, B., Ogihara, T., Mashiko, M. et al. (2009). *J. Am. Chem. Soc.* 131 (5): 1636–1637.
94 Zhang, Z., You, Y.Z., Wu, D.C. et al. (2015). *Polymer* 64: 221–226.
95 Yu, L., Zhang, Z., You, Y.Z. et al. (2018). *Eur. Polym. J.* 103: 80–87.
96 Zhang, Z., You, Y.Z., Wu, D.C. et al. (2015). *Macromolecules* 48 (11): 3414–3421.
97 Zhang, Z., Tan, Z.B., Hong, C.Y. et al. (2016). *Polym. Chem.* 7 (7): 1468–1474.
98 Long, Z., Mao, L., Liu, M. et al. (2017). *Polym. Chem.* 8 (37): 5644–5654.
99 Zhan, R., Pan, Y., Manghnani, P.N. et al. (2017). *Macromol. Biosci.* 17 (5): 1600433.

2

Carbon Disulfide Derived Polymers

Gulzar A. Bhat and Donald J. Darensbourg

Texas A&M University, Department of Chemistry, 3255 TAMU, College Station, TX 77843, USA

2.1 Introduction

The production of novel polymeric materials derived from CS_2 that itself can be obtained directly from elemental sulfur offers an attractive imperative in polymer chemistry [1]. The inexpensive and abundant CS_2 is industrially synthesized from natural gas and elemental sulfur at approximately 600 °C over a heterogeneous metal catalyst [2]. This procedure constitutes a significant use of large quantities of elemental sulfur, which is produced in over 60 million tons annually as a result of hydrodesulfurization processes (Scheme 2.1). However, the major use of elemental sulfur is in the synthesis of the commodity chemical, sulfuric acid, with smaller quantities being utilized in manufacturing synthetic rubber via vulcanization. In the absence of the hydrodesulfurization process, the burning of fossil fuel releases sulfur dioxide (SO_2) into the atmosphere as a pollutant, which causes respiratory problems and acid rain. Additionally, sulfur in gasoline can have deleterious effects on catalytic converters in cars.

Over one million tons of CS_2 is produced annually worldwide by the above process, with the United States being responsible for about 60 000 tons. CS_2's chief uses are for the manufacture of rayon, cellulose film, and carbon tetrachloride. Historically, these manufacturing processes were the principal sources of humans' exposure to CS_2, although small quantities of CS_2 are released from volcanic eruptions into the atmosphere. Currently, in the United States, the major source of CS_2 emissions into the environment arises from hydraulic fracturing. It has been shown that occupational workers in rayon and energy extraction industries have an increased risk for glucose metabolism interference [3]. CS_2 can be readily converted to CO_2 and is thereby a direct or indirect contributor to climate change [4].

The synthesis of copolymers from CO_2 and cyclic ethers (oxiranes and oxetanes), and analogous processes involving COS to a lesser extent, are extremely well-established catalytic pathways for the production of polycarbonates and polymonothiocarbonates [5–21]. Indeed, the coupling of CO_2 and ethylene oxide or

Sulfur-Containing Polymers: From Synthesis to Functional Materials,
First Edition. Edited by Xing-Hong Zhang and Patrick Theato.

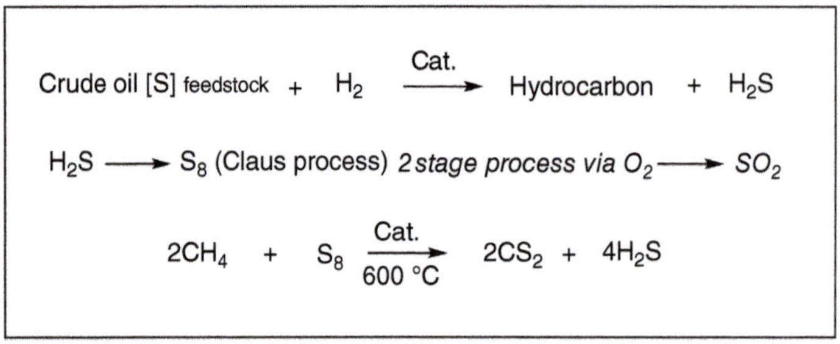

Scheme 2.1 Hydrodesulfurization to carbon disulfide.

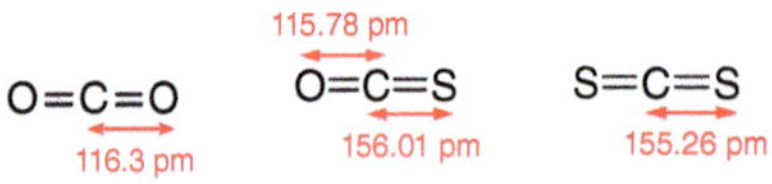

Figure 2.1 Structures and bond distances in CO_2, COS, and CS_2.

propylene oxide in the presence of protic chain transfer agents is a commercially viable process for the formation of polyols for use in polyurethane production [22–26]. Corresponding to value added products derived from greenhouse gases, polymers based on carbon disulfide provide sulfur atoms in polymer chains, thereby affording polymeric materials with numerous outstanding properties. These included high refractive indices, enhanced melting temperatures, self-healing capabilities, and metal binding sites [27–29]. This chapter will summarize the use of CS_2-based polymers derived mostly from the copolymerization of carbon disulfide, oxiranes, or thiiranes. Other minor uses of CS_2 as a chemical feedstock for the synthesis of sulfur containing polymers will be discussed.

Prior to a discussion of sulfur containing copolymers based on CS_2, it is worthwhile to compare the physical, structural, and electronic/reactivity properties of the three congeners (CO_2, COS, and CS_2). Carbon dioxide and carbonyl sulfide are gases, and unlike CO_2, COS is flammable. On the other hand, carbon disulfide is a colorless volatile organic liquid (b.p. 46.3 °C), and resembling COS is a hazardous air pollutant. All three are linear molecules consisting of double bonded carbon to oxygen or sulfur, with bond distances as indicated in Figure 2.1.

Based on structural studies of the binding of these molecules with *N*-heterocyclic olefins, the carbon disulfide adducts displayed shorter $C—CS_2$ bonds compared with their CO_2 and COS analogs [30, 31]. Furthermore, for the COS adduct, the negative charge was preferentially delocalized on the sulfur atom (Figure 2.2). The relative

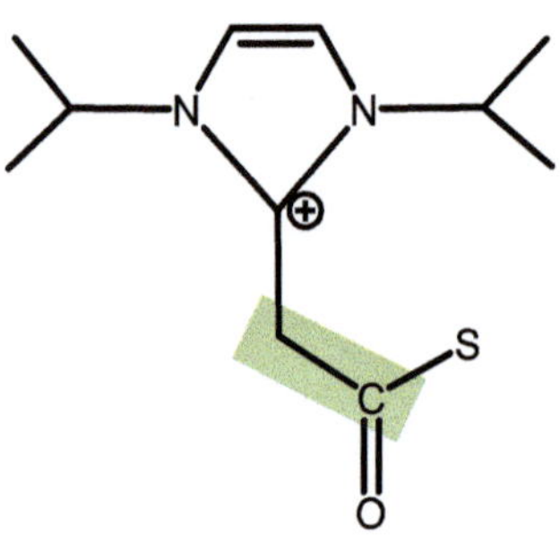

Figure 2.2 Carbonyl sulfide adduct with *N*-heterocyclic olefin.

reactivity of these triatomics with metal-oxygen bonds, a process of importance to the copolymerization of these molecules with epoxides, was shown to occur with $CS_2 > COS > CO_2$ (Eq. (2.1)) [32, 33].

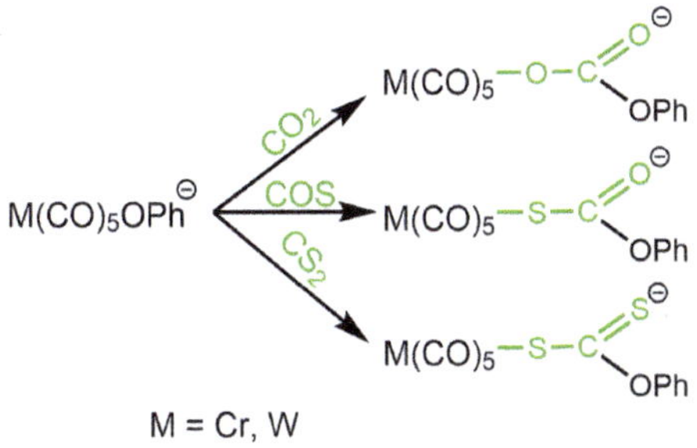

(2.1)

As stated earlier, the subject matter of this chapter will summarize the published literature on the production of polymers and their cyclic monomers from coupling reactions of CS_2 with oxiranes (epoxides) or thiiranes (episulfides). A discussion of sources of these cyclic thioethers, which serve as comonomers with CS_2, will be covered, as well as other limited polymeric materials derived from CS_2. Hence, this synopsis of sulfur containing polymers based on CS_2 ultimately, although indirectly, addresses the use of elemental sulfur as a chemical feedstock for polymer production.

2.2 Synthesis of Thiiranes (Episulfides)

Episulfide, a saturated three membered heterocycle with one sulfur atom, was the first thiirane reported by Staudinger and Pfenninger in 1916 [34]. Analogous to the related epoxides or oxiranes, such sulfur heterocycles constitute key intermediates in synthetic chemistry. The nucleophilic ring-opening of these heterocycles serves as a convenient access to sulfur-containing compounds [35]. Episulfides have also been employed in various commercial and industrial applications such as disinfectants, precursors for synthetic polymers [36], stabilizers for plastics, and as pharmacologically active substances [37].

Numerous synthetic routes have been developed to synthesize episulfides [38, 39]. In the conventional approach, episulfides are synthesized from alkenes via indirect methods of converting them first to oxygen-functionalized precursor, particularly epoxides that upon nucleophilic substitution leads to the formation of thioepoxide. The main drawback of this approach is that during the nucleophilic substitution, frequently polymerization of the thiirane occurs as a result of this side reaction selectivity is reduced [40, 41]. Catenation and sulfur extrusion severely limit the synthetic usage of this method [42]. The synthesis of thiiranes by direct interaction of alkenes with sulfur transferring agents offers attractive opportunities. This challenging concept has first been realized 55 years before and research in this direction has intensified in the last two decades [42]. The different sulfur transferring agents are broadly classified into two main categories viz stoichiometric and *metal-catalyzed* episulfidation.

The atomic sulfur is predominantly obtained by vapor-phase irradiation of COS and exists mostly in singlet state [S(^{1}D)] and triplet ground state [S(^{3}P)] [43]. The triplet atomic state reacts with alkenes to form the diradical, which upon cyclization and intersystem crossing lead to the formation of thiirane (Scheme 2.2). Due to the fast intersystem crossing, the episulfidation with S(^{3}P) occurs with retention of the alkene configuration [44]. This route of episulfidation is of little interest as the photolysis of COS, which is used for making triplet ground state [S(^{3}P)], occurs on irradiation around 254 nm for prolonged time and under these reaction conditions the thiiranes will not persist.

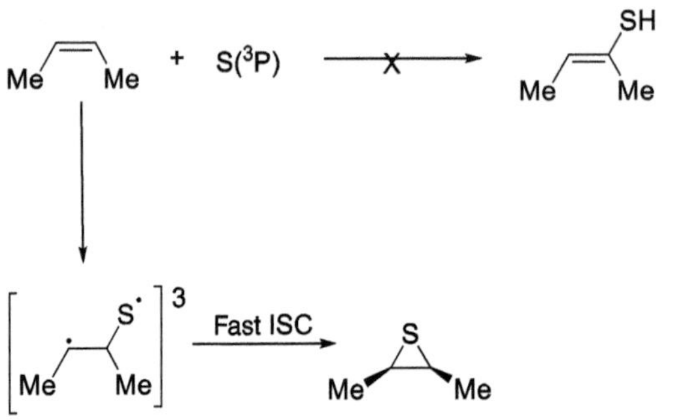

Scheme 2.2 Stereoselective addition of triplet atomic sulfur to *Z*-butene.

The catenation propensity of triplet sulfur atoms leads to S_8 formation, which further limits the usage of this approach [45]. Later it has also been observed that isothiocyanates can act as direct precursors for triplet atomic sulfur which resulted in enhancement of episulfide yields up to 45% at 76% conversion of the alkene (Scheme 2.3) [47]. In striking contrast to the triplet atomic sulfur [S(^{3}P)], the singlet state [S(^{1}D)] is unsuitable for episulfidation both in the vapor phase and in solution.

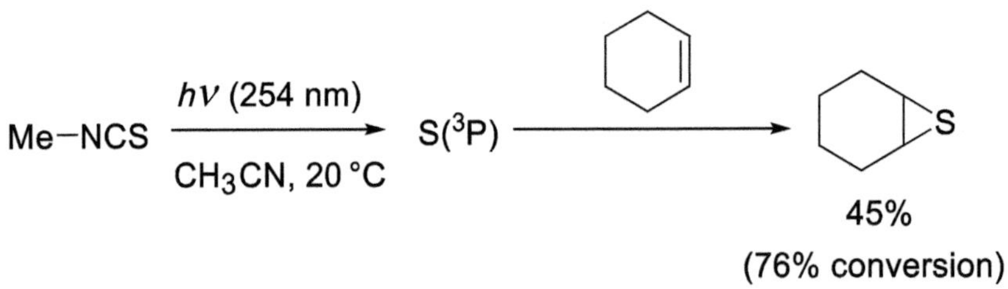

Scheme 2.3 Direct episulfidation of cyclohexene by triplet atomic sulfur. Source: Modified from Ref. [46].

Thiiranes were also synthesized in few instances from alkenes by using elemental sulfur without chemical activation [46]. Heating *syn*-bibenzonorbornenylidene (*syn*-**1**) with elemental sulfur in refluxing *o*-dichlorobenzene led exclusively to episulfide *syn*-**2** in 83% yield with retention of the configuration (Scheme 2.4) [48]. The exclusive formation of episulfide *anti*-**1** in 72% yield from *anti*-**2** under the same conditions was also observed.

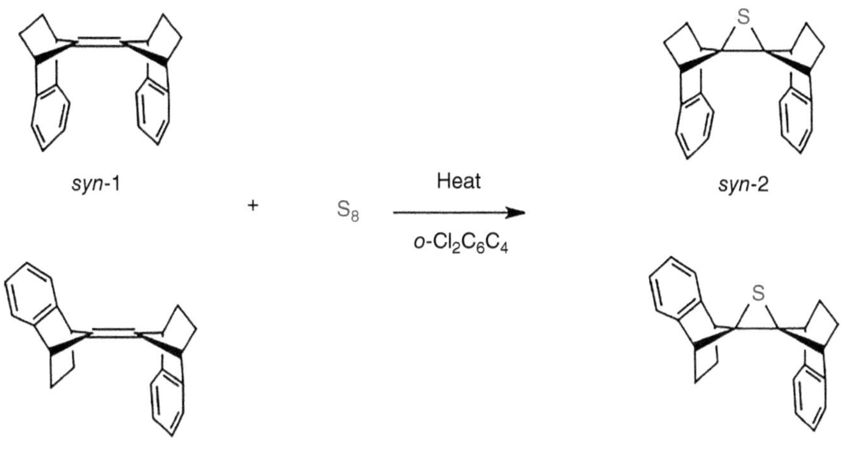

Scheme 2.4 Reaction of bibenzonorbornenylidene *syn*-**1** and *anti*-**1**with S_8. Source: Refs. [48a, b].

Under refluxing conditions alkene **3** also reacts with elemental sulfur to give both *anti*-**4** (76%) and *syn*-**4** (4%) products, along with unreacted starting alkene **3** (19%) (Scheme 2.5, entry 1) [49, 50]. Prolonged heating with excess sulfur did not improve the conversion of **3**, suggesting that the sulfuration must be reversible. Indeed, heating *syn*-**4** or *anti*-**4** product with or without elemental sulfur under the same conditions produced nearly the same product distribution.

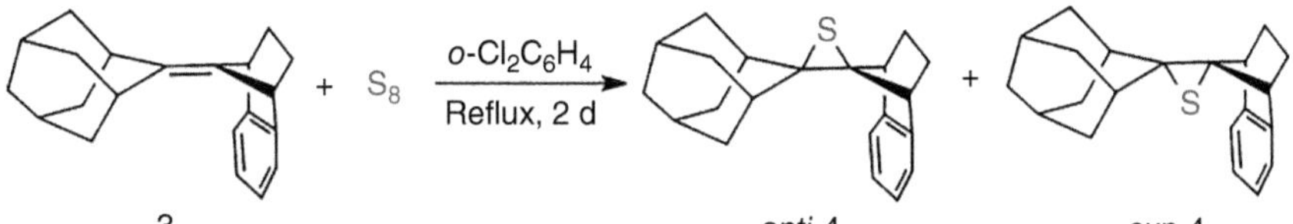

Scheme 2.5 Reaction of alkene-**3** with S_8 along with different optimization conditions. Source: Refs. [49, 50].

The elemental sulfur was also chemically activated by using different metal complexes for synthesizing episulfides. The catalytic amounts of the oxo complex **7** (7 mol%) on interaction with S_8 lead to the formation of *trans*-cyclooctene episulfide (*trans*-**6a**) and *trans*-cyclononene episulfide (*trans*-**6b**) in 69% and 87% yields, respectively (Scheme 2.6) [51].

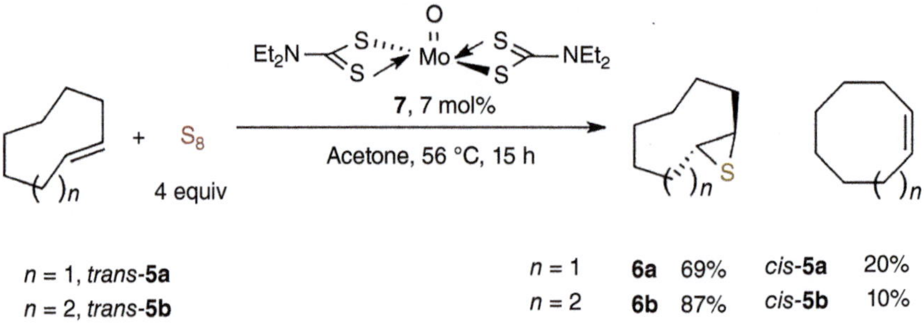

Scheme 2.6 Catalytic activation of S_8 for the conversion of alkenes to episulfides. Source: Modified from Ref. [51].

Episulfidation of strained cyclic allenes such as 1,2-cyclononadiene **10** was carried by reacting with elemental sulfur in presence of catalytic amount of the molybdenum oxo complex Mo(O)dtc_2 **7** (Scheme 2.7) [52].

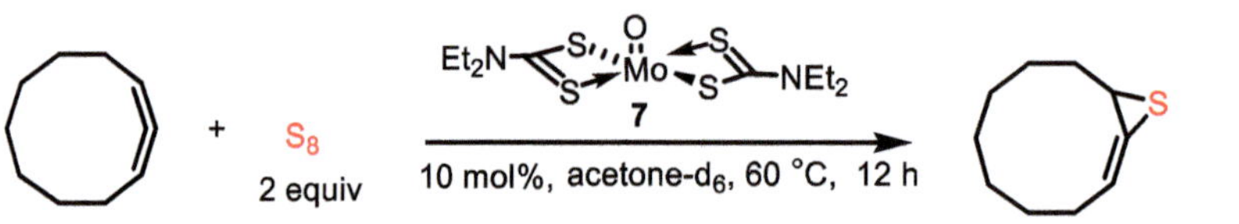

Scheme 2.7 Catalytic conversion of 1,2-cyclononadiene to the episulfide derivative. Source: Based on Ref. [52].

Recently a rhodium complex generated *in situ* from $RhH(PPh_3)_4$, dppe, and 4-tolylacetylene was found to catalyze the episulfidation of substituted norbornenes to their corresponding exo thiiranes under mild thermal conditions (Scheme 2.8) [53]. The substrate scope was further extended to *trans*-cyclooctene as well as cyclic allenes.

2.3 Copolymerization Reactions

These reactions will be categorized into two classes: (i) CS_2 and episulfides, and (ii) CS_2 and epoxides.

2.3.1 Copolymerization of Carbon Disulfide and Episulfides

Introducing sulfur atoms inside the polymer backbone imparts them with attractive features like optical and thermal properties, chemical resistance, ability to recognize heavy-metals, etc. The coupling of episulfide and CS_2 was first reported by Soga et al. in 1970s by performing the copolymerization of ethylene sulfide (ES) [54] or propylene sulfide (PS) [55] and CS_2 using $CdEt_2$, $ZnEt_2$ and $Hg(SBu)_2$ as catalysts. Besides the low catalytic activity of these catalysts the polymers produced

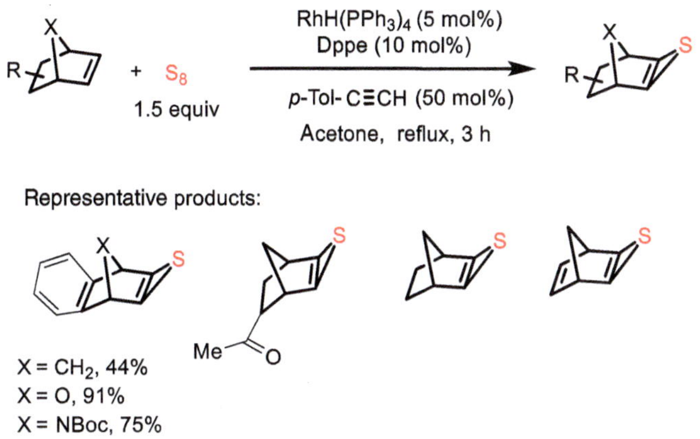

Scheme 2.8 Catalytic conversion of substituted norbornenes to their corresponding exo thiiranes. Source: Based on Ref. [53].

viz poly([thio]tricarbonate-*co*-thioether) have 93% content of thioether linkages. The first perfectly alternating poly[thio]tricarbonates with high number-average molecular weights (M_n = 11 000–20 300 g mol^{-1}) from the coupling of CS_2 and episulfides was reported by Nozaki and coworkers in 2007 [56]. The coupling of PS with CS_2 using (salen)CrX as catalyst lead to the formation of poly(propylene trithiocarbonate) and cyclic propylene trithiocarbonate (Scheme 2.9).

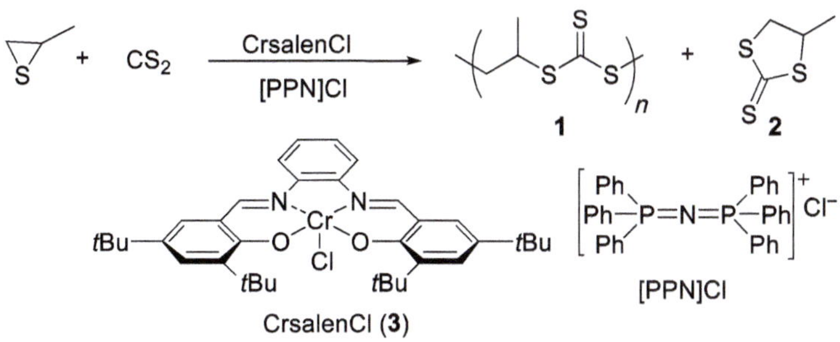

Scheme 2.9 Copolymerization of propylene sulfide with CS_2 using [PPN]Cl and CrSalenCl.

This copolymerization process was optimized to provide completely alternating polymers with high selectivity and molecular weights. Episulfides are known to polymerize following an anionic mechanism, accordingly the optimization is achieved by using different ratios of PS/initiator. Initially using thiol and DBU (1,8-diazabicyclo[5.4.0]undec-7-ene) together lead to the formation of only cyclic propylene trithiocarbonate (**2**), whereas copolymerization in the presence of bis(triphenylphosphoranylidene) ammonium chloride ([PPN]Cl) (0.20 mol%, $[PPN]^+ = [Ph_3P{=}N{=}PPh_3]^+$) for five hours yielded the completely alternating copolymer, poly(propylene trithiocarbonate) (**1**) with low selectivity (Table 2.1). It has also being observed that using tetrabutylammonium chloride produces the polymer, whereas no polymer formation is observed in the presence of phosphonium salt $[Ph_3PMe]Cl$.

Table 2.1 Copolymerization of PS with CS_2 by using [PPN]Cl/complex **3** system.[a)]

Run	PS/CS_2	Initiator/catalyst	Yield of 1 + 2 (%)[b)]	1/2[b)]	M_n (g mol^{-1})	M_w/M_n[c)]
1	1/1	[PPN]Cl	16	37/63	11 600	1.4
2	1/1	Bu_4NCl	5	42/58	nd[d)]	nd[d)]
3	1/1	[PPN]Cl/**3**	60	80/20	20 300	1.4
4	1/0.5	[PPN]Cl/**3**	44	48/52	11 000	1.6
5	1/0.7	[PPN]Cl/**3**	52	71/29	15 400	1.6
6	1/2	[PPN]Cl/**3**	53	88/12	18 200	1.4
7[e)]	1/2	[PPN]Cl/**3**	90	92/8	42 600	1.3

a) Reaction conditions: PS (10 mmol), [PPN]Cl or Bu_4NCl (0.20 mol%), [complex **3** (0.20 mol%)], CS_2 at 25 °C for 5 h.
b) Determined on the basis of ^{1}H NMR spectroscopy of the crude product by using dodecane as an internal standard.
c) Determined by gel permeation chromatography using a polystyrene standard.
d) nd, Not determined.
e) Premixed complex from [PPN]Cl and complex **3** was used.

The rate and selectivity of copolymerization were increased by adding a metal salen complex viz (salph)CrCl complex (**3**) [(salph)H_2) = bis-(3,5-di-*tert*-butylsalicylidene)-1,2-diaminobenzene]. The metal complex **3** was proved to be the most effective catalyst for selectively producing the copolymers in combination with [PPN]Cl (Table 2.1). The use of metal complex **3** alone does not copolymerize PS with CS_2. The selectivity of copolymer/cyclic trithiocarbonate(**1**/**2**) also depends on the molar ratio of PS and CS_2 viz increasing the proportion of CS_2 tends to enhance the copolymerization reaction. It has been observed that use of 2 equiv of CS_2 to PS is optimal for both high selectivity and good catalytic activity as shown in Table 2.1.

The alternating copolymers obtained were highly soluble in CS_2, whereas they were slightly soluble in $CHCl_3$ and THF. The chain sequence of the polymers was confirmed by ^{1}H and ^{13}C NMR spectroscopy, and no sequence was assigned to thioether linkages confirming the alternating nature of the polymers. The head to tail regioregularity of the polymers was confirmed by recording their ^{13}C NMR. The ^{13}C NMR signal of each carbon splits into two or three peaks (Figure 2.3b), which is a characteristic feature of atactic polymers. To further illustrate this a comparable polymer from enantio-enriched PS (90% ee) is also synthesized and its ^{13}C NMR gives sharp and almost undivided peaks for each of the carbons as expected for isotactic polymers. (Figure 2.3a).

The mechanistic pathway for this copolymerization is not clear, but it is speculated that it follows the same route as the copolymerization of CO_2 and epoxides as shown in Scheme 2.10. The chromium complex works as a Lewis acid for activating the episulfides and both of the chloride ions in the catalyst are capable of initiating the polymerization.

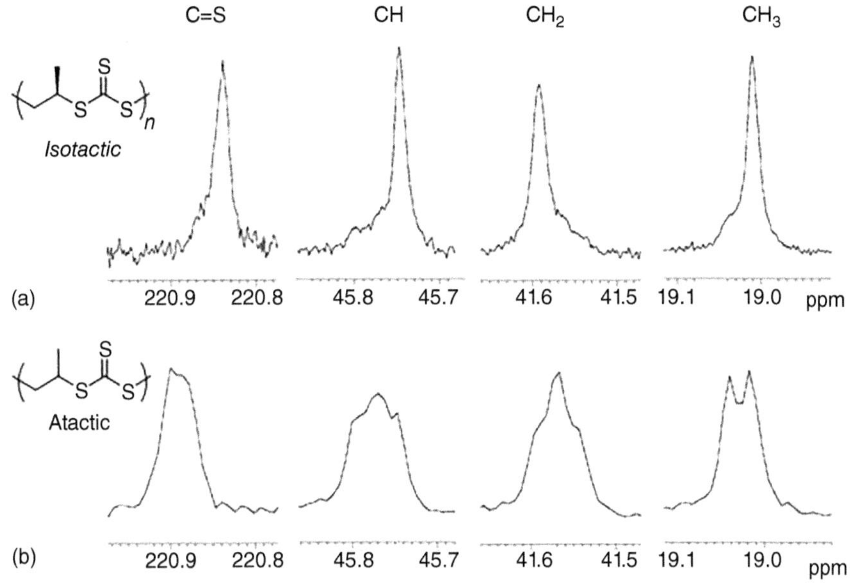

Figure 2.3 Comparison of ^{13}C NMR signals of poly(propylene trithiocarbonate)s from (a) (*R*)-propylene sulfide (90% ee) and (b) *rac*-propylenesulfide (in $CDCl_3 + CS_2$, 125 MHz).

Although Nozaki's catalyst/cocatalyst system exhibited efficient alternating copolymerization of episulfide with CS_2, both still had the limitation of by-product formation of [thio]tricarbonate. Recently, Xingmei et al. reported an asymmetrical Cr(III)-bis-schiff-base complex [Cr(L)Cl] (Table 2.2) as catalyst for the coupling of PS and CS_2 to form poly[trithio]carbonate selectively [57]. It has been observed that the pre-activation of catalyst system has dramatic effect on the product formation. Both the cyclic and polymeric products were formed using this catalyst along with [PPN]Cl as cocatalyst, but it was observed that the conversion of PS and selectivity for poly[trithio]carbonate formation were strongly dependent on the reaction conditions and concentration of catalyst/cocatalyst as shown in Table 2.2.

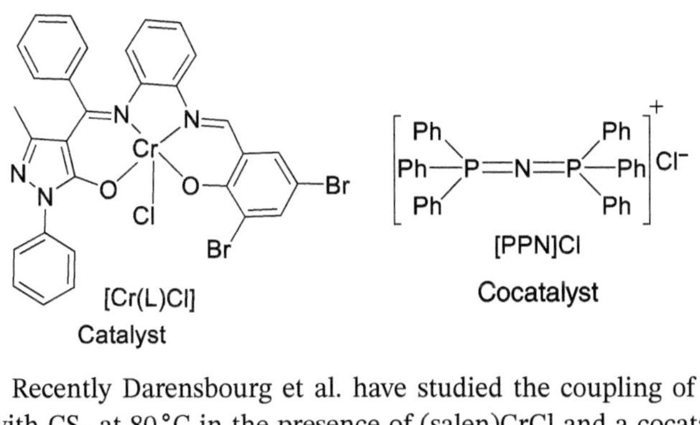

Recently Darensbourg et al. have studied the coupling of cyclopentene sulfide with CS_2 at 80 °C in the presence of (salen)CrCl and a cocatalyst (Table 2.3) [58]. This process was shown to be very ineffective, with only limited formation of

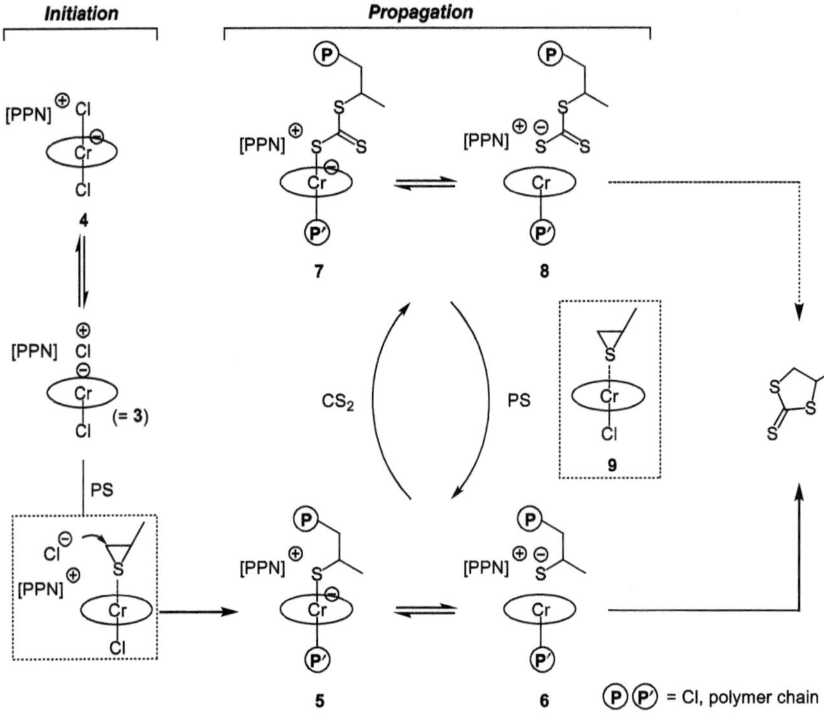

Scheme 2.10 Plausible reaction mechanism of copolymerization of propylene sulfide with CS_2 using [PPN]Cl as cocatalyst and CrsalenCl as catalyst.

cis-cyclopentene trithiocarbonate (entire 1–3), along with a trace quantity of the desulfurization product, cyclopentene. Upon increasing the temperature to 110 °C for 24 hours, ~20% and 40% conversion to *cis*-cyclopentene trithiocarbonate and to cyclopentene, respectively, were observed (entries 4–5). In the absence of cocatalyst, cyclopentene was the major product at 36% (entry 6). Cyclopentene sulfide was unreactive with CO_2 under similar reaction conditions. The lack of activity of cyclopentene sulfide is unique to the cyclopentyl derivative, for cyclohexene sulfide [59] and PS [56] both undergo reactions with CS_2 to afford poly(trithiocarbonate) and cyclic trithiocarbonate under similar conditions.

Of the cyclic thiocarbonates produced, the *trans*-cyclopentene trithiocarbonate was identified by X-ray crystallography (Figure 2.4). This trans structure is consistent with that previously reported in the literature [60]. In studies of the coupling reaction of CS_2 and cyclopentene oxide (CPO), gas chromatography–mass spectrometry (GC-MS) provided a useful method for identifying the nature of the various cyclic materials produced. The method of analysis demonstrated that eight of the possible twelve isomers of cyclic[thio]carbonates were afforded by the coupling of CS_2 and CPO.

The relative enthalpies of each of the possible cyclopentene [thio]carbonate species were calculated, with only a slight energy difference (1–2 kcal mol^{-1}) observed between the cis-boat and cis-chair conformations (Figure 2.5). In general,

Table 2.2 Optimization of different reaction conditions used for the copolymerization of PS with CS_2.

Entry	CS_2:PS:1:[PPN]Cl (molar ratio)	Temp (°C)	Time (h)	Conv.[a)] (%; PS)	Polym/Cyclic[b)]	M_n (g mol^{-1})	$Đ$[c)]
1	1000 : 500 : 0 : 0	60	10	—	—	—	—
2	1000 : 500 : 1 : 0	60	10	—	—	—	—
3	1000 : 500 : 0 : 1	60	10	12	11/89	587	2.08
4	1000 : 500 : 1 : 1	60	10	96	82/18	9762	1.56
5	1000 : 500 : 3 : 1	60	10	>99	88/12	6456	1.81
6	1000 : 500 : 1 : 3	60	10	>99	45/55	7540	1.92
7	2000 : 500 : 1 : 1	60	10	90	87/13	6740	1.95
8	500 : 500 : 1 : 1	60	10	58	58/42	4234	1.72
9	1000 : 500 : 1 : 1	60	10	45	51/49	3721	1.84
10	1000 : 500 : 1 : 1	60	10	82	84/16	9840	1.49
11	1000 : 500 : 1 : 1	60	10	98	69/31	7893	1.79
12	1000 : 500 : 1 : 1	60	5	84	83/17	5345	1.45
13	1000 : 500 : 1 : 1	60	20	>99	75/25	7500	1.62

a) Conversion of PS is determined by ^{1}H NMR spectrum of each coupling product.
b) The molar ratio of polymeric and cyclic product is determined by ^{1}H NMR spectrum of the aliquot with dodecane as internal standard.
c) $Đ = M_w/M_n$ is determined by gel permeation chromatography.

Table 2.3 Coupling reactions of cyclopentene sulfide with CS_2 [58].

Entry	Reaction loading[a)]	Cocat.	Temp (°C)	*t* (h)	Cyclopentene sulfide[b)] (%)	*cis*-Cyclopentene trithiocarbonate[b)] (%)	Cyclopentene[b)] (%)
1	1 : 2 : 500 : 1000	PPNCl	80	6	98.8	0.9	0.3
2	1 : 2 : 500 : 1000	$PPNN_3$	80	6	99.1	0.8	0.1
3[c)]	1 : 2 : 500 : 1000	n-Bu_4NN_3	80	12	99.1	0.4	0.5
4[c)]	1 : 2 : 250 : 500	PPNCl	110	24	39	19	42
5[c)]	1 : 2 : 250 : 1000	PPNCl	110	24	39	21	39
6[c)]	1 : 0 : 250 : 1000	—	110	24	56	0.7 (6.3[d)])	36
7[c)]	0 : 2 : 250 : 1000	PPNCl	110	24	86	8	6
8[c)]	0 : 0 : 250 : 500	PPNCl	110	24	96	—	4

All reactions performed in 10 mL stainless steel reactors.
a) Reaction loading refers to the relative equivalents of (salen)CrCl:cocatalyst:cyclopentene sulfide:CS_2 loaded into the reactor.
b) Relative product yield from crude ^{1}H NMR.
c) Toluene solvent employed; total volume of CS_2 + toluene = 1.0 mL.
d) Unidentified coupling products.

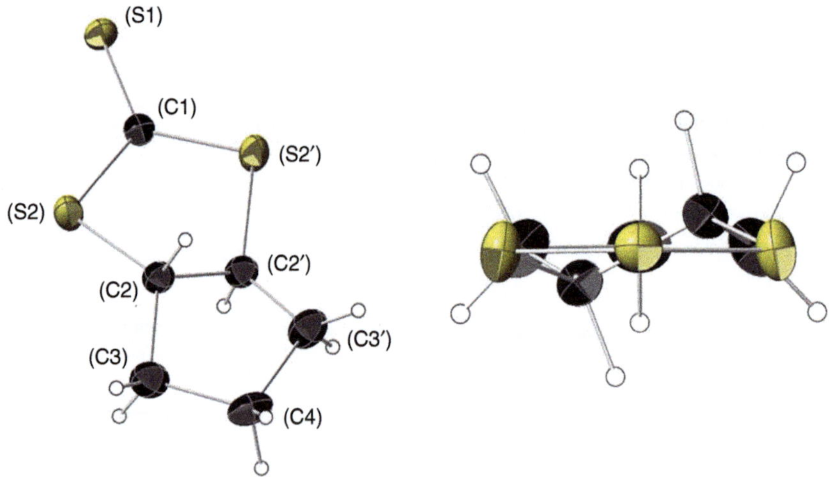

Figure 2.4 Thermal ellipsoid representation of *trans*-cyclopentene trithiocarbonate with ellipsoids at 50% probability surfaces. Inset image is looking down the S1–C4 plane to show torsion caused by the trans-configuration. S2–C2–C2′–S2′ angle = 46.44°.

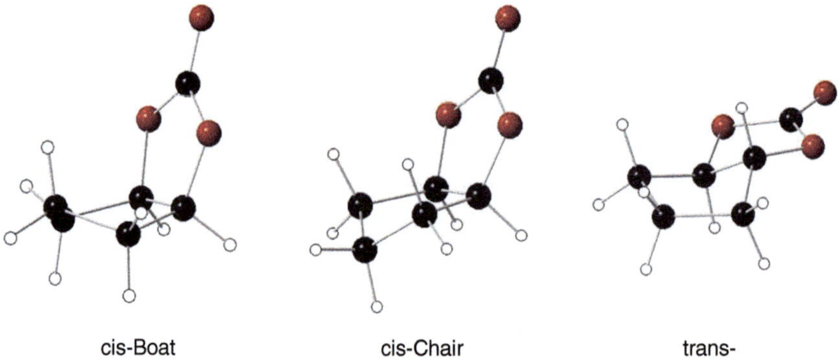

Figure 2.5 Calculated local minima for cyclopentene carbonate species: cis-boat, cis-chair, and trans-.

the cis isomers are lower in energy than their trans counterparts (Table 2.4). In the instance of cyclopentene carbonate that difference is especially large at ~20 kcal mol^{-1} [61]. This is consistent with the fact that only the *cis*-cyclopentene carbonate has been isolated and characterized, whereas all attempts to prepare the trans isomer have been unsuccessful. For the cyclopentene trithiocarbonate analogs the cis isomer is more stable by only ~5 kcal mol^{-1} because the larger sulfur atoms provide less strain in the two fused 5-membered rings. The mixed O/S species also display a preference for the cis isomer. As would be anticipated, there is a large preference (10–25 kcal mol^{-1}) for carbon–oxygen double bond formation. Table 2.4 summarizes the relative stabilities of the various cyclic materials.

Table 2.4 Calculated relative enthalpies of cyclic [thio]carbonates.

Cyclic [thio]carbonate family	Isomer[a)]	Calculated relative enthalpy (kcal mol^{-1})[b)]
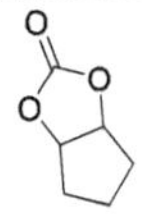	Cis-boat	0.0
	Cis-chair	2.5
	Trans-	19.2
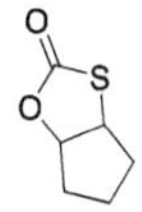	Cis-boat	0.0
	Cis-chair	1.0
	Trans-	10.3
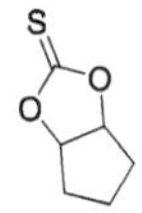	Cis-boat	14.7
	Cis-chair	17.5
	Trans-	34.7
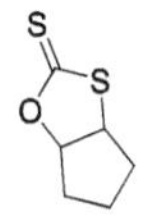	Cis-boat	11.7
	Cis-chair	12.8
	Trans-	22.9
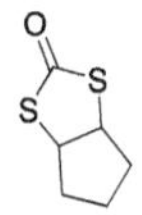	Cis-boat	0.0
	Cis-chair	0.3
	Trans-	4.1
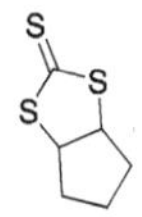	Cis-boat	0.0
	Cis-chair	0.1
	Trans-	5.0

a) As defined in Figure 2.5.
b) CBS-QB3, gas phase.

2.3.2 Copolymerization of Carbon Disulfide and Epoxides

Sulfur-rich polymers can also be synthesized by the copolymerization of epoxides and carbon disulfide. This copolymerization dates back to when zinc cobalt double metal cyanides (DMCs) were first utilized to facilitate the one-pot copolymerization of CS_2/PO [62]. The molecular weight and $Đ$ of these polymers were found to be in the range of 1.2–5.4 kg mol^{-1} and 1.24–3.50, respectively. Upon optimization of the reaction, it was revealed that the copolymerization is dependent on temperature, CS_2/PO molar ratio, and double-metal cyanide complex (DMCC) loadings. The optimal conditions for the copolymerization were found to be a temperature of 100 °C with a CS_2/PO molar ratio of 1.5–2.0 and a DMCC loading of 3.2×10^{-4} g catalyst per g reactants. One interesting aspect of this coupling process was the observation of a high degree of sulfur/oxygen exchange that occurred under the reaction conditions. Multiple S_nO_{3-n} combinations were noted in both the copolymer and cyclic by-products as confirmed by Fourier-transform infrared spectroscopy (FT-IR) and ^{13}C NMR spectroscopies.

Thus, it is quite important to study this O/S exchange reaction in detail in order to establish the conditions, which might lead to selective polymer formation. This catalyst system, (salen)CrCl/[PPN]Cl (salen = *N*,*N'*-bis(3,5-di-tertbutylsalicylidene)-1, 2-ethylenediimene, [PPN]Cl = bis(triphenylphosphoranylidene) ammonium chloride), is extensively used for the copolymerization of carbon dioxide and cyclohexene oxide (CHO), hence that knowledge should be transferable towards an understanding of the copolymerization of carbon disulfide and CHO [10]. Initially coupling of CS_2 and CHO at 50 °C was successful, which prompted further investigations [59]. It was apparent that the O/S exchange observed by Qi and coworkers was also occurring with the (salen)CrCl catalyst system. From the IR and ^{13}C NMR spectral data, a wide variety of "carbonyl functional" groups are observed in the polymer product even at 50 °C. After determining that the (salen)CrCl catalyst system would be effective for the copolymerization of CS_2 and CHO, a study of the effects of temperature and CS_2 loading was undertaken.

The effect of temperature on the coupling reaction was first investigated over a 40 °C range, maintaining the CS_2 loading at 1 equiv. Since the (salen)CrCl based catalyst system has shown activity at temperatures lower than 80 °C for the coupling of CO_2 and epoxides, this implies that the coupling of CS_2 and epoxides should occur at a lower temperature due to the more electrophilic character of the carbon center of CS_2 versus that of CO_2. The effect of temperature for the overall coupling reaction of CS_2 with CHO was studied and the turnover frequencies (TOFs) were tabulated in Table 2.5. It is evident from this data that the optimal reaction temperature is 50 °C, which is well below the 110 °C used in the Zn–Co DMC catalyst system using propylene oxide.

As earlier discussed, both copolymer and cyclic products observed from each of the above reactions exhibited O/S atom exchange, which was confirmed by both solution FT-IR and ^{13}C NMR spectroscopy. In the FT-IR spectrum peaks corresponding to both carbonate (1743 cm^{-1}) and monothiocarbonate (1711 cm^{-1}) were observed (Figure 2.6), but because of numerous overlapping vibrational modes in the trithiocarbonate region, it was not possible to definitively assign a band primarily due to the C=S stretching vibration. The ^{13}C NMR spectrum of copolymers shows resonances for the trithiocarbonate (220 ppm), monothiocarbonate (168 ppm), and carbonate (154 ppm) in all the polymer samples (Figure 2.6). The very weak resonance for the dithiocarbonate was observed at 212 ppm.

The copolymerization reaction of CHO and CO_2 catalyzed by the (salen)CrCl/[PPN]Cl system shows little propensity for the formation of a cyclic product, cyclohexylene carbonate at 80 °C. On the other hand, the analogous process between cyclohexene sulfide and CS_2 using the same catalyst system leads almost exclusively (96%) to the cyclic material at 50–60 °C. Thus a potential pathway for sulfur/oxygen atom exchange could involve ring-opening of the initially formed cyclic thiocarbonate accompanied by COS and subsequently CO_2 extrusion. If the reaction is monitored by *in situ* infrared spectroscopy, the formation of both poly(carbonate) (~1750 cm^{-1}) and poly(thiocarbonate) (~1710 cm^{-1}) is observed from the onset of the reaction; nevertheless there is an initiation period which implies that the O/S exchange process occurs prior to and during the polymerization

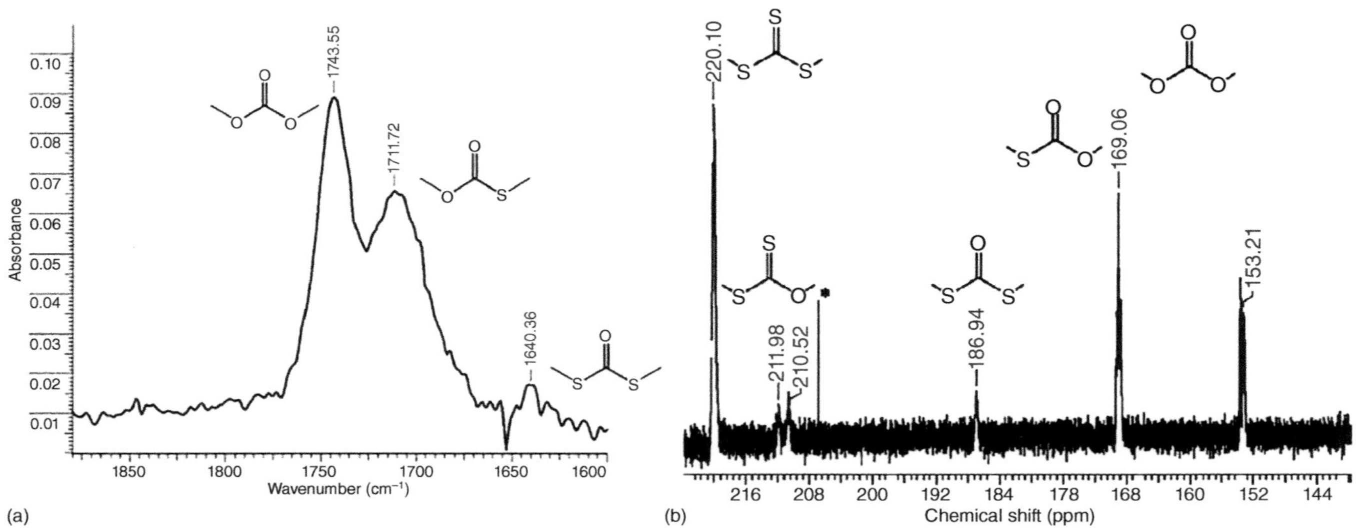

Figure 2.6 (a) FT-IR spectrum of polymer formed from the copolymerization of CS_2 and CHO at 50 °C. (b) ^{13}C-NMR spectrum of the carbonyl region of polymer product from the coupling of CS_2 and CHO at 50 °C. Peak marked with asterisk is acetone.

Table 2.5 Copolymerization of cyclohexene oxide with CS_2 utilizing the (salen)CrCl/[PPN]Cl catalyst system[a].

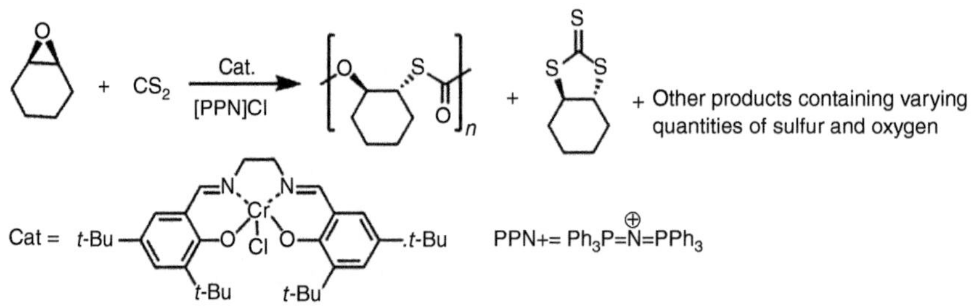

T (°C)	Mass polymer (g)	Mass cyclic (g)	Conversion (%)[b]	TOF[c] (h^{-1})	M_w[d] (kg mol^{-1})	$Đ$
40	4.68	2.27	40.6	77.6	15.3	1.19
50	7.50	3.01	61.4	124.4	34.8	1.31
60	2.24	9.35	67.7	37.2	15.2	1.48
80	1.85	–	–	30.6	3.03	1.26

a) 50 mg (salen)CrCl, 50 mg [PPN]Cl, 10 mL CHO, 6.1 mL CS2, 4 h.
b) Determined by isolated combined weights of copolymer and cyclic products.
c) (Mol epoxide consumed)/(mol Cr hr^{-1}).
d) Determined by GPC analysis.
Theoretical molecular weight equals 83 600.

process. The initial thoughts about the much greater tendency for cyclohexene sulfide and CS_2 to provide cyclic product are due to the better metal binding ability of cyclohexene sulfide compared with CHO, whereby the former can more readily displace the growing polymer chain leading to a back-biting pathway for cyclic product formation.

Besides the effect of temperature, the ratio of the monomers influences the product distribution and potentially the extent of O/S exchange. The coupling studies of CS_2 and PS revealed that 2 to 1 ratio of CS_2 to comonomer exhibited the best activity in terms of polymer selectivity, higher molecular weights, and narrow $Đ$. It was initially thought that this trend would also hold for the (salen)CrCl system with CHO but upon varying the carbon disulfide loading, it was apparent that the optimal loading of CS_2 for catalytic activity was 1 equiv (Table 2.6). It was further observed that the optimal CS_2 loading for copolymer selectivity was 2 equiv. It is also worth noting here that the $Đ$ values for all the polymerizations are narrow (1.2–1.5) and the polymers produced have relatively predictable molecular weights.

Darensbourg and coworkers have also investigated the catalytic coupling of carbon disulfide with an alternative alicyclic epoxide, cyclopentene oxide (CPO), employing (salen)CrCl in the presence of onium salts [58]. It is noteworthy that these two alicyclic epoxide monomers, cyclopentene, and CHOs show different reactivity patterns with CO_2 and CS_2 under the influence of same catalytic system.

Table 2.6 Copolymerization of carbon disulfide and cyclohexene oxide.[a)]

Equiv. CS_2	Mass polymer (g)	Mass cyclic (g)	Conversion (%)[b)]	TOF[c)]	M_w (kg mol^{-1})[d)]	Đ
0.5	2.97	3.02	69. 9	78.8	14.7	1.29
1.0	6.11	3.61	56.8	131.8	26.8	1.25
2.0	3.84	1.42	30.7	59.8	17.2	1.19
4.0	3.56	3.96	42.4	35.8	13.3	1.17

a) 4.3×10^{-3} M (salen)CrCl, 1 equiv [PPN]Cl, 10 mL CHO, 50 °C, 4 h.
b) Determined by isolated combined weights of copolymer and cyclic products.
c) Mol epoxides consumed/(mol Cr h), calculations based on monomer molecular weight of copolymer of 167.9 as determined by elemental analysis.
d) Determined by GPC analysis.

The catalytic coupling of CS_2 with CPO was initially carried out using (salen)CrCl and bis(triphenylphosphine)-iminium azide ($PPNN_3$) as the catalyst and cocatalyst, respectively, which resulted in the formation of a variety of polymeric and cyclic [thio]carbonate materials (Scheme 2.11).

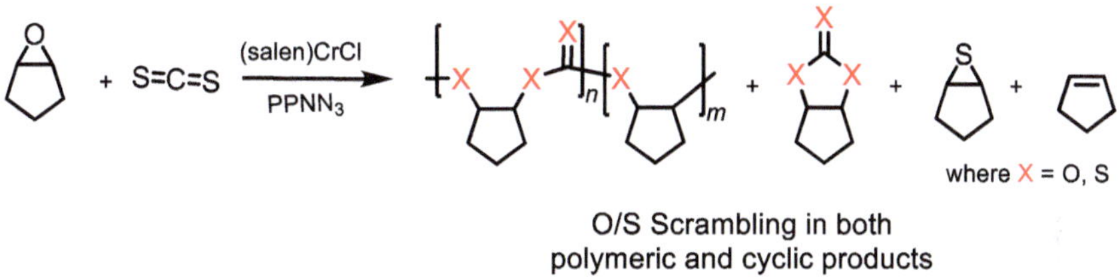

Scheme 2.11 Copolymerization of cyclopentene oxide with CS_2 utilizing the (salen)CrCl/[PPN]N_3 catalyst system.

The importance of using $PPNN_3$ as cocatalyst is due to its distinct ν_{N3} bond stretching frequency that undergoes drastic changes on perturbation of chemical environment that can be monitored by infrared spectroscopy. Due to the overlapping nature of the ^{1}H NMR and FT-IR spectral features of the products, assignments were challenging (Table 2.7).

The peak intensity of –O–(C=O)–S– polymer peak at 1712 cm^{-1} is maximized after ~four hours, after which it decays away, presumably due to backbiting or rearrangement reactions. The formation of CO_2 was observed in the reaction by the appearance of band around 2338 cm^{-1}, especially when a limiting amount of CS_2 is employed in the reaction. In order to determine the possibility of formation of ether linkages during the copolymerization reaction, microanalysis of pure polymeric samples was performed. The hydrogen atoms are set at 8.00/repeat unit because this value is consistent for both poly[thio]carbonate and poly[thio]ether linkages (Figure 2.7). It has been observed that experimentally carbon content is significantly below than that of the expected value 6.0 for [thio] carbonate linkages, indicating

Table 2.7 Reaction of cyclopentene oxide with CS_2 using (salen)CrCl and $PPNN_3$ in a 300 mL stainless steel Parr reactor.

Entry	Equivalents $PPNN_3$	Equivalents CS_2	P_{CO2} (MPa)	Polymer repeat unit[a)]	T_g[b)] (°C)
1	2	1500	0	$C_{5.69}H_8S_{1.56}O_{0.76}$	14.4
2	1	1500	0	$C_{5.58}H_8S_{1.15}O_{0.88}$	9.6
3	2	750	0	$C_{5.22}H_8S_{0.80}O_{1.22}$	−17.6
4	2	750	0.7	$C_{5.81}H_8S_{1.08}O_{1.81}$	32.3

Reaction conditions 9.64 g cyclopentene oxide (10.0 mL, 115 mmol, 1500 eq), 0.0483 g (salen)CrCl (0.076 mmol, 1 eq), and the appropriate amounts of $PPNN_3$, CS_2, and CO_2 pressure, where applicable. 80 °C, 24 h.

a) H set at 8.00, and other elements were calculated from elemental analysis of the purified polymer sample.

b) Midpoint T_g from the second heating cycle utilizing differential scanning calorimetry.

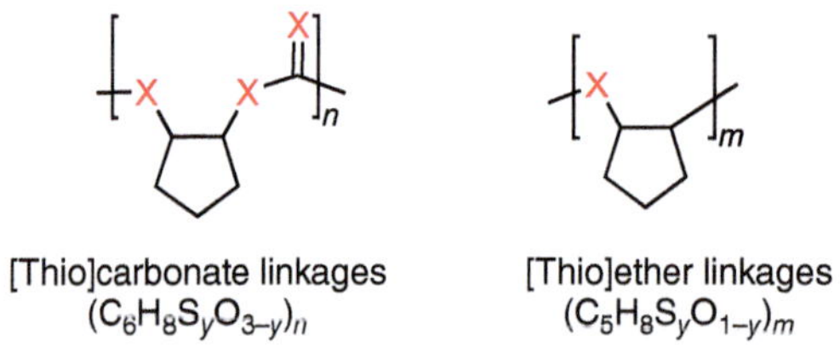

Figure 2.7 Repeat units for [thio]carbonate and [thio]ether linkages. X = O or S.

production of large amount of ether or thioether linkages in the polymer backbone. When limiting amount of CS_2 is employed, only 5.22 carbons exist in the idealized repeat unit, indicating that the polymer contains mostly [thio]ether linkages (entry 3). The decrease of [thio]carbonate content is also mirrored by a decrease in T_g, from +14.4 to −17 °C. When the reaction mixture is charged with 0.7 MPa of CO_2, both the [thio]carbonate content and T_g greatly increase, with 5.81 carbons/repeat unit and T_g = 32.3 °C.

Notably, ^1H and ^{13}C NMRs ($CDCl_3$) of the crude reaction mixtures each clearly display the formation of cyclopentene sulfide (peaks at 3.28, 30.0 ppm, respectively) and cyclopentene (peaks at 5.76, 32.4 ppm, respectively). At the end of the 24-hour reaction period, the CPO is nearly completely consumed, as evident by the disappearance of the peaks at 3.39 and 27.1 ppm in the ^1H and ^{13}C NMRs, respectively.

During this copolymerization reaction of CPO and CS_2 in presence of $[PPN]N_3$, there is also a side reaction of azide with CS_2 to form a 5-membered heterocyclic anion azidodithiocarbonate, $CS_2N_3^-$ in small quantities, a process that is well-defined in the literature (Scheme 2.12). **NOTE:** *Azidodithiocarbonate salts and derivatives are potentially explosive, and appropriate safety precautions should be taken.* Upon heating, this anionic heterocycle will breakdown to dinitrogen, thiocyanate, and elemental sulfur.

As a test to confirm the presence of the $CS_2N_3^-$ anionic heterocycle, tetrabutylammonium azide, n-Bu_4NN_3, was dissolved in toluene at ambient temperature,

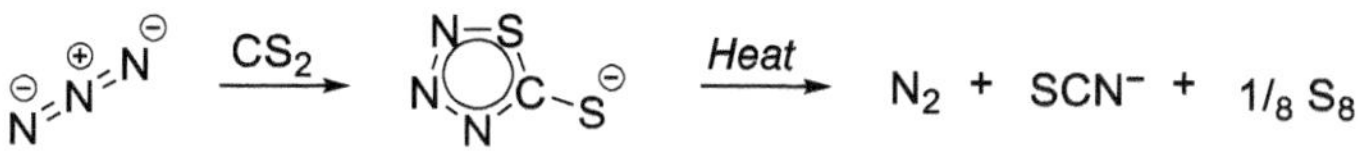

Scheme 2.12 Reaction of azide with carbon disulfide to yield the azodithiocarbonate anion.

and the characteristic free N_3^- peak was observed at 2000 cm^{-1} [63]. Upon addition of 5 equiv of CS_2, the N_3^- peak immediately disappeared, and no distinct peaks were observed, consistent with reports that $CS_2N_3^-$ complexes have only weak infrared stretching patterns [64]. $CS_2N_3^-$ salts and compounds were known to be thermally unstable [65], and so the reaction was heated to the typical reaction temperature, 80 °C, to observe any potential decomposition. Indeed, almost immediately upon heating a sharp peak appeared at 2055 cm^{-1}, which indicates the formation of n-Bu_4NSCN [66]. As such, the $CS_2N_3^-$ heterocycle is not long-lived in the reaction medium, instead decomposing to thiocyanate, SCN^-.

Due to the unclear $N_3^-/CS_2N_3^-/SCN^-$ nature of the cocatalyst in the previous studies, it was desirable to investigate the copolymerization reaction of CS_2 and CPO using the well-defined PPNCl cocatalyst. Initial ATR-FTIR monitoring of a reaction between CPO and CS_2 revealed that polymer production peak viz –O–(C=O)–S– at 1712 cm^{-1} was achieved at six hours and further prolongation of reaction results in drops off quickly, presumably due to depolymerization.

The polymerization was further optimized using different ratios of CPO and CS_2 keeping the reaction time constant at six hours as illustrated in Table 2.8. The highest polymer selectivity and yield was obtained when equal ratio of CS_2 and CPO were employed (Table 2.8, entry 4). When the amount of CS_2 was limiting, only trace amounts of polymer and cyclic [thio]carbonates were produced (entry 5). Similarly, both polymer and cyclic production were inhibited by a large excess of CS_2, presumably due to dilution of the reaction (entry 1). The addition of 0.7 MPa CO_2 prevented significant polymer growth, though a relatively large amount (1.50 g) of cyclic material was produced (entry 6). Though other mixed O/S [thio]carbonates were produced, NMR and GC-MS confirmed that *cis*-cyclopentene carbonate formed from the coupling of CPO and CO_2 was the major product for this reaction, highlighting the large difference in reactivity for CPO with CO_2 and CS_2.

Microanalysis of multiple polymeric samples give a consistent repeat unit of $C_6H_8S_{1.6}O_{13}$ indicating that there is very little [thio]ether content in the backbone of the polymer (Table 2.8, entries 2 and 4). The polymer is thus oxygen-enriched from the expected $(C_6H_8S_2O)_n$ repeat unit. It was though that the increased time of the 24 hour $[PPN]N_3$ reactions led to backbiting reactions, the loss of CS_2, CO_2, and COS, and corresponding increase in ether and thioether linkages in the polymer backbone. Similar observations were previously reported for the decomposition of polycarbonates [67]. Additionally, no *trans*-cyclopentene trithiocarbonate was observable in any of the produced cyclic materials as confirmed by GC-MS, which is starkly different from the 24-hour $[PPN]N_3$ reactions in which *trans*-cyclopentene trithiocarbonate was typically the major cyclic product. This

Table 2.8 Reaction of cyclopentene oxide with CS_2 with (salen)CrCl and PPNCl in a 25 mL stainless steel Parr reactor.

Entry	Equivalents CS_2 [a)]	Mass polymer (g) [b)]	Mass cyclic carbonates (g)	Polymer repeat unit [c)]
1	2.5	0.20	0.22	nd [d)]
2	2.0	1.11	0.91	$C_{5.92}H_8S_{1.70}O_{1.43}$
3	1.5	1.27	1.26	$C_{5.97}H_8S_{1.65}O_{1.24}$
4	1.0	2.04	0.83	$C_{5.96}H_8S_{1.64}O_{1.32}$
5	0.5	Trace	0.49	nd
6 [e)]	1.0	0.17	1.50	nd

Reaction conditions: 24.2 mg (salen)CrCl (1 eq), 43.9 mg PPNCl (2 eq), 5.0 mL cyclopentene oxide (1500 eq), 80 °C, six hours in a 25 mL stainless steel reactor.

a) Relative to cyclopentene oxide.
b) Mass of purified polymer sample.
c) H set at 8.00 and other elements were calculated from elemental analysis of the purified polymer sample.
d) Not determined.
e) 0.7 MPa CO_2 added

indicates that *trans*-cyclopentene trithiocarbonate likely forms from late-stage polymer backbiting reactions after CS_2 has been largely consumed.

The produced poly(cyclopentene [thio]carbonate)s are semicrystalline with T_g near 50 °C and T_m near 125 °C. The cyclic products viz cyclopentene sulfide and cyclopentene are again observed in the ^{1}H and ^{13}C NMRs of the crude reaction mixtures, though significant amounts of unreacted CPO also exist in each case. Notably, ^{13}C NMR revealed that the major [thio]carbonate component was actually at 211 ppm, caused by the unscrambled –S–(C=S)–O– polymeric product (Figure 2.8) [59]. The next largest contributor to the polymeric backbone is –O–(C=O)–S–, with corresponding peaks at 169 ppm and 1712 cm^{-1} in the ^{13}C NMR and FT-IR, respectively. Notably, a small amount of the all-oxygen polycarbonate can be seen both in the NMR (~153 ppm) and IR spectra (1750 cm^{-1}), and other minor contributions from the trithiocarbonate (~220 ppm) and dithiocarbonate (~193 ppm) can also be observed by NMR.

2.3.2.1 Scrambling Mechanism for Sulfur/Oxygen Atoms

During the catalytic copolymerization of CPO with CS_2, scrambling of the oxygen and sulfur atoms has been observed in both polymeric and cyclic [thio]carbonate samples. In the literature, several similar mechanisms involving the ring-opening of preformed cyclic carbonates have been proposed. In Endo's 2001 investigation of cyclic trithiocarbonate formation utilizing a titanium catalyst, scrambling was observed in the presence of excess CS_2 [68]. In 2009 Darensbourg group proposed that the insertion of CO_2, COS, and CS_2 is reversible, and it is this reversibility that brings about the scrambling [59]. A similar mechanism has been proposed by North for the coupling of propylene oxide and CS_2 in which PS is first eliminated and then

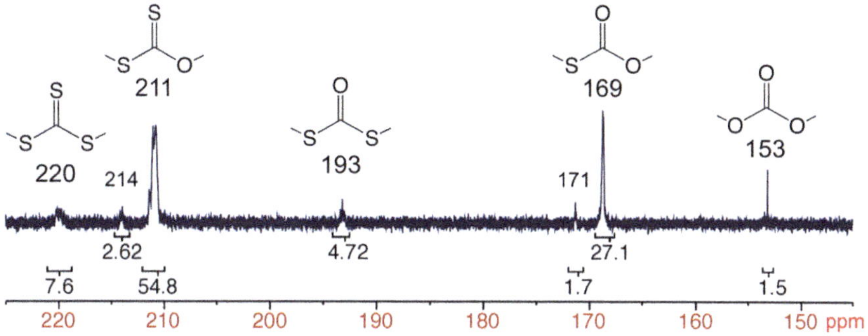

Figure 2.8 Carbonate region of the ^{13}C NMR of poly[thio]carbonate from Table 2.2, entry 2. $CDCl_3$, 125 MHz. Source: Modified from Ref. [59].

Table 2.9 Conditions needed for O/S scrambling utilizing a purified mixture of cyclic [thio]carbonate materials in the presence of (salen)CrCl and *n*-Bu_4NCl.

Entry	Added reagent[a)]	Time (h)	Scrambling observed?[b)]
1	None[c)]	48	No
2	100 psi CO_2[c)]	48	No
3	1 eq CS_2[c)]	48	No
4	1 eq cyclopentene oxide[d)]	16	Yes

a) Average MW of cyclic materials arbitrarily assigned to be 160 g mol^{-1} for calculations. Catalyst loading held at (salen)CrCl:*n*-Bu_4NCl:cyclic mix = 1 : 2 : 50. 80 °C.
b) Scrambling observed by changes in the ^1H NMR and GC-MS traces of the starting material and reaction products.
c) 0.875 M solution in toluene.
d) 1.4 M solution.

re-enters the mechanism as a new reagent [69]. Tetrahedral spiro intermediates have been proposed and isolated by Barbero et al. for propylene oxide/sulfide scrambling systems utilizing $HBF_4 \cdot Et_2O$ [70], but these are not directly observable by NMR or FT-IR in this investigation.

Several test reactions were run in order to determine the conditions necessary to induce scrambling for reactions involving CPO, (salen)CrCl, and an onium salt cocatalyst (Table 2.9). In three separate reactions, toluene solutions containing a mix of cyclic cyclopentene [thio]carbonates were heated to 80 °C with (salen)CrCl and *n*-Bu_4NCl in the presence of CO_2, CS_2, and a control with no other added reagents. After 48 hours, no scrambling was observed by ^1H NMR or GC-MS for any of the cases (entries 1–3). Conversely, a solution of the same cyclic [thio]carbonates, catalyst, and cocatalyst heated to 80 °C in the presence of CPO displayed distinct changes in composition, indicating scrambling had taken place (entry 4). Further investigations revealed that the presence of (salen)CrCl catalyst was necessary to induce scrambling, though added *n*-Bu_4NCl cocatalyst increased the activity (Table 2.10).

Table 2.10 Conditions needed for O/S scrambling utilizing *trans*-cyclohexene trithiocarbonate and cyclohexene oxide.

Entry	Catalyst	Cocatalyst	Scrambling observed?[a]
1	—	—	No
2	(Salen)CrCl	—	Yes, slight
3	—	n-Bu_4NCl	No
4	(Salen)CrCl	n-Bu4NCl	Yes

a) Analyzed by *in situ* ATR-FTIR spectroscopy, with scrambling being assigned by the growth of cyclohexene carbonate at 1809 cm^{-1}.

Thus catalyst is required to activate and ring-open the epoxide, and the more nucleophilic alkoxide then attacks the [thio]carbonyl to induce scrambling (Scheme 2.13). The trithiocarbonate species can be produced from the coupling of the ring-opened thiirane produced during the scrambling process and CS_2. In addition to the pathways outlined in Scheme 2.13, it is possible that cyclopentene sulfide is produced via the elimination of the ROC(=S)O anion from (ROC(=S)O)C5H8-S$^-$. It is unlikely that cyclopentene sulfide will be separately ring-opened to undergo coupling. The possibility of spiro intermediates has not yet been ruled.

2.3.2.2 NMR Scrambling Studies

Besides being studied by ATR-FTIR, NMR spectroscopy was employed to monitor the scrambling reactions. Figure 2.9 shows the time lapsed ^{13}C NMRs for the reaction of 4 equiv CPO with 1 equiv *trans*-cyclopentene trithiocarbonate utilizing (salen)CrCl and n-Bu_4NCl at 80 °C in toluene-d_8. After 15 hours, the growth of cyclopentene sulfide is clearly visible with peaks at 41.1, 29.0, and 18.3 ppm. Growth of *cis*-cyclopentene carbonate can also be clearly observed by 22.5 hours. After 48 hours, faint traces of cyclopentene can be observed at 130.7 ppm. Notably, CPO's signal at 56.3 ppm broadens during the course of the reaction, likely because the epoxide is undergoing reversible ring-opening and ring-closing processes.

For completeness, the reaction of cyclopentene sulfide with *cis*-cyclopentene carbonate at 80 °C was also monitored by ^{13}C NMR (Figure 2.10). The desulfurization product cyclopentene (32.7, 23.2 ppm) is the only observable product after 48 hours reaction time, indicating that either the catalyst is incapable of ring-opening the episulfide or that the thiolate is not a strong enough nucleophile to attack the carbonate. A proposed mechanism for these observed epoxide/cyclic scramblings can be seen in Scheme 2.14. It is expected that a mixture of the mechanisms proposed in Schemes 2.13 and 2.14 are both operational during the course of the reaction.

2.3.2.3 Mixed Species Scrambling

O/S scrambling is not limited to epoxides and episulfides of the same backbone connectivity. Indeed, when *trans*-cyclohexene trithiocarbonate was reacted with CPO, (salen)CrCl, and n-Bu_4NCl at 80 °C, six cyclic [thio]carbonate materials

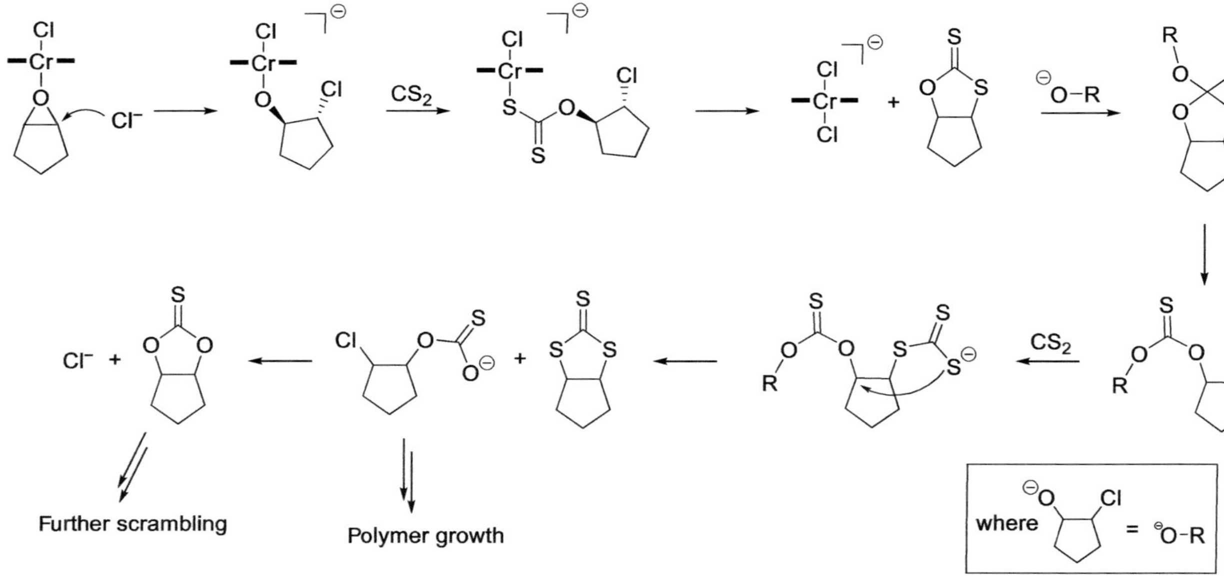

Scheme 2.13 Proposed mechanism for scrambling in the coupling of cyclopentene oxide and CS_2.

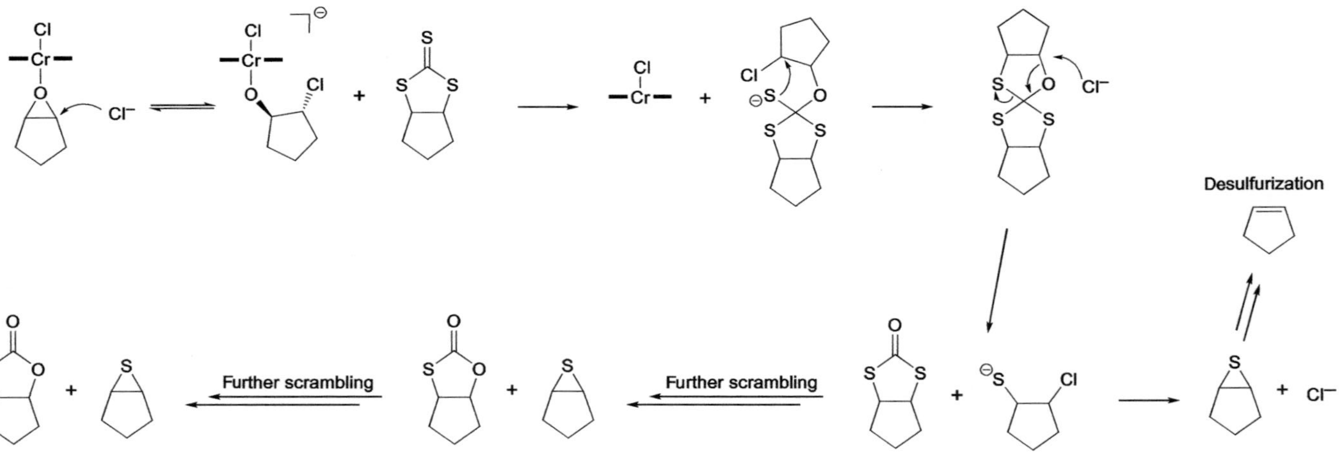

Scheme 2.14 Proposed mechanism for observed scrambling between cyclopentene oxide and *trans*-cyclopentene trithiocarbonate.

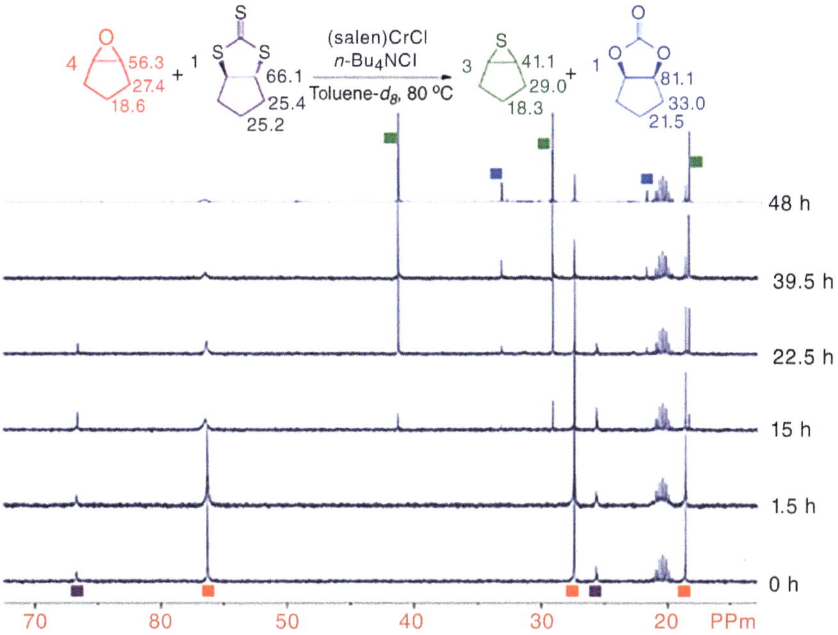

Figure 2.9 Time-lapsed ^{13}C NMR of the reaction between 4 equiv of cyclopentene oxide (red) and 1 equiv *trans*-cyclopentene trithiocarbonate (purple) to produce cyclopentene sulfide (green) and *cis*-cyclopentene carbonate (blue) in toluene-d_8. Broadening of cyclopentene oxide's carbon at 56.3 ppm is likely due to ring-opening and ring-closing of the epoxide.

were observed by ^{13}C NMR and GC-MS including *cis*-cyclopentene carbonate and residual *trans*-cyclohexene trithiocarbonate. Mixed species cyclopentene thiocarbonate (SO_2), cyclohexene thiocarbonate (SO_2), and cyclohexene dithiocarbonate (S_2O) were observed, though their exact connectivity cannot be commented on. Cyclohexene sulfide was also clearly produced during the course of the reaction. The reaction mixture was concentrated *in vacuo* prior to GC-MS and NMR analysis, so it was unclear whether cyclopentene sulfide was also generated. However, production of this thermodynamic sink is likely to have occurred.

2.4 Other Related CS_2-Based Polymers

Similar to other sulfur containing polymers, polythioureas are a promising polymeric material due to their intrinsic ability to show outstanding properties such as self-healing, high refractive index, high dielectric constant, and coordinating tendency to heavy metal ions like mercury. Because of the large dipole moment (5.4 Debye) of thiourea ($S{=}C(NH_2)_2$), polythiourea-based polymers are known as order-disorder type ferroelectric materials. The ferroelectric transitions in polythiourea are due to their "incommensurate phase," which arises mainly due to the dipole–dipole interactions rather than hydrogen bonding. Polythioureas having

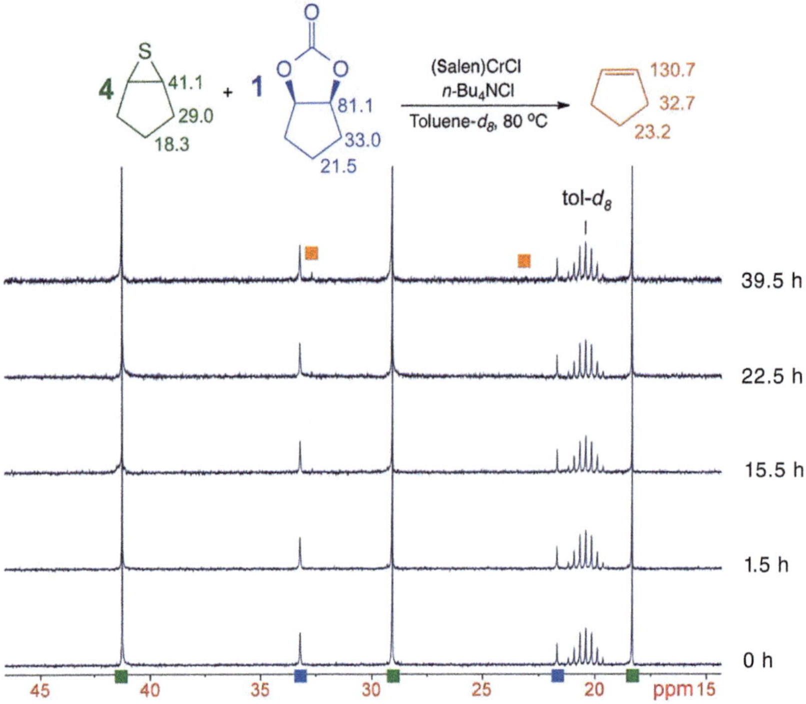

Figure 2.10 Time-lapsed ^{13}C NMR of the reaction between 4 equiv of cyclopentene sulfide (green) and 1 equiv *cis*-cyclopentene carbonate (blue) in toluene-d_8. Cyclopentene (orange) is the first observable product.

larger dipole–dipole interactions in the backbone will have a polar structure which ultimately leads to ferroelectric polar aggregates in these polymers.

As a result of their limited methods of synthesis examples of versatile polythioureas are relatively scarce. In the late nineties Yamazaki and coworkers have employed CS_2 for the construction of polythioureas at very low temperature and pressure in the presence of amines and pyridine (Scheme 2.15) [71, 72]. The reaction conditions revealed that polycondensation was almost completed within six hours at 40 °C.

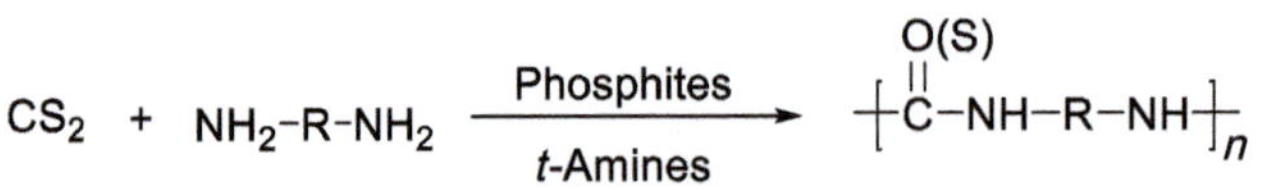

Scheme 2.15 Synthesis of polythioureas from polycondensation of CS_2 and diamines. Source: Refs. [71, 72].

The formation of polythioureas were confirmed from FT-IR spectra in which absorption bands at 1530 cm^{-1} due to $\nu_{C=S}$ were observed. Microanalysis of the polymers was consistent with the assigned structures and hence further excludes the possibility of incorporating phosphorus moiety into the polymer backbone

Table 2.11 Polycondensation of carbon disulfide with various diamines by using diphenyl phosphite in pyridine.[a)]

Diamine	Yield	η in h[b)]	Elemental analysis[c)]			
			C (%)	H (%)	N (%)	S (%)
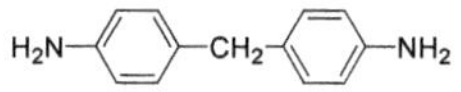	100	0.18	68.61	5.12	10.95	—
			(69.99)	(5.03)	(11.66)	—
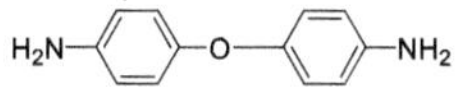	100	0.25	62.96	4.03	10.97	13.49
			(64.44)	(4.16)	(11.56)	(13.23)
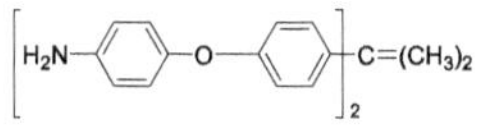	78	0.15	—	—	—	—
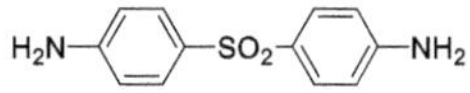	17	0.07	—	—	—	—
	100	0.12[d)]	—	—	—	—
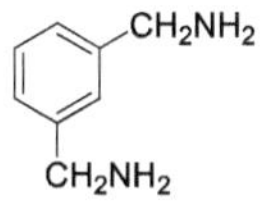	0	—	—	—	—	—

a) The reaction **was** carried out **at 40 °C** for **6** h.
b) Measured in DMSO at **30 °C.**
c) Calculated values in parentheses.
d) The polymer **was** insoluble in DMSO, and the viscosity **was** measured in sulfuric acid at 30 °C.

due to side reactions. It has also been observed that presence of phosphorous compounds is important during the polycondensation of CS_2 with diamine. Out of the different phosphorous compounds employed diphenyl or triaryl phosphites, especially triphenyl, tri-*m*-tolyl, or tri-*p*-tolyl phosphites, showed greater effect upon yield and molecular weight of polymer. Phosphorous acid and its alkyl esters did not give any promising results for this polycondensation reaction.

The effect of different diamines in the polycondensation reaction with carbon disulfide using diphenyl phosphite in pyridine is summarized in Table 2.11. Aromatic diamines having higher basicity generally produced polymers of higher viscosity in quantitative yields, whereas aromatic diaamines of lower basicity such as 4,4′-diaminodiphenyl sulfone gave only a 17% yield of low-viscosity polymer. The diamine of extremely high basicity such as *m*-xylylenediamine did not afford any polymer at all.

Later, Tasaka et al. reported for the first time ferroelectric properties of aliphatic polythioureas having an odd number of CH_2 groups in the linear aliphatic backbone [73]. These polythioureas have polar structure with ferroelectric packing similar to the structure of ferroelectric odd-nylon [74]. The polythioureas were synthesized by reacting aliphatic diamines like pentamethylenediamine, heptamethylenediamine, and nonamethylenediamine with carbon disulfide by heating at 120 °C for two to four hours. Although the structure of these polythioureas depends on the reaction conditions, generally side reactions occur during the polymerization resulting

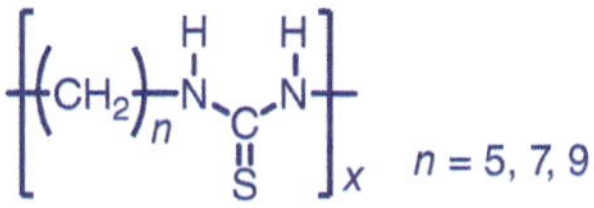

Figure 2.11 Chemical structure of aliphatic polythioureas.

Table 2.12 Glass transition temperature and dielectric relaxation strength at the primary relaxation region (140 °C) for aliphatic polythioureas.

Sample	T_g (°C)	$\Delta\varepsilon$
Polythiourea 5	55	77
Polythiourea 7	52	54
Polythiourea 9	48	48

in structural disorder, which imparts amorphous nature to these polymers. The polythioureas having odd number of CH_2 groups in the aliphatic backbone were chosen for studying their ferroelectric properties due to their unique polar packing (Figure 2.11).

These polythioureas were characterized by different spectroscopic and analytical methods, e.g. IR spectroscopy was used to understand the hydrogen bonding interactions. Generally, polymers like polyamides, polyureas, or polyurethanes show more than two stretching bands for C=S or N–H groups, with the N–H stretching mode in these polythioureas being observed at 3454 cm^{-1} (non-hydrogen bonded) and 3300 cm^{-1} (hydrogen bonded). These polymers were amorphous in nature with glass temperature (T_g) around 50 °C as evidenced by X-ray and DSC techniques as tabulated in Table 2.12 along with the dielectric relaxation strength.

Further dielectric and ferroelectric properties of these aliphatic polythioureas were studied. Temperature-dependent dielectric relaxation as obtained from Cole–Cole plots at 140 °C was observed in all the polymeric samples. The remanent polarization and pyroelectric constant in polythiourea 9 were found to be 24.6 mC m^{-2} and 20 μC m^{-2} K^{-1} at 60 °C, respectively. It has been concluded that the ferroelectric behavior of these amorphous polythioureas arises due to inter-chain hydrogen bonding interactions.

More recently Shuang Wu et al. reported a novel strategy for the synthesis of polythioureas by employing a catalyst free copolymerization of CS_2 with diamines [75]. The reactions occurs at mild temperatures (20–80 °C) in the absence of any inert/anhydrous atmosphere. Similar types of polythioureas were also recently reported by Aida and coworkers by employing the polycondensation of 1,8-diamino-3,6-dioxaoctane (**DA1**) and 1,1′-thiocarbonyldiimidazole (Scheme 2.16), but their method of preparation was very tedious and requires inert/anhydrous conditions besides having a serious purification problem for removing the imidazole by-product [76].

In order to optimize the reaction conditions for the catalyst free copolymerization of CS_2 with diamines, reactions were carried out at different temperatures in DMF as

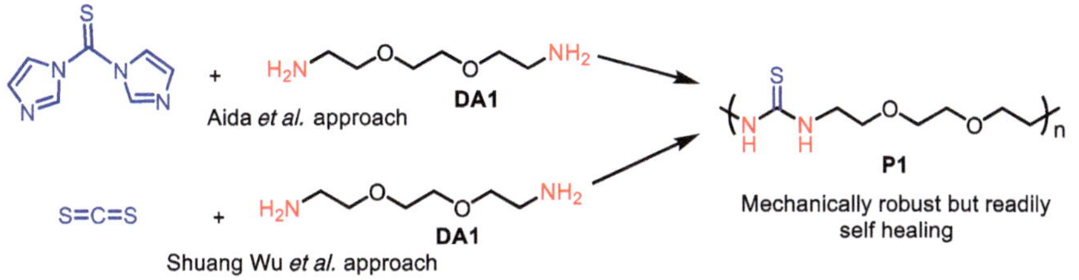

Scheme 2.16 Synthetic approaches to the self-healing polythiourea (**P1**).

Table 2.13 Catalyst-free copolymerization of CS_2 and DA1 under different reaction conditions.[a)]

Entry	DA1:CS_2 (molar ratio)	[DA1] (mol L^{-1})	T (°C)	Time (h)	Yield of P1 (%)	M_w[b)] (kg mol^{-1})	$Đ$[b)] (M_w/M_n)
1	1 : 5	1.5	20	5	82	23.8	1.40
2	1 : 5	1.5	45	5	96	25.5	1.46
3	1 : 5	1.5	60	5	90	18.8	1.48
4	1 : 5	1.5	80	5	75	12.1	1.45
5	1 : 5	1.5	100	5	70	7.0	1.18
6	1 : 2	1.5	45	5	73	9.1	1.33
7	1 : 3	1.5	45	5	81	18.8	1.47
8	1 : 10	1.5	45	5	58	8.9	1.08
9	1 : 5	2	45	5	73	18.8	1.53
10	1 : 5	1	45	5	76	19.5	1.68
11	1 : 5	0.8	45	5	0	—	—
12	1 : 5	0.5	45	5	0	—	—
13	1 : 5	1.5	45	1	62	13.2	1.66
14	1 : 5	1.5	45	3	75	18.8	1.48
15	1 : 5	1.5	45	12	90	22.7	1.64

a) Carried out in DMF as solvent, 1 mL **DA1** (6.67 mmol) was added.
b) M_w and $Đ$ were determined by GPC in DMF based on polystyrene standard samples.

a solvent. On increasing the temperature from 0 to 100 °C, the yields and molecular weights of polymer **P1** increased up to 45 °C followed by a steady decrease at higher temperatures. It was noted that during this copolymerization feed ratio of DA1 and CS_2 was set at 1 : 5, with the concentration of DA1 maintained at 1.5 M and a reaction time of five hours. The optimum reaction temperature was chosen as 45 °C, which was close to the boiling point of CS_2, and this condition efficiently produced **P1** in 96% yield with a M_w of 25.5 kg mol^{-1} and polydispersity of 1.46 (Table 2.13).

The copolymer **P1** is characterized by ^1H and ^{13}C NMR spectroscopy, which strongly supports its formation, further to confirm its alternating structure matrix

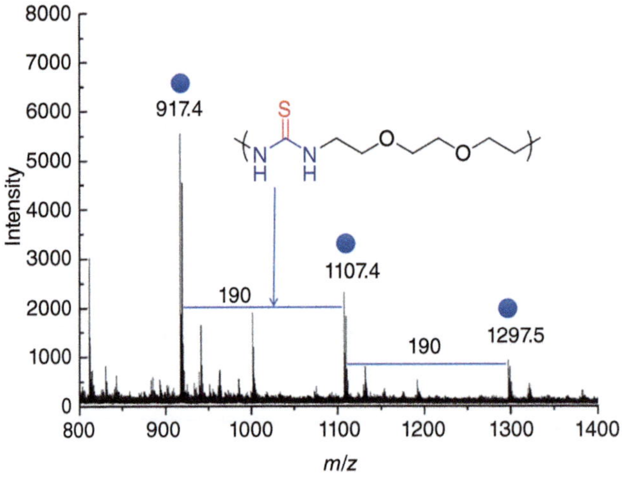

Figure 2.12 MALDI-TOF mass spectrum of the **P1**.

assisted laser desorption time-of-flight mass spectrometry (MALDI-TOF-MS) was recorded as depicted in Figure 2.12. The MALDI spectrum shows a group of equidistant peaks at m/z value of 917.4, 1107.4, and 1297 (peaks marked with blue circles), which are separated from each other by 190 m/z a mass value equivalent to repeating unit of thiourea unit.

The self-healing property of polymer **P1** was further examined by preparing a rectangular sheet of dimensions 2 × 10 × 20 mm [75]. The sheet was prepared by heating the sample in a Teflon mold at 140 °C and allowing it to cool naturally. The sheet was further cut into two pieces using a hot knife. The two pieces were pressed together after cooling for 10 minutes. The merged sheets were found to withstand a load of 200 g as shown in Figure 2.13. Further studies of this self-healing property of polythioureas are expertly described by Tian et al. [77].

The scope of this economical method of preparing polythioureas was further extended by employing a pool of different primary and secondary diamines (Scheme 2.17a). Twenty-five commercially available primary and secondary diamines whose structures are given in Scheme 2.17b are tested for the polycondensation with CS_2, and the results are shown in Table 2.14. It has been observed that all the diamines give the corresponding polythiourea polymers except the **DA25.**

It has also been observed that the linear aliphatic diamines (**DA2–DA5**) show good conversions by giving the polymers **P2–P5** in very good yields (92–98%) with high molecular weights (M_w 15.3–56.2 kg mol^{-1}). The alicyclic diamines (**DA6–DA8**) also exhibit good reactivities but the molecular weights of these polymers were low (3.4–5.8 kg mol^{-1}). The aromatic diamines with one benzene ring (**DA9–DA12**) also show good conversions whereas aromatic diamines (**DA13–DA16**), with two benzene rings display less reactivity with lower molecular weight of the resulting polymers (2.2–4.1 kg mol^{-1}).

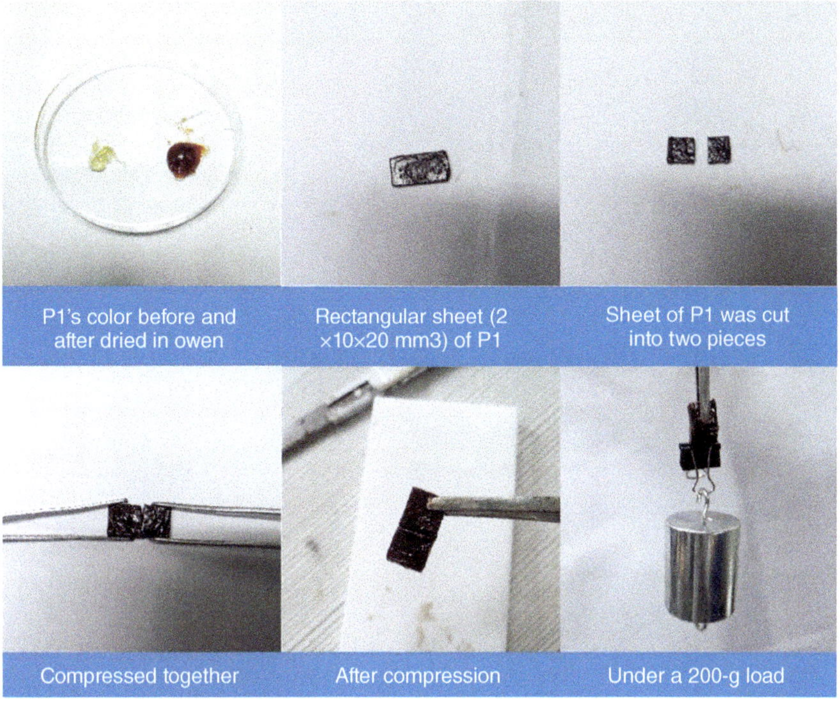

Figure 2.13 Systematic procedure for self-healing test of **P1**. Source: Reproduced from Ref. [75] with permission of American Chemical Society.

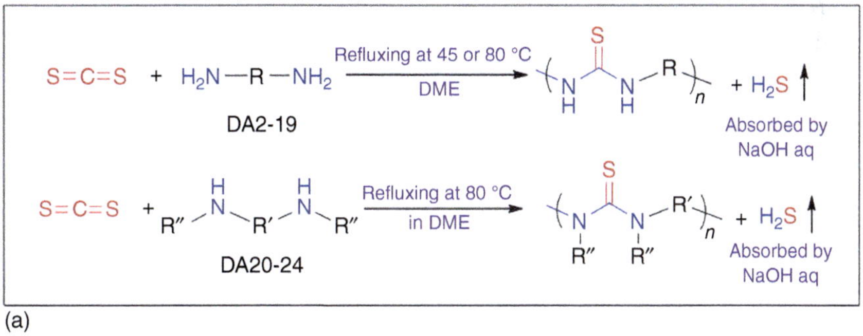

Scheme 2.17 (a) Catalyst-free copolymerization of CS_2 with different primary and secondary diamines. (b) Structures of diamine monomers.

It is well-documented in the literature that materials having thiourea functionality have strong affinity for binding heavy metal ions like mercury [77]. Inspired from these reports polythiourea polymer **P4** whose synthesis is highly cost effective was employed for Hg^{2+} extraction from water. Experimentally the solid sample of polymer **P4** was stirred with aqueous solution of $HgCl_2$ at ambient temperature for 12 hours (Figure 2.14). In the beginning the polymer **P4** remained suspended in the solution but with progress gradually it started to precipitate out at the bottom

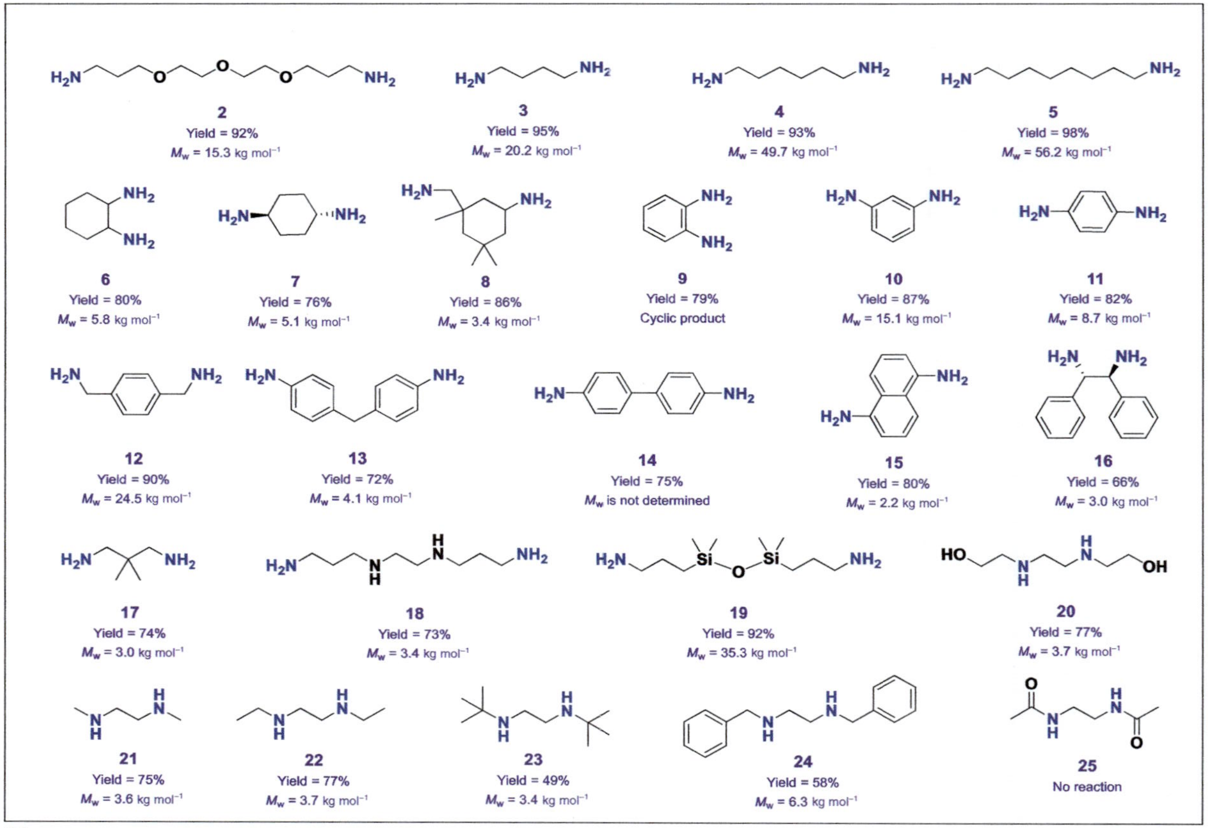

(b)

Scheme 2.17 *(Continued.)*

Table 2.14 Catalyst-free polymerization of CS_2 with different diamines.[a)]

Entry	Monomer	T (°C)	Yield (%)	M_w[b)] (kg mol^{-1})	$Đ$[b)] (M_w/M_n)	T_g (°C)	$T_{d,5\%}$ (°C)
1	**DA2**	45	92	15.3	1.75	−13	246
2	**DA3**	45	95	20.2	1.30	46	247
3	**DA4**	45	93	49.7	1.81	67	246
4	**DA5**	45	98	56.2	1.63	52	286
5	**DA6**	80	80	5.8	1.14	1	179
6	**DA7**	80	76	5.1	1.11	66	169
7	**DA8**	80	86	3.4	1.01	—	223
8	**DA9**	80	79	—	—	—	330
9	**DA10**	80	87	15.1	1.54	—	175
10	**DA11**	80	82	8.7	1.24	—	252
11	**DA12**	80	90	24.5	1.54	132	173
12	**DA13**	80	72	4.1	1.03	132	272
13	**DA14**	80	75	—	—	—	274
14	**DA15**	80	80	2.2	1.02	—	314
15	**DA16**	80	66	3.0	1.07	13	207
16	**DA17**	80	74	3.0	1.05	79	192
17	**DA18**	80	73	3.4	1.05	—	254
18	**DA19**	80	92	35.3	1.54	21	257
19	**DA20**	80	77	3.7	1.01	75	208
20	**DA21**	80	75	3.6	1.02	—	123
21	**DA22**	80	77	3.2	1.08	—	107
22	**DA23**	80	49	3.4	1.02	—	195
23	**DA24**	80	58	6.3	1.28	70	241
24	**DA25**	80	0	—	—	—	—

a) Carried out in DMF as solvent, diamine:CS_2 (molar ratio) = 1 : 5, concentration of diamine is 1.5 M, and the reactions carried out for 12 h.
b) M_w and $Đ$ are determined by GPC in DMF based on polystyrene standard samples.

due to the higher density of the Hg^{2+} polymer complex. After filtration and centrifugation, the concentration of Hg^{2+} was determined by employing inductively coupled plasma atomic emission spectroscopic technique (ICP-AES). It has been observed that on addition of 5 mg of polymeric sample **P4** into 4 mL $HgCl_2$ solution [Hg^{2+}], concentration reduces sharply from 20 to 6.56 mg l^{-1} after 12 hours of stirring thus absorbing 67.2% of total Hg^{2+}content. On addition of higher proportion of **P4** (30 mg), the absorption efficiency reached 99.9% (Figure 2.14) thus confirming that **P4** is an efficient Hg^{2+} absorbent.

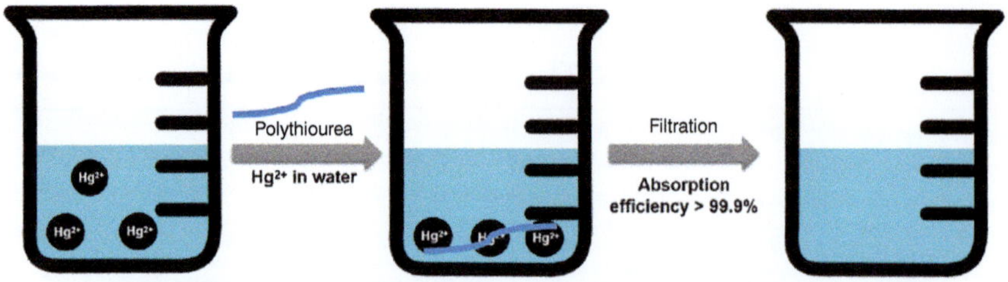

$Mass_{P4}$ (mg)	$[Hg^{2+}]_0$ (mg l^{-1})	$[Hg^{2+}]$ (mg l^{-1})	Absorption efficiency (%)
5	20.00	6.56	67.2
10	20.00	1.21	93.9
20	20.00	0.05	99.8
30	20.00	<0.01	>99.9

Figure 2.14 Schematic diagram of facile mercury absorption process. The mercury absorption efficiency of solid powder of **P4** with 4 mL of Hg^{2+} solution (20 mg l^{-1}).

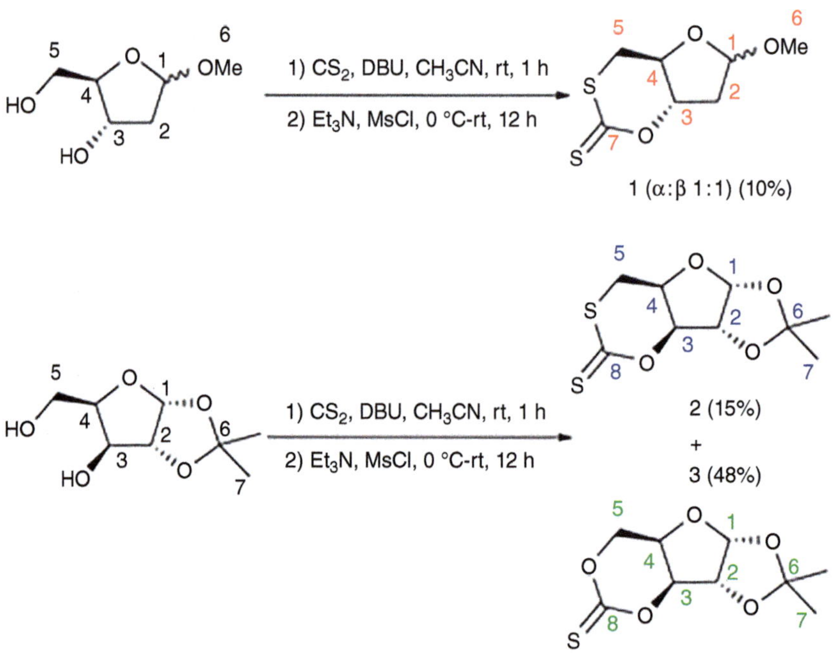

Scheme 2.18 Synthesis of monomers **1–3**. Source: Modified from Ref. [78]. Licensed Under CC BY 3.0.

Recently, Buchard and coworkers reported the synthesis of novel cyclic xanthate and thionocarbonate monomers from sugar diols by using cyclothiocarbonation of 2-deoxy-D-ribose- and D-xylose-derived diols with CS_2 as shown in Scheme 2.18 [78].

The monomers **1–3** were further polymerized using ring opening polymerization at room temperature by employing 1,5,7-triazabicyclo[4.4.0]dec-5-ene (TBD) as an organocatalyst, 4-methylbenzyl alcohol as initiator, and 1 mol L^{-1} initial concentration of monomer (Table 2.15).

From Table 2.15 it is evident that polymerization of **1** occurs rapidly reaching >99% conversion in less than 15 minutes at different monomer: initiator: catalyst feed ratios (entries 1–3), whereas polymerization of **2** was slightly slower with 86% monomer conversion after 15 minutes and >99% after two hours due to less ring strain of this xanthate compared to **1** (entries 5–8 in Table 2.15). Polymerization of **3** occurs even slower with monomer conversion of 66% after two hours (entries 8–14 in Table 2.15). The polymerization of monomers **1–3** were also studied by

Table 2.15 Ring-opening polymerization of **1–3.**[a)]

Entry	M	[M]:[C]:[1][b)]	Time (h)	Conv.[c)]	Mn (calc)[d)]	Mn (SEC)[e)]	Đ[f)]
1	1	25 : 1 : 1	0.25	>99	5.3	5.1	1.6
2	1	50 : 1 : 1	0.25	>99	10.4	5.2	1.8
3	1	100 : 1 : 1	0.25	>99	20.7	11.3	2.2
4	2	25 : 1 : 1	0.25	86	5.5	6.0	1.5
5	2	50 : 1 : 1	0.25	87	10.9	5.3	2.2
6	2	50 : 1 : 1	2	>99	12.5	12.6	1.5
7	3	100 : 1 : 1	0.25	86	21.5	15.7	1.7
8	3	25 : 1 : 1	0.25	42	2.6	5.2	1.5
9	3	50 : 1 : 1	0.25	40	4.8	7.4	1.3
10	3	50 : 1 : 1	2	66	7.8	6.7	1.2
11	3	50[g)]: 1 : 1	0.25	52	6.2	7.5	1.5
12	3	50[g)]: 1 : 1	0.5	61	7.2	10.0	1.3
13	3	50[g)]: 1 : 1	3	62	7.3	10.3	1.3
14	3	100[g)]: 1 : 1	0.25	64	10.3	10.6	1.5

a) Polymerizations carried out at room temperature in anhydrous CH_2Cl_2 solvent with initial [M]0 = 1 mol L^{-1} (M = monomer), unless stated otherwise; all entries correspond to separate experiments.
b) C is the catalyst, TBD, and I is the initiator, 4-MeBnOH.
c) Conversion measured by ^{1}H NMR determined by relative integration of the anomeric proton in the ^{1}H NMR spectrum in $CDCl_3$.
d) In 103 g mol^{-1}, calculated as Mr(I) + (Mr(monomer) × [monomer]0/[I]0 × conv/100%).
e) In 103 g mol^{-1}, estimated by SEC (RI detector) versus polystyrene standards with THF eluent.
f) $Đ = M_n/M_w$. G [M]0 = 1.58 mol.
g) $[M]_0 = 1.58$ mol L^{-1}.

size exclusion chromatography (SEC) in THF to confirm their number M_n and $Đ$. Polymers with molecular weight and polydispersities up to 11 300 g mol^{-1} ($Đ$ 2.2) and 15 700 g mol^{-1} ($Đ$ 1.7) could be achieved from monomers **1** and **2,** respectively. There was limited agreement between the predicted and SEC obtained M_n with high polydispersities. The ring opening polymerization of monomer **3** was better controlled with narrower dispersities and better agreement with SEC obtained and theoretical calculated M_n. MALDI-TOF-MS also revealed both linear and cyclic topologies further supporting to the limited control over the polymerization of these monomers. No O/S scrambling was observed in the polymerization of these monomers as revealed by $^{13}C(^{1}H)$ NMR analysis.

2.5 Concluding Remarks

In this chapter, the essential aspects of the synthesis of copolymers derived from the coupling of CS_2 and various monomers, namely, epoxides, episulfides, diamines, and sugar diols, have been summarized. In addition, methods are described for the synthesis of various thiiranes (episulfides), monomers needed for copolymerization with CS_2 to afford exclusively polytrithiocarbonates. During the addition of CS_2 to epoxides or COS to thiiranes, O/S scrambling reactions are prominent. Hence, mechanistic considerations of this process are presented in some detail. As stated in the introduction, the preparation/characterization/utilization of sulfur containing polymers is the subject of much research activity. The use of CS_2, which can be produced from the large quantities of elemental sulfur that is available from hydrodesulfurization, offers an excellent monomer to expand the range of polymeric materials containing sulfur in their main chains.

Fundamental studies of the type described in this chapter hopefully will provide the necessary foundation, as well as stimulus, to enable others to develop the large-scale production of valuable sulfur-containing polymeric materials for various uses.

Acknowledgments

The authors' original research on the utilization of carbon disulfide as a source for polymer synthesis has been funded over the years by the National Science Foundation and the Robert A. Welch Foundation of Texas under grant A-0923.

References

1 Chung, W.J., Griebel, J.J., Kim, E.T. et al. (2013). The use of elemental sulfur as an alternative feedstock for polymeric materials. *Nat. Chem.* 5 (6): 518.
2 The Editors of Encyclopaedia Britannica. Carbon disulfide. 2. https://www.britannica.com/science/carbon-disulfide (accessed 2019).

3 Rich, A.L., Patel, J.T., and Al-Angari, S.S. (2016). Carbon disulfide (CS_2) interference in glucose metabolism from unconventional oil and gas extraction and processing emissions. *Environ. Health Insights* 10: 51–57. https://doi.org/10.4137/EHI.S31906.

4 Rich, A.L. and Patel, J.T. (2015). Carbon disulfide (CS_2) mechanisms in formation of atmospheric carbon dioxide (CO_2) formation from unconventional shale gas extraction and processing operations and global climate change. *Environ. Health Insights* 9: 35–39. https://doi.org/10.4137/EHI.S15667.

5 Darensbourg, D.J. and Holtcamp, M.W. (1995). Catalytic activity of zinc (II) phenoxides which possess readily accessible coordination sites. Copolymerization and terpolymerization of epoxides and carbon dioxide. *Macromolecules* 28 (22): 7577–7579.

6 Coates, G.W. and Moore, D.R. (2004). Discrete metal-based catalysts for the copolymerization of CO_2 and epoxides: discovery, reactivity, optimization, and mechanism. *Angew. Chem. Int. Ed.* 43 (48): 6618–6639.

7 Sugimoto, H. and Inoue, S. (2004). Copolymerization of carbon dioxide and epoxide. *J. Polym. Sci. Part A: Polym. Chem.* 42 (22): 5561–5573.

8 Darensbourg, D.J., Mackiewicz, R.M., Phelps, A.L., and Billodeaux, D.R. (2004). Copolymerization of CO_2 and epoxides catalyzed by metal salen complexes. *Acc. Chem. Res.* 37 (11): 836–844.

9 Chisholm, M.H. and Zhou, Z. (2004). New generation polymers: the role of metal alkoxides as catalysts in the production of polyoxygenates. *J. Mater. Chem.* 14 (21): 3081–3092.

10 Darensbourg, D.J. (2007). Making plastics from carbon dioxide: salen metal complexes as catalysts for the production of polycarbonates from epoxides and CO_2. *Chem. Rev.* 107 (6): 2388–2410.

11 Klaus, S., Lehenmeier, M.W., Anderson, C.E., and Rieger, B. (2011). Recent advances in CO_2/epoxide copolymerization—New strategies and cooperative mechanisms. *Coord. Chem. Rev.* 255 (13–14): 1460–1479.

12 Kember, M.R., Buchard, A., and Williams, C.K. (2011). Catalysts for CO_2/epoxide copolymerisation. *Chem. Commun.* 47 (1): 141–163.

13 Lu, X.-B. and Darensbourg, D.J. (2012). Cobalt catalysts for the coupling of CO_2 and epoxides to provide polycarbonates and cyclic carbonates. *Chem. Soc. Rev.* 41 (4): 1462–1484.

14 Lu, X.-B., Ren, W.-M., and Wu, G.-P. (2012). CO_2 copolymers from epoxides: catalyst activity, product selectivity, and stereochemistry control. *Acc. Chem. Res.* 45 (10): 1721–1735.

15 Darensbourg, D.J. and Wilson, S.J. (2012). What's new with CO_2? Recent advances in its copolymerization with oxiranes. *Green Chem.* 14 (10): 2665–2671.

16 Paul, S., Zhu, Y., Romain, C. et al. (2015). Ring-opening copolymerization (ROCOP): synthesis and properties of polyesters and polycarbonates. *Chem. Commun.* 51 (30): 6459–6479.

17 Trott, G., Saini, P., and Williams, C. (2016). Catalysts for CO_2/epoxide ring-opening copolymerization. *Philos. Trans. R. Soc. A Math. Phys. Eng. Sci.* 374 (2061): 20150085.

18 Poland, S.J. and Darensbourg, D.J. (2017). A quest for polycarbonates provided *via* sustainable epoxide/CO_2 copolymerization processes. *Green Chem.* 19 (21): 4990–5011.

19 Kozak, C.M., Ambrose, K., and Anderson, T.S. (2018). Copolymerization of carbon dioxide and epoxides by metal coordination complexes. *Coord. Chem. Rev.* 376: 565–587.

20 Kamphuis, A.J., Picchioni, F., and Pescarmona, P.P. (2019). CO_2-fixation into cyclic and polymeric carbonates: principles and applications. *Green Chem.* 21 (3): 406–448.

21 Luo, M., Zhang, X.-H., and Darensbourg, D.J. (2016). Poly (monothiocarbonate)s from the alternating and regioselective copolymerization of carbonyl sulfide with epoxides. *Acc. Chem. Res.* 49 (10): 2209–2219.

22 Darensbourg, D.J. (2019). Chain transfer agents utilized in epoxide and CO_2 copolymerization processes. *Green Chem.* 21 (9): 2214–2223.

23 Covestro.(2015) Carbon dioxide is revolutionizing plastics production. https://www.co2-dreams.covestro.com/en (accessed 12 December 2018).

24 Tullo, A.H. (2016). Aramco buys Novomer's CO_2-based polyols unit. *Chem. Eng. News* 94 (45): 14–14.

25 Ok, M.-A. and Jeon, M. (2012). Properties of poly(propylene carbonate) produced via SK Energy's Greenpol Technology. ANTEC-Conference Proceedings, Society of Petroleum Engineers, Richardson, TX, 2012, 2134–2139.

26 Econic Technologies Ltd. Nether Alderley.(2017) www.econic-technologies.com (accessed 2019).

27 Griebel, J.J., Namnabat, S., Kim, E.T. et al. (2014). New infrared transmitting material via inverse vulcanization of elemental sulfur to prepare high refractive index polymers. *Adv. Mater.* 26 (19): 3014–3018.

28 Kleine, T.S., Diaz, L.R., Konopka, K.M. et al. (2018). One dimensional photonic crystals using ultrahigh refractive index chalcogenide hybrid inorganic/organic polymers. *ACS Macro Lett.* 7 (7): 875–880.

29 Sun, X., Shi, W., Ma, F. et al. (2014). Thymine-covalently decorated, AIEE-type conjugated polymer as fluorescence turn-on probe for aqueous Hg^{2+}. *Sensors Actuators B Chem.* 198: 395–401.

30 Wang, Y.-B., Wang, Y.-M., Zhang, W.-Z., and Lu, X.-B. (2013). Fast CO_2 sequestration, activation, and catalytic transformation using N-heterocyclic olefins. *J. Am. Chem. Soc.* 135 (32): 11996–12003.

31 Wang, Y.-B., Sun, D.-S., Zhou, H. et al. (2015). CO_2, COS and CS_2 adducts of N-heterocyclic olefins and their application as organocatalysts for carbon dioxide fixation. *Green Chem.* 17 (7): 4009–4015.

32 Darensbourg, D., Mueller, B., Bischoff, C. et al. (1991). Investigations into the steric influences on the reaction mechanism of CO_2 insertion into metal-oxygen bonds. COS activation as a model for CO_2. *Inorg. Chem.* 30 (10): 2418–2424.

33 Darensbourg, D.J., Sanchez, K.M., Reibenspies, J.H., and Rheingold, A.L. (1989). Synthesis, structure, and reactivity of zerovalent group 6 metal pentacarbonyl aryl oxide complexes. Reactions with carbon dioxide. *J. Am. Chem. Soc.* 111 (18): 7094–7103.

34 Staudinger, H. and Pfenninger, F. (1916). Über die Einwirkung von Schwefeldioxyd auf diphenyl-diazomethan. *Ber. Dtsch. Chem. Ges.* 49 (2): 1941–1951.

35 Tarbell, D.S. and Harnish, D.P. (1951). Cleavage of the carbon-sulfur bond in divalent sulfur compounds. *Chem. Rev.* 49 (1): 1–90.

36 Mackinney, H. (1960). Polythiirane-rubber compositions. Patent 2, 962, 457, Chem. Abstr, 1961, p. 6009.

37 Schirmeister, T. (2000). (S)-Thiirancarboxylic acid as a reactive building block for a new class of cysteine protease inhibitors. *Bioorg. Med. Chem. Lett.* 10 (23): 2647–2651.

38 Procter, D., Penning, T., Akonas, L. et al. (1999). Kelatorphan and related analogs: potent and selective inhibitors of leukotriene A_4 hydrolase. *J. Chem. Soc., Perkin Trans. 1* 5: 641–667.

39 Fan, W. and Katritzky, A. (1996). *Comprehensive Heterocyclic Chemistry II*, vol. 4, 1–126. Oxford: Elsevier Science.

40 Goodman, L. and Baker, B. (1959). Potential anticancer agents. 1 XXII. The conversion of cyclopentene oxide to cyclopentene sulfide. *J. Am. Chem. Soc.* 81 (18): 4924–4926.

41 van Tamelen, E.E. (1951). The formation and ring-opening of alkene sulfides. *J. Am. Chem. Soc.* 73 (7): 3444–3448.

42 Sander, M. (1966). Thiiranes. *Chem. Rev.* 66 (3): 297–339.

43 Kazemi, F., Kiasat, A., and Ebrahimi, S. (2003). LiBF4: a mild and efficient catalyst for conversion of oxiranes to thiiranes with thiourea. *Synth. Commun.* 33 (4): 595–600.

44 Joseph, J., Gosavi, R.K., Otter, A. et al. (1990). Stereospecificity of the $S(^3P_J)$ + butene-2 reaction and the NMR spectra of the 1,2-dimethylthiiranes: an experimental and theoretical study. *J. Am. Chem. Soc.* 112 (24): 8670–8678.

45 Gunning, H. and Strausz, O. (1966). The reactions of sulfur atoms. *Adv. Photochem.* 4: 143–194.

46 Ashitani, T. and Nagahama, S. (1999). Direct episulfidation of caryophyllene and humulene. *Nat. Prod. Lett.* 13 (3): 163–167.

47 Nau, W.M., Bucher, G., and Scaiano, J. (1997). Absolute rate constants for the reactions of sulfur (^3P_J) atoms in solution. *J. Am. Chem. Soc.* 119 (8): 1961–1970.

48 (a) Nakayama, J. and Ito, Y. (1989). The reaction of acenaphthylene with elemental sulfur. *Sulfur Lett.* 9: 135. (b) Nakayama, J., Ito, Y., and Mizumura, A. (1992). Sulfuration of dibenzobarralene and adamantylidene adamantane. *Sulfur Lett.* 14: 247.

49 Bartlett, P.D. and Ghosh, T. (1987). Sulfuration of the norbornene double bond. *J. Organomet. Chem.* 52 (22): 4937–4943.

50 Sugihara, Y., Kobiki, A., and Nakayama, J. (2009). Synthesis and sulfuration of 2′-adamantylidene-9-benzonorbornenylidene. *Heterocycles* 78 (1): 103–115.

51 Adam, W. and Bargon, R.M. (2001). Molybdenum-catalyzed episulfidation of (E)-cycloalkenes with elemental sulfur. *Chem. Commun.* 19: 1910–1911.

52 Adam, W., Bargon, R.M., and Schenk, W.A. (2003). Direct episulfidation of alkenes and allenes with elemental sulfur and thiiranes as sulfur sources, catalyzed by molybdenum oxo complexes. *J. Am. Chem. Soc.* 125 (13): 3871–3876.

53 Arisawa, M., Ichikawa, T., and Yamaguchi, M. (2015). Synthesis of thiiranes by rhodium-catalyzed sulfur addition reaction to reactive alkenes. *Chem. Commun.* 51 (42): 8821–8824.
54 Soga, K., Imamura, H., Sato, M., and Ikeda, S. (1976). Copolymerization of ethylene sulfide and carbon disulfide. *J. Polymer Sci. Polymer Lett. Ed.* 14 (3): 677–684.
55 Soga, K., Sato, M., Imamura, H., and Ikeda, S. (1975). Copolymerization of carbon disulfide and episulfides. *J. Polymer Sci. Polymer Lett. Ed.* 13 (3): 167–171.
56 Nakano, K., Tatsumi, G., and Nozaki, K. (2007). Synthesis of sulfur-rich polymers: copolymerization of episulfide with carbon disulfide by using [PPN] Cl/(salph) Cr (III) Cl system. *J. Am. Chem. Soc.* 129 (49): 15116–15117.
57 Zhang, X., Su, P., Liu, Y. et al. (2017). Coupling of propylene sulfide (PS) and carbon disulfide (CS_2) catalyzed by the asymmetrical Cr(III)-*bis*-Schiff-base complex. *Inorg. Chem. Commun.* 85: 59–61.
58 Darensbourg, D.J., Wilson, S.J., and Yeung, A.D. (2013). Oxygen/sulfur scrambling during the copolymerization of cyclopentene oxide and carbon disulfide: selectivity for copolymer vs cyclic [thio]carbonates. *Macromolecules* 46 (20): 8102–8110.
59 Darensbourg, D.J., Andreatta, J.R., Jungman, M.J., and Reibenspies, J.H. (2009). Investigations into the coupling of cyclohexene oxide and carbon disulfide catalyzed by (salen) CrCl. Selectivity for the production of copolymers vs. cyclic thiocarbonates. *Dalton Trans.* 41: 8891–8899.
60 Iqbal, S.M. and Owen, L.N. (1960). 212. Dithiols. Part XXI. Dithiols from cyclic trithiocarbonates. *J. Chem. Soc.*: 1030–1036.
61 Darensbourg, D.J., Yeung, A.D., and Wei, S.-H. (2013). Base initiated depolymerization of polycarbonates to epoxide and carbon dioxide co-monomers: a computational study. *Green Chem.* 15 (6): 1578–1583.
62 Zhang, X.-H., Liu, F., Sun, X.-K. et al. (2008). Atom-exchange coordination polymerization of carbon disulfide and propylene oxide by a highly effective double-metal cyanide complex. *Macromolecules* 41 (5): 1587–1590.
63 Darensbourg, D.J. and Moncada, A.I. (2008). Mechanistic insight into the initiation step of the coupling reaction of oxetane or epoxides and CO_2 catalyzed by (salen)CrX complexes. *Inorg. Chem.* 47 (21): 10000–10008.
64 Crawford, M.-J., Klapötke, T.M., Mayer, P., and Vogt, M. (2004). CS_2N_3-containing pseudohalide species: an experimental and theoretical study. *Inorg. Chem.* 43 (4): 1370–1378.
65 Sommer, F. (1915). Über den Imidcharakter der Stickstoff-wasserstoffsäure I. Die Synthese der Azido-dithio-kohlensäure. *Ber. Dtsch. Chem. Ges.* 48 (2): 1833–1841.
66 Bailey, P.J., Bell, N.L., Nichol, G.S. et al. (2012). Insertion and substitution chemistry at the boron fourth position in charge-neutral zwitterionic tripodal tris(methimazolyl) borate ligands. *Inorg. Chem.* 51 (6): 3677–3689.
67 Darensbourg, D.J., Bottarelli, P., and Andreatta, J.R. (2007). Inquiry into the formation of cyclic carbonates during the (salen) CrX catalyzed CO_2/cyclohexene oxide copolymerization process in the presence of ionic initiators. *Macromolecules* 40 (21): 7727–7729.

68 Motokucho, S., Takeuchi, D., Sanda, F., and Endo, T. (2001). Synthesis of cyclic trithiocarbonates from cyclic ethers and carbon disulfide catalyzed by titanium complex. *Tetrahedron* 57 (33): 7149–7152.

69 Clegg, W., Harrington, R.W., North, M., and Villuendas, P. (2010). A bimetallic aluminum(salen) complex for the synthesis of 1, 3-oxathiolane-2-thiones and 1, 3-dithiolane-2-thiones. *J. Organomet. Chem.* 75 (18): 6201–6207.

70 Barbero, M., Degani, I., Dughera, S. et al. (1996). Convenient procedure for converting 1,3-dithiolane-2-thiones into 1,3-dithiolan-2-one. *J. Chem. Soc., Perkin Trans. 1*: 289–294.

71 Yamazaki, N., Higashi, F., and Iguchi, T. (1974). Polyureas and polythioureas from carbon dioxide and disulfide with diamines under mild conditions. *J. Polymer Sci. Polymer Lett. Ed.* 12 (9): 517–521.

72 Yamazaki, N., Iguchi, T., and Higashi, F. (1975). Studies on reactions of the N-phosphonium salts of pyridines. XIII. Direct polycondensation reaction of carbon dioxide or disulfide with diamines under mild conditions. *J. Polymer Sci. Polymer Lett. Ed.* 13 (4): 785–795.

73 Tasaka, S., Ohishi, K., and Inagaki, N. (1995). Ferroelectric behavior in aliphatic polythioureas. *Ferroelectrics* 171 (1): 203–210.

74 Lee, J.W., Takase, Y., Newman, B.A., and Scheinbeim, J.I. (1991). Ferroelectric polarization switching in nylon-11. *J. Polym. Sci. B Polym. Phys.* 29 (3): 273–277.

75 Wu, S., Luo, M., Darensbourg, D.J., and Zuo, X. (2019). Catalyst-free construction of versatile and functional CS_2-based polythioureas: characteristics from self-healing to heavy metal absorption. *Macromolecules* 52 (22): 8596–8603.

76 Yanagisawa, Y., Nan, Y., Okuro, K., and Aida, T. (2018). Mechanically robust, readily repairable polymers via tailored noncovalent cross-linking. *Science* 359 (6371): 72–76.

77 Tian, T., Hu, R., and Tang, B.Z. (2018). Room temperature one-step conversion from elemental sulfur to functional polythioureas through catalyst-free multicomponent polymerizations. *J. Am. Chem. Soc.* 140 (19): 6156–6163.

78 López-Vidal, E.M., Gregory, G.L., Kociok-Köhn, G., and Buchard, A. (2018). Polymers from sugars and CS_2: synthesis and ring-opening polymerisation of sulfur-containing monomers derived from 2-deoxy-d-ribose and d-xylose. *Polym. Chem.* 9 (13): 1577–1582.

3

Carbonyl Sulfide Derived Polymers

Cheng-Jian Zhang, Jia-Liang Yang, Xiao-Han Cao, and Xing-Hong Zhang

Zhejiang University, MOE Key Laboratory of Macromolecular Synthesis and Functionalization, Department of Polymer Science and Engineering, Hangzhou 310027, China

3.1 Introduction

3.1.1 Overview of COS

Carbonyl sulfide, COS or SCO (CAS: 463-58-1), also known as carbon oxysulfide, carbon oxide sulfide, or carbon monoxide monosulfide, was observed in 1867 when it was prepared by the reaction of carbon monoxide (CO) and sulfur vapors at a dull red heat [1]. A review paper written by Bruno et al. in 2002 demonstrated the synthesis of COS, its basic physical properties, and chemical reactions [2].

COS is a key intermediate of the atmospheric sulfur cycle and the most abundant sulfur-containing gas in the troposphere (Figure 3.1). In the age of industry, COS is massively released from the consumption of fossil fuels and finally converted to hydrogen sulfide (H_2S) and elemental sulfur (S_8). COS is also an air pollutant, causing sulfur aerosols, fog, haze, and acid rain and damaging of the ozonosphere through a series of photochemical reactions [3, 4]. Therefore, intensive study on COS for its capture and utilization has great significance.

The major physical properties of COS have been well summarized in the previous review [2]. COS is a colorless gas at ambient pressure and temperature, which has unpleasant rotten egg odors, except when it is very pure. COS is somewhat soluble in water (0.80 mL mL^{-1} at 13.5 °C, 1 atm) and toluene (15.0 mL mL^{-1} at 22.0 °C, 1 atm). It is very soluble in potassium hydroxide (KOH) aqueous solutions where its hydrolysis is accelerated. In pure water, COS is slowly hydrolyzed to H_2S. COS does not cause corrosion to many metals such as copper, aluminum, carbon steel, monel, stainless steels, brass, and so on. However, the compatibility is sharply reduced in the presence of moisture, which is commonly observed with many acidic gases. Synthetic polymers such as Teflon and PVC are very compatible with COS.

Sulfur-Containing Polymers: From Synthesis to Functional Materials,
First Edition. Edited by Xing-Hong Zhang and Patrick Theato.

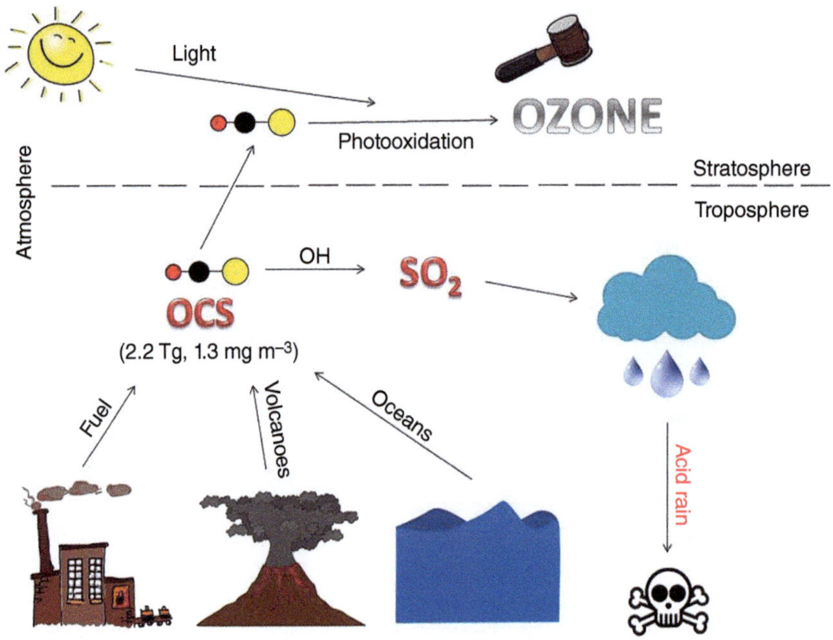

Figure 3.1 Carbonyl sulfide (COS) cycle in nature and the utilization of COS to sulfur-containing polymers.

3.1.2 Preparation of COS

COS can be produced by several preparation methods, as introduced in the following.

3.1.2.1 Direct Reaction of CO and Sulfur

The first reported synthesis of COS was by the reaction of CO with sulfur vapors at a dull red heat [1].

$$CO + S_x \rightarrow COS$$

Sita and coworkers reported a metal-catalyzed on-demand production of COS from CO and S_8. The group 6 molybdenum cyclopentadienyl amidinate bis(carbonyl) complex was applied as the catalyst for the production of COS from CO and S_8 under near-ambient conditions (e.g. 10 psi, 25 °C) [5].

3.1.2.2 Using Carbon Disulfide (CS_2)

COS can also be obtained by the reaction of oxidizing agents on CS_2, and as a byproduct during the preparation of CS_2 from carbonaceous matter and SO_2. It may be formed by heating amides and CS_2 together in a sealed tube, by the action of carbonyl chloride on sulfides, or by the action of concentrated sulfuric acid (H_2SO_4) on allyl thiocyanates [1]. Recently, our group disclosed a new route to COS from the oxygen/sulfur exchanges of CS_2 and epoxides (Figure 3.2). The process can be

Figure 3.2 Coupling of CS_2 and epoxides to poly(thioether)s and COS.

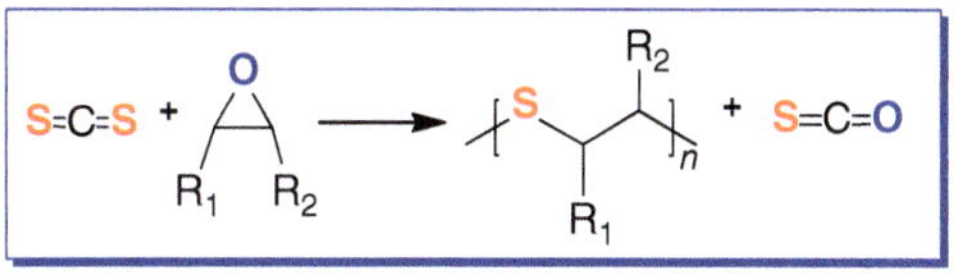

conducted in ambient air, and poly(thioether)s can also be produced [6]. More details are shown in Section 3.5.2.

3.1.2.3 Laboratory Preparation of COS

A common synthetic procedure in laboratory involves the reaction of H_2SO_4 with potassium thiocyanate (KCNS) solution at 30 °C, as follows:

$$KCNS + 2H_2SO_4 + H_2O \longrightarrow KHSO_2 + NH_4HSO_4 + COS$$

The subsequent purification process involves the removal of the side products CO_2, SO_2, H_2S, CS_2, hydrogen cyanide (HCN), and formaldehyde (CH_2O). To complete the purification, the product gas is firstly bubbled into a mixture of saturated aqueous cupric sulfate and concentrated H_2SO_4 (50% v/v) solution to remove H_2S. The gas is subsequently went through the aqueous KOH solution fast to remove all remaining acid gases. An extraction with a solution of aniline in ethanol eliminates CS_2, and finally, further passed through the concentrated H_2SO_4 yields anhydrous COS.

3.1.3 Environmental and Safety Considerations

COS is a trace component of Earth's atmosphere, which is emitted from oils and trees, volcanoes and hot springs, and chemical processes in soils and sediments. It is also considerably released in emissions from combustion processes, such as the combustion of fossil fuel and biomass, as well as gasification and petrochemical processes. COS is the most abundant sulfur compound in the atmosphere with the concentration of c. 500 ppm. The total global sources and sinks were estimated about six times less than those of H_2S and twice those of CS_2 [7]. Similar to CS_2, H_2S, and dimethyl sulfide, COS is implicated in the generation of atmospheric SO_2 with a possible global climate influence.

The lifetime of COS with respect to the oxidation by hydroxyl radicals is greater than nine years [8]. Consequently, COS can be photodissociated to CO and S ($COS \rightarrow CO + S$) and then sulfate particles (Figure 3.1) when transported into the stratosphere. It therefore becomes a factor of making the Junge layer that has an influence on the Earth's climate and can scatter the incoming solar radiation. Vertical profiles indicate that the COS concentration can be up to an altitude of 17 km. Quantum yield measurements for CO production in the photodissociation of COS show that the contribution of COS to lower stratospheric sulfate aerosol layer is much larger than previously thought. COS can also be removed from the atmosphere by hydrolysis, dissolution, oxidation in atmospheric water, and dry deposition on the earth. But oxidation leads to SO_2 and hydrolysis produces H_2S. The modest solubility of COS in water makes scavenging by rain only a minor factor in its removal from the atmosphere. COS can also be metabolized by plants, however, the COS uptake by plants is slower than those for SO_2, H_2S, and CH_3SH [9, 10]. Therefore, it is very important to prohibit the releasing of COS in concerning of our environment.

Special caution is to be taken when working with COS as it is considered irritant to the eyes, skin, nose, lungs, and throat, causing cough and sneezing, because of its ability to generate H_2S via hydrolysis [11]. A low to moderate COS exposure may cause painful and severe conjunctivitis, increase the production of tears (lachrymation), and increase sensitivity to profuse salivation, light, diarrhea, nausea, headache, and mental confusion. Exposure to high concentrations can cause sudden collapse, blurred vision, tremors, and tachycardia. Prolonged exposure leads to coma, unconsciousness, and death due to the respiratory paralysis. The concentrations of 0.1% (v/v) can produce death within two hours or less. It should be noted that COS is also a flammable gas. Health concerns from consumer activist groups point out that analyses of the volatile COS gases evolved from a number of polymer samples commonly used in domestic applications show that CS_2 and COS are emitted anywhere from room temperature to 150 °C in part per million (mass/mass) quantities [12]. Such polymers include the materials used in the fabrication of the nipples for baby feeding bottles. This led to the suggestion of imposed tolerance limits on the toxicity of COS.

3.1.4 COS Chemistry

COS is stable at ambient conditions, but it can undergo decomposition, hydrolysis, oxidation, and reduction to produce reactive compounds such as H_2S and S_8. These reactions as well as the reactions with ammonia and amines are important in the removal of COS from other gases [1].

3.1.4.1 Dissociation

COS undergoes thermal decomposition at high temperatures (600–900 °C) by either of two reactions:

$$2COS \rightarrow 2CO + 2S$$

$$2COS \rightarrow CO_2 + CS_2$$

Dissociation of the first reaction is rapid and reaches a maximum conversion of 64% at 900 °C.

3.1.4.2 Hydrolysis

The reaction of COS with water follows a unimolecular kinetic order and is catalyzed by many substances and ions, especially the hydroxyl ion.

$$COS + H_2O \rightarrow CO_2 + H_2S$$

This reaction is relatively slow, because both H_2S and CO_2 are acidic and COS is relatively stable towards acids. COS exists indefinitely in an aqueous solution with 50 wt% of H_2SO_4. In carbonated mineral waters it decomposes to H_2S after a long

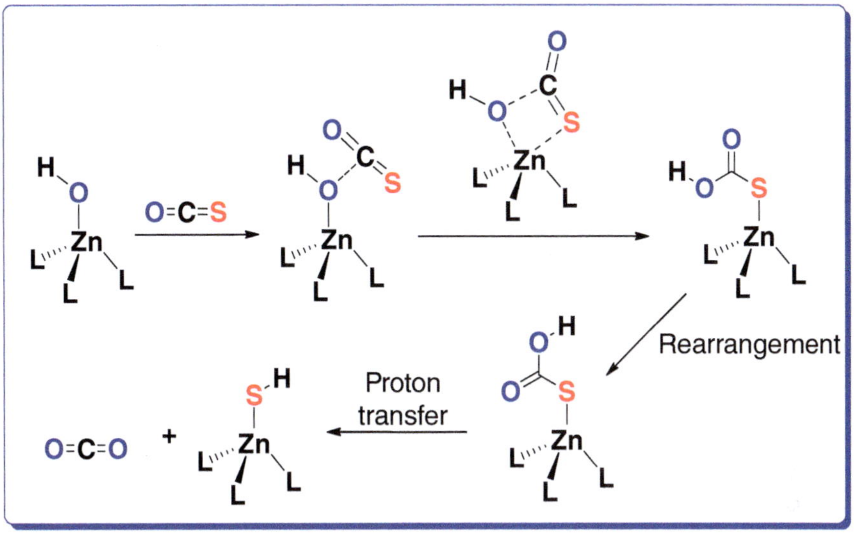

Figure 3.3 Mechanism of fixing COS by the CA Compound $[L_3Zn{-}OH]^+$, with L as histidine (CA center structure), or ligand (generally, amine).

storage in the closed container. Although hydrolysis can be accelerated by alkaline, COS is only slowly hydrolyzed by strong alkaline reagents (e.g. NaOH).

It should note that, human carbonic anhydrase (HCA) zinc enzyme can accelerate the hydrolysis of COS to H_2S in organisms. Many studies on HCA models such as $[L_3Zn{-}OH]^+$ for activating COS were reported for understanding this process [13–15], as shown in Figure 3.3. In the context, OH^- acts as a nucleophile and attack COS, through a four-membered ring transition state, then via rearrangement and proton transfer steps, a CO_2 molecule was produced along with an $L_3Zn{-}SH$. This process may be helpful to the oxygen/sulfur exchange reaction (O/S ER) (Sections 3.2.1 and 3.4).

3.1.4.3 Oxidation

COS is highly flammable, burning with a slightly luminous blue flame. However, it is less readily oxidized than CS_2. A mixture of COS/O_2 (1/1.5, v/v) burns with a small explosion, while the combustion of COS/O_2 with a volume ratio of 1/7.5 occurs without explosion. The critical explosion limits for the mixture of COS and O_2 at temperatures between 190 and 408 °C.

$$2COS + 3O_2 \rightarrow 2SO_2 + 2CO_2$$

In addition, bromine in alkaline solutions readily oxidizes COS [1].

$$COS + 4Br_2 + 12KOH \rightarrow K_2CO_3 + K_2SO_4 + 8KBr + 6H_2O$$

3.1.4.4 Reduction

COS can be reduced by hydrogen according to the equation:

$$COS + H_2 \rightarrow CO + H_2S$$

The COS content of gases from the petroleum cracking is lowered from 1000 to <0.04 ppm by the hydrogenation reaction under a low pressure, which is just slightly greater than the vapor pressure of the hydrocarbon. Cuprous sulfide is the basic ingredient in the catalyst and it is believed to function as follows:

$$COS + Cu_2S \rightarrow CO + 2CuS$$

$$2CuS + H_2 \rightarrow Cu_2S + H_2S$$

$$COS + H_2 \rightarrow CO + H_2S$$

Manganese oxide or manganese ore catalyst is effective as long as a large proportion has not been converted to manganese sulfide. The lifetime of a catalyst containing a mixture of iron oxide and sodium carbonate may be increased by pretreatment with hydrogen at 500 °C.

3.1.4.5 Reaction with SO_2

Elemental sulfur is formed when COS is mixed with an excess of SO_2 and passed over red-hot pumice or brick at 400–425 °C by the catalyst prepared from blast furnace slag containing about 55% aluminum oxide.

$$2COS + SO_2 \rightarrow 2CO_2 + 3S$$

The reaction of COS with SO_2 was also investigated at 600–800 °C in the presence of oxides of iron, aluminum, or titanium on pumice and porcelain supports. It proceeds nearly to completion with iron and titanium catalysts, ceasing when the gas phase contains 0.6% COS and 0.3% SO_2.

3.1.4.6 Reaction with Ammonia and Amines

Ammonia and many types of organic amines can react readily with COS. These reactions are of commercial importance in removing COS from gases. Ammonia gas and ammonia solutions absorb COS with the formation of ammonium thiocarbamate:

$$COS + 2NH_3 \rightarrow NH_2COSNH_4$$

If an ammonia solution is used, evaporation decomposes it to urea and H_2S. The first reaction is slow, while decomposition to urea is much more rapid. When COS and ammonia react in the absence of large amounts of water and at least one of them is in the liquid state, urea is obtained directly [1].

3.1.4.7 Miscellaneous Reactions

COS reacts extremely slow with dry alcohol to form a mercaptan:

$$COS + C_2H_5OH \rightarrow CO_2 + C_2H_5SH$$

Though, when aqueous KOH solution is used, COS is rapidly consumed:

$$COS + 2KOH \rightarrow KSCOOK + H_2O$$

$$COS + KOH + C_2H_5OH \rightarrow H_2O + KSCOOC_2H_5$$

The reaction is so fast that alcoholysis does not occur.

3.1.4.8 The Claus Reaction

The awareness of the importance of COS in industry has recently been on the rise as a result of the environmental concerns, as discussed in the aforementioned section. For instance, in the Claus reaction, highly pure sulfur can be prepared in large scale by high temperature oxidation of H_2S according to the sequence of Eqs. (3.1, 3.2), which leads to net Eq. (3.3)

$$3H_2S + 3/2O_2 \rightarrow 2H_2S + SO_2 + H_2O \quad (3.1)$$

$$2H_2S + SO_2 \rightarrow 3S + 2H_2O \quad (3.2)$$

$$3H_2S + 3/2O_2 \rightarrow 3S + 3H_2O \quad (3.3)$$

Thus, first, the combustion of H_2S leads to the formation of SO_2 and water (Eq. (3.1)). The formed SO_2 reacts with more H_2S to yield water and condensed sulfur (Eq. (3.2)). However, the combustion process also involves the formation of CO_2, which reacts with H_2S to yield COS as the main product, as well as CS_2, according to Eqs. (3.4, 3.5)

$$CO_2 + H_2S \rightarrow COS + H_2O \quad (3.4)$$

$$CO_2 + 2H_2S \rightarrow CS_2 + 2H_2O \quad (3.5)$$

The formed COS is a polluting byproduct in the Claus reaction because it undergoes the hydrolysis to yield CO_2 and H_2S (Eq. (3.6)).

$$COS + H_2O \rightarrow CO_2 + H_2S \quad (3.6)$$

As a result, additional costly treatments must be employed to remove the COS downstream from the Claus converter. These include the Beavon sulfur removal process, in which the catalytic reduction with hydrogen or reducing gas mixtures convert all sulfur content to H_2S. In general, the hydrolysis products, CO_2 and H_2S, are much easier to remove from the refinery gas streams than COS.

In this section, we have revisited the chemical and physical properties of COS, and the preparation methods in industry and in the laboratory, COS chemistry and the concerns of COS on the environment, and safety. One of big industrial issues is the removal of naturally occurring COS in liquefied petroleum gas (LPG); in addition, COS is very promising to use as an agricultural fumigant for eliminating pests. In very recent years, high pure COS has been used as a special gas for electronics industry. Based on the recent endeavors of the authors' group [4, 16–18] and other groups [19, 20], COS is very potential to be a sustainable and low-cost one-carbon (C1) monomer for synthesizing a variety of sulfur-containing polymers, which is also the main topic of this chapter.

3.2 Metal Catalysts for COS Copolymerization

Since COS is a sulfur-containing analogue of CO_2 with the similar structure, its copolymerization with cyclic ethers (epoxides and oxetanes) also shares the same elementary issues with the CO_2/epoxides copolymerization. Because the utilization

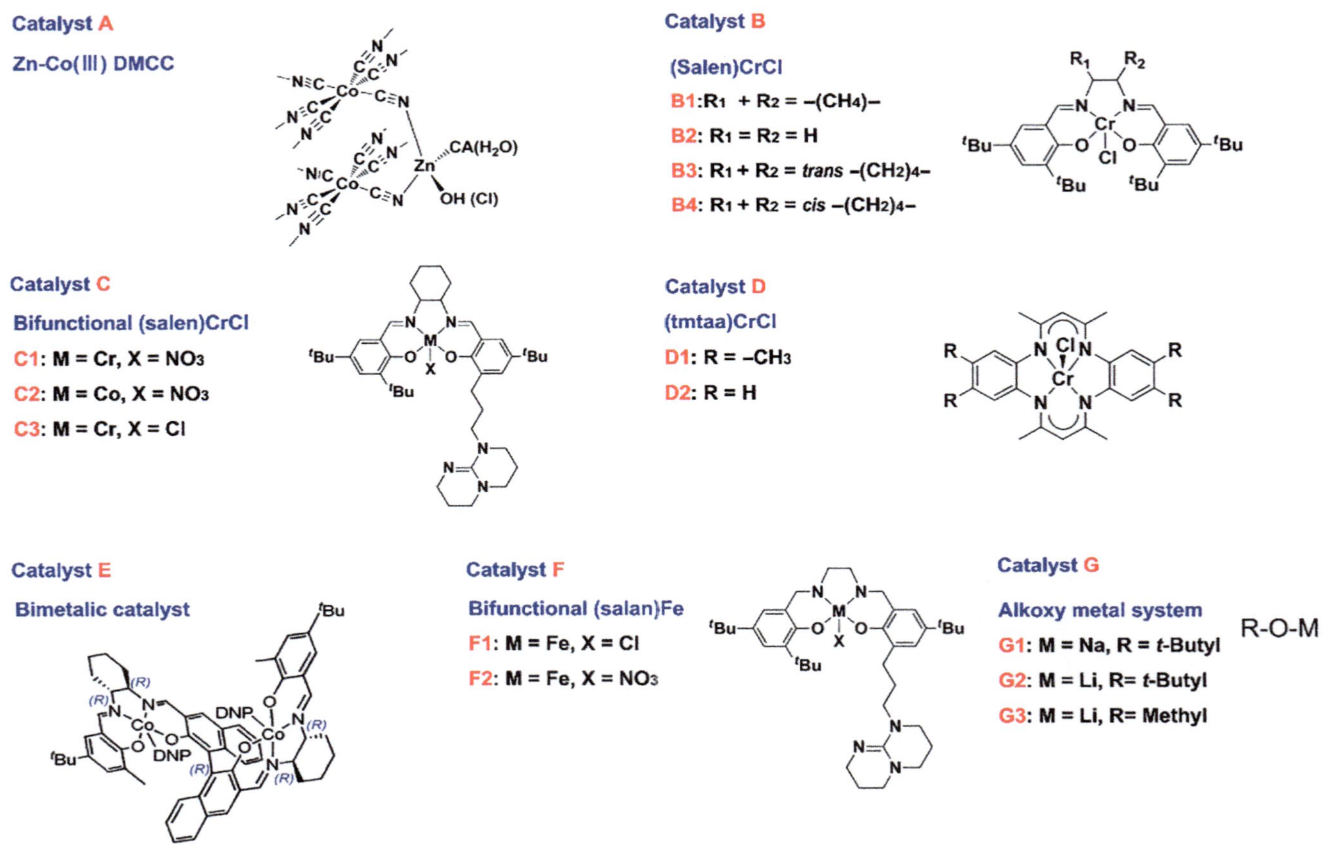

Figure 3.4 Metal catalysts and cocatalysts reported for the COS/epoxide copolymerization.

of CO_2 as a C1 monomer has already made considerable headway [15, 17, 21–32], some of the developed metallic catalyst systems for CO_2 copolymerization could be directly used for the COS/epoxides copolymerization or had a hint for its catalyst design. Up to date, several metallic catalyst systems have been discovered and widely used for meditating the copolymerization of COS with cyclic ethers (epoxides and oxetane), including zinc-cobalt(III) double-metal cyanide complex (Zn-Co(III) DMCC, catalyst A in Figure 3.4) [33], (salen)chromium complexes [16, 19, 34], iron complexes (Fe(III)-salan (*N,N*-disubstituted bis(aminophenoxide))) [35] and alkoxy metal salts [6, 36]. These catalysts can be divided into two categories, heterogeneous catalysts such as Zn-Co(III) DMCC that may be beneficial for industrial applications and homogeneous complexes (**Catalysts B–G** in Figure 3.4) accompanied with organic bases that present excellent control for the copolymerization and provides a mechanistic understanding (Figures 3.5 and 3.6).

3.2.1 Zinc-Cobalt(III) Double-Metal Cyanide Complex

DMCC is an inorganic coordinated complex with a three-dimensional network, in which the inner metal M^{II} is linked with the external metal M^{I} by several cyano-bridges (M^{II}-CN-M^{I}, generally, M^{I} is the divalent metal ions such as Zn^{2+}, Fe^{2+}, Co^{2+}, Ni^{2+}, etc., M^{II} is the transition metal ions such as Fe^{2+}, Fe^{3+}, Co^{2+}, Co^{3+}, Ni^{2+}, etc.). The external metal M^{I} on the surface of the catalyst is generally considered to be the active site owing to its unsaturated coordination structure. The active site of Zn-Co(III) DMCC is proposed as shown in Figure 3.4 (catalyst A), in which Zn^{2+} is a tetrahedral coordinated structure with two cyano anions, one OH^- and a complexing agent (e.g. *t*-BuOH). Because it is structurally a metal-cyanide complex with the typical coordination structure, it is abbreviated as DMCC and shows a coordinated anion mechanism for C1 and epoxide-involved copolymerizations [37–39].

Zn-Co(III) DMCC was discovered to be a highly active catalyst for the ring-opening polymerization (ROP) of epoxides, especially propylene oxide (PO) [38, 40]. It is still industrially used to produce poly(propylene oxide) (PPO)

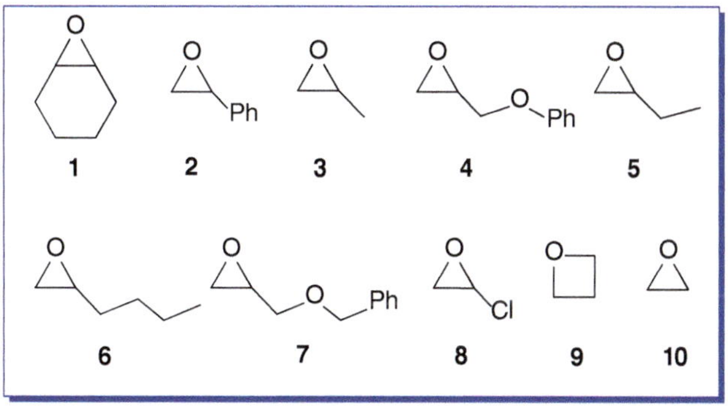

Figure 3.5 Reported cyclic ethers including epoxides and oxetane for the copolymerization with COS.

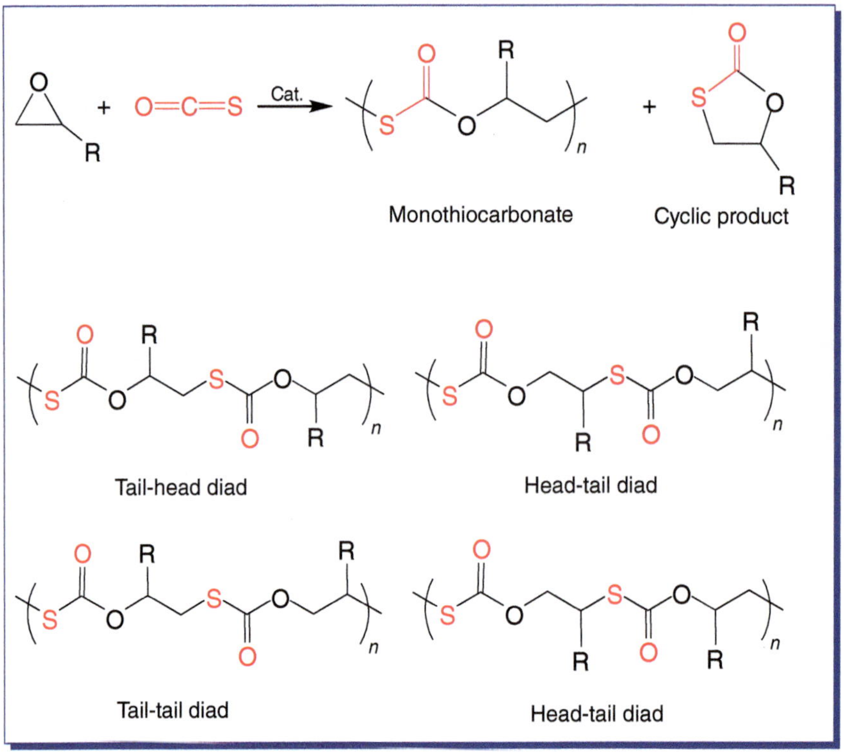

Figure 3.6 The regioselectivity of the copolymerization of COS with monosubstituted epoxides.

polyols with moderate and high molecular weights and low unsaturation degrees, which are widely used for making soft polyurethanes. Further, it has been found to be an effective catalyst for C1-involved copolymerizations, such as CO_2/epoxide copolymerization [39], epoxide/CS_2 copolymerization [17], and epoxide/COS copolymerization [15].

In 2008, Zhang's group firstly disclosed the concept of oxygen/sulfur atom exchange for explaining the complicated unit structures of the CS_2/PO copolymer, resulting from the synthesis using the Zn-Co(III) DMCC catalyst (Figure 3.7) [41]. The GC-MS and $^1H(^{13}C)$ NMR spectra showed that CO_2, COS, and PS were generated during the copolymerization of CS_2 and PO. These mixed C1 monomers including CO_2, COS, and CS_2, cross-coupled with PO and PS in random order. Meanwhile, PO and PS could also ring-opened and copolymerized with each other. As a result, the generated copolymers had monothiocarbonate (possibly from COS/PO, CO_2/PS) as the main repeating unit, while always containing contaminant ether, thioether, dithiocarbonate, and carbonate units. Considerable amounts of cyclic (thio)carbonates were also produced. The copolymerization of CS_2 with CHO catalyzed by Zn-Co(III) DMCC showed the similar result [33]. After that, Darensbourg et al. reported the O/S ERs in the copolymerization of CS_2 and CPO catalyzed by (salen) CrCl accompanied with organic salts [42]. The exchange of

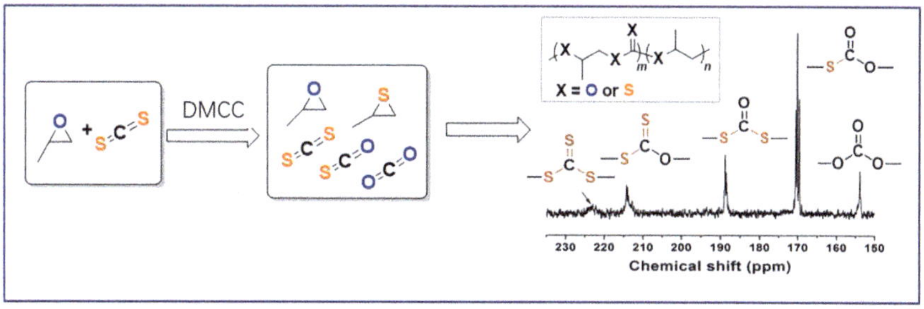

Figure 3.7 Atom-exchange coordination polymerization of CS_2 with PO in the presence of Zn-Co(III) DMCC. The ether and thioether units are omitted. Oxygen/sulfur exchange reaction (O/S ER) is proposed. Source: Modified from Ref. [41].

sulfur and oxygen atoms was observed in both copolymers and cyclic products. *In situ* FT-IR and ^{1}H NMR monitoring of the coupling reactions revealed an initiation period followed by formation of copolymers enriched in oxygen content, and concomitantly cyclic thiocarbonates enriched in sulfur content (see detail in Chapter 2).

The observation on O/S ER intrigued us to investigate the copolymerization of COS with epoxides. In 2014, we described the first example of alternating copolymerization of COS with CHO via Zn-Co(III) DMCC [33], which afforded poly(cyclohexene monothiocarbonate) with an alternating degree of up to 93%. M_ns of this copolymer was 6.5–25.0 kg mol^{-1} with dispersities of $Đ = 1.6–2.1$ (e.g. entry 1 in Table 3.1). The productivity of the catalyst reached 970 g polymer/g catalyst (TOF: 97 h^{-1}) within five hours and the conversion of CHO was up to 95%. O/S ER could be largely suppressed in tetrahydrofuran even at 110 °C, resulting in 90% of monothiocarbonate linkages, 2% of carbonate linkages, and 8% of ether linkages in the polymer chain. The monothiocarbonate unit account was 98% in the total alternating linkages. The obtained poly(cyclohexene monothiocarbonate) is highly transparent and colorless, exhibits good solubility in various solvents such as dichloromethane and tetrahydrofuran, and has a high T_g of 112 °C and an initial decomposition temperature of 214 °C.

Aim to study the steric and electronic effects on the copolymerization of COS with styrene oxide (SO, **2** in Figure 3.6), Zn-Co(III) DMCC was used and showed a high TOF of 394 h^{-1} for COS/SO copolymerization [44]. Of interest, the phenyl group of SO possesses both a large steric hindrance and an electron-withdrawing ability. As shown in Figure 3.8, the thiocarbonate growing anion can attack the methylene carbon (C_β) of SO or it can attack the electron-poor methine carbon (C_α) because of the strong electronic-withdrawing ability of the phenyl group. Since Zn-Co(III) DMCC is not sterically encumbering, ring opening of SO occurred predominantly at C_α site, producing poly(styrene monothiocarbonate) with 92% of linkage 2 (M_n: 24.7 kg mol^{-1}; $Đ$: 2.10, entry 2, Table 3.1). Therefore, the chain growth is driven by the electronic effect of phenyl group of SO, making Zn-Co(III) DMCC a regioselective catalyst for the COS/SO copolymerization.

Table 3.1 Representative copolymerization of COS with cyclic ethers.[a)]

Entry	Epoxide[b)]	Catalyst system[c)]	T (°C)	TOF (h^{-1})[d)]	Polymer selectivity[e)]	M_n[f)] ($kg\,mol^{-1}$)	$Đ$[f)] M_w/M_n	T-H[g)] (%)	T_g[h)] (°C)	Refs.
1	**1**	**A**	110	940	>99	24.7	2.10		112.4	[43]
2	**2**	**A**	50	394	>99	11.5	2.01	84	ND	[44]
3	**3**	**B1**/PPNCl	25	288	99	24.4	1.26	99	ND	[16]
4	**3**	**B2**/PPNCl	25	310	99	21.9	1.35	99	ND	[16]
5	**3**	**B3**/PPNCl	25	332	99	25.3	1.41	98	22.4	[16]
6	**3**	**C1**	80	271 000	>99	58.8	1.26	ND	30.0	[19]
7	**1**	**B2**/PPNCl	40	325	>99	12.3	1.10		114.6	[45]
8	**1** + **3**	**B2**/PPNCl	40	323	>99	22.3	1.16		55.9	[45]
9	**4**	**B3**/PPNCl	20	7 300	>99	22.6	1.12	98	ND	[46]
10	**4**	**B3**/PPNCl	20	500	>99	40.8	1.20	99	44.9	[46]
11	**1**	**C1**	80	260	>99	12.3	1.30			[19]
12	**1**	**C1**	80	15 800	>99	45.5	1.23		105.5	[19]
13	**4**	**C1**	80	48 800	>99	123.0	1.36	ND	14.1	[19]
14	**2**	**B3**/PPNCl	20	83	>99	77.2	1.18	78	78.4	[44]
15	**2**	**D1**/PPNCl	30	36	84	32.8	1.15	52	ND	[44]
16	**2**	**D2**/PPNCl	30	40	93	41.6	1.18	50	ND	[44]
17	**5**	**C1**	80	25 700	>99	124.0	1.34	ND	15.0	[19]
18	**6**	**C1**	80	21 500	>99	190.0	1.38	ND	2.9	[19]
19	**7**	**C1**	80	53 900	>99	135.0	1.38	ND	13.7	[19]
20	**8**	**C1**	25	988	98	0.8	1.88	ND	ND	[47]
21	**9**	**B1**/PPNCl	80	44	>99	13.9	2.19	ND		[48]
22	**10**	**C1**	80	75 600	>99	53.8	1.36	ND	5.0	[49]
23	**3**	**C2**	25	34	99	11.5	1.21	ND	ND	[19]
24	**3**	**C2**	80	103	60	4.3	1.33	ND	ND	[29]
25	**1**[i)]	**E**	25	<1	-	-	-			[34]
26	**1**	**E**	25	270	>99	51.2	1.22			[45]
27	**3**	**F1**	25	1 040	94	33.3	1.15	ND	ND	[35]
28	**3**	**F2**	25	700	>99	22.3	1.14	ND	ND	[35]
29	**3**	**F1**	80	7 290	67	48.9	1.28	ND	ND	[35]
30	**3**	**G1**	25	3	>99	5.2	1.13	>99	ND	[36]
31	**3**	**G2**	25	70	93	77.4	1.22	>99	ND	[36]
32	**3**	**G3**	25	23	93	6.8	1.05	>99	ND	[36]

a) These results are selected from the reported articles without occurrence of oxygen/sulfur exchange reaction (O-S ER, discussed in the follows) based on ^{13}C NMR spectra; monothiocarbonate linkage >99% from ^{1}H NMR spectra. ND, not determined.
b) If not mentioned, the epoxide is racemic (Figure 3.5).
c) See Figure 3.4.
d) Turnover frequency (TOF) = (mol epoxide consumed)/(mol metal hour).
e) Determined by ^{1}H NMR spectroscopy.
f) Determined by gel permeation chromatography in THF calibrated with polystyrene standards.
g) Determined by ^{13}C NMR spectroscopy, see Figure 3.6.
h) Glass transition temperature determined by differential scanning calorimetry.
i) The copolymerization was conducted without cocatalyst.

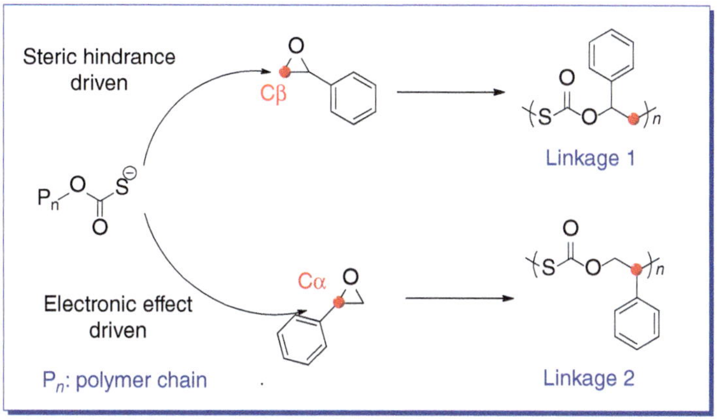

Figure 3.8 Regioselectivity in the copolymerization of SO and COS.

3.2.2 Catalysts Centered with Chromium (Cr)

In 2013, we reported the first example of the COS/PO copolymerization using **Catalyst B1–B3** (Figure 3.4) in conjunction with PPNCl as the cocatalyst (entries 3–5 in Table 3.1) [16]. A series of binary and bifunctional salen metal complexes (**Catalysts C, E,** and **F**, Figure 3.4; **entries 6–14, 17–29** in Table 3.1) were then developed for COS/epoxide copolymerization with improved performance. **Catalyst B1–B3** showed high TOFs of 288–332 h^{-1} for the COS/PO copolymerization. The resulting poly(propylene monothiocarbonate) (PPMTC) possessed M_n of 25.3 kg mol^{-1}, $Đ$ of 1.41, and alternating degree of up to 100% because O/S ER was totally suppressed at 25 °C. The T–H linkage was determined to be 98–99% based on the ^{13}C NMR spectra. The ligand difference in **B1–B3** had no observable effect on the activity or selectivity of copolymerization. Note that O/S ER occurred and >20 mol% cyclic thiocarbonates were produced when the temperature was higher than 50 °C, which can be ascribed to the chain transfer reaction to water being favored at higher reaction temperatures causing the generation of Cr–OH species. Similar to the Zn–OH bond in HACII, as shown in Figure 3.3, we proposed that Cr–OH could trigger the O/S ER. Except for O-S ER, the other situation worth noting is the regiochemistry control of COS/PO copolymerization involving the two asymmetric monomers. The attacking manner of the propagating species of –OC(=O)S^- to the carbon of CH_2 or CH of PO will produce four kinds of consecutive monothiocarbonate diads: head-to-tail (H–T), tail-to-head (T–H), tail-to-tail (T–T), and head-to-head (H–H). Note that a T–H diad is different from a H–T diad due to the asymmetric monothiocarbonate structure. All predicted units in the copolymer and cyclic (thio)carbonates are listed in Figure 3.6. Types of ligands and cocatalyst had effect on the copolymerization result. Catalyst with ligands of trans-$(CH_2)_4$-connection (**B3**) had the highest activity in the same reaction condition.

The (salen)CrCl complex could also be applied to the copolymerization of COS with other epoxides. The binary **B2**/PPNCl catalyst for the COS/CHO copolymerization provided a fully alternating poly(cyclohexene monothiocarbonate) with a M_n of

12.3 kg mol^{-1} and *Đ* of 1.13 at 40 °C for three hours without the observation of any cyclic products and the occurrence of O/S ER (entry 7, Table 3.1). The TOF was up to 325 h^{-1}. The COS/CHO/PO terpolymerization catalyzed by the binary **B2**/PPNCl system afforded terpolymers with a 100% of polymer selectivity and a TOF of 323 h^{-1} at 40 °C for three hours (entry 8, Table 3.1) [46]. This terpolymer possessed a M_n of 22.3 kg mol^{-1} and narrow *Đ* of 1.16. Using the binary **B3**/PPNCl system, phenyl glycidyl ether (PGE), as a commercially available monomer, could copolymerize with COS in a highly active and selective manner [45], affording copolymers with >99% of copolymer selectivity, 100% of monothiocarbonate linkages, and 96–98% of T–H diad content (entries 9–10, Table 3.1). As compared with our early work [16], this copolymerization process exhibited an unprecedented activity at room temperature, where the conversion of PGE was up to 73% (TOF: 7300 h^{-1}) within six minutes. The copolymer had a M_n of 40.8 kg mol^{-1} and *Đ* of 1.2. The bifunctional catalyst centered with Cr could also meditate the copolymerization of COS with CHO at 80 °C, providing polymonothiocarbonate with M_n of 45.5 kg mol^{-1}, *Đ* of 1.23, and T_g of the copolymer reached 105.5 °C (entries 11–12, Table 3.1). Remarkably, a bifunctional (salen)$CrNO_3$ complex **C1** (Figure 3.4) exhibited the TOF as high as 48 800 h^{-1} for COS/PGE copolymerization at 80 °C, providing a fully alternating copolymer with a high M_n of 123.0 kg mol^{-1} (entry 13, Table 3.1).

The binary **B3**/PPNCl system exhibited high activity and selectivity towards the copolymerization of COS with SO at 20 °C. The resultant poly(styrene monothiocarbonate) had a high M_n of 77.2 kg mol^{-1} and *Đ* of 1.18 (entry 14, Table 3.1) [44]. Meanwhile, the copolymer selectivity was > 99%, and O/S ER was totally suppressed. Other catalysts, i.e. **A** and tetramethyltetraazaannulene chromium chloride [(tmtaa)CrCl] (**D1** and **D2** in Figure 3.4), were also employed for the COS/SO copolymerization (entries 15 and 16, Table 3.1). Catalysts **D1** and **D2** were less active than **B3** with the copolymer selectivity being 84% and 93% for **D1** and **D2**, respectively. Similar to catalyst **B3**, O/S ER did not occur. Employing catalyst **B**, ring-opening of SO was observed to be 88% selective at the methylene carbon. That is, the reaction is driven by the steric hindrance, where ring-opening occurs preferentially at the less congested carbon center (Figure 3.8). On the other hand, upon employing catalyst **A**, where the ligands around the active zinc site are not sterically encumbering, ring-opening of SO occurs predominantly at the methine carbon site (92% at methine carbon); i.e. the reaction is electronically driven.

Bifunctional (salen)$CrNO_3$ complex **C1** (Figure 3.4) showed a remarkable TOF of 271 000 h^{-1} for the COS/PO copolymerization (entry 6, Table 3.1), without observing O-S ER even at an elevated temperature of 80 °C, providing a fully alternating PPMTC with a M_n of 58.8 kg mol^{-1} and a *Đ* of 1.26 [19]. Such bifunctional chromium complex combined the metal Lewis acid center and a sterically hindered organic base group in a molecule, the synergistic effect is clearly exhibited, contributing the selectively production of corresponding poly(thiocarbonate)s with completely alternating structure and exceedingly high activity, high molecular weight, and narrow dispersity. This is the highest TOF value (271 000 h^{-1}, entry 6, Table 3.1) observed in the copolymerization of COS when the bifunctional chromium complex was used. Synergy between the metal center and the cocatalyst is the key to improve its catalytic performance. This bifunctional catalyst showed a TOF of 260 h^{-1} for COS/CHO copolymerization (entry 9, Table 3.1). Of interest, the addition of a small amount of PO resulted in rapid conversion of CHO with a TOF of 15 800 h^{-1} (entry 10, Table 3.1). In addition, **C1** was also expanded to the copolymerization of COS with the monosubstituted epoxides 5–7 (entries 17–20, Table 3.1) with high TOFs (988–53 900 h^{-1}), afforded copolymers with high molecular weights (0.8–190.0 kg mol^{-1}).

The copolymerization of symmetric cyclic ethers (ethylene oxide (EO), oxetane (OX)) catalyzed by either the (salenCr)Cl (**B1**)/PPNCl system or the bifunctional catalyst **C1** provided semicrystalline copolymers (entries 21–22, Table 3.1) [47, 48].The detail discussion is summarized in Section 3.6.

3.2.3 Catalyst Centered with Iron (Fe) and Cobalt (Co)

Other earth abundant metal elements were found to be also effective in the copolymerization of COS with epoxides. Iron and cobalt are of the most ideal metals for catalysis owing to their low-cost and environmentally benign, especially for iron. As described, (salen)Cr complexes bearing a sterically hindered organic base (**C1**) were excellent for catalyzing the COS/epoxides copolymerization. When the Cr^{3+} was replaced by Co^{3+}, i.e. **C2** in Figure 3.4, rather low activity was exhibited at ambient temperature (entry 23, Table 3.1). The elevation of the reaction temperature to 80 °C could improve TOF but caused a decrease of the copolymer selectivity (entry 24 Table 3.1) [19]. Similarly, the (salen)Fe complex was found to be inefficient in the same reactions [35]. To solve the problem, a salan (*N,N*-disubstituted bis(aminophenoxide)) ligand was designed and synthesized, which possesses greater donor character than the corresponding salen ligand. When 1,5,7-triazabicyclo[4.4.0]dec-5-ene (designated as TBD) is introduced to the framework as the base site, the resultant bifunctional catalyst (**F**, Figure 3.4) had improved copolymer selectivity even at a low catalyst loading. With a catalyst loading of 0.05 mol%, **F1** exhibited a high activity of 1040 h^{-1} with a copolymer selectivity of 94% at ambient temperature (entry 27, Table 3.1). The nucleophilicity of the axial anion of iron(III) complexes had a significant influence on the COS/PO copolymerization. Although the variation of the axial anion from Cl^- to less nucleophilic NO^{3-} decreases the TOF values from 1040 to 700 h^{-1}, an enhanced copolymer selectivity of >99% was obtained at ambient temperature (entry 28, Table 3.1). The resultant PPMTC had a fully alternating structure. Remarkably, a TOF of 7290 h^{-1} was achieved along with an alternating degree of 99% at 80 °C under a **F1** loading of 0.01 mol%, while the copolymer selectivity was sacrificed and only 67% (entry 29, Table 3.1).

Notably, enantiopure binaphthol-linked binuclear Co(III) complexes (**E**, Figure 3.4) could meditate the asymmetric alternating copolymerization of a series of achiral *meso*-epoxides including CHO, CPO, 4,4-dimethyl-3,5,8-trioxa-bicyclo-[5.1.0] octane (CXO), and 3,4-epoxytetrahydrofuran (COPO) with COS. CHO was chosen as the model monomer of *meso*-epoxides for testing the catalytic activity and enantioselectivity of enantiopure binuclear metal complexes (**Catalyst E**)

[34]. It was found that the cocatalyst played an important role in meditating the copolymerization, the lacking of which resulted in no activity (entry 25, Table 3.1). Meanwhile, R,R,R,R,R conformer (**Catalyst E**) was effective and enantioselective in catalyzing COS/CHO copolymerization, with TOF value of 270 h^{-1}, and the resultant poly(monothiocarbonate)s have main-chain chirality and the isotacticity of >99%, and molecular weights up to 51.2 kg mol^{-1} with Ðs around 1.2 (entry 26, Table 3.1). A DFT calculation was carried out for understanding the mechanism, using the methyl monothiocarbonate ($CH_3OC(O)S^-$) as the model for the growing anion. One CHO is coordinated with one of two Co centers inside the catalytic cleft, and the methyl monothiocarbonate anion coordinated the neighbored Co centers then attacks this CHO, leading to the ring-opening reaction. And two DNP ions existed in the exo positions as the counterions. The relative energy of the transient state resulted from ring-opening of CHO by **Catalyst E** was the lowest among all the tested Co(III) complex, consistent to the experiment observations.

3.2.4 Alkoxy Metal Salts

Typically transition metal complexes exhibited high activity and accessed high molecular weight samples than those derived from earth-abundant main group metal complexes, but the latter are cheap and environment-friendly. Accounting for these advantages, alkali metal alkoxides were used in catalyzing the copolymerization of COS with epoxides [36]. When *t*-BuONa was used as the catalyst for the COS/PO copolymerization, a relative low TOF of 4 h^{-1} was observed, while the copolymerization had the PPMTC selectivity of >99% with fully alternating chain structure. M_ns of the resulted copolymer were up to 5.2 kg mol^{-1} with Ð of 1.13 (entry 30, Table 3.1). When the metal ion was replaced by potassium, the activity of the catalyst system was enhanced without sacrificing the selectivity. TOF value for the catalyst of lithium *tert*-butoxide and lithium methoxide were 70 and 23 h^{-1}, respectively (entries 31–32, Table 3.1). Different from other metallic catalysts, alkali metal alkoxides are simple, low cost, and effective, meanwhile the copolymerization of COS/PO was proceeded with anionic copolymerization mechanism. The order of the activity of alkali metal alkoxides follows $Li^+ > Na^+ > K^+$, decreasing with increasing of the size of the metal cations that is adversely related to their solvation ability. The success of alkali metal alkoxides for COS/epoxide copolymerization indicated an anionic mechanism, which is one of the motivations of seeking organocatalysts for COS-based polymers.

3.3 Organocatalysts for COS Copolymerization

Although a series of metal(Zn, Cr, Fe) complexes have achieved a big success in catalyzing the copolymerization of COS with epoxides, the metal residues are inevitably brought into the polymer products, resulting in potential toxicity or deep coloration that will limit the applications of these COS-derived polymers in optical materials or package materials for food and medicine. On the other hand, the production of metal-OH bond via chain transfer reaction of growing species is the key reason to the O/S ER especially at high reaction temperatures, leading to poor selectivity of the repeating unit. Inspired by the fact of CO_2 activation by the frustrated Lewis pairs (FLPs) [50], we have investigated the organic Lewis pair for catalyzing the copolymerization of COS with epoxides. We have observed the triethyl borane (TEB)/organic base and TU/organic base systems are effective to this copolymerization [51]. Since the organocatalytic polymerization is an emerging area in the field of polymer chemistry very recent years, we herein discuss the organocatalytic polymerization of COS with epoxides in detail.

3.3.1 TEB-Organic Base Pair

The first example of organic Lewis pair-catalyzed copolymerization of COS with epoxides was disclosed by the collaboration of our group with the Darensbourg group in 2017 (Figure 3.9). The use of the TEB/organic bases was motivated by two

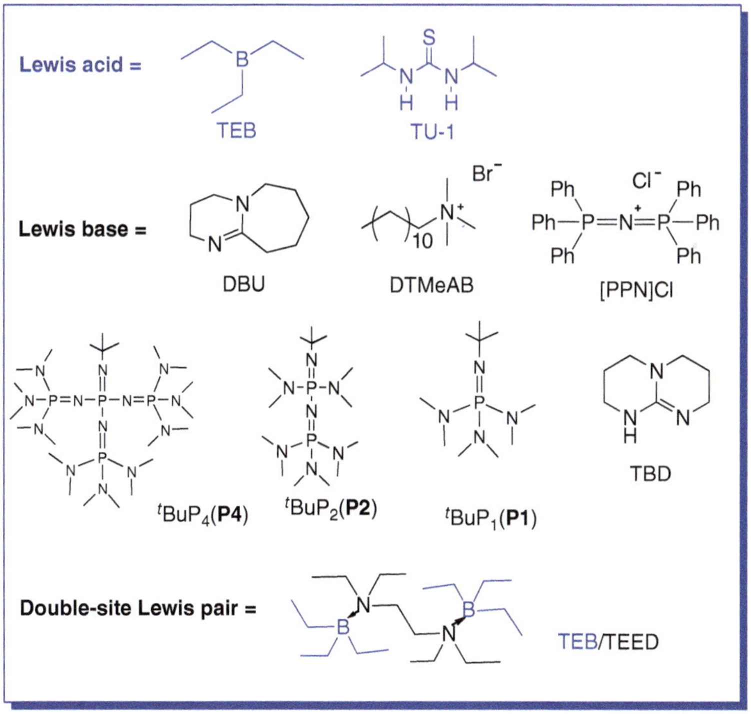

Figure 3.9 Illustration of metal-free Lewis pairs reported for copolymerization of COS and epoxides. TEB: triethyl borane; TU-1: diisopropyl thiourea; DBU: 1,8-diazabicyclo[5.4.0] undec-7-ene; DTMeAB: dodecyltrimethylammonium bromide; [PPN]Cl: bis(triphenylphosphine)iminium chloride; TBD: 1,5,7-triazabicyclododecene; ^{t}Bu-P_2: **P2**, 1-tert-Butyl-2,2,4,4,4-pentakis(dimethylamino)-2λ^5,4λ^5-catenadi(phosphazene); and ^{t}Bu-P_1: **P1**, tert-butylimino-tris(dimethylamino) phosphorene; TEED: *N*,*N*,*N′*,*N′*-tetraethylethylenediamine; ^{t}Bu-P_4: **P4**, 1-*tert*-butyl-4,4,4-tris(dimethyla mino)-2,2-bis[tris(dimethylamino) phosphoranyli denamino]-2λ^5,4λ^5 catenadi (phosphazene).

points. One, the ligand-stabilized metal catalysts are commonly utilized for the alternating copolymerization of epoxides with C1 including CO_2, CS_2, and COS, with a widely accepted coordination-insertion mechanism [23, 26, 52]. Second, the FLPs have also been developed to activate small molecules such as CO_2 [50] and catalyze the polymerizations of vinyl monomers [53]. In a general mode of metal catalysis for polymerizations, the metal complex essentially acts as Lewis acid to activate epoxides, and the organic bases such as amidine, guanidine, or quaternary ammonium salts are often to serve as the Lewis bases to enhance the catalytic activity. It is reasonable to find the organic Lewis acid to replace the metal complex to develop metal-free catalyst for COS/epoxide copolymerization.

As shown in Figure 3.9, the combination of TEB with neutral organic bases (e.g. DBU and TBD) and quaternary onium salts (e.g. DTMeAB, PPNCl, NBu_4Cl, PPh_4Br, and PPh_4Cl) was effective to yield the alternating copolymerization of COS with PO at room temperature, as summarized Table 3.2 (entries 1–7). The produced PPMTC exhibited a strictly alternating structure, i.e. the alternating degree of >99%, as revealed by the ^1H NMR spectra. The copolymer selectivities of all samples were >90% and up to >99% for the TEB/DBU (1/0.5, entry 1 in Table 3.2) pair and the TEB/DTMeAB (1/0.5, and 1/1) pair. Remarkably, all of the generated PPMTCs had greater than 99% T-H diad linkage content (entries 1–7, Table 3.2). The ^{13}C NMR result suggested that PO was ring-opened at the less sterically hindered methylene (CH_2) carbon by the attacking of thiocarbonate anion ($OC(=O)S^-$), exhibiting a highly regioselective copolymerization. Clearly, the nucleophilicity of the Lewis base formed the cations from these Lewis bases strongly impacted on the copolymerization. For example, large cations (NBu_4Cl *vs* PPh_4Cl), together with strongly nucleophilic anions (Br^- *vs* Cl^-), benefit the increase of the activities. Of special note, the TEB/DTMeAB pair was effective at producing high molecular weight copolymers, i.e. 87.1–92.5 kg mol^{-1}. The catalysis of these TEB/organic base pair for the COS/PO copolymers is similar to the typical free radical polymerization of acrylate monomers by free radical initiator. The molecular weight distribution of the resultant copolymers was up to 1.8.

The TEB/organic base pairs were successfully utilized to catalyze the copolymerizations of COS with other two epoxides (CHO and PGE). For instance, the COS/CHO copolymers with perfect alternating structures were produced using TBD, DBU, and DTMeAB as Lewis bases combined with TEB (entries 8–10, Table 3.2). Because of bulky structure of CHO, the production of cyclic carbonate was completely inhibited, and the copolymer selectivity was more than 99%. However, the TEB/organic base pairs exhibited low activities towards the copolymerization of PGE with COS at 25 °C (entries 11–12, Table 3.2). Especially, the catalysis of the TEB/DBU pair for the PGE/COS copolymerization afforded 90% of cyclic thiocarbonate. Clearly, the structure of epoxides has strongly impacted on the copolymerization results.

The O/S ER (see the detail in Section 3.4) was completely inhibited by the TEB organic base pairs at room temperature, in agreement with our original expectation that the chain transfer reaction to the proton compounds could be significantly depressed under the organocatalysis. We thus investigated the effect

Table 3.2 Representative reported copolymerization of COS with epoxides via organocatalyst.[a)]

Entry	Epoxide[b)]	Catalyst system[c)]	Epoxide/ catalysts	T (°C)	TOF (h^{-1})[d)]	Polymer selectivity (%)[e)]	M_n[f)] (kg mol^{-1})	$Đ$[f)]	T-H[g)] (%)	Refs.
1	**3**	TEB/DBU	1000/1/0.5	25	90	>99	26.3	1.2	>99	[51]
2	**3**	TEB/TBD	1000/2/1	25	100	94	24.3	1.6	>99	[51]
3	**3**	TEB/PPNCl	1000/2/1	25	86	96	22.4	1.6	>99	[51]
4	**3**	TEB/DTMeAB	1000/2/1	25	119	94	87.1	1.2	>99	[51]
5	**3**	TEB/PPh_4Cl	1000/2/1	25	76	96	14.4	1.8	>99	[51]
6	**3**	TEB/PPh_4Br	1000/2/1	25	104	96	26.8	1.6	>99	[51]
7	**3**	TEB/NBu_4Cl	1000/2/1	25	61	92	14.4	1.8	>99	[51]
8	**1**	TEB/DBU	500/1/0.5	80	51	>99	56.5	1.2	-	[51]
9	**1**	TEB/TBD	500/1/0.5	80	41	>99	36.3	1.4	-	[51]
10	**1**	TEB/DTMeAB	1000/1/1	80	102	>99	24.2	1.3	-	[51]
11	**4**	TEB/DTMeAB	250/1/0.5	25	19	88	13.0	1.5	-	[51]
12	**4**	TEB/DBU	250/1/0.5	25	14	10	7.9	1.3	-	[51]
13	**1**	TEB/TEED	500/1/0.5	60	22 500	>99	46.6	1.3	>99	[54]
14	**1**	TEB/TEED	500/1/0.5	40	8 900	87	80.8	1.4	>99	[54]
15	**1**	TEB/TEED	500/1/0.5	0	8 000	88	81.0	1.4	>99	[54]
16	**3**	TU-1/**P4**	4000/5/1	25	112	98	9.9	1.1	>99	[55]
17	**3**	TU-1/**P2**	2000/1/1	25	35	>99	19.3	1.1	>99	[55]
18	**3**	TU-1/**P1**	2000/1/1	25	1	>99	9.0	1.2	>99	[55]
19	**3**	TU-1/DBU	2000/1/1	25	22	94	11.4	1.2	>99	[55]
20	**3**	TU-1/MTBD	2000/1/1	25	17	97	12.0	1.2	>99	[55]
21	**1**	TU-1/**P4**	250/1/1	40	5	96	13.4	1.2	-	[55]
22	**4**	TU-1/**P4**	4000/1/1	25	52	>99	17.8	1.2	-	[55]

a) These results are selected from the reported articles without occurrence of oxygen-sulfur exchange reaction (O-S ER, discussed in the follows) based on ^{13}C NMR spectra; monothiocarbonate linkage >99% from ^{1}H NMR spectra.
b) If not mentioned, the epoxide is racemic (Figure 3.6).
c) See Figure 3.9.
d) Turnover frequency (TOF) = (mol epoxide consumed)/(mol metal hour).
e) Determined by ^{1}H NMR spectroscopy.
f) M_w/M_n, determined by gel permeation chromatography in THF calibrated with polystyrene standards.
g) Determined by ^{13}C NMR spectroscopy.

of reaction temperature on the COS/PO copolymerization reaction. The TEB/DBU and TEB/DTMeAB pairs afforded fully alternating and regioselective copolymers over a range of reaction temperatures. However, elevating the reaction temperature led to the observation of the O-S ER products and low copolymer selectivity. Comparatively, the copolymerization carried out at 70–80 °C by the TEB/DBU pair led to 100% production of cyclic monothiocarbonate and dithiocarbonate as a result

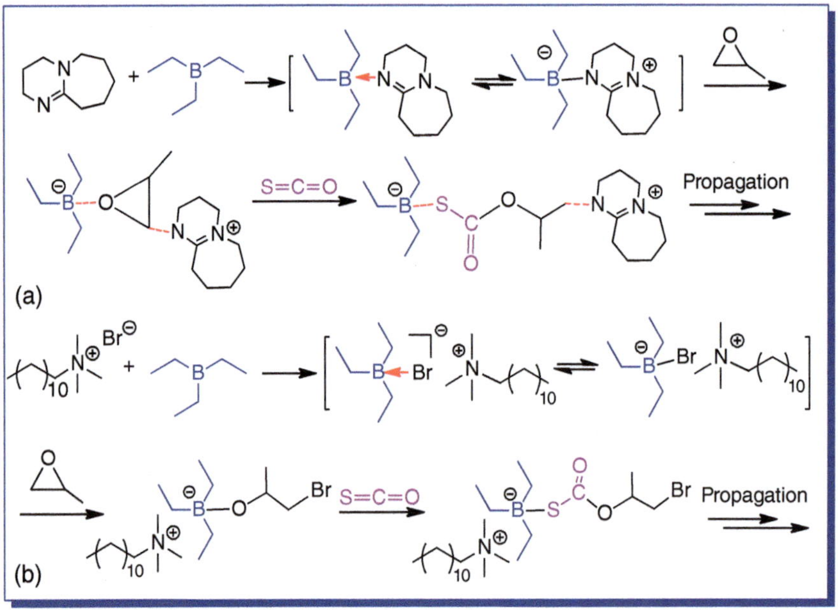

Figure 3.10 The formation of intramolecular TEB/DBU pair (a) and intermolecular TEB/DTMeAB pair (b) for activating COS and PO.

of O/S ERs. While the TEB/DTMeAB pair provided a copolymer selectivity of 84/16 for COS/PO copolymerization even at 80 °C owing to the effect of steric hindrance by the larger cation in DTMeAB as a result of less ion-pairing.

The mechanism of the TEB-organic base pair-catalyzed COS/epoxide copolymerization reaction was proposed for understanding the process. Taking the TEB/DBU (and DTMeAB) pairs as the instances, we traced the interactions of these pairs with the monomers by ^{1}H NMR spectroscopy. The result suggested the formation of the intermediates of the [DBU-PO-COS-TEB] adduct and the $[TEB\text{-}COS\text{-}PO\text{-}Br]^{-}[N(CH_3)_3C_{12}H_{25}]^{+}$ adduct, respectively, in the TEB/DBU pair and TEB/DTMeAB pair-catalyzed systems (Figure 3.10). The as-formed tetrahedral boron anion after COS insertion are proposed as the initiating species for the copolymerization process. As indicated, such species are stabilized by the bulky cations, i.e. an *intramolecular* cation for the TEB/DBU pair and an *intermolecular* cation the TEB/DTMeAB pair in Figure 3.10.

It should note that the COS/PO copolymerization catalyzed by these Lewis pair exhibited an induction period of around 1–2 hours in the absence of protonic compounds (ROH), suggesting the formation of such initiating species is rate-determining. The unpredictable molecular weights of the resultant copolymers imply less initiation efficiency of these Lewis pairs for the copolymerization. In the presence of ROH, such as polyethylene glycol (PEO) with one or two end -OH groups, PPMTC-*b*-PEO, and PPMTC-*b*-PEO-b-PPMTC were achieved with perfectly alternating degree and regioregularity using a variety of TEB/organic base pairs. Such copolymerization showed a living character and could be carried out up to 80 °C without observing O/S ER. An improved TOF of 240 h^{-1} for COS/PO polymerization was achieved. The molecular weight of resultant block copolymers (BCPs) could be mediated by varying the feed ratio of PEOs and monomers with narrow dispersity of 1.1.

There are two points that should be kept in mind for this metal-free copolymerization process. First, no polymerization was observed upon replacing TEB with the bulkier, less acidic $(C_6F_5)_3B$ or $(C_6H_5)_3B$. That is, a proper acidity/alkalinity and matched size of the Lewis pair are necessary for effective copolymerization of PO and COS. Second, different to the TEB-quaternary onium salt pair-catalyzed CO_2/PO copolymerization reaction [56], neural organic bases coupled with TEB were effective catalyst systems for the COS/PO copolymerization.

3.3.2 Dual-Site Lewis Pair

Aiming to further improve the activity of the TEB/base Lewis pair for the COS/epoxide copolymerization, we recently investigated the structure of organic bases. As the proposed mechanism in Figure 3.10, amidine, guanidine, or quaternary onium salts may have single base sites for activating monomers and initiating the copolymerization. Low initiator efficiency and the feature of R_i (rate of initiation) < R_p (rate of propagation) lead to unpredictable molecular weights and broad molecular weight distributions in the absence of chain initiators. In addition to using chain initiators [57] as discussed earlier, we have reported a new type of dual-site LPs using tertiary diamines as the Lewis bases for the COS/epoxide copolymerization process based on the hypothesis that dual sites or multiple-sites Lewis pairs may increase the interaction probability of LA, LB, and monomers for instant initiation, because they can increase the local concentration of the activated monomers [54].

A series of dual-site LPs using the combination of simple tertiary diamine, such as TEED (*N,N,N′,N′*-tetraethylethylenediamine), TMED (*N,N,N′,N′*-tetramethylethylenediamine), and TEPD (*N,N,N′,N′*-tetraethyl-1,3-propanen diamine) with TEB are very active to the COS/PO copolymerization. Noteworthy, under the catalysis of the TEB/TEED pair for the COS/PO copolymerization (entries 13–15, Table 3.2), the conversion of PO reached 77% within 1.0 minute at 60 °C, and thus the TOF was up to 22 500 h^{-1}, ranking as an *exceedingly high activity*. High molecular weight copolymers were obtained with M_ns of 46.6–81.0 kg mol^{-1} and $Đ$ of 1.3–1.4 (Table 3.2). Both the alternating selectivity and regioselectivity of such dual Lewis pair are >99%. And the generation of cyclic products was greatly inhibited (<1%).

The TEB/TEED Lewis pair led to the COS/PO copolymerization with remarkably living features. All the samples with the [PO]/[LP] molar ratio of more than 200/1 afforded fully alternating PPMTCs with a copolymer selectivity of >99%. M_ns of the resultant PPMTCs ($Đ$: 1.09–1.15) increased linearly with the increase of the [PO]/[LP] molar ratios and well agreed with the calculated values on the supposition that one TEED molecule produced two chains (Figure 3.11a). This result suggests that the initiator efficiency for the TEB/TEED pair was near

>99%. Moreover, a linear relationship ($R_2 > 0.99$) of M_ns with the conversion of PO was observed with narrow Đs (1.11–1.12), as shown in Figure 3.11b. As compared with the above mono-site pair-catalyzed COS/PO copolymerization, the well-controlled copolymerization feature of dual-site pair can be attributed to the fast initiation rate.

As compared with the previous C1/epoxide copolymerizations [4] that require an excess of C1 monomers (such as CO_2 and COS) to ensure a fully alternating structure, the dual-site pair-catalyzed COS/PO copolymerization system is very special, because this copolymerization could proceed when the molar ratio of COS/PO was less than 1. As shown in Figure 3.11c, M_ns of the resultant PPMTCs increased in a linear fashion and exhibited narrow Đs (1.09–1.18) with increasing the feeding ratios of COS/PO, which is also demonstrated that the chain initiation was directly related to COS. Based on this distinct character of the TEB/TEED pair for the COS/PO copolymerization, the chain extension reaction was performed by turning on the paused copolymerization through charging with COS four times in a one pot manner (Figure 3.11d). M_ns of the resultant PPMTCs increased corresponding with the COS addition. The final resultant PPMTC (M_n: 68.9 kg mol^{-1}; Đ: 1.19) had a perfect alternating structure without observing any polyether linkages in the ^{1}H NMR spectra. Such on–off character and chain extension of the copolymer by stepwise feeding of COS also strongly supported the living nature of the TEB/TEED pair-catalyzed COS/PO copolymerization. Meanwhile, this system is very promising in large-scale production of PPMTC because step-by-step supplementation of COS can easily avoid the heat aggregation.

Of significance, a BCP of PPMTC-*b*-PPO was also synthesized in the case of COS/PO molar ratio of 200/500, by increasing the dosage of TEB. The copolymerization reaction was conducted in a one-pot reaction for 12 hours, with the resulting copolymer featuring an M_n of 24.6 kg mol^{-1} with a narrow Đ of 1.11, which was proved to be a diblock copolymer of PPMTC-*b*-PPO (Figure 3.11e). The formation of PPMTC was fast, while the formation of PPO block that was generated by the ROP of PO was relatively slow. Herein, the huge difference in the activity of forming two blocks guaranteed the perfect block structure of PPMTC-*b*-PPO. Note that, it is important to use an excess of TEB for activating PO in the second stage [58]. This result again indicates that COS was necessary in the chain initiation stage of copolymerization catalyzed by the TEB/TEED pair.

Based on the aforementioned results, the mechanism for the TEB/TEED pair-catalyzed COS/PO copolymerization is proposed, as shown in Figure 3.12. It is a dual zwitterion copolymerization with two Lewis base sites, the structure (**a**) is formed owing to the interaction between the Lewis acid and the Lewis base. With the alternating insertion of PO and COS, (**a**) is converted to the initiation species (**b**) and goes through a propagation process. The product (**c**) and TEED after termination step were observed by MALDI-TOF-MS spectrum and GC-MS spectra, respectively.

The dual-site Lewis pair exhibited exceedingly high activity without sacrifice of structural selectivity, representing the highest activity for C1 copolymerization via metal-free catalysis. Of significance, this catalyst system is very simple and

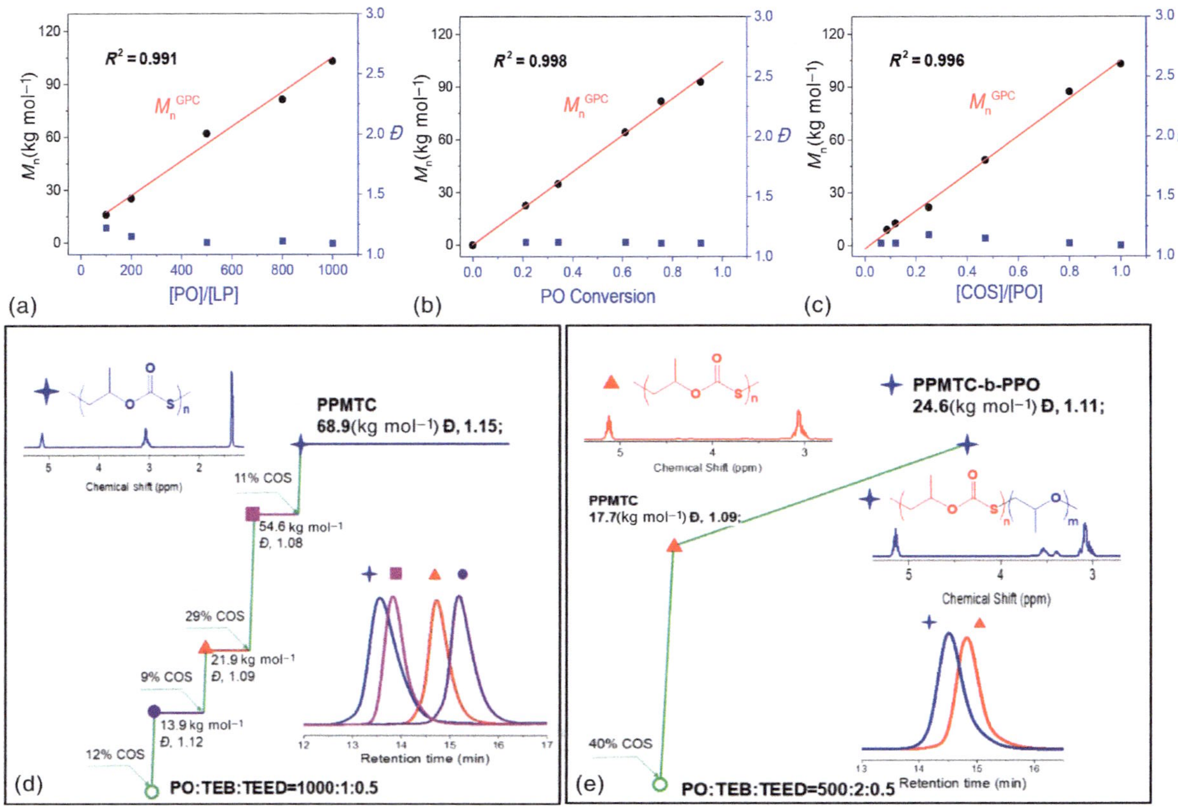

Figure 3.11 Living and on–off character of COS/PO copolymerization catalyzed by the TEB/TEED pair. (a) The effect of the [PO]/[TEED] (1/0.5) ratio on the M_ns of PPMTCs, 60 °C. (b) Linear plot of the M_n of PPMTCs versus PO conversion, [TEB]/[TEED]/[PO]/[COS] = 1/0.5/1000/1000, 60 °C. (c) Linear plot of the M_n of the copolymer versus the [COS]/[PO] ratios, [TEB]/[TEED]/[PO] = 1/0.5/1000, 60 °C. (d) On–off copolymerization of COS and PO. GPC traces of the original PPMTCs and products formed by three consecutive on–off reactions by charging particular amounts of COS. (e) GPC traces and ^{1}H NMR spectra of PPMTCs and PPMTC-*b*-PPO, [TEB]/[TEED]/[PO]/[COS] = 2/0.5/500/200, 60 °C. Source: Reprinted with permission from Ref. [54]. Copyright Royal Society of Chemistry.

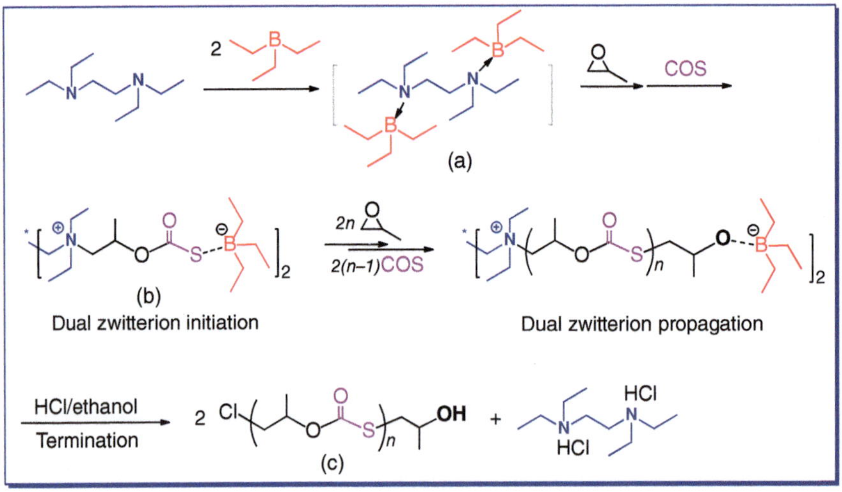

Figure 3.12 The proposed mechanism of COS/PO copolymerization catalyzed by the TEB/TEED pair.

effective and thus provides a bright future for organocatalytic polymerizations. For TEB-participated C1 copolymerization, the formation of organic coordination bonds, the N → B bond in the initiation stage and of O → B bond in chain growth stage are the key to the copolymerization process. In essence, such organic coordination bonds are typical supramolecular interactions, similar to interactions of traditional metal complexes with organic bases as cocatalysts or growing anions. It is worthwhile to study the organocatalytic C1 polymerization process by other types of supramolecular interactions. Although TEB is easily hydrolyzed after copolymerization, it is toxic and spontaneous combustion in ambient air. Therefore, the development of other effective metal-free and boron-free catalyst systems for COS/epoxide copolymerizations is highly desired.

3.3.3 Thiourea-Organic Base Pairs

In this section, a *living* anionic copolymerization of COS with epoxides is described using commonly available TUs and a variety of neutral organic Lewis bases (LB: **P4**, **P2**, **P1**, DBU, and 7-methyl-1,5,7-triazabicyclo[4.4.0]dec-5-ene (MTBD)) [55]. The development of this TU/LB catalyst system was based on the hypothesis that the cooperative catalytic process of Lewis pairs composed of TU and base would afford the hydrogen bonding interaction for activating and stabilizing the alcohol initiator/chain end.

The copolymerization of COS with PO catalyzed by the TU/LB pairs was successfully carried out at 25 °C. The generated PPMTCs had perfect alternating degree, > 99% of T-H diad content, well controlled molecular weights, and narrow molecular weight distributions. TOFs of these Lewis pairs were as high as 112 h^{-1}. The living copolymerization manner was revealed by the kinetic studies and chain extension

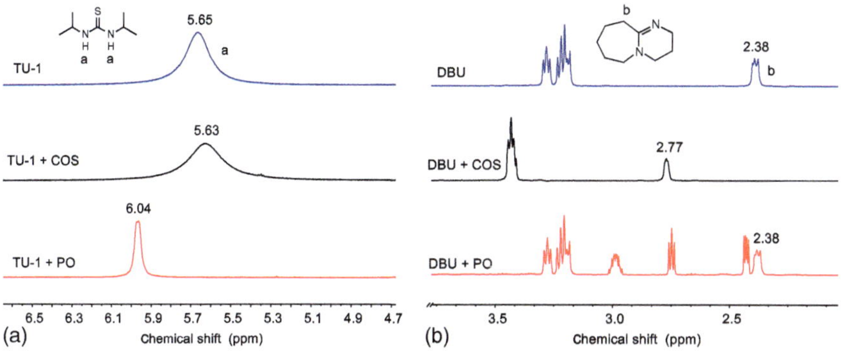

Figure 3.13 The specific recognition of TU-1 to PO and DBU to COS in TU-1/DBU pair via hydrogen-bonding interaction. (a) ^{1}H NMR spectra of TU-1, TU-1/COS (1/excess), TU-1/PO (1/60), (b) ^{1}H NMR spectra of DBU, DBU/COS (1/excess), DBU/PO (1/1), 0.5 M TU-1 (DBU) in $CDCl_3$. The recognition of supramolecular interactions of LA(LB) with the substrates. Source: Reprinted with permission from Ref. [55]. Copyright Springer Nature.

reactions. The available thioureas were selected as TU-1, while the corresponding Lewis bases were selected as **P4**, **P2**, **P1**, DBU, and MTBD (entries 16–22, Table 3.2). The stronger the basicity of the LBs, the higher TOFs of the COS/PO copolymerization. It was observed that the TOF values decreased with the basicity orders of **P4** > **P2** > DBU ≥ MTBD > **P1**. Of special interest, the MTBD/TU-1 pair-catalyzed COS/PO copolymerization provided a copolymer with a high M_n of 98.4 kg mol^{-1} and a narrow $Đ$ of 1.14 after 72 hours, suggesting that the growing species featured a quite long lifespan of the active species, thus minimizing the chain transfer reaction. Therefore, the occurrence of O/S ER was completely inhibited, which was also confirmed by our results.

The catalysis of the TU-1/LB pair has a clear cooperative effect on the COS/PO copolymerization, as revealed by the kinetic studies and physical organic chemistry (Figure 3.13). The apparent rate constant (k_{app}) of the plots of PO conversion *versus* time is well proportional to the concentration of [TU-1] + [DBU] in the absence or presence of BnOH, which suggests that the TU-1/DBU pair behaved as a discreet catalyst species (Figure 3.14, species (***a***)). In addition, k_{app} of the TU-1/**P2** pair was determined to be $13.6 \pm 0.3 \times 10^{-2}$ h^{-1} and higher than that of the TU-1/DBU pair ($5.7 \pm 0.8 \times 10^{-2}$ h^{-1}) for the COS/PO copolymerization under the same conditions owing to the stronger basicity of **P2**. This result is agreed with the determined binding constants via the ^{1}H NMR dilution, 10.3 ± 2.5 for TU-1/DBU pair and 21.2 ± 2.2 for TU-1/**P2** pair in equilibrium in $CDCl_3$ at 25 °C. To probe the supramolecular interaction of the TU/LB pair with the monomers, the combining capabilities of TU-1, DBU, COS, and PO in $CDCl_3$ were also further investigated using ^{1}H NMR spectra. As shown in Figure 3.13, TU-1 (NH from 5.65 to 6.04 ppm) only activated PO and did not interact with COS in $CDCl_3$ (Figure 3.13a). In contrary, DBU could solely activate COS owing to a clear deshielding effect of protons of NCH_2 of DBU (from 2.38 to 2.77 ppm) and appearance of chemical equivalent protons of NCH_2 of DBU rather than activate PO (Figure 3.13b). The specific recognition of TU-1 to PO

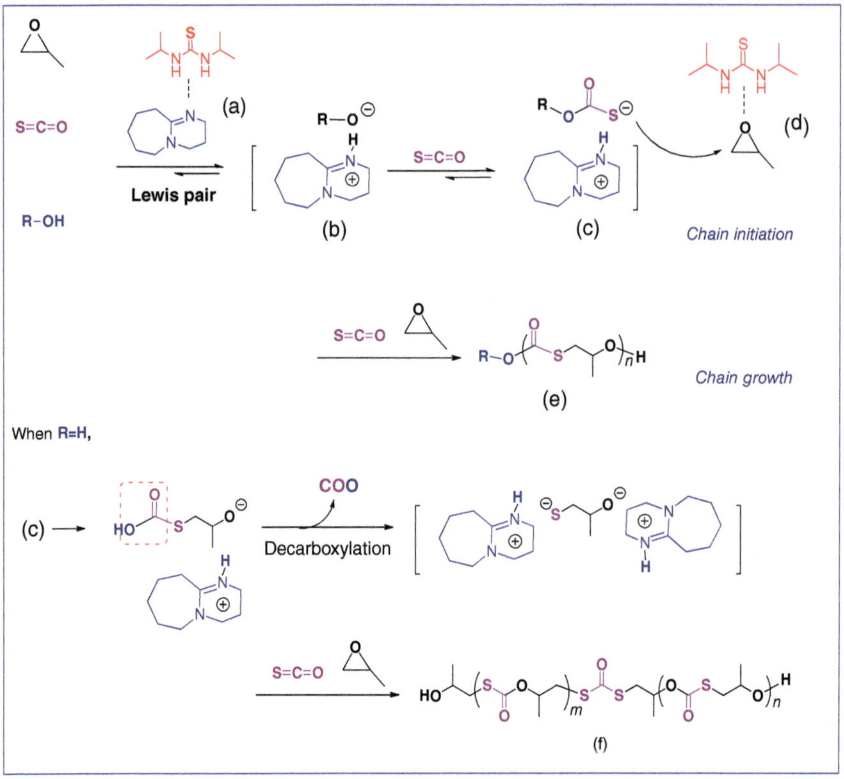

Figure 3.14 Proposed anionic copolymerization mechanism. Chain initiation and growth routes are presented upon the cooperative catalysis of TU-1/DBU pair in the presence of BnOH and H_2O. Species (*1*)–(*4*) were revealed by ^{1}H NMR spectra, the copolymer distributions (*5*) and (*6*) were evidenced by MALDI-TOF-MS spectra. Source: Reprinted with permission from Ref. [55]. Copyright Springer Nature.

and DBU to COS in TU-1/DBU pair was a typical supramolecular interaction that promoted the copolymerization cooperatively [59].

Based on the copolymerization results and the characterization results of the Lewis pairs with monomers, it was proposed that the TEB/LB pair-catalyzed COS/PO copolymerization is an anionic process, as shown in Figure 3.14. In the presence of BnOH as the initiator, a possible species of [BnOC(=O)S$^-$…DBUH$^+$](***b***) is formed via the deprotonation of BnOH by DBU, the species (***c***) is then generated when COS approaches. Since PO is activated by TU-1 through H-bond [species (***d***)], the chain growth is accelerated. In the case of absence of BnOH, traces of water can also initiate the copolymerization via the similar route with BnOH, leading to the formation of the –S(O=C)–OH end group, which after a decarboxylation process results in a –SH end group that can then initiate the copolymerization owing to its the stronger acidity. A copolymer with two secondary –OH end groups and one dithiocarbonate unit (***f***) was formed, which was clearly revealed by MALDI-TOF-MS spectrum.

Noteworthy, many types of thioureas and Lewis pairs can be used for this copolymerization, and such TU/LB pair can also expand to the copolymerization of COS with other epoxides, including PGE and CHO (entries 18–19, Table 3.2). It is a well-controlled synthesis of perfectly alternating and regioregular poly(monothiocarbonate)s from COS and epoxides under mild conditions, thus a promising alternative to get metal-free well-defined sulfur-containing polymers with high activity and selectivity. Note that, the previously reported catalysts for COS/epoxide copolymerization are screened from those that could well catalyze the CO_2/epoxide copolymerization. However, the TU/LB pair is the first example of metal-free catalyst that is active to the COS/epoxide copolymerization, while it did not catalyze the CO_2/epoxide copolymerization. The inherent mechanism may guide us to find new metal-free catalysts for C1 copolymerization.

3.3.4 Supramolecular Anion

As discussed in Sections 3.3.1–3.3.3, in the process of the TEB/LB and TU/LB Lewis pairs for the copolymerizations of COS with epoxides, TEB is combined with the growing anion by an O → B coordination bond, and TU is linked with the growing anion with the hydrogen bond. Both supramolecular interactions of the incorporated TEB or TU with the growing anion are key to the chain growth. In contrast, the growing species (e.g. oxyanion) in most of the traditional anionic polymerization processes does not have such a Lewis acid that can be potentially regulated. In our system, the basicity and nucleophilicity of these growing anions capped with TEB or TU can also be mediated by the types, structures of these Lewis acids, as well as the related supramolecular interactions. That is, the performance of the growing oxyanions can be tailored by altering these neutral Lewis acids. Indeed, these organic Lewis pairs have shown improved performance for catalyzing the COS/epoxide copolymerization, as compared with the metal catalysts as disclosed in Section 3.2. As a result, the "supramolecular anion" is generalized as the strategy for developing metal-free polymerization catalysts, by introducing a neutral Lewis acid that could insert into the growing ion pair and combine with the anion by supramolecular interactions (Figure 3.15) [60]. Herein, we advance this method and hope to move it forward because many recently reported studies have demonstrated the successful attempts of "supramolecular anion" for the polymerization of a variety of monomers.

This strategy has succeeded in the alternating copolymerization of a variety of epoxides and cyclic anhydrides (Figure 3.15c). Polyesters with completely alternating structure were synthesized by the combination of organoboranes including TEB, tributylborane (TBB) and Lewis bases such as DBU, TBD, **P2**, **P1**, and quaternary onium salts. The effect of the acidity, type, and size of Lewis pairs on the catalytic activity and selectivity of forming polyesters are investigated by several groups [61–65]. Of significance, the undesirable transesterification and etherification could be effectively inhibited even in the case of complete conversion of the cyclic anhydride. The reason can be ascribed to the formation of the tetracoordinate bond-carboxylate (or alkoxide) anions with the O → B coordination bond during the

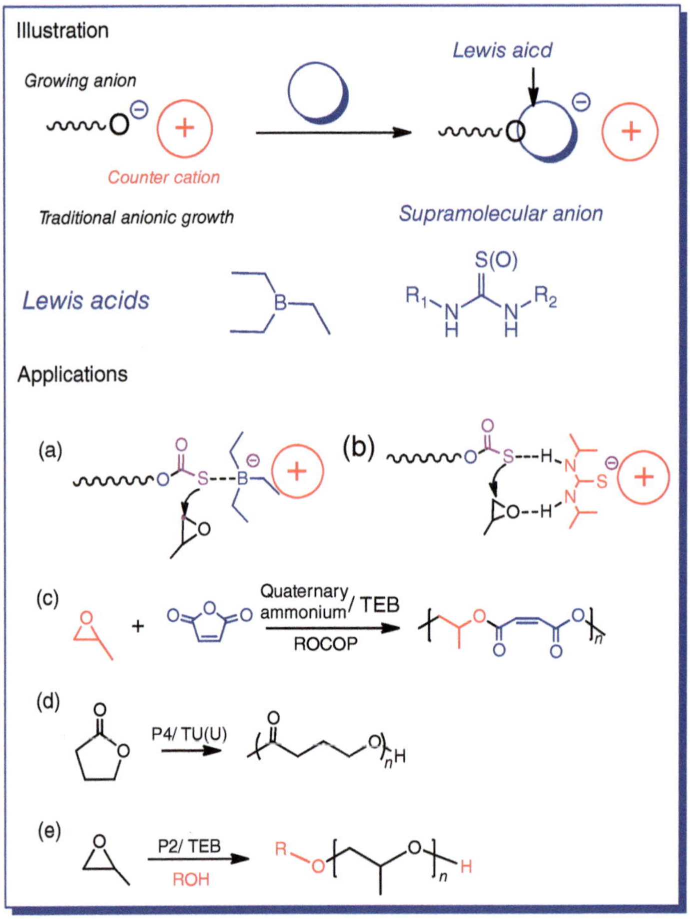

Figure 3.15 Illustration and applications of "supramolecular anion" for anionic polymerizations.

chain growth. These Lewis pairs are highly active, with a TOF of up to 303 h^{-1} for the copolymerization of PO with phthalic anhydride at 80 °C. A BCP polyester with narrow *Đ* of 1.05 was obtained by sequential addition of cyclic anhydrides. Li and coworkers have recently reported that the TEB/LB pair-catalyzed copolymerization of epoxide with cyclic anhydride, exhibiting the highest TOF of 330 h^{-1} at 110 °C, the resultant polyesters exhibited narrow *Đ* of <1.20, fully alternating structure, and the ester content of >99% [62]. Very recently, aimed to improve the activity of the copolymerization of PO with succinic anhydride (SA), a biorenewable cyclic anhydride for biodegradable aliphatic polyesters was employed and we investigated the effect of the types of Lewis bases, organoboranes, and their feed ratios on the activity and selectivity of the copolymerization [65]. The result showed that the catalysis of TEB/**P1**(1/1) for the PO/SA copolymerization at 30–80 °C provided polyesters with the ester content of >99%. When the molar ratio of TEB/**P1** increased to 4/1 at 80 °C, TOF reached 128 h^{-1}, ranking the highest activity for the

PO/SA copolymerization to date. In these examples, the TEB/LB pair exhibited better catalytic performance than some traditional metal complexes.

The "supramolecular anion" concept had a big success in solving the ROP of γ-butyrolactone (γ-BL), a biomass-derived cyclic ester with non-strained five-membered ring and is thermodynamically stable monomer. In 2016, Hong and Chen et al. reported the first example of ROP of γ-BL by using La and Y metal complexes as the catalysts at −40 to −60 °C [66]. Immediately after, the same research team reported the first example of metal-free ROP access to γ-BL, using the strongest superbase **P4** as the catalyst. Both processes produced mixed linear and macrocyclic poly(γ-butyrolactone)s (PγBLs) with the M_ns of 26.7–30.0 kg mol^{-1} [67]. Li and coworkers disclosed the selective synthesis of linear PγBLs with M_ns of up to 22.9 kg mol^{-1}, using a new cyclic trimeric phosphazene base [68]. Based on the result of **P4**-catalzyzed ROP of γ-BL, we applied the *supramolecular anion* strategy by using (thio)ureas as hydrogen bond donors on the consideration that the basicity of the growing anion may be reduced by the incorporated Brønsted acid, leading to the inhibition of the backbiting reaction of such anion (Figure 3.15d) [69]. Meanwhile, ionized (thio)ureas could also effectively activate γ-BL. As expected, the intramolecular transesterification was eliminated, because the combination of (T)U with the growing anion through supramolecular H-bond interactions. As a result, linear poly(γ-BL)s with high M_ns of 64.3 kg mol^{-1} were selectively produced. The similar methods was also reported for controlled ROP of γ-BL by Meng, Li, and coworkers [70, 71].

It should be noted that the idea of tailoring the growing anions is an old story. The supramolecular interactions of the Lewis pairs with the growing anions had been utilized in the ROP of lactides, lactones, and cyclic carbonates [72–74]. The first example of TU/LB pair-catalyzed ROP was reported by Waymouth and coworkers in 2005 [75]. The catalyst system exhibited exceptional selectivity in the ROP of lactide for the synthesis of well-defined poly(lactide)s with narrow dispersity. After that, a family of the hydrogen-bond catalyst systems including (thio)ureas with different structures and bases (e.g. DBU, DBN, MTBD, **P4**, **P2**, **P1**, alkali alkoxide, and so on) have been developed and prompted the development of the synthetic polymer chemistry [76]. Very recently, Tao and Wang reported a bifunctional thiourea linked to a pyridine group as the intramolecular Lewis pair for selective ROP of *O*-carboxyanhydrides, allowed the synthesis of stereoregular poly (mandelic acid), which is a good example of *supramolecular anion* that successfully retain the configuration of *O*-carboxyanhydride monomer [77].

This strategy was also successfully employed in some *old* polymerizations, for example, the ROP of PO [58] (Figure 3.12e) and EO [78], as well as the corresponding BCPs. The industrial production of PPO commonly uses the basic alkali metal derivatives as the catalysts. One of the big problem is that vast undesired chain transfer reactions to PO lead to low M_ns and difficulty in synthesizing polyether-based BCPs [79]. Surprisingly, our attempt of using TEB combined with organic Lewis bases (e.g. phosphazene base, amidine, and guanidine) provided a fast and living metal-free ROP of PO. High molecular weight (M_n up to 156 kg mol^{-1}) PPOs were produced [58]. In this instance, excess TEB either masked the growing oxyanion and thus reduced its basicity and nucleophilicity or activated PO. This process has also exhibited a truly living behavior, as evidenced by the kinetic results of fast initiation and growth, unobserved chain transfers to PO, easy-to-handle chain extension reactions. A variety of PPO-based BCPs with narrow *Đ*s (<1.1) had also been synthesized. Simultaneously, Zhao and coworkers reported ROP of EO [78] and a variety of polyether-polyester BCPs using TEB/base catalyst system. Extremely high catalytic efficiency (TOF: up to 6000 h^{-1}) is exhibited and the catalyst loading can be minimized to 44 ppm. PEO products are noncytotoxic towards many cell lines even without purification. The success of ROP of PO (EO) by using TEB/base catalyst system indicates the control of structure and properties of growing anion is a powerful method to controlled polymerization.

In short, the *supramolecular anion* strategy is expected to be a useful tool for chemists in some challenging and uncontrolled anionic polymerizations involving heteroatom (O, N, and S)-containing monomers. There is still a lot of space for seeking and designing organocatalysts for these polymerizations.

3.4 Oxygen-Sulfur Exchange Reaction

We have also proposed O/S ER in Section 3.2.1 for explaining the observation of low repeating unit selectivity during the copolymerization of CS_2 with PO via the catalysis of Zn-Co(III) DMCC catalyst [41]. Based on the idea of inhibiting O/S ER, we have developed a variety of catalysts, including metal complexes and organocatalysts for the controlled copolymerization of COS with epoxides, as described in the Sections 3.2 and 3.3, affording a family of sulfur-containing polymers, i.e. poly(monothiocarbonate)s. Being different from the CO_2/epoxide copolymerization, O/S ER is a unique but very common phenomenon in the copolymerization of sulfur-containing C1 monomers (COS and CS_2) with epoxides (Figure 3.16). Herein, O/S ER can be simply defined as the exchange reaction of a sulfur atom in CS_2 (COS) with the oxygen atom in epoxides generates COS (CO_2) and episulfide under the specific catalytic conditions. Three proposed routes are summarized for causing the O/S ER process, as follows (Figure 3.16).

The first route is metal-hydroxide (M-OH) bond catalysis when metal complexes are used for the copolymerization of COS (CS_2) with epoxides, as shown in Figure 3.16. For metal catalysts, a credible cause of the generation of O/S ERs is that the metal ions have a high Lewis acidity, the chain transfer reaction of growing species to protonic compounds (trace water) affords M-OH, which can react with CS_2 (COS) to generate COS (CO_2) and M-SH via the similar process of CS_2 (COS) with human carbonic anhydride II with Zn-OH bond (HCAII, a widely existing zinc enzyme) [15, 80–82]. Note that the generated M-OH (and M-SH) can also initiate the ring-opening reaction of epoxides in many cases. The second route is the rearrangement of the tetrahedral spiro intermediate (Figure 3.16) that was for the first time proposed by Barbero et al. [83] and observed by the Darensbourg group in the process of (salen)Cr complex-catalyzed copolymerization of CS_2 with epoxides using NMR and FT-IR spectra for the interpretation of the scrambling

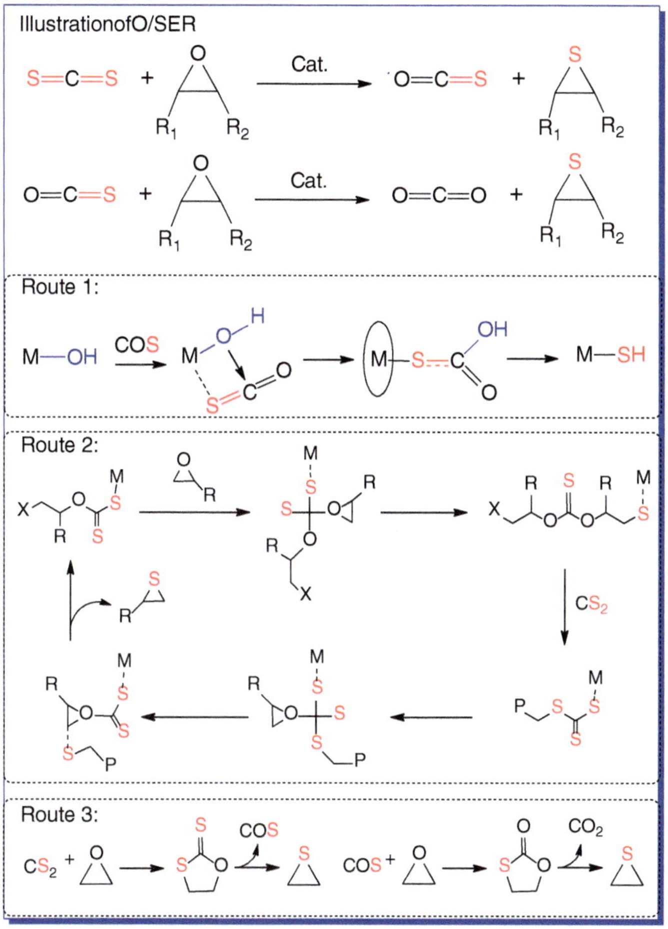

Figure 3.16 Illustration of O/S ER and involved mechanisms.

process. This mechanism can reasonably explain the generation of a variety of repeating units during the copolymerization, which is intensively discussed in Chapter 2. A joint work of the Darensbourg group and our group [84] investigated the copolymerization reaction of OX and CS_2 via (salen)CrCl/onium salt catalyst system. Because five different polymer linkages and two cyclic products could be clearly determined by ^1H and ^{13}C NMR spectroscopy, we could trace the scrambling of oxygen and sulfur atoms during the coupling process under different conditions. It was observed that reaction temperature could clearly promote O/S exchange process. However, the sole use of (salen)CrCl led a low O/S ER, the onium salt not only accelerated the copolymerization reaction but also promoted O/S ER process. This system could also be explained by the proposed tetrahedral rearrangement mechanism. These endeavors to understand O/S ERs are aimed to prevent these uncontrolled reactions as much as possible. Limited success has been made in synthesizing copolymers with inhibiting O/S ER in recently years. For example,

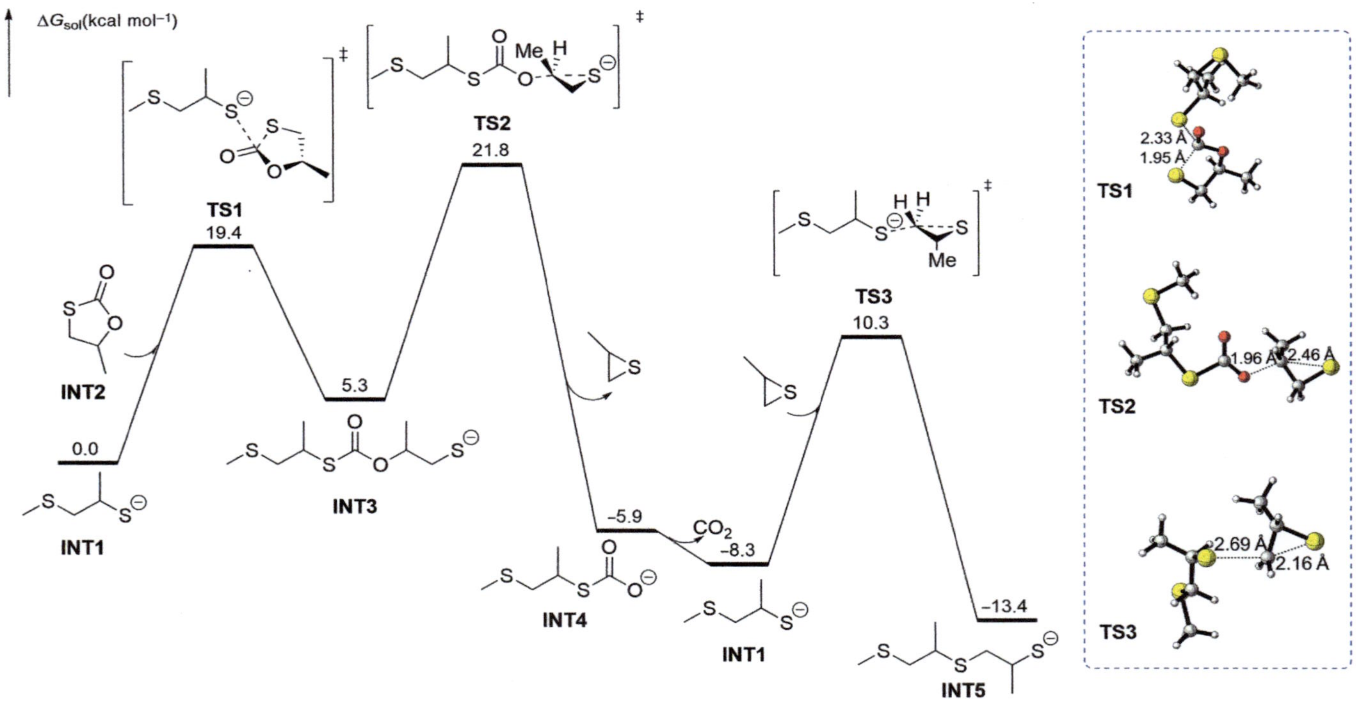

Figure 3.17 Free energies associated with the chain growth of ROP of 5-methyl-1,3-oxathiolan-2-one. Source: Reprinted with permission from Ref. [87]. Copyright American Chemical Society.

Werner and coworkers reported an alkoxide-initiated regioselective coupling of CS_2 with terminal epoxides, afforded alternating polythiocarbonates with monothiocarbonate and trithiocarbonate linkages [85]. Recently, we found that this process could be further utilized as a tool to synthesize sulfur-rich copolymers with monothiocarbonate and trithiocarbonate units via the catalysis of the TEB/organic base system [86].

On the contrary, is there a way to utilize O/S ER to produce well-defined polymers? Obviously, the idea is bold due to the uncontrollability of the O/S ERs but very promising for producing novel structures. As such, we revealed a new O/S ER mechanism, i.e. the third route in Figure 3.16, in which the cycloaddition of CS_2 (COS) with epoxide is the first step, followed by the dethiocarboxylative (decarboxylative) ring-opening of the formed five-membered cyclic thiocarbonates to form COS (CO_2) and episulfide [6, 87]. The O/S ER process was driven by successively regioselective elementary reactions and spontaneously releasing of CO_2 (COS), leading to the sulfur atom of COS transferring to poly(thioether)s, which was well demonstrated by DFT studies. As shown in Figure 3.17, the initiation of polymerization has an activation free energy of 19.4 kcal mol^{-1}, leading to the thiol anion **INT1**, which attacks the carbonyl group of **INT2** via **TS1** in a regioselective manner. The C—S bond is cleaved and **INT3** is formed. Subsequent intramolecular attack via **TS2** generates an episulfide molecules and **INT4**, which then undergoes a spontaneous decarboxylation, releasing CO_2 and generating the thiol anion **INT1**. **INT1** attacks the generated episulfide via **TS3** in a regioselective step, leading to the enchainment. Based on the free energy profile of chain growth, the rate-limiting step is the intramolecular nucleophilic attack via **TS2** with a 21.8 kcal mol^{-1} overall barrier (**INT1** to **TS2**). In total, continuously regioselective ring-opening reaction of the cyclic carbonate and ROP of episulfide, plus the entropy-driven decarboxylation process, O/S ER was achieved, resulting in the sulfur atom of COS transferring to poly(thioether). We utilized this route to synthesize polythioethers, as discussed in the next section.

3.5 Utilization of O/S ER for Poly(thioether)s

3.5.1 COS-Epoxides Route to Poly(thioether)s

The essential part of this subchapter is to discuss the development of simple and practical catalysts for synthesizing poly(thioether)s from COS by utilizing O/S ER. Poly(thioether), a kind of functional polymer, is very potential to the fabrication of optical materials, energy conversion materials for alkaline metal storage and rechargeable batteries [88–90]. The previous reported methods to poly(thioether)s are routinely by the ROP of episulfide and thiol-ene click polymerization of dithiols and dienes (Figure 3.18) [90, 91]. These monomers are difficult to synthesize in a large scale and are very reactive and thus required special storage. Moreover, the production of episulfide generates large amounts of waste water, and dithiol is toxic and has unpleasant smell and poor shelf life due to the oxidization. As a result, it is highly desirable to prepare poly(thioether)s using readily available, scalable raw chemicals. However, in fact, the synthesis of poly(thioether)s remains limited.

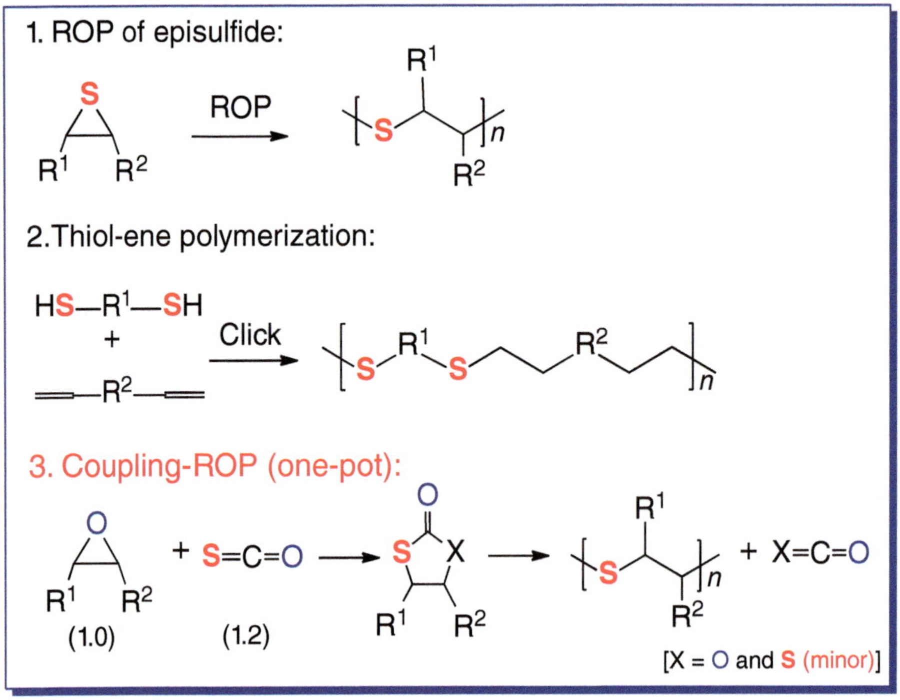

Figure 3.18 Synthetic methods of poly(thioether)s: (1) ROP of episulfide; (2) thiol-ene click polymerization; and (3) the reaction of COS and epoxides using organic bases as the catalysts/initiators via one-pot/two-step (80 °C/120 °C) or one-pot/one-step (120 °C). Source: Refs. [90, 91].

The utilization of O/S ER for synthesizing poly(thioether)s from COS and epoxides is motivated as follows: is it possible that a five-membered cyclic thiocarbonate can be ring-opened via the similar route to the ROP of five-membered cyclic carbonates with releasing CO_2 at high temperatures (>100 °C)? Our initial screening experiments gave us a positive answer. The reaction of COS with PO (PO/COS molar ratio was 1/1.2) in the presence of DBU at 80 °C for 48 h afforded two cyclic thiocarbonates, 5-methyl-1,3-oxathiolan-2-one, and 4-methyl-1,3-dithiolan-2-one with a molar ratio of 83/17, after releasing the generated gas (CO_2 and a little COS), the mixture was then heated in a sealed autoclave at 120 °C for two hours, two cyclic thiocarbonates were completely converted to poly(thioether) with an M_n of 14.8 kg mol^{-1} [87]. The aforementioned two-step reaction was also handled in a closed system without venting of gas. This result guided us to disclose a new mechanism for O/S ER (route 3 in Figure 3.16). We have also attempted to accelerate the coupling reaction of PO with COS to produce cyclic thiocarbonates by activating epoxide and using TU as a hydrogen-bond donor. As expected, the apparent rate constant (K_{obs}) of the coupling reaction catalyzed by the DBU/TU pair was 19.7×10^{-3} min^{-1} and c. 18 times larger than the use of single DBU (K_{obs} : 1.1×10^{-3} min^{-1}). Note that excess loading of COS ([COS]/[PO] ≥ 2) only led to the generation of two cyclic thiocarbonates, that is, the ROP process of two cyclic thiocarbonates was completely inhibited. The resulting poly(thioether) was pure without observing any thiocarbonate units or other linkages, as revealed by the ^{1}H NMR spectrum (Figure 3.19a). Of interest, the resultant poly(thioether) was highly regioselective, with c. 96% of H-T diad content based on the ^{13}C NMR spectrum (Figure 3.19b).

This method can be expanded to a variety of commercially available organic bases, including TBD, 1,5-diazabicyclo[4.3.0]non-5-ene (DBN), MTBD, dimethyl aminopyridine (DMAP), and PPNCl in the presence of TU, and many other epoxides such as 1,2-epoxybutane, 1,2-epoxypentane, 1,2-epoxyhexane, 1,2-epoxyheptane, 1,2-hexyloxirane, 1,2-decyloxirane, CHO, 1,2-epoxy-4-vinylcyclohexane, glycidol benzyl ether, and SO. The generated poly(thioether)s had M_ns among 2 100 to 17 300 g mol^{-1} with Đs of 1.4–1.6. Correspondingly, T_gs of these poly(thioether)s were from -45.8 to 53.4 °C and strongly affected by the substituents of the epoxides, which makes them a potential precursor for elastomers. It is hopeful that this method enables the broad synthesis of a family of poly(thioether)s. As discussed earlier, this route underwent the coupling reaction of COS with epoxides, followed by ROP of the generated mixed cyclic thiocarbonates with releasing CO_2 and a little bit of COS. Organic bases first catalyzed the coupling reaction and then initiated the ROP, therefore a route of killing two birds with one stone.

3.5.2 CS_2-Epoxides Route to Poly(thioether)s

CS_2 is a low-cost chemical with annual industrial production of around 1.2 million tons in the world [92]. The utilization of CS_2 as a feedstock for producing sulfur-containing polymers is of great interest in polymer chemistry, however, the success is rather limited. Nozaki and coworkers reported the copolymerization of CS_2 with PS, no O/S ER occurred because no epoxides were used. The copolymerization was performed well in the presence of (salen)Cr complex with PPNCl (Figure 3.20), or even in the presence of PPNCl, which is a good example of organocatalytic copolymerization [93]. The copolymerization of CS_2 and epoxides has been reported, as discussed in the Chapter 4, written by Prof. Darensbourg and his coworkers. However, the current CS_2/epoxide copolymers contain complex and random units including thiocarbonate, carbonate, and even ether and thioether (Figure 3.20, see Sections 3.2 and 3.4) due to severe O/S ER. Indeed, the O/S ERs have not been effectively inhibited so far in the copolymerization process of CS_2 and epoxides.

Inspired by the aforementioned COS-epoxides route to poly(thioether)s [87], our group developed a coupling reaction process of CS_2 and epoxides to synthesize poly(thioether)s and COS using organic bases as the catalysts (and initiators), as summarized in Table 3.3 [6]. The CS_2/PO coupling reaction was catalyzed by DBU in an opened reaction system, the [CS_2]/[PO]/[DBU] feeding ratio was optimized to be 1.5/1/0.01. PO was completely consumed at 120 °C within 12 hours, affording poly(propylene sulfide) (PPS) and trithiopropylene carbonate (TTPC) ([thioether unit]/[TTPC]: 0.8/0.2, see entry 1 in Table 3.3). The yield of the purified PPS was 3.8 g (73 %) with an M_n of 1800 g mol^{-1} and a Đ of 1.5. After an investigation of the impact of reaction temperatures and solvents on CS_2/PO coupling reaction

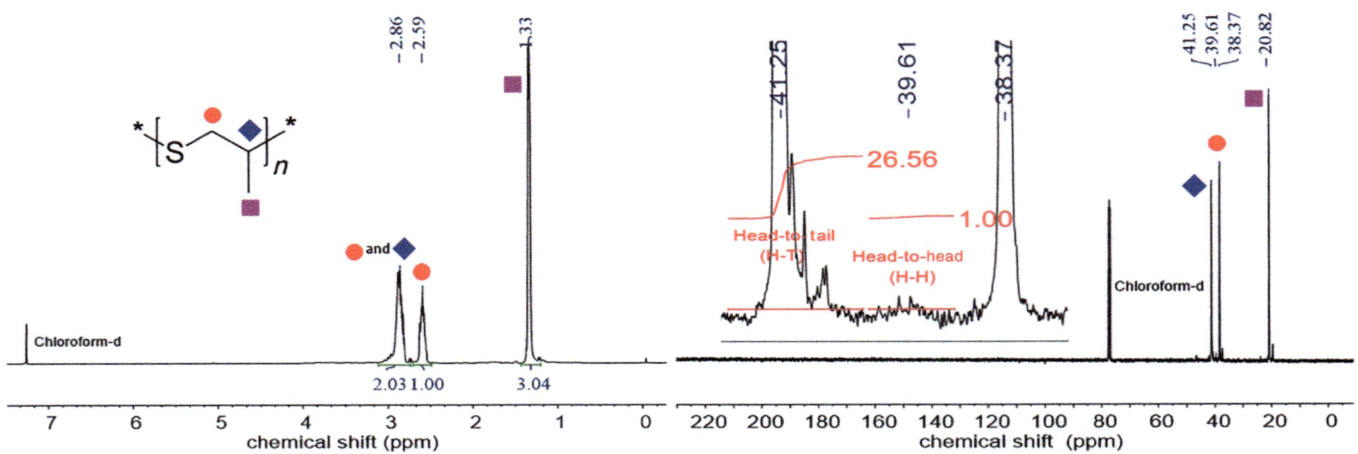

Figure 3.19 (a) ^{1}H NMR and (b) ^{13}C NMR of the produced poly(propylene sulfide) (in $CDCl_3$). Source: Reprinted with permission from Ref. [87]. Copyright American Chemical Society.

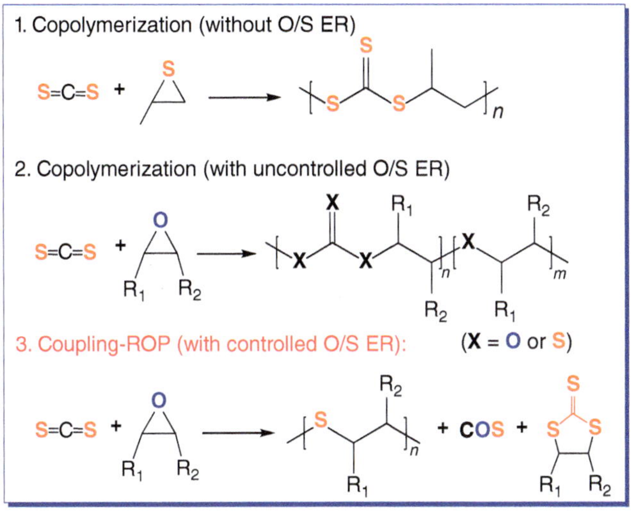

Figure 3.20 Illustration of (1) copolymerization of CS_2 and episulfides to sulfur-rich copolymers, (2) copolymerization of CS_2 and epoxides to copolymers with complex and random units, (3) coupling of CS_2 and epoxides to poly(thioether)s and COS.

in the presence of DBU, the optimum temperature for producing PPS is about 120 °C, and less polar solvent benefits the selective production of PPS. A variety of common organic bases were effective to the CS_2/PO coupling reaction, including TBD, MTBD, and DBN (entries 2–4 in Table 3.3). This process was also successfully extended to several kinds of epoxides (entries 6–10, Table 3.3). As compared with the COS route, this process is much cheaper, without using high-pressure autoclaves. However, the CS_2-based poly(thioether)s have a deeper color and heavier odor than the COS-based one, and the cyclic trithiocarbonate was produced as the byproduct.

It is of great importance that COS was also generated in this process, which is method of synthesizing COS with the potential in large-scale production in industrial. As shown in Figure 3.21, gas chromatography time-of-flight mass spectrum (GC-TOF-MS) of the released gas dissolved in THF suggested the production of COS (mass: 60.07 g mol^{-1}), and the sole chemical shift at 153.36 ppm appears in the ^{13}C NMR spectrum. Since CS_2 and epoxides are commercially available and lost cost, this method can be easily performed in the laboratory for preparing pure COS.

Mechanistically, the exchange of sulfur in CS_2 with oxygen in PO is accomplished through two consecutive reactions, i.e. cycloaddition of CS_2 to PO and *dethiocarboxylative* ROP of DTPC, forming PPS and COS, as shown in Figure 3.22. The first step involved the cycloaddition of CS_2 to PO to form dithiopropylene carbonate (DTPC, **C′**), which is catalyzed by DBU via the zwitterionic DBU-CS_2 adduct. The second step is that the *dethiocarboxylative* ROP of DTPC generates PPS via a PS intermediate with releasing COS, which is an entropy-favorable process. The *dethiocarboxylative* ROP of DTPC is very fast, because we did not observe the coexistence of DTPC and PPS from ^{1}H NMR spectra of all crude products. Indeed,

Table 3.3 Coupling reaction of epoxides with CS_2 using organic bases [6].[a)]

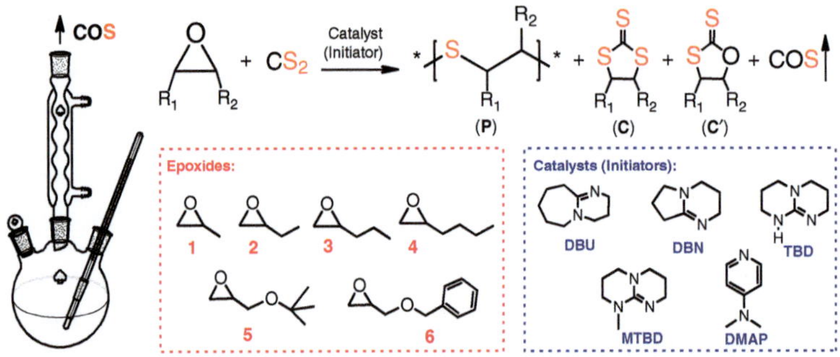

Entry	Epo.	Cat. (Ini.)	P:C:C′[b)]	Yield (g, %)[c)]	M_n (g mol^{-1})[d)]	$Đ$ (M_w/M_n)[d)]
1	**1**	DBU	80 : 20 : 0	3.8 (73)	1800	1.5
2	**1**	TBD	91 : 9 : 0	4.4 (83)	1900	1.5
3	**1**	DBN	74 : 26 : 0	3.2 (61)	1400	1.4
4	**1**	MTBD	83 : 17 : 0	3.7 (69)	1700	1.5
5[e)]	**1**	DMAP	0 : 43 : 57	0	–	–
6	**2**	DBU	54 : 46 : 0	2.0 (41)	2300	1.4
7	**3**	DBU	91 : 9 : 0	3.8 (80)	3100	1.5
8	**4**	DBU	76 : 24 : 0	2.7 (59)	5100	1.2
9	**5**	DBU	70 : 30 : 0	2.4 (47)	4400	1.4
10	**6**	DBU	60 : 40 : 0	2.6 (44)	4100	1.3

a) Reactions were run at 120 °C in 5.0 mL TOL (epoxide: 5.0 mL; with the epoxide/CS_2/catalyst molar ratio of 100/150/1) in a 25 mL flask with reflux device, 12 hours, epoxide was totally consumed determined by ^{1}H NMR spectroscopy.
b) The molar ratio of thioether unit to cyclic thiocarbonates (trithiopropylene carbonate (TTPC, **C**) and dithiopropylene carbonate (DTPC, **C′**)) in the final products, determined by ^{1}H NMR spectroscopy.
c) Yield of poly(thioether), determined by gravimetry.
d) Determined by gel permeation chromatography in THF, calibrated with polystyrene standards.
e) PO conversion was 23%.

the mechanism of CS_2 route is more complicated than the COS-route. One of the obvious differences is that the production of TTPC via the coupling reaction of CS_2 with *in situ* generated PS. TTPC could also ring-opened to PPS with a low conversion rate.

This section has disclosed a chemoselective coupling reaction of CS_2 with epoxides for producing poly(thioether)s and COS in one pot by utilizing the O/S ERs. This is a robust, metal-free, and efficient approach to polythioethers that can be operated under ambient conditions. The production of COS from CS_2 is very meaningful because pure COS is highly desired in electronics industry and key to the production of COS-based copolymers in a large scale.

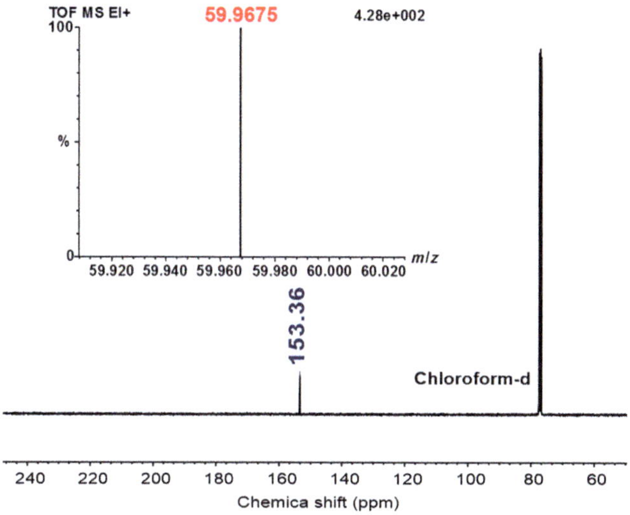

Figure 3.21 ^{13}C NMR spectrum in $CDCl_3$ and GC-TOF-MS (insert part) of the released gas. Source: Reprinted with permission from Ref. [6]. Copyright American Chemical Society.

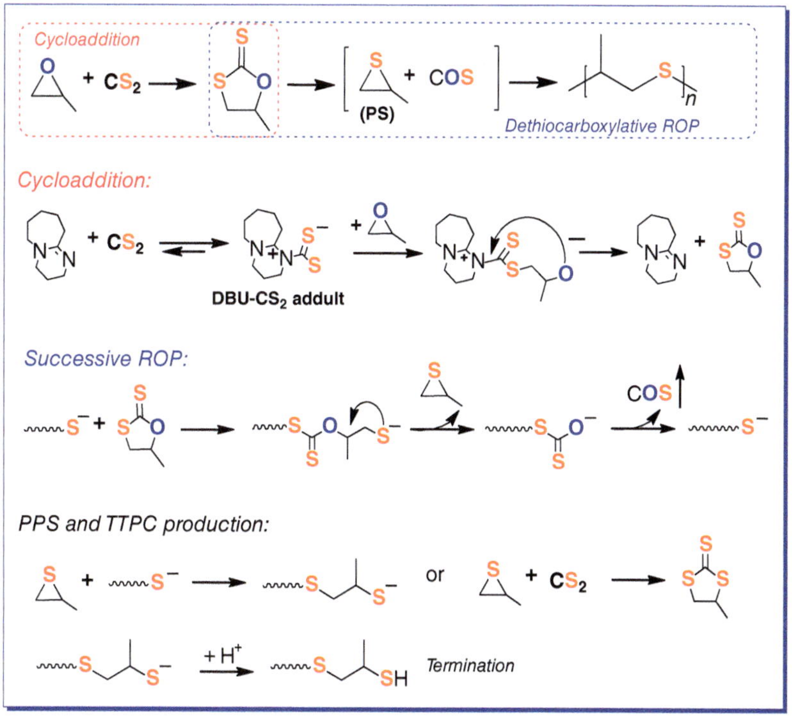

Figure 3.22 Illustration of the mechanism of the CS_2/PO reaction towards COS and PPS.

3.5.3 Repurposing Poly(monothiocarbonate)s to Poly(thioether)s

In Section 3.5.1, the key step of COS-route to poly(thioether) is the ROP of cyclic monothiocarbonate via a *decarboxylative* process. In consideration that the cyclic monothiocarbonate is often generated during the copolymerization of COS with epoxides, we were inspired to explore a method of using poly(monothiocarbonate) (PMTC) as the precursor for synthesizing poly(thioether). In view of the method of recycling polymers, the previous works focused on the *polymer-to-monomer* method, as shown in Figure 3.23a. The proposed PMTC to poly(thioether) represents a *polymer-to-polymer* method for recycling a polymer. In this section, we discussed a rapid transformation of PMTC into poly(thioether) using organic bases, which is of significant to the upcycling of PMTC and meets the requirements of sustainable chemistry [94].

Organic bases such as DBU were effective to degrade PPMTC (***1*** in Figure 3.23) with an M_n of 110.0 kg mol^{-1} in 1,2,4-trichlorobenzene (TCB) at 25 °C ([PPMTC]:[DBU] = 50:1). PPMTC was completely degraded within four hours, and 5-methyl-1,3-oxathiolan-2-one (***2***) was produced. PS (***3***) and PPS (***4***) were detected with a ***2***/***3***/***4*** molar ratio of 57/10/33 (numbers of repeated units). The resultant PPS was regioregular with a 93% of H-T diad content. Remarkably, the transformation of PPMTC into PPS is very fast, it can be completely converted to PPS within 15 minutes at 100 °C. In total, many other commercially available organic bases including TBD, MTBD, DBN, and 1-methylimidazole (NMI) are effective to this transformation process. The stronger basicity of the applied organic bases, the faster the transformation of PPTMC to PPS. This method can also expand to repurpose other PMTCs such as semicrystalline poly(ethylene monothiocarbonate) (PEMTC) to poly(ethylene sulfide) at 140 °C, and amorphous poly(allyloxy benzene monothiocarbonate) to poly(allyloxy benzene sulfide) at 100 °C.

^{1}H NMR spectroscopy and GPC were used to trace the process under the catalysis of DBU in TCB at 100 °C, as shown in Figure 3.24. PPMTC rapidly degraded, as revealed by a dramatic decrease of M_n from 110.0 to 5.8 kg mol^{-1} within two minutes,

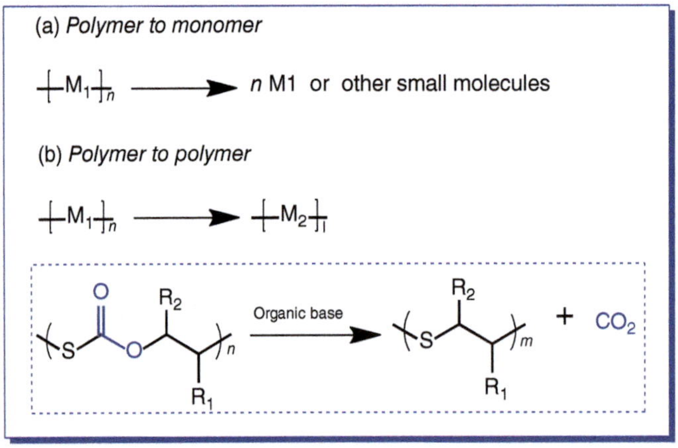

Figure 3.23 Different methods for the recycling of polymers.

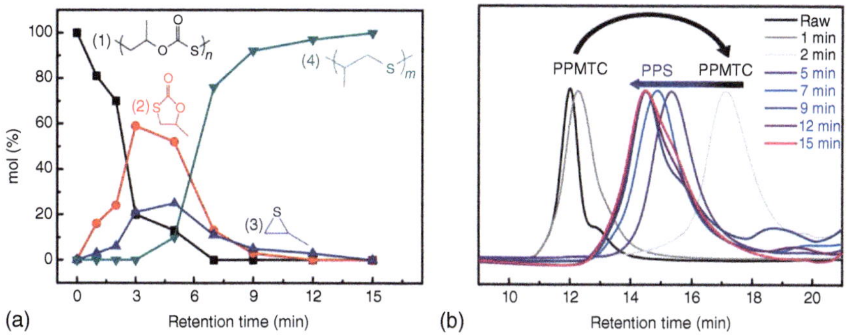

Figure 3.24 (a) The variations in the amounts of PPMTC (*1*), 5-methyl-1,3-oxathiolan-2-one (*2*), propylene sulfide (*3*) and PPS (*4*) found during the degradation of PPMTC. (b) GPC curves of the products removed from the reaction at different times. Source: Reprinted with permission from Ref. [94]. Copyright Royal Society of Chemistry.

and very low levels remained at three minutes (undetectable by GPC), ***2*** and ***3*** were simultaneously produced and reached to the maximum value of concentration at three and four minutes, respectively, while PPS was still undetected. Afterwards, PPS was rapidly produced from c. 3 to 15 minutes, and ***2*** and ***3*** were rapidly consumed (Figure 3.24a), as displayed by the GPC curves in Figure 3.24b.

This section has demonstrated a new pathway to poly(thioether)s by repurposing PMTCs via organic bases. In contrast, a pioneering work [95] reported the transformation of bisphenol A PCs into value-added poly(aryl ether sulfone)s using a carbonate salt as the catalyst and bis(aryl fluorides) as the polymerizable agent. The popular polymer-to-monomer approaches for chemically recycling polymers often required sizable energy inputs, additional operations such as separation and purification of monomers for subsequent polymerization, as such the polymer-to-polymer method is a simple one-pot route and a thermal-stable polymer is produced.

3.6 Crystalline COS-Derived Polymers

As with the CO_2-derived polymers, most of reported COS-derived polymers are amorphous. Up to now, it is a challenge to synthesize crystalline COS-derived polymers with good thermal and mechanical properties. It is well-established that the introduction of sulfur in place of oxygen atoms in the backbone of polycarbonates (PCs) can enhance the thermal properties [96]. That is, some structural analogue of amorphous CO_2-derived copolymers, poly(thiocarbonate)s, can be crystalline. Researchers often prepare crystalline COS-derived copolymers by one of three synthetic routes: (1) regioselective copolymerization of COS with chiral epoxides; (2) copolymerization of COS with symmetric cyclic ethers (OX and EO); (3) asymmetric alternating copolymerization of COS with achiral *meso*-epoxides (Figure 3.25).

For the first route, high regioselectivity of the copolymerization is required to ensure the stereoregularity of the PMTCs. Epicholorohydrin (ECH) is an ideal

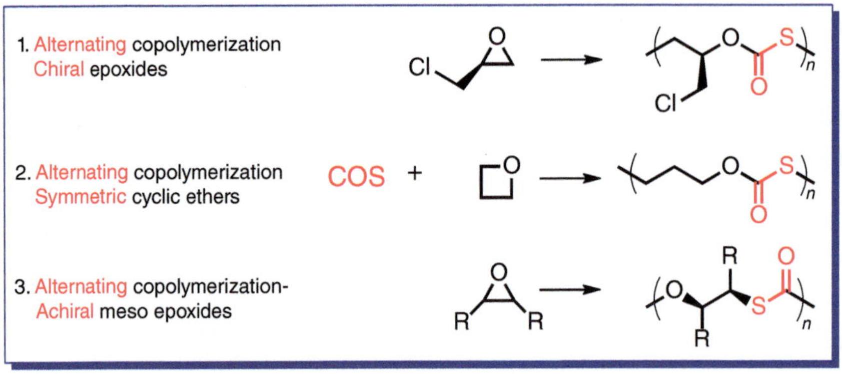

Figure 3.25 Synthetic routes of crystalline COS-derived copolymers.

epoxide in the coupling reaction with CO_2 as a result of its high reactivity and low cost, whereas the cyclic carbonate rather than the PC usually is the main product [97]. Highly regioregular ring-opening of enantiopure ECH with CO_2 could produce a typical semicrystalline polymer with a T_g of 42 °C and a T_m of 108 °C [98]. As for the copolymerization of COS and ECH, however, the regio- and stereoselectivity may be more complicated due to the asymmetric structure of COS. When a single-site bifunctional chromium catalyst (**Catalyst C1**) was used as the catalyst, the copolymerization of COS with ECH provided fully alternating PMTC with high regioselectivity at low reaction temperature [47]. As the enantiomer was used, so the chain configuration was isotactic, leading to the crystallinity of the final copolymer. The T_m of that PMTC is 96.7 °C with a T_g of 15.6 °C. Influenced by the chain transfer reaction to ECH itself in the copolymerization process, M_n of the copolymer was as low as 3.1 kg mol^{-1} even that the copolymerization temperature was −25 °C. Note that this method is also confined by the use of chiral epoxides.

Compared with the route 1, the copolymerization of COS with symmetric cyclic ethers is a more convenient way to synthesize crystalline COS-derived copolymers. Rather than the expensive chiral monomers, OX and EO are employed to copolymerize with COS [47, 48]. Meanwhile, because the cyclic ether is symmetric molecules, the regioselectivity in this copolymerization system is not as complex as that in the copolymerization of two asymmetric monomers. As a result, the requirement on the catalyst design is also less strict. The commercial catalyst systems of the (salen)Cr/PPNCl system and the TEB/LB Lewis pair (Section 3.3.1) could successfully catalyze the copolymerization of OX with COS, providing poly(trimethylene monothiocarbonate) (PTMMTC) with fully alternating structure and a T_m of up to 127.5 °C. At 80 °C, the copolymerization provided semicrystalline PMTC with M_n of 13.9 kg mol^{-1}. And this reaction could be conducted in a wide range of temperatures from 40 to 130 °C with copolymer selectivity up to 99 %. M_n of this copolymer could reach 55.7 kg mol^{-1} with a $Đ$ of 1.76 (Table 3.1, entry 21). Similarly, the copolymerization of EO with COS also provided semicrystalline copolymer with a T_m of up to128.2 °C by the catalyst centered with Cr (Catalyst **C**).

This process provides an economical and convenient route to synthesize crystalline COS-derived polymers with high molecular weight, using commercially available monomers of EO and OX, which is derived from 1,3-propanediol. Therefore, these two linear aliphatic PMTCs are expected to become a new class of general-purpose polymer materials.

Recently, an enantiopure binaphthol-linked dinuclear Co(III) complex (**Catalyst E1**) has been developed to meditate the asymmetric alternating copolymerization of achiral *meso*-epoxides with COS [34]. Asymmetric copolymerization of *meso*-epoxides (such as, CHO, CPO, CXO, and COPO) and nucleophilic reagents is a powerful synthetic strategy for preparing optically active polymers with main-chain chirality, as introduced in Section 3.2.3. Such bimetallic synergistic effect had been successfully used in catalyzing the copolymerization of epoxides with either CO_2 or anhydrides [30]. Its utilization in COS/epoxides copolymerization provided crystalline sulfur-containing copolymers with high T_ms of 141–232 °C.

3.7 COS-Derived Block Polymers

The synthesis of BCPs with a simple method is always highly desired but a challenging issue in polymer chemistry, because BCPs can provide high-performance polymeric materials [99, 100]. A series of COS-based BCPs have been developed recently by using three routes, (1) group transformation, (2) macroinitiator, and (3) sequential addition, as shown in Figure 3.26.

We reported the synthesis of polystyrene-*block*-poly(propylene monothiocarbonate)-*block*-polystyrene (PS-*b*-PPMTC-*b*-PS) by the group transformation [101]. First, α-OH, ω-OH PPMTC with an M_n of 3.7 kg mol^{-1} and a $Đ$ of 1.06 was prepared using binary (salen)CrCl/PPNCl system and water as the chain transfer agent (route 1 in Figure 3.26). Then, the α-OH, ω-OH PPMTC was transformed into the brominated PPTMC by the esterification reaction of PPMTC diol with 2-bromoisobutyryl bromide. The resultant brominated PPTMC initiated the ATRP of styrene, PS-*b*-PPMTC-*b*-PS with a M_n of 10.8 kg mol^{-1} and $Đ$ of 1.08 was achieved.

The second approach for synthesizing COS-derived BCPs is sequential monomer addition (route 2 in Figure 3.26) [47]. A typical example is the synthesis of a triblock copolymer of PEMTC-*b*-PPMTC-*b*-PEMTC, reported by Ren, Lu, and coworkers. 1,2-Propanediol was used as the chain transfer agent and stepwise addition of PO and EO was carried out, PPMTC block was firstly formed, and then the crystalline PEMTC block was formed. In addition, the author's group reported the synthesis of PPMTC-*b*-PPO depended on the kinetic difference of COS/PO copolymerization and ROP of PO in the presence of TEB/LB pair, as described in Section 3.3.2.

The third approach is of using polymer with end –OH group as the initiator (i.e. macroinitiator, see Section 3.3.1). PPMTC-*b*-PEO and PPMTC-*b*-PEO-*b*-PPMTC with narrow $Đ$s of 1.1–1.3 could be obtained by the copolymerization of COS and PO using the TEB/LB pair as the catalyst and PEO with one or two end –OH groups as the macroinitiators [57]. Of note, the initiating efficiency of these PEOs was up to

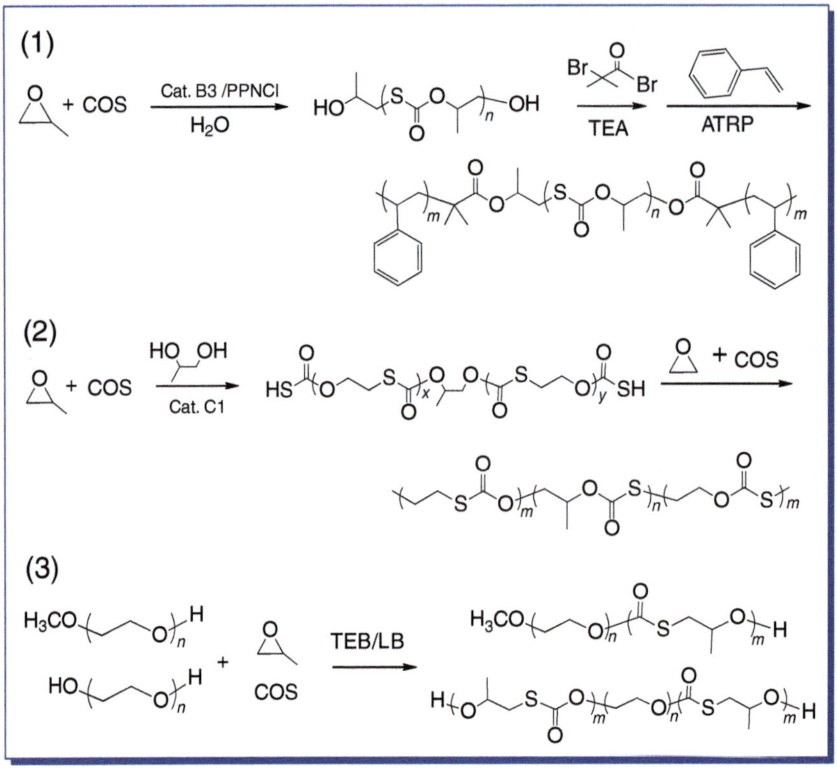

Figure 3.26 Illustration of synthetic method to COS-based BCPs. (1) Group transformation, (2) sequential addition, and (3) macroinitiator. Source: Based on Ref. [47].

100%; therefore, the molecular weights of the resultant BCPs can be well regulated. In the next section, the application of PPMTC-*b*-PEO as the polymer matrix for a new type of solid polymer electrolytes (SPEs) will be introduced.

3.8 Properties of COS-Derived Polymers

COS-derived polymers have excellent thermal, optical, and crystalline properties, and most of them are highly transparent colorless solids [51]. This section focuses on the structure–property relationship and applications of reported COS-derived polymers.

3.8.1 Thermal Properties

The T_g and thermal decomposition temperature (T_d) are the basic thermal properties of a polymer material. As mentioned earlier, COS-derived polymers include PMTCs and poly(thioether)s. Firstly, the T_g of the reported PMTCs from COS/epoxide copolymerization vary from 2.9 to 162.0 °C (Figure 3.27) [4, 34]. The

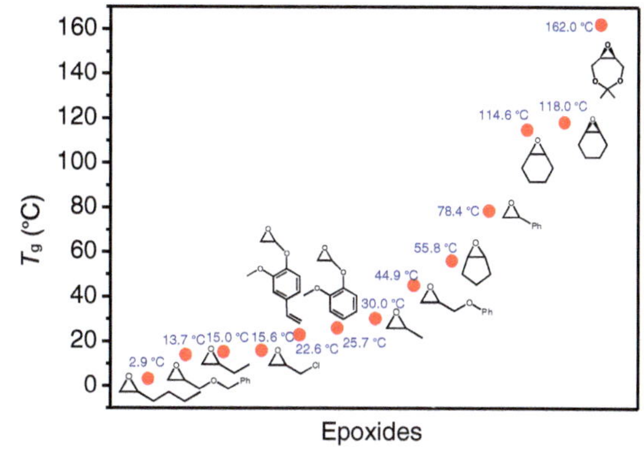

Figure 3.27 T_gs of the PMTCs from COS/epoxide copolymerization.

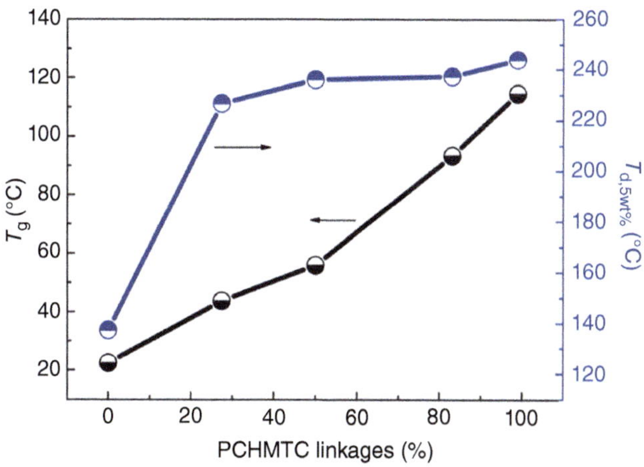

Figure 3.28 Plots of the T_g and $T_{d,5wt\%}$ values versus COS/CHO linkages percentages in the COS/PO/CHO terpolymers. Source: Reprinted with permission from Ref. [45]. Copyright Royal Society of Chemistry.

structure of the epoxide has a strong impact on T_g of the PMTCs. For instance, COS/1,2-epoxyhexane copolymer has the lowest reported T_g (2.9 °C) due to the long butane group. By contrast, the COS/CXO copolymer possesses the highest T_g (162.0 °C) resulting from its rigid heterocyclic structure. Interestingly, the T_g of PMTCs can be regulated by ternary copolymerization. We found that T_gs of the COS/PO/CHO terpolymer could be tuned from 22.4 to 114.6 °C by changing the feeding ratio of PO and CHO (Figure 3.28) [45]. In addition, T_gs of the COS/eugenol-based glycidyl ether (EGE)/guaiacol-based glycidyl ether (GGE) terpolymer were in the range of 16.2 to 18.3 °C and lower than that of the COS/EGE copolymer (22.6 °C) and COS/GGE copolymer (25.7 °C) [102].

The T_d of the polymer determines its maximum operating temperature. COS/epoxide copolymers possess high $T_{d,5wt\%}$ over 200 °C, for examples, 243.9 °C for the COS/CHO copolymer, 206.7 °C for the COS/PGE copolymer, and 252.1 °C for the COS/EGE copolymer [4]. For the terpolymerization of COS, PO, and CHO, introduction of CHO as the third monomer, T_d values of the resultant terpolymers increased with increasing the content of the incorporation of the CHO-COS units. It is worth noting that the $T_{d,5wt\%}$ of COS/PO copolymer we previously reported was rather low (137 °C), which may be attributed to the presence of the residual catalysts (**B3**/PPNCl) in the copolymer [16]. Recently, we found that COS/PO copolymer derived from organic catalysis (TEB/DTMeAB) possessed higher $T_{d,5wt\%}$ of 200 °C [96]. In contrast to traditional plastics like PE, PS, and PC, COS-derived polymers presents moderate thermal properties, similar to the CO_2/epoxide copolymers.

As mentioned in Section 3.5.1, poly(thioether)s synthesized by one-pot reaction of COS (or CS_2) and epoxides also have excellent thermal properties [6, 87]. As shown in Table 3.4, T_gs of these poly(thioether)s are in the range of −58.4 to 53.4 °C, like the PMTCs, strongly depending on the substituents of the epoxides. For example, T_g values of poly(thioether)s with monosubstituted alkyl groups decrease with increasing the alkyl chain length, while poly(thioether)s with large steric groups show increased T_g values. In addition, these poly(thioether)s possess high $T_{d,5wt\%}$s of up to 257 °C. These poly(thioether)s have higher thermal properties than the polysulfide rubber, and thus can be used as elastomers or plastics for various purposes.

3.8.2 Crystalline Properties

Most of the COS-derived copolymers are amorphous and highly soluble in common organic solvents. As shown in Figure 3.29, amorphous COS/PO copolymer possesses the same level of transparency as poly(methyl methacrylate) (PMMA). Ren and coworkers reported the first example of semicrystalline COS-derived copolymer from the regioselective copolymerization of COS and enantiopure ECH [47]. The crystallization behavior of the COS/ECH copolymer with an M_n of 3.1 kg mol^{-1} can be largely attributed to the stereochemistry of the adjacent group locations, exhibiting a T_m of 96.7 °C with the melting enthalpy (ΔH_m) of 49.8 J g^{-1}.

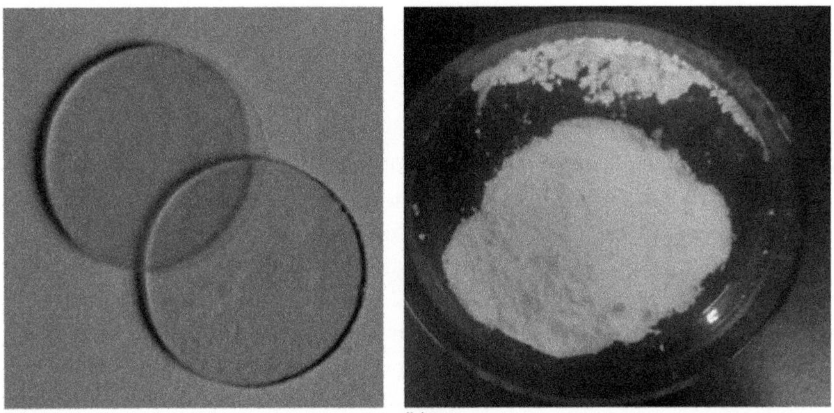

(a) (b)

Figure 3.29 Images of COS-derived polymers: Left is amorphous COS/PO copolymer; Right is the crystalline COS/OX copolymer. Courtesy of Xing-Hong Zhang.

Table 3.4 Thermal properties of poly(thioether)s synthesized by one-pot reaction of COS (or CS_2) and epoxides.

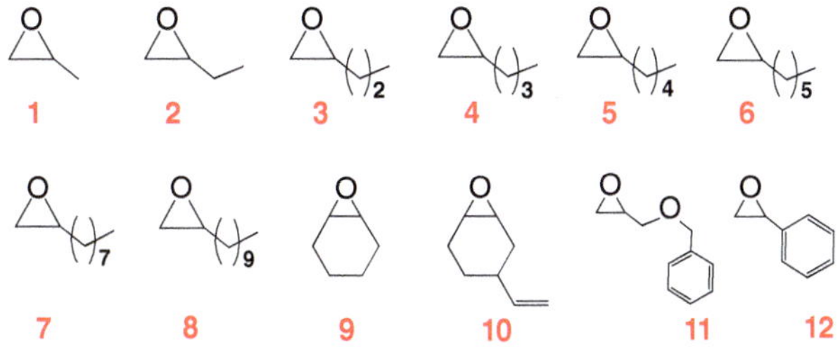

Sample	Epoxides	M_n (kg mol^{-1})	$Đ$	T_g/T_d (°C)
COS/epoxide	1	16.2	1.5	−38.3/240
	2	10.7	1.5	−45.7/235
	3	9.8	1.5	−46.9/238
	4	17.3	1.6	−49.4/257
	5	12.4	1.5	−52.0/248
	6	13.5	1.5	−56.3/251
	7	12.3	1.5	−58.4/253
	8	10.8	1.6	−45.8/254
	9	2.1	1.4	53.4/224
	10	3.4	1.4	48.7/217
	11	6.4	1.6	−14.2/255
	12	2.3	1.5	23.9/227
CS_2/epoxide	1	1.8	1.5	−32.5/–
	2	2.3	1.4	−42.0/–
	3	3.1	1.5	−44.4/–
	4	5.1	1.2	−49.2/–
	11	4.1	1.3	−8.2/–

The synthesis of high molecular weight crystalline COS-derived polymers without incorporating chiral carbon centers is a great significant challenge. Semicrystalline PTMMTC [48] and PEMTC [49] have been reported by simple synthetic routes. It is hence of practical significance to study their crystallization behavior and crystalline morphology, which are helpful for choosing suitable processing conditions of these two polymers.

We had comparatively investigated the crystallization behavior and spherulite morphology of PTMMTC and PEMTC [96]. The T_m of PTMMTC is located at 133 °C with a ΔH_m of 74.4 J g^{-1}, while the T_m of PEMTC is 125 °C and its ΔH_m is 41.5 J g^{-1} (Figure 3.30). Moreover, the equilibrium melting temperature of PTMMTC (147.9 °C) is also higher than that of PEMTC (134.6 °C). These differences can be

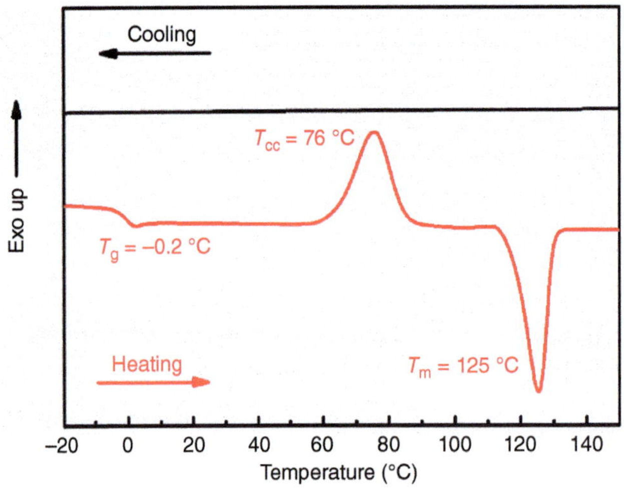

Figure 3.30 The cooling and second-run heating DSC traces of PEMTC [96].

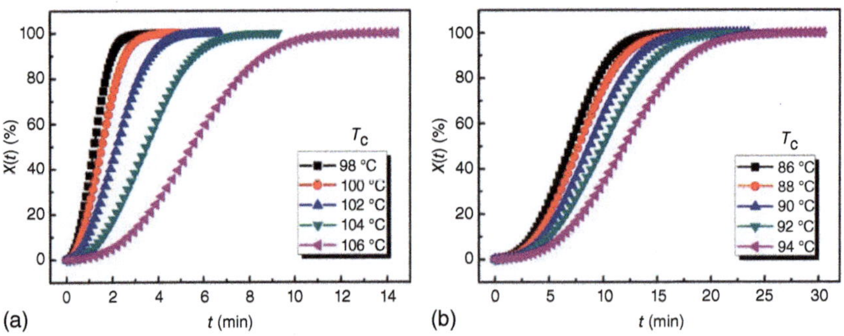

Figure 3.31 Plots of relative crystallinity as a function of crystallization time for (a) PTMMTC and (b) PEMTC. Source: Reprinted with permission from Ref. [96]. Copyright Elsevier.

attributed to the *odd–even* effect of polymeric chain. When the PMTC chains adopt a planar *zig–zag* conformation, the interchain interaction can take place for all carbonyl groups in PTMMTC, while only half of the carbonyl groups in PEMTC are involved the interchain interaction. Figure 3.31 shows the curves for evolution of relative crystallinity with crystallization time, which is analyzed by the Avrami equation. It is found that crystallization rate constant (K) decreases but half crystallization time ($t_{1/2}$) increases with increasing T_c, indicating that crystallization rate of both PMTCs decreases as T_c increases (Table 3.5). In addition, the values of Avrami exponent (n) for PTMMTC and PEMTC are between 2.0 and 3.0, indicating the formation of spherulites with less perfect building lamellae. As shown in Figure 3.32, PTMMTC tends to form the banded spherulites, while PEMTC tends to form a non-banded spherulites. In addition, PTMMTC exhibited a larger average linear spherulitic growth rate (G) than PEMTC at same degree of supercoiling ($\Delta T = T_m^0 - T_c$) (Figure 3.33). It was observed that the Gs of both

Table 3.5 Isothermal crystallization parameters of PTMMTC and PEMTC [96].

Sample	T_c (°C)	n	K (min^{-n})	$t_{1/2}$ (min)
PTMMTC	98	2.22	0.45	1.22
	100	2.15	0.27	1.55
	102	2.17	0.13	2.21
	104	2.41	0.03	3.46
	106	2.63	0.01	5.50
PEMTC	86	2.60	4.5E-3	6.98
	88	2.68	2.9E-3	7.75
	90	2.80	1.6E-3	8.75
	92	2.78	1.2E-3	9.93
	94	2.88	5.5E-4	11.97

PMTCs increased steadily with the increase of ΔT in the studied T_c range, which was consistent with the results of K and $1/t_{1/2}$. Based on the Hoffman–Lauritzen model, the regime III–regime II transition temperatures of PEMTC and PTMMTC were located around 100 and 115 °C, respectively.

As mentioned earlier, Ren et al. has described the synthesis of chiral COS-derived polymers by the alternating copolymerization of achiral *meso*-epoxides with COS (Table 3.6) [34]. These copolymers showed crystallization properties. The COS/CPO and COS/CXO copolymers possess high melting peaks at 157 °C (ΔH_m = 42.4 J g^{-1}) and 221 °C (ΔH_m = 21.7 J g^{-1}), respectively. By contrast, the COS/COPO copolymer has two melting peaks at 215 and 232 °C, with the whole ΔH_m of 26.5 J g^{-1}. The degree of isotacticity of these copolymers has a great influence on its crystalline behavior. As shown in Figure 3.34, the COS/CHO copolymer with a 90% isotacticity was a typical semicrystalline polymer with a T_m of 189 °C (ΔH_m = 27.3 J g^{-1}). However, only a single T_g was observed at 118 °C for the copolymer with a 75% isotacticity, and an improved T_m of up to 201 °C (ΔH_m = 33.3 J g^{-1}) for the highly isotactic copolymers (>90%). This phenomenon is well consistent with the XRD results. The highly isotactic COS/CHO copolymers have four evident diffraction peaks, demonstrating their semicrystalline properties.

3.8.3 Optical Properties

Unlike CO_2-derived polymers, the presence of sulfur atoms in the backbone of COS-derived polymers endows them with excellent optical properties. The refractive index (n_d, n value at the wavelength of 587.6 nm) of two common COS-derived copolymers of COS/PO and COS/CHO are 1.524 and 1.548, respectively, and higher than a typical commercial optical plastic, PMMA (n_d: 1.49) (Table 3.7) [45]. The

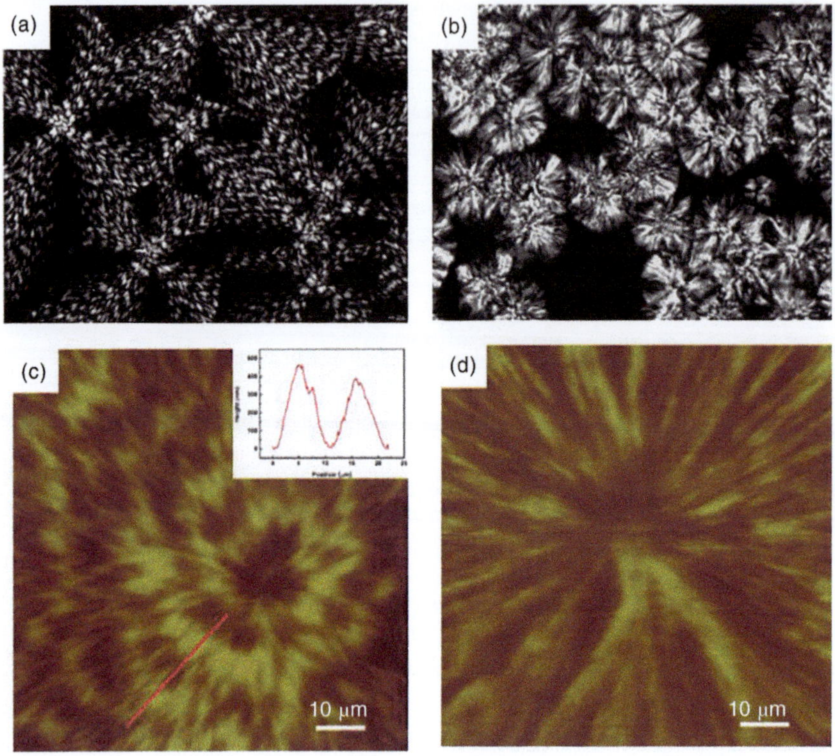

Figure 3.32 POM micrographs (a,b) and AFM height images (c,d) for spherultes of PTMMTC (a,c) and PEMTC (b,d). Source: Reprinted with permission from Ref. [96]. Copyright Elsevier.

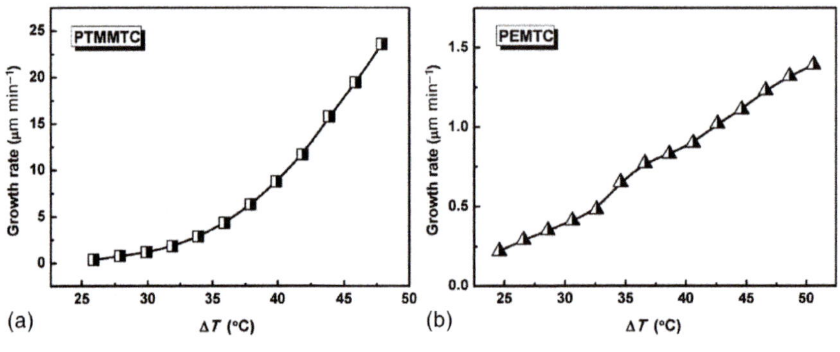

Figure 3.33 Spherulitic growth rate as a function of ΔT for (a) PTMMTC and (b) PEMTC [96].

COS/PGE copolymer pending with side phenyl group has n_d of 1.602, which is higher than the n_ds (1.58–1.59) of PS and aromatic PC [46].

The Abbe number (V_d) of a polymer is also a key parameter for its optical properties, and the V_d of a visual optical material should be around 30–60. The V_d of the COS/PO copolymer is as low as 15.9, meant a severe chromatic dispersion

Table 3.6 Copolymerization of COS and various *meso*-epoxides [34].

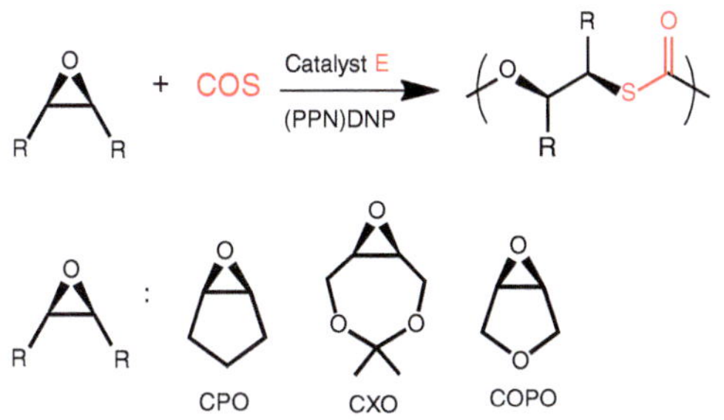

Copolymer	M_n (kg mol^{-1})	$Đ$	T_g/T_m (°C)
COS/CPO	10.4	1.32	–/141
COS/CXO	31.2	1.32	162/221
COS/COPO	46.3	1.30	–/232

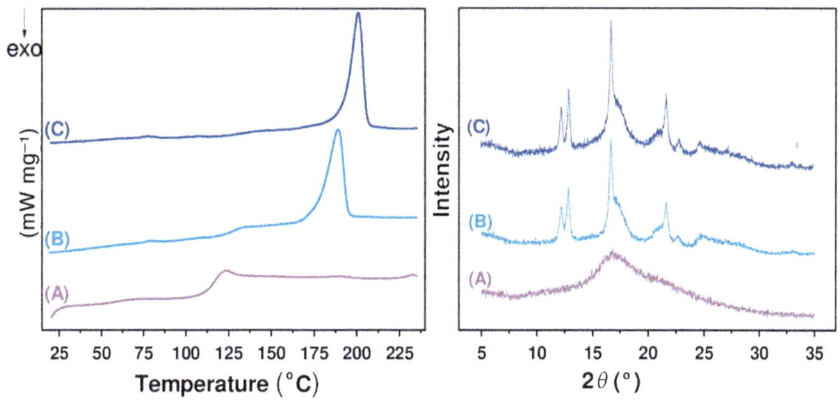

Figure 3.34 DSC thermograms (a) and powder XRD profiles (b) of COS/CHO copolymers with various degrees of isotacticity: (A) 75%, (B) 90%, and (C) >99% [34].

of this material. The V_d value of the COS/CHO copolymer is 32.2 and higher than those of PS (31) and aromatic PC (30). Of interest, we observed that V_d values of the COS/CHO/PO terpolymers could be regulated by changing the PO/CHO feeding ratios during the terpolymerization and were 32.1–43.1, the n_d values of these COS/CHO/PO terpolymers remained at high values of 1.552–1.556. Especially, the terpolymers with equivalent CHO/PO feed ratio exhibited a maximum of V_d 43.1, which is more suitable for application as a visual optical material. This result is ascribed the polarizability change of the C—S bond in each repeated unit owing to the random terpolymerization. After that, we have also reported the COS-derived polymers with aryl structure through the copolymerization of COS

Table 3.7 The refractive index and Abbe number of the COS-derived polymers [45].

Copolymer	Refractive index (n_d)	Abbe number (V_d)
COS/PO	1.524	15.9
COS/CXO	1.530	43.0
COS/CHO	1.548	32.2
COS/CPO	1.560	21.0
COS/EGE	1.588	32.5
COS/PGE	1.602	–
COS/GGE	1.605	40.4
COS/CHO/PO	1.552–1.556	32.1–43.1
COS/EGE/GGE	1.589–1.592	34.5–37.7
PMMA	1.490	58.0
Aromatic PC	1.580	30.0
PS	1.590	31.0

with EGE and GGE [102]. The n_ds of the COS/EGE and COS/GGE copolymers are 1.588 and 1.605, respectively. By the similar terpolymerization process, the n_ds and V_ds of COS/EGE/GGE terpolymers are in the range of 1.589–1.592 and 34.5–37.7, respectively. Therefore, for COS-based copolymers, it is a good way to balance the n_d and V_d values by terpolymerization. Recently, Ren and coworkers reported the synthesis of highly stereoregular COS-derived polymers with main-chain chirality by asymmetric alternating copolymerization of COS and achiral *meso*-epoxides [34]. Like COS/CHO copolymer with alicyclic structure, the COS/CPO copolymer shows good optical properties with n_d of 1.56 and V_d of 21. Notably, COS/CXO copolymer with heterocycle structure possesses a higher n_d of 1.53 and a higher V_d of 43. In addition, a sulfur-rich copolymers with monothiocarbonate and trithiocarbonate units from the copolymerization of CS_2 and PGE via the catalysis of the TEB/organic base system had a n_d of 1.79 [86], representing the highest n_d value from COS(CS_2)-derived polymers. Poly(thioether)s, as introduced in Section 3.5.1, synthesized by one-pot reaction of COS and epoxides also have high n_ds of 1.620 (PO) and 1.639 (SO) (Figure 3.35) [87]. Note that the amorphous COS-based copolymers synthesized by organocatalyst are often highly transparent and colorless. As a result, COS-derived polymers are a new type of optical materials with tailored optical properties.

3.8.4 Electronic Properties

Structurally, the monothiocarbonate repeating unit (–S–C(=O)–O–) of COS-derived polymers is very promising for application as the host matrix for Li^+ ion transport. We have investigated the application of COS-derived polymers for the fabrication

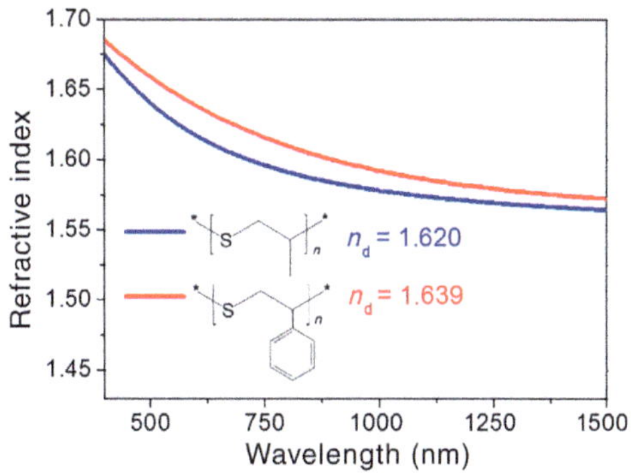

Figure 3.35 Refractive indices of the COS-derived poly(thioether)s [87].

of SPEs by using lithium bis-(trifluoromethanesulfonyl)imide (LiTFSI)-doped PPMTC-*b*-PEO BCPs [103]. In this system, both PPMTC and PEO blocks could associate with Li^+ ions. As shown in Figure 3.36a, the ether stretching band of PEO/LiTFSI electrolytes at 1114 cm^{-1} shift to lower wavenumber after addition of LiTFSI due to the complexation with Li^+ ions. In addition, the stretching band (1714 cm^{-1}) of the carbonyl in the PPMTC/LiTFSI electrolytes splits two peaks, and the new absorption peak (1668 cm^{-1}) becomes gradually obvious with the increase of the $r = [Li^+]/([EO]+[PMTC])$ (Figure 3.36b). Of interest, addition of LiTFSI induces the microphase separation of the PPMTC-*b*-PEO BCPs with a nearly homogeneous structure (Figure 3.37). Moreover, stronger microphase separation usually occurs at moderate doping ratio ($r = 1/12$ and $1/6$), but weaker microphase separation is observed at lower and higher doping ratios ($r = 1/24$ and $1/3$). As shown in Figure 3.37d, a sharp scattering peak with a second-order peak ($q = 2q^*$, q^* is the scattering vector of the primary peak) is observed for $PPMTC_{59}$-*b*-PEO_{44}/LiTFSI electrolyte, confirming the formation of an ordered lamellae (LAM) structure. The extent of microphase separation of $PPMTC_n$-*b*-PEO_{44} BCPs does not monotonically increase with r, which is attributed to the differential complexation of Li^+ ions with the PPMTC and PEO blocks.

In BCP electrolytes with a single conductive phase, the microphase-separated morphology has a remarkable influence on ionic conductivity. However, PPMTC-*b*-PEO BCPs with double conductive phases shows a weak correlation between the conductivity and microphase-separated structure. Being different from PEO/LiTFSI electrolytes, the conductivity of PPMTC-*b*-PEO/LiTFSI electrolytes increases as the amount of LiTFSI rises owing to the increase in the number of charge carriers (Figure 3.38) [104]. It was found that a small grain size and a diffused interphase structure were favorable to the improvement of the ionic conductivity. What's more interesting, the ionic conductivity of PPMTC-*b*-PEO/LiTFSI electrolytes (2×10^{-4} S cm^{-1}) at room temperature is higher than that of PEO/LiTFSI electrolyte with the same r (Figure 3.38d). At 100 °C, the ionic conductivity of the electrolytes can even reach 1.5×10^{-3} S cm^{-1}.

In order to inhibit the dendrite formation, high mechanical strength is essential for SPEs. As shown in Figure 3.39a, the storage moduli (G') values of the PPMTC-*b*-PEO/LiTFSI electrolyte are approximately one to four

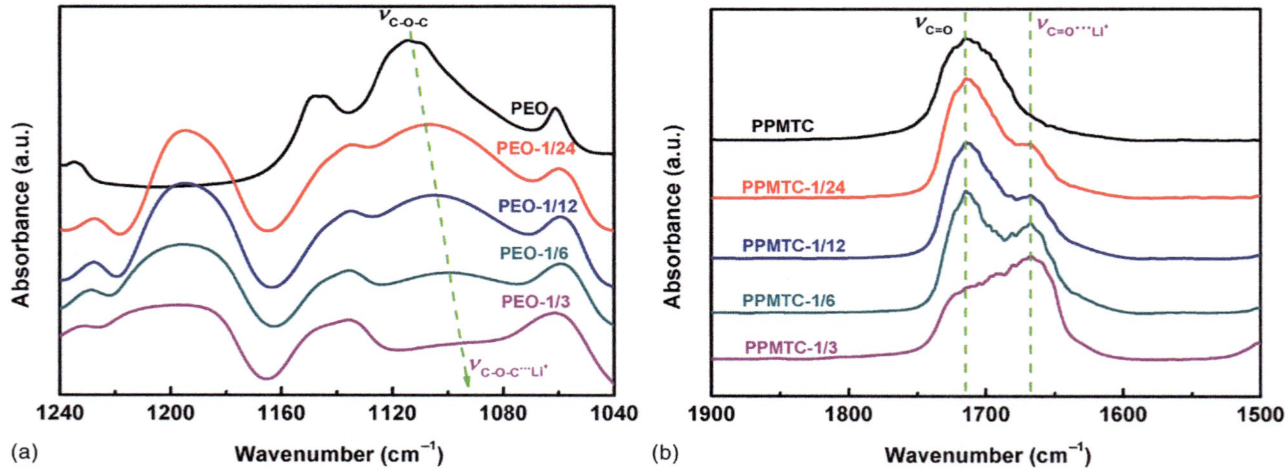

Figure 3.36 FTIR characterizations of (a) PEO/LiTFSI and (b) PPMTC/LiTFSI electrolytes. Source: Reprinted with permission from Ref. [103]. Copyright Elsevier.

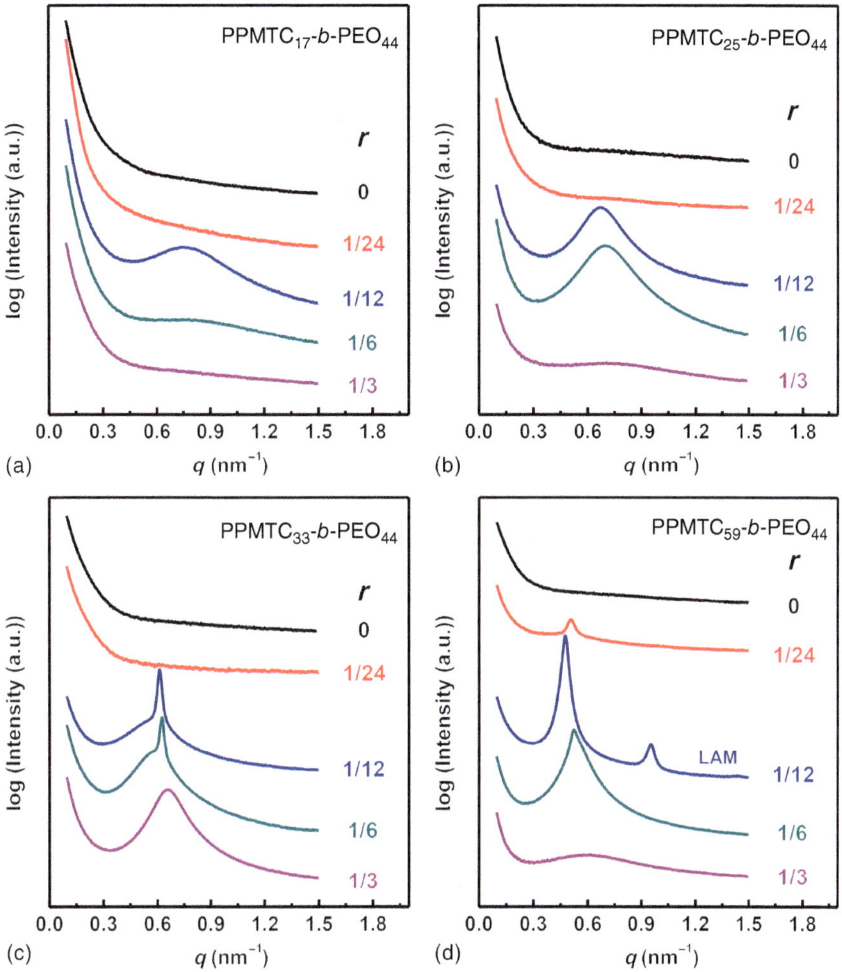

Figure 3.37 SAXS profiles of PPMTC$_n$-*b*-PEO$_{44}$/LiTFSI electrolytes at 40 °C. Source: Reprinted with permission from Ref. [103]. Copyright Elsevier.

orders of magnitude higher than that of the PEO/LiTFSI electrolytes. Because the microphase separation can improve the mechanical properties of polymer electrolytes, the G' of PPMTC-*b*-PEO/LiTFSI electrolytes are also higher than that of the neat PPMTC-*b*-PEO BCPs (Figure 3.39b). In addition, the PPMTC59-*b*-PEO44/LiTFSI-1/12 electrolyte with ordered LAM structure exhibited the highest G's ($\approx 10^5$ Pa at 30 °C and $\approx 10^4$ Pa at 90 °C) among all electrolytes. As a result, the tradeoff between the ionic conductivity and storage moduli of these SPEs can be achieved by microphase-separated double conductive nanophases structure.

Mechanical properties of COS-derived polymers have also been introduced, although the related reports are rare. The crystalline COS/epoxide copolymers possess high tensile strength. For example, PTMMTC exhibits the tensile behavior of

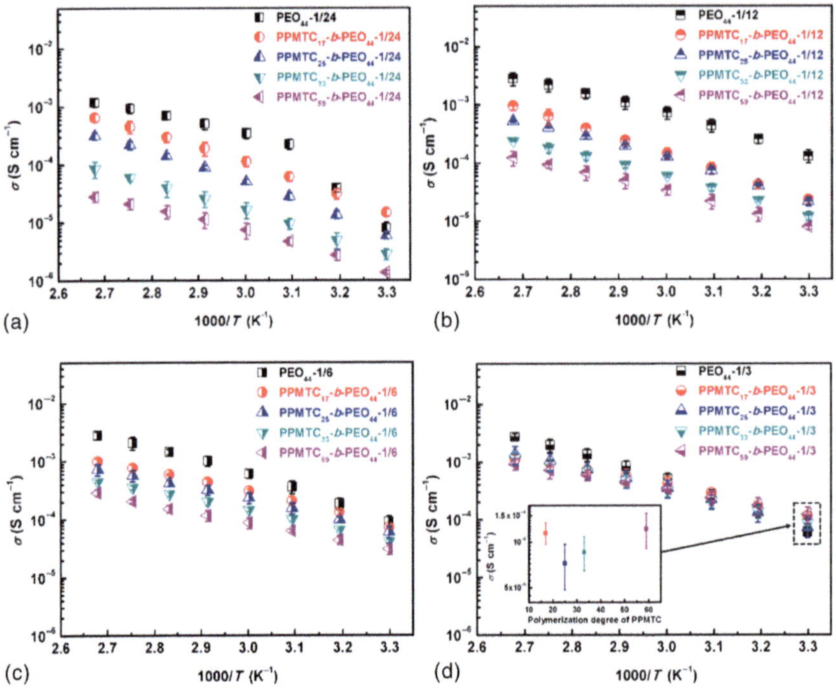

Figure 3.38 Ionic conductivity of PPMTC$_n$-*b*-PEO$_{44}$/LiTFSI electrolytes. Source: Reprinted with permission from Ref. [104]. Copyright Wiley.

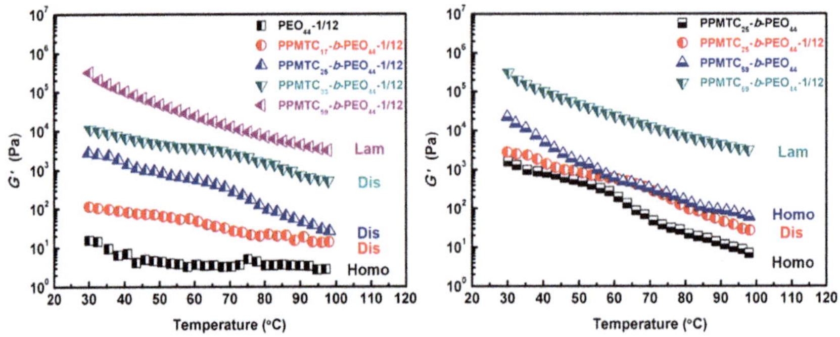

Figure 3.39 Storage moduli of (a) PEO$_{44}$-1/12 and PPMTC$_n$-*b*-PEO$_{44}$-1/12 electrolytes and (b) PPMTC$_n$-*b*-PEO$_{44}$ BCPs electrolytes during heating. Source: Reprinted with permission from Ref. [104]. Copyright Wiley.

typical crystalline polymers. After yielding, necking is observed in wide strain range, followed by breaking. The tensile strength of the PTMMTC (M_n: 13.0 kg mol^{-1}) is 34.3 MPa with a strain at break of 430% (Figure 3.40). Another example reported by Ren et al. that a triblock PEMTC-*b*-PPMTC-*b*-PEMTC copolymer had an ultimate tensile strength of (11.2 ± 0.1) MPa and a strain at break of (575 ± 52)% [49]. By contrast, the PPMTC (M_n: 10.3 kg mol^{-1}) had a strain at break greater than 1000%

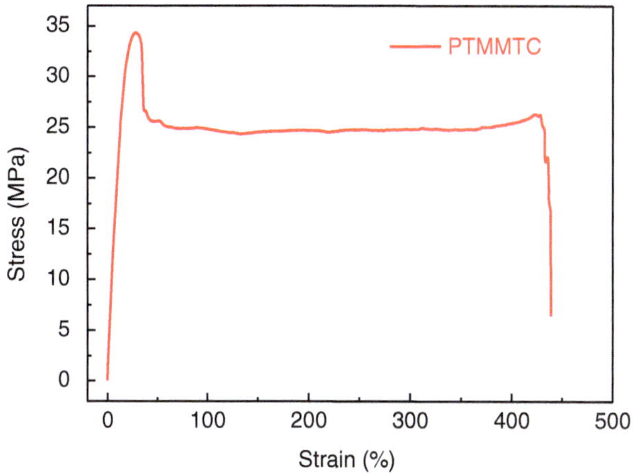

Figure 3.40 Stress–strain curve of PTMMTC.

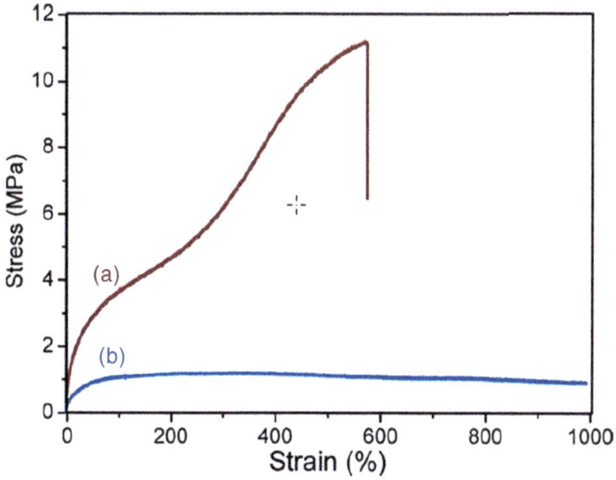

Figure 3.41 Stress–strain curves of (a) the PEMTC-*b*-PPMTC-*b*-PEMTC copolymer and (b) PPMTC. Source: Reprinted with permission from Ref. [49]. Copyright American Chemical Society.

strain, while the largest tensile strength is only (1.6 ± 0.1) MPa (Figure 3.41). These polythiocarbonates derived from COS is one kind of potentially degradable polymer that could be used the alternate of polyolefin in application.

3.9 Summary and Outlook

Since 2013, many efforts have been made for developing a variety of COS-based polymers. These materials are excellent candidates for applications as structural or functional materials because of their unique structure and derived properties such as thermal, mechanical, optical, and electrical properties. In this chapter, we have comprehensively summarized fundamental research activities on the synthesis of sulfur-containing polymers from the polymerization of COS with epoxides (and OX). The catalytic approaches including metal or metal-free systems are essentially introduced with the mechanistic investigation results for providing the understanding to the readers on this subject.

It should be noted again, the conceptualization of O/S ER, a unique phenomenon in the copolymerization process of COS(CS_2) with epoxides, has guided us to be successful in synthesizing a variety of new COS-derived copolymers by either inhibiting it or using it. The O/S ER created mixed monomers in the reaction system, providing chances to prepare polymers with diverse structures. The other important proposal is the *supramolecular anion* strategy for polymerization of many types of monomers with heteroatoms (e.g. O, S, N). The developed polymerizations introduced in this chapter show a clear trend. It is worthwhile to revisit this strategy because supramolecular chemistry has progressed dramatically providing theories and results that can be utilized in designing new catalysts. The stereoselective and sequence-controlled polymerization by organocatalysts is a new challenge in polymer chemistry.

Finally, as introduced in this chapter, COS can be produced in large scale from the current resource system in a low-cost manner and the resultant polymers may find applications in biomaterials, packaging materials, optical, and electrical materials. Future efforts can be directed to the development of robust and versatile catalytic systems that can be accepted by industry for producing sulfur-containing polymers.

References

1 Ferm, R.J. (1957). The chemistry of carbonyl sulfide. *Chem. Rev.* 57: 621–640.

2 Svoronos, P.D.N. and Bruno, T.J. (2002). Carbonyl sulfide: a review of its chemistry and properties. *Ind. Eng. Chem. Res.* 41: 5321–5336.

3 Notni, J., Schenk, S., Protoschill-Krebs, G. et al. (2007). The missing link in COS metabolism: a model study on the reactivation of carbonic anhydrase from its hydrosulfide analogue. *ChemBioChem* 8: 530–536.

4 Luo, M., Zhang, X.-H., and Darensbourg, D.J. (2016). Poly(monothiocarbonate)s from the alternating and regioselective copolymerization of carbonyl sulfide with epoxides. *Acc. Chem. Res.* 49: 2209–2219.

5 Farrell, W.S., Zavalij, P.Y., and Sita, L.R. (2015). Metal-catalyzed “on-demand” production of carbonyl sulfide from carbon monoxide and elemental sulfur. *Angew. Chem. Int. Ed.* 54: 4269–4273.

6 Zhang, C.-J. and Zhang, X.-H. (2020). Chemoselective coupling of CS_2 and epoxides for producing poly(thioether)s and COS via oxygen/sulfur atom exchange. *Macromolecules* 53: 233–239.

7 Watts, S.F. (2000). The mass budgets of carbonyl sulfide, dimethyl sulfide, carbon disulfide and hydrogen sulfide. *Atmos. Environ.* 34: 761–779.

8 Khalil, M.A.K. and Rasmussen, R.A. (1984). Global sources, lifetimes and mass balances of carbonyl sulfide (OCS) and carbon disulfide (CS_2) in the earth's atmosphere. *Atmos. Environ. (1967)* 18: 1805–1813.

9 Kluczewski, S.M., Brown, K.A., and Bel, J.N.B. (1985). Deposition of carbonyl sulphide to soils. *Atmos. Environ. (1967)* 19: 1295–1299.

10 Taylor, G.E., McLaughlin, S.B., Shriner, D.S., and Selvidge, W.J. (1983). The flux of sulfur-containing gases to vegetation. *Atmos. Environ. (1967)* 17: 789–796.

11 Gosselin, R.E. (1984). Hydrogen sulfide. *Clin. Toxicol. Commer. Prod.*: 198–202.

12 Houriet, R. and Louvier, D. (1999). Emission of toxic sulfur gases from polymers coming in contact with food products and with infants. *Analusis* 27: 369–372.

13 Schenk, S., Kesselmeier, J., and Anders, E. (2004). How does the exchange of one oxygen atom with sulfur affect the catalytic cycle of carbonic anhydrase? *Chem. A Eur. J.* 10: 3091–3105.

14 Schenk, S., Notni, J., Köhn, U. et al. (2006). Carbon dioxide and related heterocumulenes at zinc and lithium cations: bioinspired reactions and principles. *Dalton Trans.*: 4191–4206.

15 Li, Y., Zhang, Y.-Y., Liu, B., and Zhang, X.-H. (2018). HCAII-inspired catalysts for making carbon dioxide-based copolymers: the role of metal-hydroxide bond. *Chin. J. Polymer Sci.* 36: 139–148.

16 Luo, M., Zhang, X.-H., Du, B.-Y. et al. (2013). Regioselective and alternating copolymerization of carbonyl sulfide with racemic propylene oxide. *Macromolecules* 46: 5899–5904.

17 Luo, M., Li, Y., Zhang, Y.-Y., and Zhang, X.-H. (2016). Using carbon dioxide and its sulfur analogues as monomers in polymer synthesis. *Polymer* 82: 406–431.

18 Zhang, C.-J. and Zhang, X.-H. (2019). Recent progress on COS-derived polymers. *Chin. J. Polymer Sci.* 37: 951–958.

19 Ren, W.-M., Liu, Y., Xin, A.-X. et al. (2015). Single-site bifunctional catalysts for COX (X = O or S)/epoxides copolymerization: combining high activity, selectivity, and durability. *Macromolecules* 48: 8445–8450.

20 Folsom, T.M., Bhat, G.A., Rashad, A.Z., and Darensbourg, D.J. (2019). Approach for introducing a single metal complex into a polymer chain: metallo-chain transfer agents in CO_2 or COS/epoxide copolymerization processes. *Macromolecules* 52: 5217–5222.

21 Sakakura, T., Choi, J.-C., and Yasuda, H. (2007). Transformation of carbon dioxide. *Chem. Rev.* 107: 2365–2387.

22 Mikkelsen, M., Jørgensen, M., and Krebs, F.C. (2010). The teraton challenge. A review of fixation and transformation of carbon dioxide. *Energ. Environ. Sci.* 3: 43–81.

23 Darensbourg, D.J. (2007). Making plastics from carbon dioxide: salen metal complexes as catalysts for the production of polycarbonates from epoxides and CO_2. *Chem. Rev.* 107: 2388–2410.

24 Cokoja, M., Bruckmeier, C., Rieger, B. et al. (2011). Transformation of carbon dioxide with homogeneous transition-metal catalysts: a molecular solution to a global challenge? *Angew. Chem. Int. Ed.* 50: 8510–8537.

25 Liu, Q., Wu, L., Jackstell, R., and Beller, M. (2015). Using carbon dioxide as a building block in organic synthesis. *Nat. Commun.* 6: 5933.

26 Coates, G.W. and Moore, D.R. (2004). Discrete metal-based catalysts for the copolymerization of CO_2 and epoxides: discovery, reactivity, optimization, and mechanism. *Angew. Chem. Int. Ed.* 43: 6618–6639.

27 Lu, X.-B. and Darensbourg, D.J. (2012). Cobalt catalysts for the coupling of CO_2 and epoxides to provide polycarbonates and cyclic carbonates. *Chem. Soc. Rev.* 41: 1462–1484.

28 Kember, M.R., Buchard, A., and Williams, C.K. (2011). Catalysts for CO_2/epoxide copolymerisation. *Chem. Commun.* 47: 141–163.

29 Cheng, M., Moore, D.R., Reczek, J.J. et al. (2001). Single-site β-diiminate zinc catalysts for the alternating copolymerization of CO_2 and epoxides: catalyst synthesis and unprecedented polymerization activity. *J. Am. Chem. Soc.* 123: 8738–8749.

30 Lu, X.-B., Ren, W.-M., and Wu, G.-P. (2012). CO_2 copolymers from epoxides: catalyst activity, product selectivity, and stereochemistry control. *Acc. Chem. Res.* 45: 1721–1735.

31 Darensbourg, D.J., Mackiewicz, R.M., Phelps, A.L., and Billodeaux, D.R. (2004). Copolymerization of CO_2 and epoxides catalyzed by metal salen complexes. *Acc. Chem. Res.* 37: 836–844.

32 Li, Y., Zhang, Y.-Y., Hu, L.-F. et al. (2018). Carbon dioxide-based copolymers with various architectures. *Prog. Polym. Sci.* 82: 120–157.

33 Luo, M., Zhang, X.-H., Du, B.-Y. et al. (2014). Alternating copolymerization of carbonyl sulfide and cyclohexene oxide catalyzed by zinc-cobalt double metal cyanide complex. *Polymer* 55: 3688–3695.

34 Yue, T.-J., Ren, W.-M., Chen, L. et al. (2018). Synthesis of chiral sulfur-containing polymers: asymmetric copolymerization of meso-epoxides and carbonyl sulfide. *Angew. Chem.* 130: 12852–12856.

35 Gu, G.-G., Yue, T.-J., Wan, Z.-Q. et al. (2017). A single-site iron(III)-salan catalyst for converting COS to sulfur-containing polymers. *Polymers* 9: 515.

36 Zhang, C.-J., Yang, J.-L., Hu, L.-F., and Zhang, X.-H. (2018). Anionic copolymerization of carbonyl sulfide with epoxides via alkali metal alkoxides. *Chin. J. Chem.* 36: 625–629.

37 Sun, X.-K., Zhang, X.-H., Chen, S. et al. (2010). One-pot terpolymerization of CO_2, cyclohexene oxide and maleic anhydride using a highly active heterogeneous double metal cyanide complex catalyst. *Polymer* 51: 5719–5725.

38 Zhang, X.-H., Wei, R.-J., Sun, X.-K. et al. (2011). Selective copolymerization of carbon dioxide with propylene oxide catalyzed by a nanolamellar double metal cyanide complex catalyst at low polymerization temperatures. *Polymer* 52: 5494–5502.

39 Wei, R.-J., Zhang, X.-H., Du, B.-Y. et al. (2013). Highly regioselective and alternating copolymerization of racemic styrene oxide and carbon dioxide via heterogeneous double metal cyanide complex catalyst. *Macromolecules* 46: 3693–3697.

40 Sun, X.-K., Zhang, X.-H., Wei, R.-J. et al. (2012). Mechanistic insight into initiation and chain transfer reaction of CO_2/cyclohexene oxide copolymerization catalyzed by zinc-cobalt double metal cyanide complex catalysts. *J. Polym. Sci. A Polym. Chem.* 50: 2924–2934.

41 Zhang, X.-H., Liu, F., Sun, X.-K. et al. (2008). Atom-exchange coordination polymerization of carbon disulfide and propylene oxide by a highly effective double-metal cyanide complex. *Macromolecules* 41: 1587–1590.

42 Darensbourg, D.J., Wilson, S.J., and Yeung, A.D. (2013). Oxygen/sulfur scrambling during the copolymerization of cyclopentene oxide and carbon disulfide: selectivity for copolymer vs cyclic [thio]carbonates. *Macromolecules* 46: 8102–8110.

43 Zhang, X., Huang, Y., Liu, F. et al. (2009). Copolymerization of carbon disulfide and cyclohexene oxide with a double-metal cyanide complex catalyst. *Acta Polym. Sinica*: 546–552.

44 Luo, M., Zhang, X.-H., and Darensbourg, D.J. (2015). An examination of the steric and electronic effects in the copolymerization of carbonyl sulfide and styrene oxide. *Macromolecules* 48: 6057–6062.

45 Luo, M., Zhang, X.-H., Du, B.-Y. et al. (2015). Well-defined high refractive index poly(monothiocarbonate) with tunable Abbe's numbers and glass-transition temperatures via terpolymerization. *Polymer Chem.* 6: 4978–4983.

46 Luo, M., Zhang, X.H., and Darensbourg, D.J. (2015). Highly regioselective and alternating copolymerization of carbonyl sulfide with phenyl glycidyl ether. *Polymer Chem.* 6: 6955–6958.

47 Yue, T.-J., Ren, W.-M., Liu, Y. et al. (2016). Crystalline polythiocarbonate from stereoregular copolymerization of carbonyl sulfide and epichlorohydrin. *Macromolecules* 49: 2971–2976.

48 Wu, H.-L., Yang, J.-L., Luo, M. et al. (2016). Poly(trimethylene monothiocarbonate) from the alternating copolymerization of COS and oxetane: a semicrystalline copolymer. *Macromolecules* 49: 8863–8868.

49 Ren, W.-M., Yue, T.-J., Li, M.-R. et al. (2017). Crystalline and elastomeric poly(monothiocarbonate)s prepared from copolymerization of COS and achiral epoxide. *Macromolecules* 50: 63–68.

50 Stephan, D.W. (2015). Frustrated Lewis pairs. *J. Am. Chem. Soc.* 137: 10018–10032.

51 Yang, J.-L., Wu, H.-L., Li, Y. et al. (2017). Perfectly alternating and regioselective copolymerization of carbonyl sulfide and epoxides by metal-free Lewis pairs. *Angew. Chem. Int. Ed.* 56: 5774–5779.

52 Darensbourg, D.J., Moncada, A.I., Choi, W., and Reibenspies, J.H. (2008). Mechanistic studies of the copolymerization reaction of oxetane and carbon dioxide to provide aliphatic polycarbonates catalyzed by (Salen)CrX complexes. *J. Am. Chem. Soc.* 130: 6523–6533.

53 Hong, M., Chen, J., and Chen, E.Y.X. (2018). Polymerization of polar monomers mediated by main-group Lewis acid–base pairs. *Chem. Rev.* 118: 10551–10616.
54 Yang, J.-L., Wang, H.-L., Hu, L.-F. et al. (2019). A double-site Lewis pair for highly active and living synthesis of sulfur-containing polymers. *Polymer Chem.* 10: 6555–6560.
55 Zhang, C.-J., Wu, H.-L., Li, Y. et al. (2018). Precise synthesis of sulfur-containing polymers via cooperative dual organocatalysts with high activity. *Nat. Commun.* 9: 2137.
56 Zhang, D., Boopathi, S.K., Hadjichristidis, N. et al. (2016). Metal-free alternating copolymerization of CO2 with epoxides: fulfilling "green" synthesis and activity. *J. Am. Chem. Soc.* 138: 11117–11120.
57 Yang, J.-L., Cao, X.-H., Zhang, C.-J. et al. (2018). Highly efficient one-pot synthesis of COS-based block copolymers by using organic Lewis pairs. *Molecules* 23: 298.
58 Zhang, C.-J., Duan, H.-Y., Hu, L.-F. et al. (2018). Metal-free route to precise synthesis of poly(propylene oxide) and its blocks with high activity. *ChemSusChem* 11: 4209–4213.
59 Wan, X.H. and Wang, X.H. (2018). Metal-free access to precise synthesis of sulfur-containing polymers. *Acta Polym. Sinica*: 773–775.
60 Zhang, C.-J. and Zhang, X.-H. (2019). Organocatalytic polymerization. *Sci. China Chem.* 62: 1087.
61 Hu, L.-F., Zhang, C.-J., Wu, H.-L. et al. (2018). Highly active organic Lewis pairs for the copolymerization of epoxides with cyclic anhydrides: metal-free access to well-defined aliphatic polyesters. *Macromolecules* 51: 3126–3134.
62 Ji, H.-Y., Wang, B., Pan, L., and Li, Y.-S. (2018). Lewis pairs for ring-opening alternating copolymerization of cyclic anhydrides and epoxides. *Green Chem.* 20: 641–648.
63 Ji, H.-Y., Chen, X.-L., Wang, B. et al. (2018). Metal-free, regioselective and stereoregular alternating copolymerization of monosubstituted epoxides and tricyclic anhydrides. *Green Chem.* 20: 3963–3973.
64 Kummari, A., Pappuru, S., and Chakraborty, D. (2018). Fully alternating and regioselective ring-opening copolymerization of phthalic anhydride with epoxides using highly active metal-free Lewis pairs as a catalyst. *Polymer Chem.* 9: 4052–4062.
65 Hu, L.-F., Chen, D.-J., Yang, J.-L., and Zhang, X.-H. (2020). An investigation of the organoborane/Lewis base pairs on the copolymerization of propylene oxide with succinic anhydride. *Molecules* 25: 253.
66 Hong, M. and Chen, E.Y.X. (2016). Completely recyclable biopolymers with linear and cyclic topologies via ring-opening polymerization of gamma-butyrolactone. *Nat. Chem.* 8: 42–49.
67 Hong, M. and Chen, E.Y.X. (2016). Towards truly sustainable polymers: a metal-free recyclable polyester from biorenewable non-strained γ-butyrolactone. *Angew. Chem. Int. Ed.* 55: 4188–4193.

68 Zhao, N., Ren, C., Li, H. et al. (2017). Selective ring-opening polymerization of non-strained gamma-butyrolactone catalyzed by a cyclic trimeric phosphazene base. *Angew. Chem. Int. Ed.* 56: 12987–12990.

69 Zhang, C.-J., Hu, L.-F., Wu, H.-L. et al. (2018). Dual organocatalysts for highly active and selective synthesis of linear poly(γ-butyrolactone)s with high molecular weights. *Macromolecules* 51: 8705–8711.

70 Lin, L., Han, D., Qin, J. et al. (2018). Nonstrained γ-butyrolactone to high-molecular-weight poly(γ-butyrolactone): facile bulk polymerization using economical ureas/alkoxides. *Macromolecules* 51: 9317–9322.

71 Shen, Y., Zhao, Z., Li, Y. et al. (2019). A facile method to prepare high molecular weight bio-renewable poly(γ-butyrolactone) using a strong base/urea binary synergistic catalytic system. *Polymer Chem.* 10: 1231–1237.

72 Lohmeijer, B.G.G., Pratt, R.C., Leibfarth, F. et al. (2006). Guanidine and amidine organocatalysts for ring-opening polymerization of cyclic esters. *Macromolecules* 39: 8574–8583.

73 Dove, A.P. (2012). Organic catalysis for ring-opening polymerization. *ACS Macro Lett.* 1: 1409–1412.

74 Mespouille, L., Coulembier, O., Kawalec, M. et al. (2014). Implementation of metal-free ring-opening polymerization in the preparation of aliphatic polycarbonate materials. *Prog. Polym. Sci.* 39: 1144–1164.

75 Dove, A.P., Pratt, R.C., Lohmeijer, B.G.G. et al. (2005). Thiourea-based bifunctional organocatalysis: supramolecular recognition for living polymerization. *J. Am. Chem. Soc.* 127: 13798–13799.

76 Ottou, W.N., Sardon, H., Mecerreyes, D. et al. (2016). Update and challenges in organo-mediated polymerization reactions. *Prog. Polym. Sci.* 56: 64–115.

77 Li, M., Tao, Y., Tang, J. et al. (2019). Synergetic organocatalysis for eliminating epimerization in ring-opening polymerizations enables synthesis of stereoregular isotactic polyester. *J. Am. Chem. Soc.* 141: 281–289.

78 Chen, Y., Shen, J., Liu, S. et al. (2018). High efficiency organic lewis pair catalyst for ring-opening polymerization of epoxides with chemoselectivity. *Macromolecules* 51: 8286–8297.

79 Brocas, A.-L., Mantzaridis, C., Tunc, D., and Carlotti, S. (2013). Polyether synthesis: from activated or metal-free anionic ring-opening polymerization of epoxides to functionalization. *Prog. Polym. Sci.* 38: 845–873.

80 Smeulders, M.J., Barends, T.R.M., Pol, A. et al. (2011). Evolution of a new enzyme for carbon disulphide conversion by an acidothermophilic archaeon. *Nature* 478: 412–416.

81 Bräuer, M., Anders, E., Sinnecker, S. et al. (2000). The TpZn–OH/CS reaction: theoretical and preparative visualization of an essential bioinorganic reaction path. *Chem. Commun.*: 647–648.

82 Rombach, M. and Vahrenkamp, H. (2001). Pyrazolylborate–zinc–hydrosulfide complexes and their reactions. *Inorg. Chem.* 40: 6144–6150.

83 Barbero, M., Degani, I., Dughera, S. et al. (1996). Convenient procedure for converting 1,3-dithiolane-2-thiones into 1,3-dithiolan-2-ones. *J. Chem. Soc. Perkin Trans. 1*: 289–294.

84 Luo, M., Zhang, X.-H., and Darensbourg, D.J. (2015). An investigation of the pathways for oxygen/sulfur scramblings during the copolymerization of carbon disulfide and oxetane. *Macromolecules* 48: 5526–5532.
85 Diebler, J., Komber, H., Häußler, L. et al. (2016). Alkoxide-initiated regioselective coupling of carbon disulfide and terminal epoxides for the synthesis of strongly alternating copolymers. *Macromolecules* 49: 4723–4731.
86 Yang, J.-L., Hu, L.-F., Cao, X.-H. et al. (2020). An investigation on the production of random copolymer with monothiocarbonate and trithiocarbonate units over cyclic thiocarbonate via metal-free catalysis. *Chin. J. Chem.* 38: 269–274.
87 Zhang, C.-J., Zhu, T.-C., Cao, X.-H. et al. (2019). Poly(thioether)s from closed-system one-pot reaction of carbonyl sulfide and epoxides by organic bases. *J. Am. Chem. Soc.* 141: 5490–5496.
88 Bearinger, J.P., Terrettaz, S., Michel, R. et al. (2003). Chemisorbed poly(propylene sulphide)-based copolymers resist biomolecular interactions. *Nat. Mater.* 2: 259.
89 Napoli, A., Valentini, M., Tirelli, N. et al. (2004). Oxidation-responsive polymeric vesicles. *Nat. Mater.* 3: 183.
90 Mutlu, H., Ceper, E.B., Li, X. et al. (2019). Sulfur chemistry in polymer and materials science. *Macromol. Rapid Commun.* 40: 1800650.
91 Park, N.H., Gomes, G.d.P., Fevre, M. et al. (2017). Organocatalyzed synthesis of fluorinated poly(aryl thioethers). *Nat. Commun.* 8: 166.
92 DeMartino, A.W., Zigler, D.F., Fukuto, J.M., and Ford, P.C. (2017). Carbon disulfide. Just toxic or also bioregulatory and/or therapeutic? *Chem. Soc. Rev.* 46: 21–39.
93 Nakano, K., Tatsumi, G., and Nozaki, K. (2007). Synthesis of sulfur-rich polymers: copolymerization of episulfide with carbon disulfide by using PPN Cl/(salph)Cr(III)Cl system. *J. Am. Chem. Soc.* 129: 15116–15117.
94 Cao, X.-H., Zhang, C.-J., Yang, J.-L. et al. (2020). Repurposing poly(monothiocarbonate)s to poly(thioether)s with organic bases. *Polymer Chem.* 11: 309–314.
95 Jones, G.O., Yuen, A., Wojtecki, R.J. et al. (2016). Computational and experimental investigations of one-step conversion of poly(carbonate)s into value-added poly(aryl ether sulfone)s. *Proc. Natl. Acad. Sci.* 113: 7722–7726.
96 Cao, X.-H., Yang, J.-L., Wu, H.-L. et al. (2019). Crystallization behavior and morphology of novel aliphatic poly(monothiocarbonate)s. *Polymer* 165: 112–123.
97 Wu, G.-P., Wei, S.-H., Ren, W.-M. et al. (2011). Perfectly alternating copolymerization of CO2 and epichlorohydrin using cobalt(III)-based catalyst systems. *J. Am. Chem. Soc.* 133: 15191–15199.
98 Wu, G.-P., Xu, P.-X., Lu, X.-B. et al. (2013). Crystalline CO2 copolymer from epichlorohydrin via Co(III)-complex-mediated stereospecific polymerization. *Macromolecules* 46: 2128–2133.
99 Jeske, R.C., Rowley, J.M., and Coates, G.W. (2008). Pre-rate-determining selectivity in the terpolymerization of epoxides, cyclic anhydrides, and CO2: a one-step route to diblock copolymers. *Angew. Chem. Int. Ed.* 47: 6041–6044.

100 Bates, F.S., Hillmyer, M.A., Lodge, T.P. et al. (2012). Multiblock polymers: panacea or Pandora's box? *Science* 336: 434–440.

101 Li, Y., Duan, H.-Y., Luo, M. et al. (2017). Mechanistic study of regio-defects in the copolymerization of propylene oxide/carbonyl sulfide catalyzed by (Salen)CrX complexes. *Macromolecules* 50: 8426–8437.

102 Hu, L.-F., Li, Y., Liu, B. et al. (2017). Alternating and regioregular copolymers with high refractive index from COS and biomass-derived epoxides. *RSC Adv.* 7: 49490–49497.

103 Cao, X.-H., Yang, J.-L., Wang, R.-Y. et al. (2019). Microphase separation of poly(propylene monothiocarbonate)-b-poly(ethylene oxide) block copolymers induced by differential interactions with salt. *Polymer* 180: 121745.

104 Cao, X.-H., Li, J.-H., Yang, M.-J. et al. Simultaneous improvement of ionic conductivity and mechanical strength in block copolymer electrolytes with double conductive nanophases. *Macromol. Rapid Commun.* 41: 1900622.

4

Thiol-Based Click Polymerizations for Sulfur-Containing Polymers

Die Huang[1], Anjun Qin[1], and Ben Zhong Tang[1,2]

[1], State Key Laboratory of Luminescent Materials and Devices, Guangdong Provincial Key Laboratory of Luminescence from Molecular Aggregates, Center for Aggregation-Induced Emission, South China University of Technology, Guangzhou 510640, China
[2]Hong Kong Branch of Chinese National Engineering Research Centre for Tissue Restoration and Reconstruction, Institute for Advanced Study, Department of Chemistry, and Department of Chemical and Biological Engineering, The Hong Kong University of Science & Technology, Clear Water Bay, Kowloon 999077, Hong Kong, China

4.1 Introduction

The construction of functional polymers is an important research topic in polymer chemistry. Generally, sulfur-containing polymers exhibit various attractive features, such as high melting temperature, thermal stability, and high refractive index, suggesting their great potential in functional polymer arena [1]. As a result, sulfur-containing polymers have drawn increasing scientific attentions and industrial interests, which have been extensively investigated and found promising applications in the fields of optical materials, optoelectronic materials, biological materials, and rechargeable batteries [2–7].

Nowadays, plentiful synthetic techniques, such as polymerizations based on elemental sulfur, CS_2, and carbonyl sulfide COS, have been developed for sulfur-containing polymers with diverse structures and functional properties [8–12]. For example, Pyun and coworkers developed the inverse vulcanization of elemental sulfur and vinyl molecules for the preparation of functional chalcogenide hybrid inorganic/organic polymers with pretty high sulfur contents [13–16]. Tang and coworkers reported a series of highly efficient multicomponent polymerizations based on elemental sulfur, which can be used to produce functional poly(thioamide)s and poly(thiourea)s under mild reaction conditions [17–19]. CS_2 has been developed into a versatile monomer and widely used in homopolymerization, step polymerization, and chain polymerization for the synthesis of sulfur-containing polymers [20–23]. COS/epoxide copolymerization has also been established for the synthesis of poly(monothiocarbonate)s [24–26].

Since it was coined by Sharpless and coworkers in 2001 [27], click chemistry had served as inspiration in most of the areas [28–41]. In particular, click

Sulfur-Containing Polymers: From Synthesis to Functional Materials,
First Edition. Edited by Xing-Hong Zhang and Patrick Theato.

polymerization evolved from click reaction has been developed into a powerful polymerization technique for the construction of functional polymers with various structures owing to its excellent click characteristics, such as high efficiency, atom economy, mild reaction conditions, great functionality tolerance, and regio- and/or stereoselectivities [42–47]. Besides the famous Cu(I)-catalyzed azide-alkyne click polymerization, thiol-based click polymerization is another well-known one. Thanks to the high reactivity of thiol, which can react with a vast range of functional groups under benign conditions, thiol-based click polymerization not only holds the click features but also exhibits diversity, which has become a versatile tool for the synthesis of sulfur-containing polymer owing to the great endeavors of polymer chemists [48–54]. For example, thiol-ene click polymerization including radical-mediated click polymerization and thiol-Michael addition click polymerization has been widely applied to produce poly(thioether)s. Thiol-yne click polymerization, a complement of the former, can proceed in a bis-addition or mono-addition manner to furnish poly(thioether)s or poly(vinyl sulfide)s with different regio-/stereoregularity. Furthermore, other thiol-based click polymerizations, including thiol-epoxy, thiol-isocyanate, and thiol-halogen click polymerizations, have also been established and applied to prepare sulfur-containing polymer with diverse structures.

Over the past decade, remarkable progress has been made in the field of thiol-based click polymerizations. This chapter intends to briefly summarize the research efforts on developing thiol-based click polymerization into a more powerful technique and focus on the preparation of soluble sulfur-containing polymers with linear and hyperbranched structures. Thereinto, the most representative thiol-ene/yne click polymerizations are highlighted and introduced in detail. Some other thiol-based click polymerizations are also mentioned.

4.2 Thiol-Ene Click Polymerization

4.2.1 Radical-Initiated Thiol-Ene Click Polymerization

As a well-known thiol-based click reaction, light or heat triggered radical-initiated thiol-ene click reaction has attracted a lot of scientific interest and been systematically investigated [52]. As shown in Scheme 4.1, the radical-initiated thiol-ene reaction consists of alternating addition and hydrogen abstraction reactions. In the presence of radical initiator, a thiyl radical is generated photochemically or thermally and then attacks a vinyl group to form a new carbon radical. The carbon radical subsequently abstracts hydrogen from another thiol group, thereby furnishing an anti-Markovnikov thioether product and a new formed thiyl radical.

In spite of its widespread applications in post-functionalization of preformed polymers and construction of dendrimers and polymer networks, radical-initiated thiol-ene click reaction has also been developed into polymerization for the preparation of soluble poly(thioether)s with linear and hyperbranched structures. For example, using photo-/thermo initiated thiol-ene click polymerization of AB-type

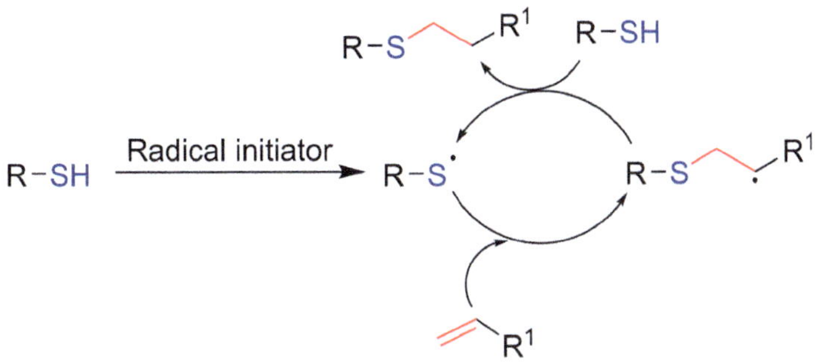

Scheme 4.1 The mechanism of radical-initiated thiol-ene click reaction.

monomers, Du Prez and coworkers prepared linear poly(thioether)s with high molecular weights (number-average molecular weights, M_n up to 40 200), which could be oxidized with hydrogen peroxide and transformed into polysulfones with improved thermal stability (Scheme 4.2) [55].

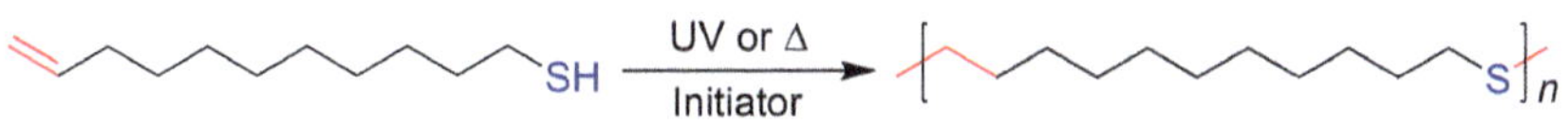

Scheme 4.2 Preparation of linear poly(thioether)s by photo-/thermoinitiated thiol-ene click polymerization of AB-type monomers. Source: Based on Ref. [55].

Considering the self-oligomerization of the AB-type monomers during storage, $A_2 + B_2$ strategy has also been applied to construct linear poly(thioether)s. As shown in Scheme 4.3, Tew and Sarapas conducted the photoinitiated click polymerization of dienes and ditiols in the presence of photoinitiator Irgacure 2959, producing a series of redox-active polymers with various and dense functional groups, including carbonate and zwitterionic phosphocholine in the backbone and pendent hydroxyl groups [56]. These poly(thioether)s could be selectively oxidized to sulfoxides or sulfones by treatment with hydrogen peroxide or *m*-chloroperoxybenzoic acid (*m*CPBA), respectively. Meanwhile, they also prepared a series of lithium polymer

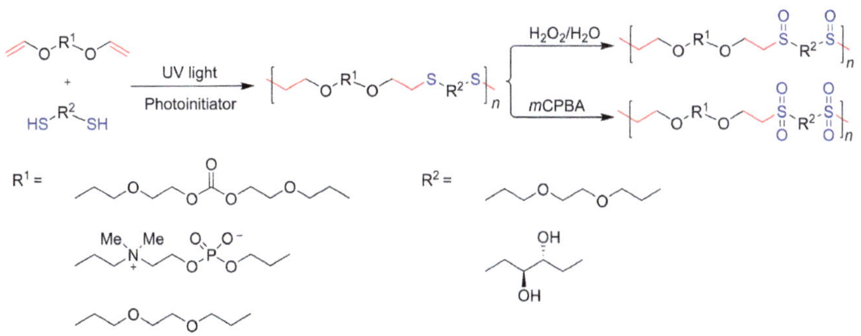

Scheme 4.3 Preparation of linear poly(thioether)s by photoinitiated thiol-ene click polymerization of dienes and dithiols.

electrolytes by photo-initiated click polymerization of dienes and ditiols [57] or bisnorbornenes and ditiols [58].

Recently, to obtain high-molecular-weight polyisobutylenes (PIBs), Kennedy and coworkers adopted low-molecular-weight PIB as macromonomer to perform the thiol-ene click polymerization [59]. As shown in Scheme 4.4, the polymerization of allyl-telechelic PIB and dithiols performed under UV radiation by using 2,2-dimethoxy-2-phenylacetophenone (DMPA) as photoinitiator or initiated thermally in the presence of 2,2′-azobis(2-methylpropionitrile) (AIBN) produced PIBs with high molecular weights.

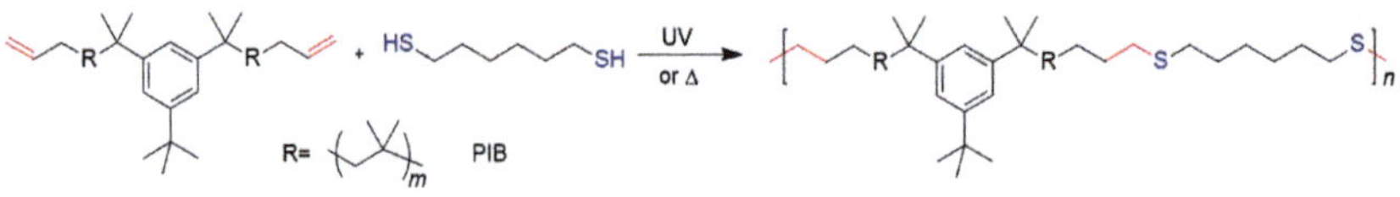

Scheme 4.4 Preparation of high-molecular-weight polyisobutylenes (PIBs) by photo-/thermoinitiated click polymerization of allyl-telechelic low-molecular-weight PIB and dithiols.

Hyperbranched poly(thioether)s have also been prepared via radical-mediated thiol-ene click polymerization. As shown in Scheme 4.5, with DMPA as photoinitiator, Feng and coworkers synthesized silicon-containing hyperbranched poly(thioether)s via an effective photoinitiated thiol-ene click polymerization of AB_n ($n \geq 2$) monomers instead of the platinum-catalyzed hydrosilylation, which is restricted by its harsh reaction conditions [60]. Unlike the former, Lin and coworkers adopted an $A_2 + B_3$ strategy to prepare silicon-containing hyperbranched poly(thioether)s [61]. As shown in Scheme 4.6, the polymerization of disiloxane-containing dienes and dithiols was carried out in THF under a UV lamp in the presence of DMPA, furnishing polysiloxane-based hyperbranched poly(thioether)s with high molecular weights (M_w up to 14 000) and good thermal stability. Through rare earth ions coordination, the obtained hyperbranched polymers could be transformed into fluorescent materials, which could be applied in cell imaging.

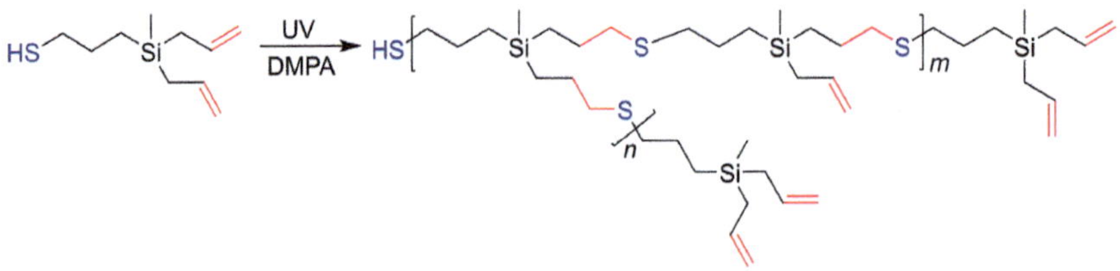

Scheme 4.5 Preparation of hyperbranched poly(thioether)s by photoinitiated thiol-ene click polymerization of AB_n ($n \geq 2$) monomers.

Using dithiol and triene ionic liquid as monomers and AIBN as initiator, Liu and coworkers successfully prepared a hyperbranched polymeric ionic liquid (*hb*-PIm^+Cl^-) with an imidazolium(Im^+)-salt backbone by thermoinitiated thiol-ene click polymerization (Scheme 4.7) [62]. Then, *hb*-$PIm^+PF_6^-$ was obtained

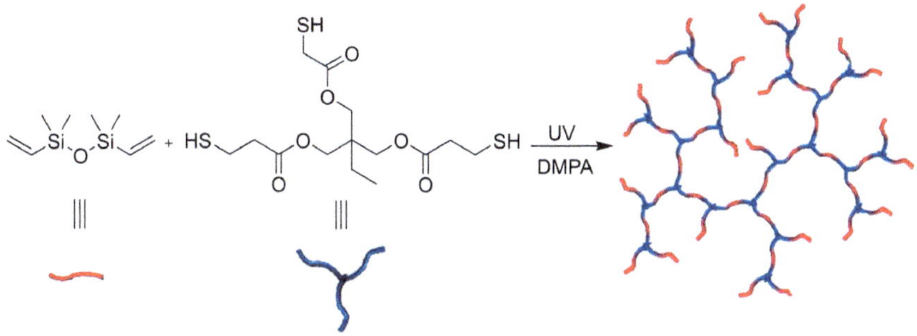

Scheme 4.6 Preparation of polysiloxane-based hyperbranched poly(thioether)s by photoinitiated click polymerization of dienes and trithiols.

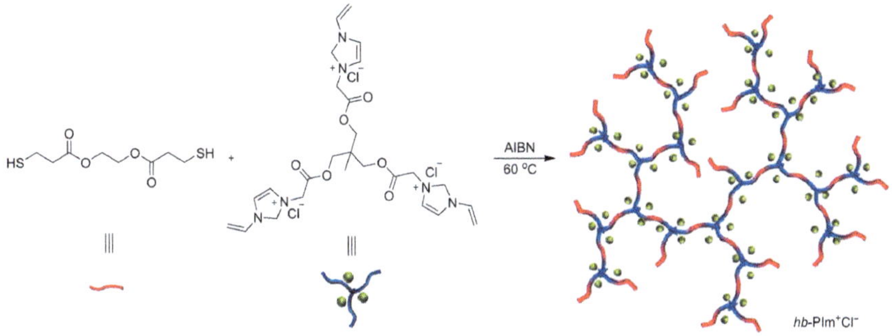

Scheme 4.7 Preparation of hyperbranched polymeric ionic liquid as anionic dye adsorbent by thermoinitiated click polymerization of dithiol and triene. Source: Modified from Ref. [62].

via the anion exchange reaction of *hb*-PIm^+Cl^- with KPF_6, which exhibits better adsorption capacity towards anionic dyes compared with cationic dyes. With Congo red as the model adsorbate, they investigated the adsorption mechanism in detail and found that the electrostatic attraction is the main driving force for the adsorption and the cavities in the hyperbranched structure provide storage sites for the interaction.

4.2.2 Thiol-Ene Michael Addition Click Polymerization

In addition to the aforementioned radical-initiated thiol-ene click reaction, Michael addition of a thiol to an electron-deficient C=C bond catalyzed by a base or a nucleophile is another important thiol-ene click reaction, including thiol-acrylate, thiol-maleimide, and thiol-vinyl sulfone systems, which has been widely applied in the area of polymeric materials [48]. For example, Dumont and Zhang prepared 5-hydroxymethylfurfural (HMF) derived linear poly(β-thioether ester) by nucleophile-catalyzed thiol-acrylate click polymerization (Scheme 4.8) [63]. The HMF derivative 2,5-furan diacylate (2,5-FDA) was polymerized with dithiols in the presence of the catalysis of dimethylphenylphosphine (DMPP) in THF at room temperature, producing poly(β-thioether ester)s with high molecular weight (M_n

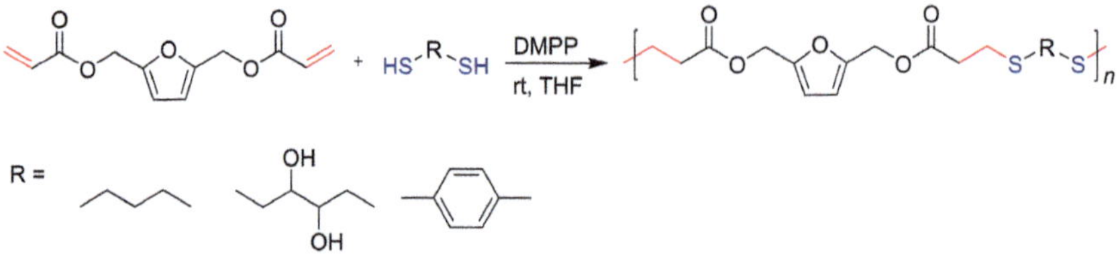

Scheme 4.8 Preparation of HMF-derived poly(β-thioether ester)s by nucleophile-catalyzed thiol-acrylate click polymerization. Source: Modified from Ref. [63].

up to 21 800). Interestingly, thanks to their contained pendant hydroxyl groups, the resultant poly(β-thioether ester)s could be applied as hot melt adhesives. Meanwhile, taking advantages of the reversibility of the Diels–Alder reaction between furan and maleimide, the furan-containing poly(β-thioether ester)s could be dynamically crosslinked with bismaleimide to produce reprocessable HMF-derived thermoset elastomers [64]. Using DMPP-catalyzed Michael addition polymerization between 2,5-FDA and dithiols, Dumont and Zhang also prepared HMF-derived poly(β-thioether ester) with thermoresponsivity [65].

Naturally derived monomers with favorable biocompatibility and renewability, such as saccharides, fatty acids, and amino acids, have captured increasing attention of polymer chemists [66]. Using dithiols and isosorbide diacrylate (iSDA) as monomers, Long and coworkers prepared sugar-based poly(β-thioether ester) via base-catalyzed thiol-acrylate click polymerization [67]. As shown in Scheme 4.9a, the polymerization carried out in $CHCl_3$ at room temperature in the presence of DBU could yield iSDA-derived homopolymers with high M_n (up to 52 300). However, these homopolymers with glass transition temperatures lower than room temperature were unsuitable for a biomedical scaffold. In order to enhance the mechanical properties of the obtained sugar-based polymers, *N,N′*-methylenebisacrylamide (MBAm) capable of engaging hydrogen bonding was applied as comonomer in the thiol-ene Michael addition click polymerization of iSDA and dithiols (Scheme 4.9b). The obtained iSDA-MBAm copolymers showed mechanical integrity at room temperature and enhanced cell attachment to film surfaces, indicating their potential application in the field of tissue engineering.

Thanks to the rigid cyclic structure of maleimide, thiol-maleimide click polymerization can be applied in synthesizing thermosetting resins with relatively high T_g [68, 69]. Inspired by the preparation of traditional thermoplastic polyurethane elastomers, Torkelson and coworkers designed and prepared segmented thermoplastic polymers by thiol-maleimide click polymerization using a soft segment + hard segment + chain extender approach [70]. As shown in Scheme 4.10, with long chain dithiols (thiol-terminated polypropylene glycol with M_w of 2500), maleimide, and small-molecule dithiols (1,2-ethanedithiol) as soft segments, hard segments, and chain extenders, respectively, the polymerization was conducted at room temperature under the catalysis of TEA, producing thermoplastic thiol-maleimide materials. When the hard segment content was 45% or 50%, the segmented thermoplastic polymers showed highly effective nanophase separation and thermal

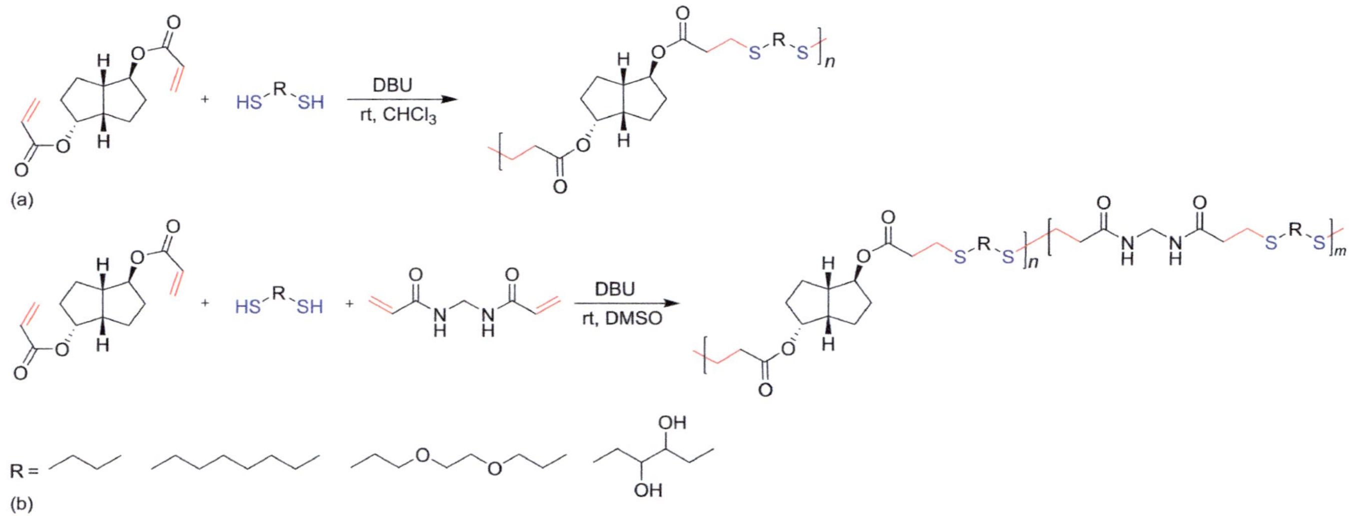

Scheme 4.9 Preparation of sugar-based homopolymers (a) and copolymers (b) by base-catalyzed thiol-acrylate click polymerization.

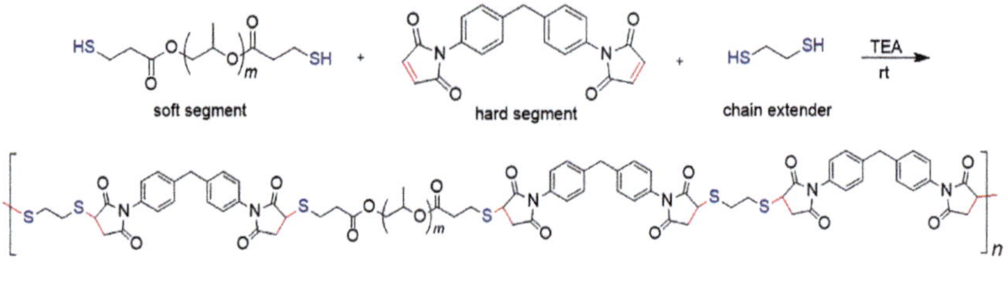

Scheme 4.10 Preparation of thiol-maleimide thermoplastic elastomer by thiol-maleimide click polymerization.

and mechanical properties associated with thermoplastic elastomer, rendering their potential as nonisocyanate-based substitutes for traditional isocyanate-based polyurethanes.

4.3 Thiol-Yne Click Polymerization

As a complement of thiol-ene click polymerization, thiol-yne click polymerization can proceed via a bis-addition process to afford polymers with high crosslink densities and high sulfur contents [71]. Meanwhile, mono-additive poly(vinyl sulfide)s with different regio-/stereoregularity can be obtained via thiol-yne click polymerization by adopting appropriate catalyst system [50]. Scheme 4.11 exhibits the mono-additive products of thiol-yne click reaction. According to the reaction mechanism, thiol-yne click polymerization can generally be sorted into three categories, including radical-initiated, base-mediated, and transition-metal catalyzed processes.

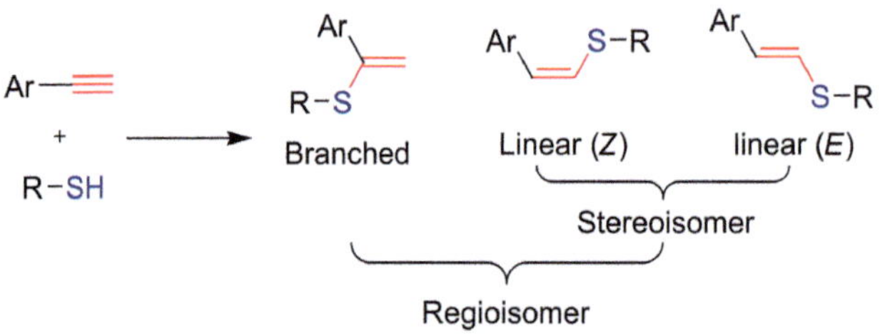

Scheme 4.11 The graphical products of the mono-addition of a thiol to an alkyne.

4.3.1 Radical-Initiated Thiol-Yne Click Polymerization

The free radical-initiated thiol-yne click polymerization, which can be triggered thermally or photochemically, has become a useful tool for the synthesis of hyperbranched polymers. Since the pioneering work of Perrier and coworkers in 2009 [72], photo-/thermo-initiated thiol-yne click polymerization of AB_2-type monomers bearing a thiol and an alkyne in which each π-bond corresponds

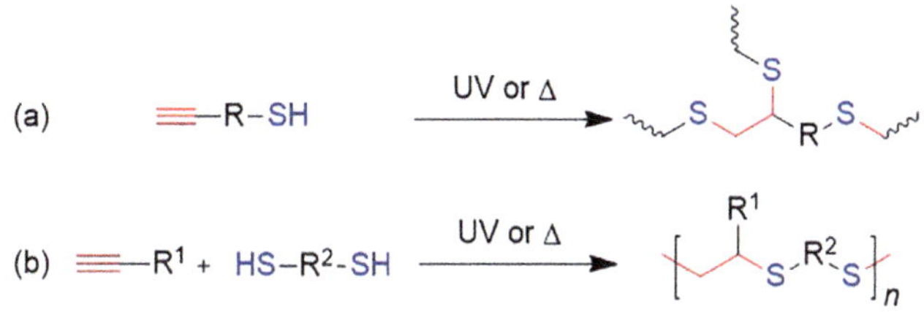

Scheme 4.12 Preparation of hyperbranched poly(thioether)s (a) and linear poly(thioether)s (b) by photo-/thermoinitiated thiol-yne click polymerization. Source: Refs. [73–77] and Refs. [80, 81] .

to a B has been systematically studied and extensively applied in the construction of functional hyperbranched poly(thioether)s with versatile properties such as temperature-responsivity, self-assembly, biodegradability, and luminescence (Scheme 4.12a) [73–77]. By slowly adding AB_2-type monomers into the multifunctional core molecules (such as B_3), hyperbranched polymers with high degree of branching and low polydispersity indices could be produced via photoinitiated thiol-yne click polymerization [78, 79]. In addition to crosslinked and hyperbranched polymers, functional linear polymers can also be produced via radical-initiated thiol-yne click polymerization of monoynes and dithiol monomers (Scheme 4.12b) [80, 81]. To avoid the tedious preparation procedure of AB_2-type monomer, Gao and coworkers developed an interesting couple-monomer method, which combined thiol-halogen or thiol-ene click reaction with radical-initiated thiol-yne click polymerization to prepare hyperbranched polymers without the risk of gelation (Scheme 4.13) [82].

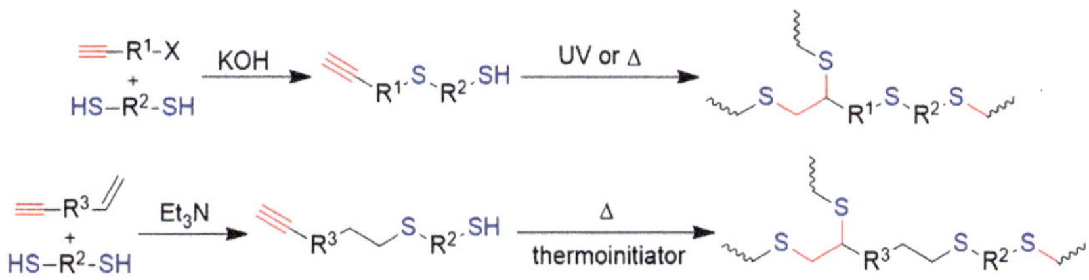

Scheme 4.13 Preparation of hyperbranched poly(thioether)s by an $A_2 + CB_2$ strategy combined sequential click chemistry and couple-monomer methodology. Source: Modified from Ref. [82].

The aforementioned radical-initiated thiol-yne click polymerizations occurred in a bis-addition manner to furnish polythioethers. By using diphenylacetylene derivatives as monomers and controlling the ratio of mercapto and ethynyl groups, Voit and coworkers successfully prepared poly(vinyl sulfide)s via thermoinitiated thiol-yne click polymerization (Scheme 4.14) [83, 84]. Using this selective radical mono-addition thiol-yne click polymerization, they prepared a series of hyperbranched poly(vinyl sulfide)s with diversiform properties such as dissolvability, thermal stability, high transparency in the visible light, high reflective indices,

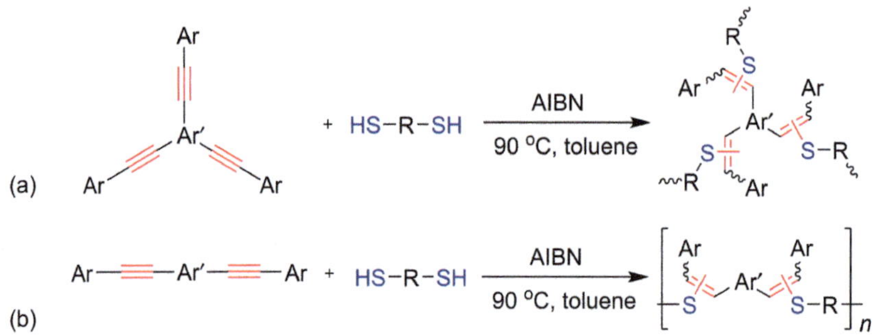

Scheme 4.14 Preparation of hyperbranched (a) and linear (b) poly(vinyl sulfide)s by thermoinitiated click polymerization of diphenylacetylene derivatives and thiols. Source: Ref. [83, 84].

and low optical dispersion, which have been applied in the construction of planar one-dimensional all-polymer photonic crystals and organic light-emitting diodes [85–88].

4.3.2 Base-Mediated Thiol-Yne Click Polymerization

Apart from free radicals, the thiol-yne click polymerization can also be initiated by enhancing the nucleophilicity of thiols under base conditions. As depicted in Scheme 4.15, Tang and coworkers developed an effective organobase-mediated thiol-yne click polymerization by using diphenylamine as the promoter [89]. The polymerization of dithiols and propiolate derivatives was carried out in the presence of diphenylamine at room temperature in *N,N*-dimethylformamide (DMF), and polymers with high-molecular-weight (M_w up to 32 300) could be produced in high yields (up to 98.2%) after 24 hours. The structural characterization indicated that the resultant polymers are anti-Markovnikov additive poly(vinylene sulfide)s with high stereoregularities (*Z*-isomer contents up to 81.4%). These poly(vinylene sulfide)s exhibited high refractive indices owing to their high sulfur contents, which

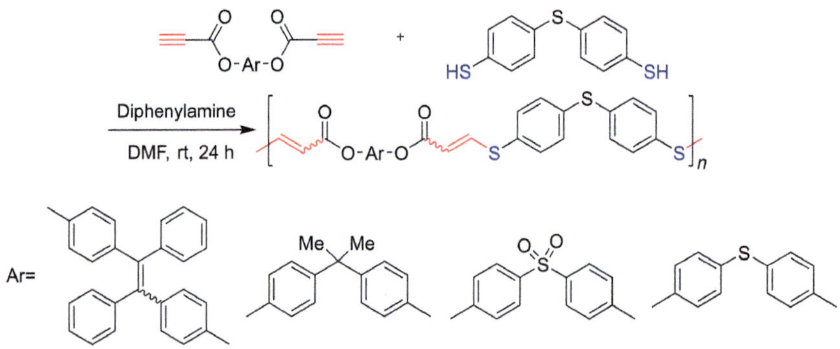

Scheme 4.15 Preparation of functional poly(vinylene sulfide)s with high *Z*-isomer contents by diphenylamine-mediated click polymerization of propiolate derivatives and dithiols.

could be further increased by coordinating with metallic species. Thanks to their photocrosslinking ability, these poly(vinyl sulfide)s could be applied in construction of fluorescent photopatterns.

Recently, Qin, Tang, and coworkers also established an inorganic base-mediated thiol-yne click polymerization for the synthesis of Z-stereoregular poly(vinylene sulfide)s [90]. As shown in Scheme 4.16, in the presence of K_3PO_4, dithiols and diynes were polymerized in N-methyl-2-pyrrolidinone (NMP) at 100 °C, solely furnishing poly(vinylene sulfide)s with high Z-isomer contents (100%) in good yields after 24 hours. Using this polymerization technique, they successfully prepared poly(vinylene sulfide)s with aggregation-induced emission (AIE) characteristics by introducing AIE-active tetraphenylethene (TPE) moieties into polymer chains. Compared with the diphenylamine-mediated one, this K_3PO_4-mediated click polymerization shows broader substrate scope and better Z-stereoselectivity.

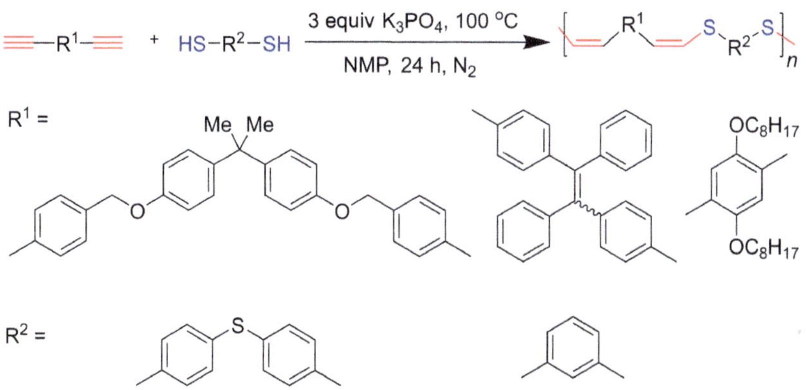

Scheme 4.16 Preparation of Z-stereoregular poly(vinylene sulfide)s (Z-isomer content of 100%) by K_3PO_4-mediated click polymerization of dithiols and diynes.

In addition to mono-additive products of poly(vinylene sulfide)s, organobase-mediated thiol-yne click polymerization has also been reported by Durmaz and coworkers for the synthesis of poly(thioether)s via a bis-addition process [91]. Using dialkyl acetylenedicarboxylates and dithiols as monomers, the click polymerization proceeded rapidly under the catalysis of 1,5,7-triazabicyclo[4.4.0]dec-5-ene (TBD) in $CHCl_3$ at room temperature, quantitatively affording poly(thioether)s with M_w as high as 50 000 in one minute (Scheme 4.17). The post-functionalization study demonstrated the versatility of the resultant poly(thioether)s.

4.3.3 Metal-Catalyzed Thiol-Yne Click Polymerization

Transition-metal catalyzed thiol-yne click reaction with regio- and stereo-selectivity has also been extensively investigated and developed for the selective synthesis of vinyl sulfides [92, 93]. Thereinto, Rh-catalyzed thiol-yne click reaction with tunable selectivity is more attractive for polymer chemists [94–96]. For example, Tang and coworkers developed a regiospecific and stereoselective thiol-yne click

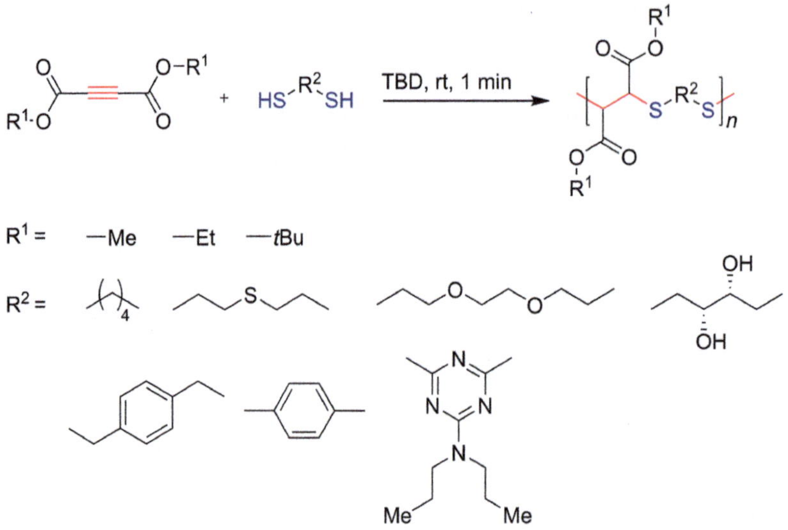

Scheme 4.17 Preparation of poly(thioether)s by TBD-catalyzed click polymerization of dialkyl acetylenedicarboxylates and dithiols.

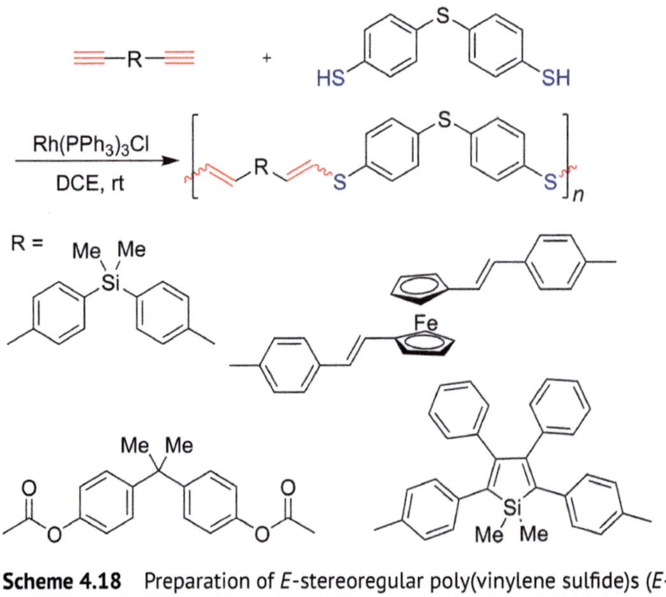

Scheme 4.18 Preparation of *E*-stereoregular poly(vinylene sulfide)s (*E*-isomer contents up to 100%) by $Rh(PPh_3)_3Cl$-catalyzed click polymerization of dithiols and diynes.

polymerization by adopting $Rh(PPh_3)_3Cl$ as catalyst [97]. The polymerization of diynes and dithiols occurred at room temperature in 1,2-dichloroethane (DCE), generating poly(vinylene sulfide)s with high M_w (up to 31 500) and high stereoregularity (*E*-isomer contents up to 100%) (Scheme 4.18). Notably, the stereostructures of the products could be modulated by adjusting the monomer addition sequence during polymerization and UV irradiation after polymerization. Owing to the great functionality tolerance of this polymerization, functional groups like ferrocene and silole could be incorporated into polymer chains to endow them with various properties such as ceramization capability and aggregation-enhanced emission features, respectively.

Adopting Rh(I)-NHC instead of $Rh(PPh_3)_3Cl$ as catalyst, Di Giuseppe, Pérez-Torrente, and coworkers developed a new Rh-catalyzed thiol-yne click polymerization with inverse regioselectivity [98]. As shown in Scheme 4.19, aromatics diynes and aliphatic dithiols were polymerized in THF at room temperature under the catalysis of RhCl(IPr)(pyridine)(η^2-coe) (IPr = 1,3-bis-(2,6-diisopropylphenyl) imidazol-2-ylidene), producing Markovnikov additive poly(vinylidene sulfide)s with high molecular weights (M_w as high as 199 000) and vinylidene content of 80%. These poly(vinylidene sulfide)s could be further chemically modified via selective hydrogenation and oxidation.

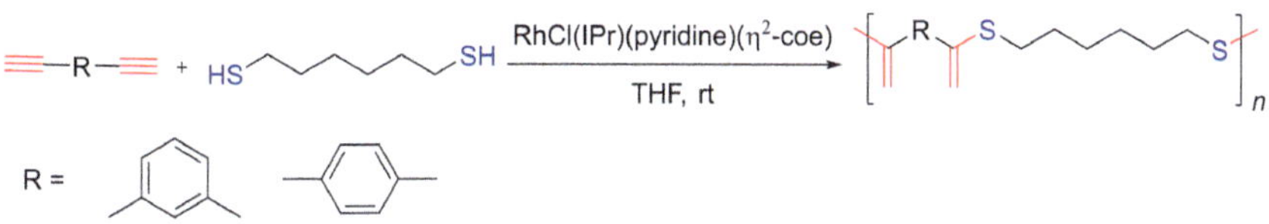

Scheme 4.19 Preparation of poly(vinylidene sulfide)s (vinylidene content of 80%) by Rh(I)-NHC-catalyzed click polymerization of dithiols and diynes.

4.3.4 Spontaneous Thiol-Yne Click Polymerization

Without the assistance of external stimuli (such as heat, light, and microwave) and additional catalyst, a spontaneous thiol-yne click polymerization has successfully established by Qin, Tang, and coworkers in 2014 [99]. As depicted in Scheme 4.20, mixing dithiols and aromatic diynes in THF and stirring at 30 °C for two hours readily produced poly(vinylene sulfide)s with M_w as high as 85 200 in excellent yields (up to 97%). The gained polymers exhibited high light refractivity owing to the high content of sulfur atoms and polarizable aromatic rings. Moreover, taking an $A_2 + B_n$ (n = 3 or 4) strategy, they also prepared soluble multifunctional hyperbranched poly(vinylene sulfide)s via this efficient thiol-yne click polymerization [100]. The obtained hyperbranched polymers contained plentiful unreacted ethynyl groups on their peripheries could be further modified by a coherent thiol-yne click reaction.

4.4 Other Thiol-Based Click Polymerizations

Taking advantage of the high reactivity of thiols, besides the aforementioned thiol-ene/yne click polymerizations, other thiol-based click polymerizations, such as thiol-epoxy (nucleophilic ring opening), thiol-isocyanate (carbonyl addition), and thiol-bromo (nucleophilic substitution) click polymerizations, have also emerged

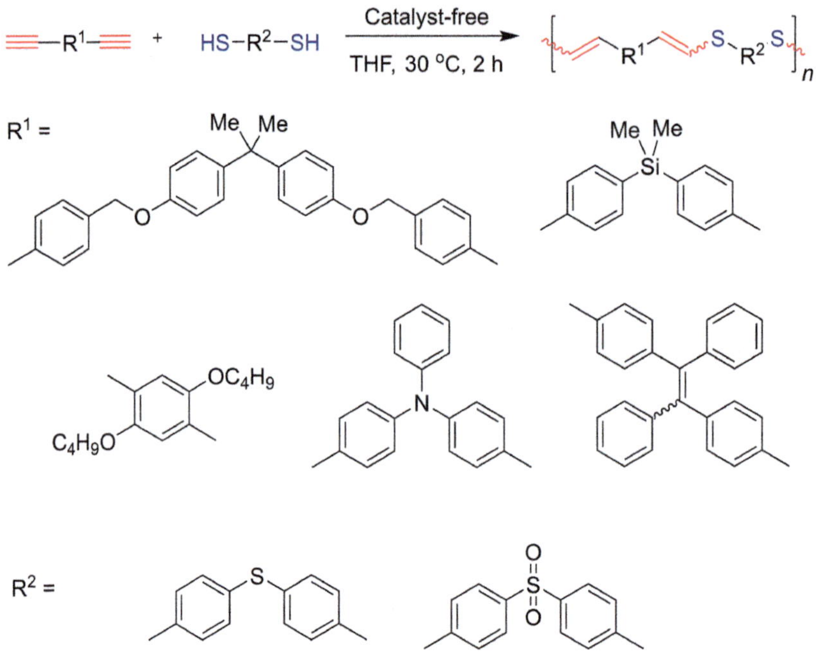

Scheme 4.20 Preparation of poly(vinylene sulfide)s by spontaneous thiol-yne click polymerization of dithiols and aromatic diynes.

and found applications in the preparation of functional sulfur-containing polymers with manifold topological structures [52].

4.4.1 Thiol-Epoxy Click Polymerization

Thiol-epoxy click reaction with very high efficiency has shown clear importance in traditional synthetic area and biological and pharmaceutical applications, which has also been applied in the industrial formation of adhesives and high-performance coatings [52]. Another attractive merit of this click reaction is the formation of a reactive hydroxyl center upon coupling reaction, revealing its great potential in polymer synthesis. As shown in Scheme 4.21, Khan and Braendle reported a thiol-epoxy click polymerization for the construction of reactive and functional polymers [101]. The polymerization of 1,4-butanediol diglycidyl ether and 1,2-ethanedithiol propagated smoothly in THF at 0 °C in the presence of lithium hydroxide monohydrate (LiOH). Linear poly(β-hydroxythio-ether)s with M_w of 10 000 and plentiful reactive hydroxyl groups were obtained, which could be post-functionalized via esterification to produce functional polymers with substituent dependent properties. Using AB-type monomer, linear poly(β-hydroxythio-ether)s with higher molecular weights (M_w of 47 700) were produced by LiOH-catalyzed thiol-epoxy click polymerization in bulk at room temperature (Scheme 4.22). Fluorescent polymer and cationic polymer could be gained via post-polymerization modification [102]. Khan and coworkers also reported that the thiol-epoxy click polymerization in

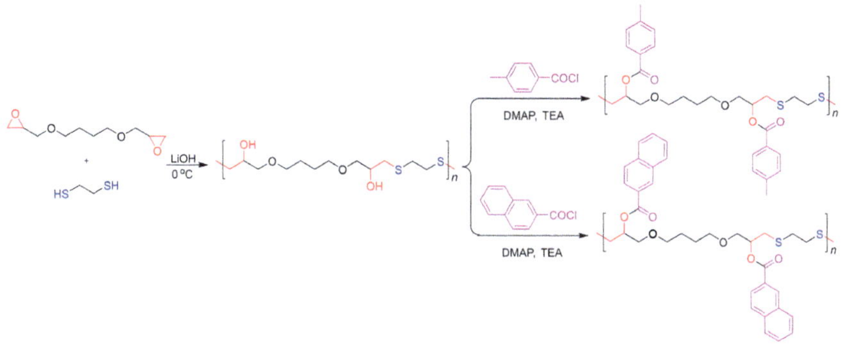

Scheme 4.21 Preparation and modification of linear poly(β-hydroxythio-ether)s by base-catalyzed click polymerization of bisepoxides and dithiols and esterification. DMAP, 4-dimethylaminopyridine. Source: Modified from Ref. [101].

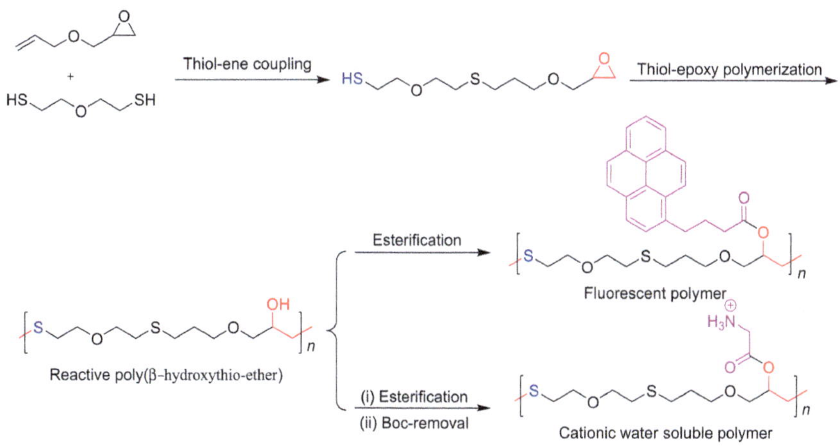

Scheme 4.22 Preparation and modification of linear poly(β-hydroxythio-ether)s by base-catalyzed thiol-epoxy click polymerization of AB-type monomer and esterification.

THF/water mixture could be triggered by a moderate catalyst such as TEA owing to the high polarity of the solvent system [103].

Through an AB_2 strategy, Khan and coworkers also prepared hyperbranched poly(β-hydroxythio-ether)s [104]. As depicted in Scheme 4.23, AB_2-type monomer bearing a thiol group and two epoxide groups was polymerized at room temperature under base conditions, furnishing hyperbranched poly(β-hydroxythio-ether)s with high degree of branching (up to 0.69). Thanks to the wide distribution of the unreacted epoxide groups and the formed hydroxyl groups during the polymerization and the first functionalization in the hyperbranched structures, these polymers could be modified via sequential thiol-epoxy click reaction and esterification to produce dual-functionalized hyperbranched polymers with potential in gene delivery applications.

Amphiphilic alternating copolymers have been successfully prepared via thiol-epoxy click copolymerization by Zhou, He, and coworkers [105]. As shown in

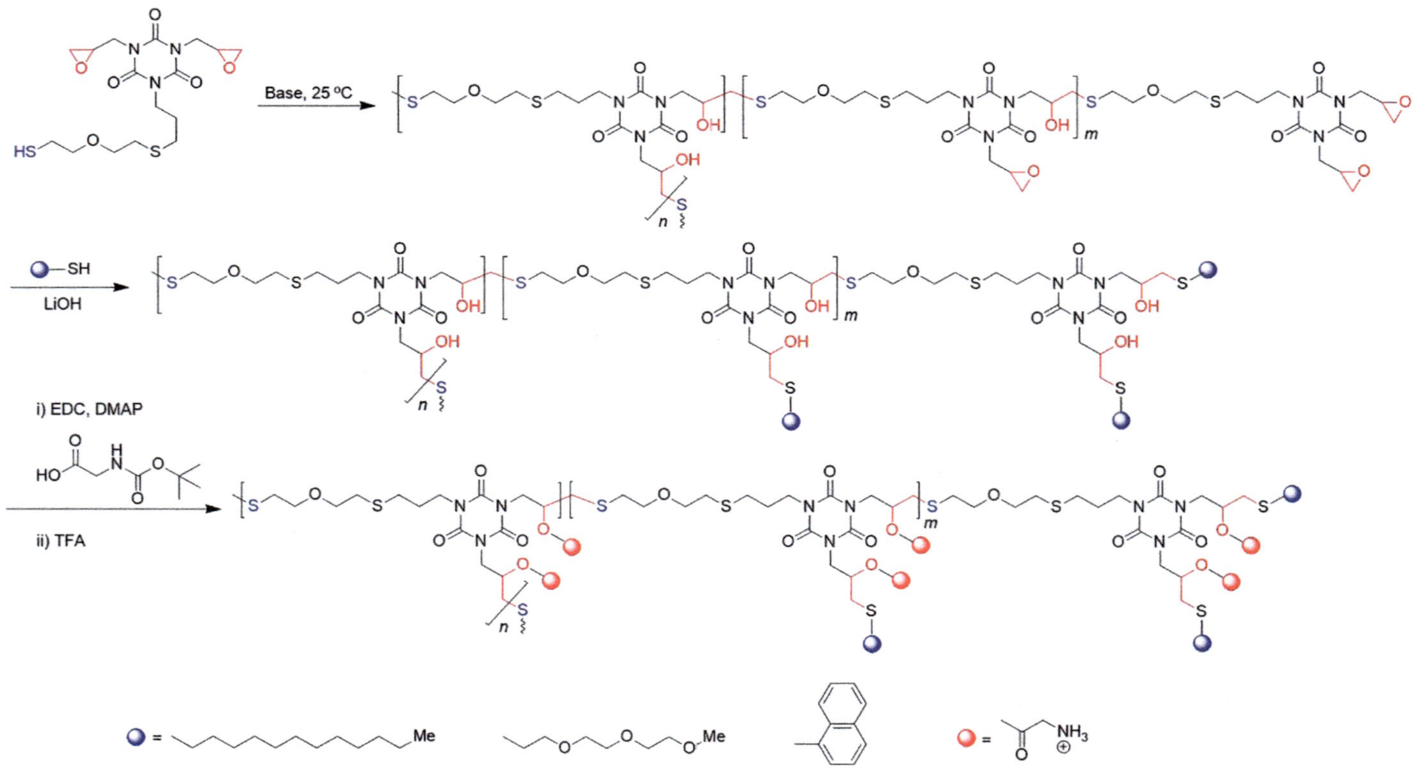

Scheme 4.23 Preparation of hyperbranched poly(β-hydroxythio-ether)s by base-catalyzed thiol-epoxy click polymerization of AB_2-type monomer and sequential modification by thiol-epoxy click reaction and esterification.

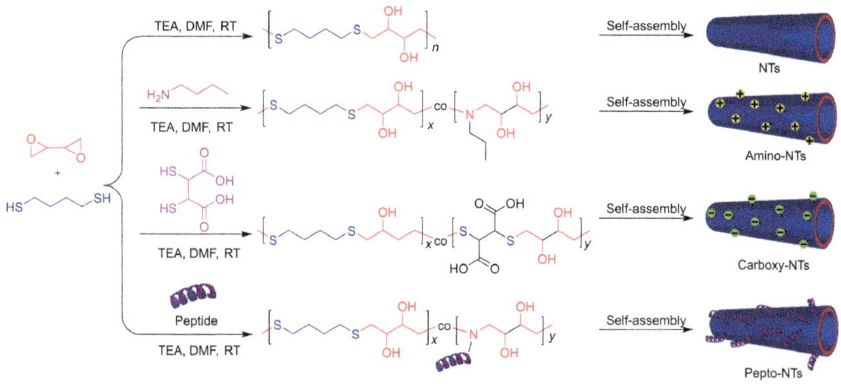

Scheme 4.24 Preparation and self-assembly of functional amphiphilic alternating copolymers by base-catalyzed thiol-epoxy click copolymerization.

Scheme 4.24, poly(β-hydroxythio-ether)s were firstly synthesized by TEA-catalyzed thiol-epoxy click polymerization in DMF at room temperature, which could self-assemble into well-defined ultrathin nanotubes (NTs). In particular, functional NTs bearing carboxy groups, amino groups, or peptides could be achieved by simple modular click polymerization with functionalized dithiols or amines. Experiments and dissipative particle dynamics simulations were applied in investigating the self-assembly mechanism, revealing a sequential self-assembly pathway including formation of small sheets, fusion of small sheets into large ribbons, rolling of ribbons, and closing of the rolled ribbons to form NTs. Functional monolayer vesicles could also be prepared via similar strategy [106].

4.4.2 Thiol-Isocyanate Click Polymerization

Thiol-isocyanate click polymerization, which can readily and efficiently furnish poly(thiourethane)s in high yields, has drawn increasing attention in the past decade and found extensively applications in the areas of self-healing materials, dental materials, shape memory materials, and high refractive index materials [107–111]. For example, Voit and coworkers successfully prepared tough thiourethane thermoplastics for fused filament fabrication via base-catalyzed thiol-isocyanate click polymerization [112]. As shown in Scheme 4.25, the polymerization of diisocyanate and dithiol propagated smoothly in DMF at room temperature in the presence of TEA for 30 minutes, affording poly(thiourethane)s with M_w of 17 200 in excellent yield (99.97%). Taking advantage of its high toughness and low melting point, the obtained poly(thiourethane)s could be applied as three-dimensional (3D) printing

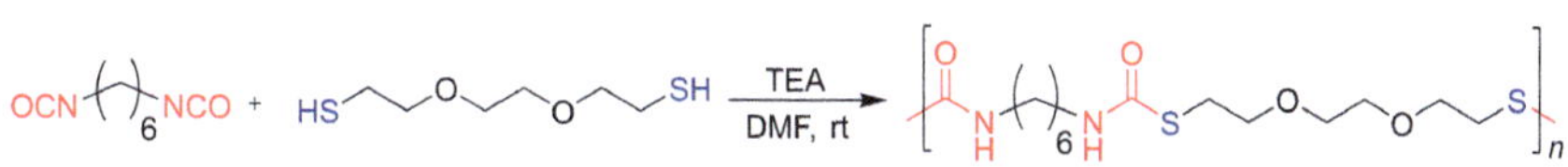

Scheme 4.25 Preparation of linear poly(thiourethane)s by base-catalyzed thiol-isocyanate click polymerization.

filament, exhibiting better tensile property than that of commercial filaments when printed.

4.4.3 Thiol-Halogen Click Polymerization

Owing to its high efficiency, simple isolation process as well as great orthogonality, thiol-halogen nucleophilic substitution click reactions such as thiol-bromo and thiol-*para*-fluorine click reactions have been widely applied in polymer chemistry, including monomers synthesis, construction of dendrimers, selective functionalization of preformed polymers, etc. [113–118]. Click polymerization derived from thiol-halogen click reaction has also been reported for the preparation of poly(thioether)s. For example, Tang and coworkers synthesized multifunctional polymers via base-catalyzed thiol-bromo click polymerization [119]. As shown in Scheme 4.26, TPE-bearing bis(alkyl bromide)s and dithiol were polymerized in a THF/TEA mixture (4 : 1, v/v) at room temperature for 24 hours, producing polymers with high molecular weights (M_w up to 36 300). The resultant polymers contained abundant sulfur atoms and aromatic rings in the structures exhibited high refractive indices. Moreover, these TPE-containing polymers integrally inherited the AIE features of TPE and could be used as highly sensitive explosive probe. Recently, thiol-bromo click polymerization has also been combined with copper-mediated reversible deactivation radical polymerization for the construction of multi-block copolymers in one-pot [120].

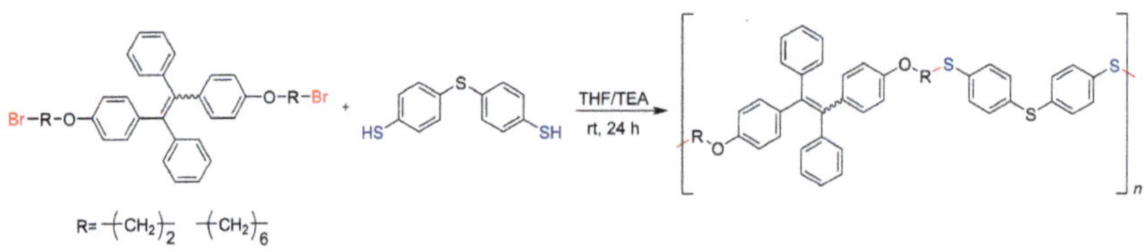

Scheme 4.26 Preparation of multifunctional poly(thioether)s by base-catalyzed thiol-bromo click polymerization.

Generally, elevated temperature and long reaction time are necessary for the assembly of metallopolymers. However, Schubert and coworkers successfully prepared metallopolymers by thiol-*para*-fluorine click polymerization in a fast and effective manner [121]. As depicted in Scheme 4.27, the polymerization of perfluorophenyl-terpyridine ruthenium complex and acetate-protected thiol was carried out in DMF at 0 °C under the catalysis of DBU, producing ruthenium(II) bisterpyridine metallopolymers with M_n as high as 89 000 in a good yield of 88%.

4.5 Conclusion

This chapter summarizes the recent advances of thiol-based click polymerizations for the preparation of sulfur-containing polymers with various structures and

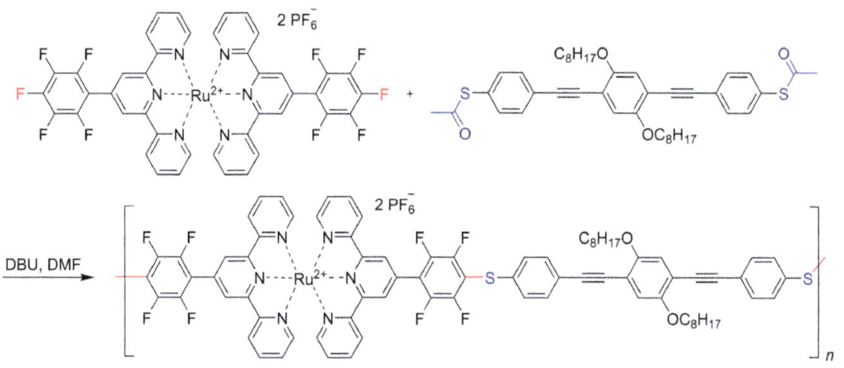

Scheme 4.27 Preparation of metallopolymer by base-catalyzed thiol-*para*-fluorine click polymerization.

diverse functional properties. Among various thiol-based click polymerizations, thiol-ene/yne click polymerization is the most prominent one, which has made a remarkable progress in the synthesis of sulfur-containing polymers. Poly(thioether)s can be offered by thiol-ene click polymerization through a free radical process or a Michael addition process. As an updated version of the former, thiol-yne click polymerization can proceed in a bis-addition or mono-addition manner to furnish poly(thioether)s or poly(vinyl sulfide)s with diverse structures. Poly(thioether)s can be obtained by the traditional radical-initiated thiol-yne click polymerization and an extremely rapid TBD-mediated click polymerization of dithiols and dialkyl acetylenedicarboxylates. The radical-initiated thiol-yne click polymerization based on diphenylacetylene derivatives occurs via a mono-addition process to afford poly(vinyl sulfide)s. Diphenylamine-/K_3PO_4-mediated and $Rh(PPh_3)_3Cl$-catalyzed thiol-yne click polymerizations with regio- and stereo-selectivity can produce Z-stereoregular and E-stereoregular anti-Markovnikov additive poly(vinylene sulfide)s, respectively. Poly(vinylene suifide)s can also be readily obtained by spontaneous thiol-yne click polymerization. Meanwhile, Markovnikov additive poly(vinylidene sulfide)s can be prepared by Rh(I)-NHC-catalyzed thiol-yne click polymerization. Besides the famous thiol-ene/yne click polymerization, some other thiol-based click polymerizations including thiol-epoxy, thiol-isocyanate, and thiol-halogen click polymerizations have also been developed for the preparation of poly(β-hydroxythio-ether)s, poly(thiourethane)s, and poly(thioether)s. Despite the great achievements of thiol-based click polymerization in sulfur-containing polymer synthesis, there are still many challenges in the field, suggesting plentiful opportunities. For example, controlled thiol-based click polymerization needs to be exploited for the preparation of sulfur-containing polymers with well-defined structures, controlled molecular weights and polydispersities. The construction of conjugated sulfur-containing polymer can be an interesting and attractive subject in the field of thiol-based click polymerization.

Acknowledgments

This work was financially supported by the National Natural Science Foundation of China (21788102, 21525417 and 21490571), the Natural Science Foundation of Guangdong Province (2019B030301003, 2019A1515110034 and 2016A030312002), and the Innovation and Technology Commission (ITC-CNERC14S01). Huang D. thanks the support from the China Postodoctoral Science Foundation (2019M662888).

References

1 Mutlu, H., Ceper, E.B., Li, X. et al. (2019). *Macromol. Rapid Commun.* 40: 1800650.
2 Zhao, F.L., Li, Y., and Feng, W. (2018). *Small Methods* 2: 1800156.
3 Kausar, A., Zulfiqar, S., and Sarwar, M.I. (2014). *Polym. Rev.* 54: 185–267.
4 Vo, C.D., Kilcher, G., and Tirelli, N. (2009). *Macromol. Rapid Commun.* 30: 299–315.
5 Jha, G.S., Seshadri, G., Mohan, A., and Khandal, R.K. (2008). *E-Polymers* https://doi.org/10.1515/epoly.2008.8.1.376.
6 Kricheldorf, H.R. and Schwarz, G. (2007). *J. Macromol. Sci. Part A: Pure Appl. Chem.* 44: 625–649.
7 Griebel, J.J., Glass, R.S., Char, K., and Pyun, J. (2016). *Prog. Polym. Sci.* 58: 90–125.
8 Kleine, T.S., Glass, R.S., Lichtenberger, D.L. et al. (2020). *ACS Macro Lett.* 9: 245–259.
9 Lim, J., Pyun, J., and Char, K. (2015). *Angew. Chem. Int. Ed.* 54: 3249–3258.
10 Abbasi, A., Nasef, M.M., and Yahya, W.Z.N. (2019). *Sustain. Chem. Pharm.* 13: 100158.
11 Ochiai, B. and Endo, T. (2005). *Prog. Polym. Sci.* 30: 183–215.
12 Zhang, C.J. and Zhang, X.H. (2019). *Chin. J. Polym. Sci.* 37: 951–958.
13 Chung, W.J., Griebel, J.J., Kim, E.T. et al. (2013). *Nat. Chem.* 5: 518–524.
14 Griebel, J.J., Nguyen, N.A., Astashkin, A.V. et al. (2014). *ACS Macro Lett.* 3: 1258–1261.
15 Griebel, J.J., Nguyen, N.A., Namnabat, S. et al. (2015). *ACS Macro Lett.* 4: 862–866.
16 Kleine, T.S., Nguyen, N.A., Anderson, L.E. et al. (2016). *ACS Macro Lett.* 5: 1152–1156.
17 Li, W., Wu, X., Zhao, Z. et al. (2015). *Macromolecules* 48: 7747–7754.
18 Tian, T., Hu, R., and Tang, B.Z. (2018). *J. Am. Chem. Soc.* 140: 6156–6163.
19 Cao, W., Dai, F., Hu, R., and Tang, B.Z. (2020). *J. Am. Chem. Soc.* 142: 978–986.
20 Luo, M., Zhang, X.H., and Darensbourg, D.J. (2015). *Polym. Chem.* 6: 6955–6958.
21 Yue, T.J., Ren, W.M., Liu, Y. et al. (2016). *Macromolecules* 49: 2971–2976.
22 Lopez-Vidal, E.M., Gregory, G.L., Kociok-Kohn, G., and Buchard, A. (2018). *Polym. Chem.* 9: 1577–1582.

23 Wu, S., Luo, M., Darensbourg, D.J., and Zuo, X. (2019). *Macromolecules* 52: 8596–8603.

24 Luo, M., Zhang, X.H., Du, B.Y. et al. (2013). *Macromolecules* 46: 5899–5904.

25 Luo, M., Zhang, X.H., and Darensbourg, D.J. (2016). *Acc. Chem. Res.* 49: 2209–2219.

26 Zhang, C.J., Wu, H.L., Li, Y. et al. (2018). *Nat. Commun.* 9: 2137.

27 Kolb, H.C., Finn, M.G., and Sharpless, K.B. (2001). *Angew. Chem. Int. Ed.* 40: 2004–2021.

28 Tiwari, V.K., Mishra, B.B., Mishra, K.B. et al. (2016). *Chem. Rev.* 116: 3086–3240.

29 Meyer, J.P., Adumeau, P., Lewis, J.S., and Zeglis, B.M. (2016). *Bioconjug. Chem.* 27: 2791–2807.

30 Holub, J.M. and Kirshenbaum, K. (2010). *Chem. Soc. Rev.* 39: 1325–1337.

31 Franc, G. and Kakkar, A.K. (2010). *Chem. Soc. Rev.* 39: 1536–1544.

32 Amblard, F., Cho, J.H., and Schinazi, R.F. (2009). *Chem. Rev.* 109: 4207–4220.

33 Parker, C.G. and Pratt, M.R. (2020). *Cell* 180: 605–632.

34 Takayama, Y., Kusamori, K., and Nishikawa, M. (2019). *Molecules* 24: 172.

35 Li, P.Z., Wang, X.J., and Zhao, Y. (2019). *Coord. Chem. Rev.* 380: 484–518.

36 Dai, Y., Chen, X., and Zhang, X. (2019). *Polym. Chem.* 10: 34–44.

37 Arslan, M., Acik, G., and Tasdelen, M.A. (2019). *Polym. Chem.* 10: 3806–3821.

38 Xu, Z. and Bratlie, K.M. (2018). *ACS Biomater. Sci. Eng* 4: 2276–2291.

39 Caneque, T., Mueller, S., and Rodriguez, R. (2018). *Nat. Rev. Chem.* 2: 202–215.

40 Yoon, H.Y., Koo, H., Kim, K., and Kwon, I.C. (2017). *Biomaterials* 132: 28–36.

41 Meghani, N.M., Amin, H.H., and Leel, B.J. (2017). *Drug Discov. Today* 22: 1604–1619.

42 Li, B., Huang, D., Qin, A., and Tang, B.Z. (2018). *Macromol. Rapid Commun.* 39: 1800098.

43 Huang, D., Liu, Y., Qin, A., and Tang, B.Z. (2018). *Polym. Chem.* 9: 2853–2867.

44 Shi, Y., Cao, X., and Gao, H. (2016). *Nanoscale* 8: 4864–4881.

45 Zheng, Y., Li, S., Weng, Z., and Gao, C. (2015). *Chem. Soc. Rev.* 44: 4091–4130.

46 Qin, A.J., Liu, Y., and Tang, B.Z. (2015). *Macromol. Chem. Phys.* 216: 818–828.

47 Qin, A., Lam, J.W.Y., and Tang, B.Z. (2010). *Chem. Soc. Rev.* 39: 2522–2544.

48 Nair, D.P., Podgorski, M., Chatani, S. et al. (2014). *Chem. Mater.* 26: 724–744.

49 Lowe, A.B. (2014). *Polym. Chem.* 5: 4820–4870.

50 Yao, B., Sun, J., Qin, A., and Tang Ben, Z. (2013). *Chin. Sci. Bull.* 58: 2711–2718.

51 Lowe, A.B. (2010). *Polym. Chem.* 1: 17–36.

52 Hoyle, C.E., Lowe, A.B., and Bowman, C.N. (2010). *Chem. Soc. Rev.* 39: 1355–1387.

53 Hoyle, C.E. and Bowman, C.N. (2010). *Angew. Chem. Int. Ed.* 49: 1540–1573.

54 Yao, Z., Dai, B., Yu, Y., and Cao, K. (2016). *Prog. Chem.* 28: 1062–1069.

55 van den Berg, O., Dispinar, T., Hommez, B., and Du Prez, F.E. (2013). *Eur. Polym. J.* 49: 804–812.

56 Sarapas, J.M. and Tew, G.N. (2016). *Angew. Chem. Int. Ed.* 55: 15860–15863.

57 Sarapas, J.M. and Tew, G.N. (2016). *Macromolecules* 49: 1154–1162.

58 Sarapas, J.M., Saijo, K., Zhao, Y. et al. (2016). *Polym. Adv. Technol.* 27: 946–954.

59 Kurnaz, E., Helvacioglu, E., Kekec, N.C. et al. (2019). *J. Polym. Sci. Part A: Polym. Chem.* 57: 1197–1208.

60 Xue, L., Yang, Z., Wang, D. et al. (2013). *J. Organomet. Chem.* 732: 1–7.

61 Zhang, Y., Zou, Y., Yang, T. et al. (2019). *Eur. Polym. J.* 112: 515–523.

62 Song, W., Liu, Y., Qian, L. et al. (2016). *Chem. Eng. J.* 287: 482–491.

63 Zhang, D. and Dumont, M.J. (2018). *Polym. Chem.* 9: 743–756.

64 Zhang, D. and Dumont, M.J. (2018). *J. Mater. Sci.* 53: 11116–11129.

65 Zhang, D. and Dumont, M.J. (2018). *Polymer* 145: 157–161.

66 Mülhaupt, R. (2013). *Macromol. Chem. Phys.* 214: 159–174.

67 Moon, N.G., Mazzini, F., Pekkanen, A.M. et al. (2018). *Macromol. Chem. Phys.* 219: 1800177.

68 Parker, S., Reit, R., Abitz, H. et al. (2016). *Macromol. Rapid Commun.* 37: 1027–1032.

69 Shibata, M. and Hashimoto, Y. (2017). *Eur. Polym. J.* 93: 561–571.

70 Jin, K., Leitsch, E.K., Chen, X. et al. (2018). *Macromolecules* 51: 3620–3631.

71 Fairbanks, B.D., Scott, T.F., Kloxin, C.J. et al. (2009). *Macromolecules* 42: 211–217.

72 Konkolewicz, D., Gray-Weale, A., and Perrier, S. (2009). *J. Am. Chem. Soc.* 131: 18075–18077.

73 Konkolewicz, D., Gaillard, S., West, A.G. et al. (2011). *Organometallics* 30: 1315–1318.

74 Barbey, R. and Perrier, S. (2013). *ACS Macro Lett.* 2: 366–370.

75 Konkolewicz, D., Poon, C.K., Gray-Weale, A., and Perrier, S. (2011). *Chem. Commun.* 47: 239–241.

76 Liu, W. and Dong, C.M. (2010). *Macromolecules* 43: 8447–8455.

77 Cai, T., Li, M., Zhang, B. et al. (2014). *J. Mater. Chem. B* 2: 814–825.

78 Cook, A.B., Barbey, R., Burns, J.A., and Perrier, S. (2016). *Macromolecules* 49: 1296–1304.

79 Hartlieb, M., Floyd, T., Cook, A.B. et al. (2017). *Polym. Chem.* 8: 2041–2054.

80 Han, J., Zheng, Y., Zhao, B. et al. (2014). *Sci. Rep.* 4: 4387.

81 Zheng, Y., Zhang, Q., Gao, Z. et al. (2018). *Macromol. Chem. Phys.* 219: 1800144.

82 Han, J., Zhao, B., Gao, Y. et al. (2011). *Polym. Chem.* 2: 2175–2178.

83 Potzsch, R., Stahl, B.C., Komber, H. et al. (2014). *Polym. Chem.* 5: 2911–2921.

84 Poetzsch, R., Komber, H., Stahl, B.C. et al. (2013). *Macromol. Rapid Commun.* 34: 1772–1778.

85 Wei, Q., Zan, X., Qiu, X. et al. (2016). *Macromol. Chem. Phys.* 217: 1977–1984.

86 Wei, Q., Pötzsch, R., Liu, X. et al. (2016). *Adv. Funct. Mater.* 26: 2545–2553.

87 Gazzo, S., Manfredi, G., Pötzsch, R. et al. (2016). *J. Polym. Sci. B Polym. Phys.* 54: 73–80.

88 Wei, Q., Poetzsch, R., Komber, H. et al. (2014). *Polymer* 55: 5600–5607.

89 Jim, C.K.W., Qin, A., Lam, J.W.Y. et al. (2010). *Adv. Funct. Mater.* 20: 1319–1328.

90 Huang, D., Liu, Y., Guo, S. et al. (2019). *Polym. Chem.* 10: 3088–3096.

91 Daglar, O., Gunay, U.S., Hizal, G. et al. (2019). *Macromolecules* 52: 3558–3572.

92 Kuniyasu, H., Ogawa, A., Sato, K. et al. (1992). *J. Am. Chem. Soc.* 114: 5902–5903.

93 Castarlenas, R., Di Giuseppe, A., Pérez-Torrente, J.J., and Oro, L.A. (2013). *Angew. Chem. Int. Ed.* 52: 211–222.

94 Ogawa, A., Ikeda, T., Kimura, K., and Hirao, T. (1999). *J. Am. Chem. Soc.* 121: 5108–5114.

95 Di Giuseppe, A., Castarlenas, R., Pérez-Torrente, J.J. et al. (2012). *J. Am. Chem. Soc.* 134: 8171–8183.

96 Yang, Y. and Rioux, R.M. (2011). *Chem. Commun.* 47: 6557–6559.

97 Liu, J., Lam, J.W.Y., Jim, C.K.W. et al. (2011). *Macromolecules* 44: 68–79.

98 Funes-Hernando, D., Hermosilla, P., Vispe, E. et al. (2018). *Polym. Chem.* 9: 1298–1302.

99 Yao, B., Mei, J., Li, J. et al. (2014). *Macromolecules* 47: 1325–1333.

100 Yao, B., Hu, T., Zhang, H. et al. (2015). *Macromolecules* 48: 7782–7791.

101 Braendle, A. and Khan, A. (2012). *Polym. Chem.* 3: 3224–3227.

102 Binder, S., Gadwal, I., Bielmann, A., and Khan, A. (2014). *J. Polym. Sci. Part A: Polym. Chem.* 52: 2040–2046.

103 JiHyeon, H., Youngson, C., Bang, J., and Khan, A. (2017). *J. Polym. Sci. Part A: Polym. Chem.* 55: 3381–3386.

104 Gadwal, I., Binder, S., Stuparu, M.C., and Khan, A. (2014). *Macromolecules* 47: 5070–5080.

105 Chen, J., Yu, C., Shi, Z. et al. (2015). *Angew. Chem. Int. Ed.* 54: 3621–3625.

106 Li, C., Chen, C., Li, S. et al. (2017). *Polym. Chem.* 8: 4688–4695.

107 Hillewaere, X.K.D., Teixeira, R.F.A., Nguyen, L.T.T. et al. (2014). *Adv. Funct. Mater.* 24: 5575–5583.

108 Reinelt, S., Tabatabai, M., Moszner, N. et al. (2014). *Macromol. Chem. Phys.* 215: 1415–1425.

109 Podgorski, M., Wang, C., and Bowman, C.N. (2015). *Soft Matter* 11: 6852–6858.

110 McBride, M.K., Podgorski, M., Chatani, S. et al. (2018). *ACS Appl. Mater. Interfaces* 10: 22739–22745.

111 Jia, Y., Shi, B., Jin, J., and Li, J. (2019). *Polymer* 180: 121746.

112 Ellson, G., Carrier, X., Walton, J. et al. (2018). *J. Appl. Polym. Sci.* 135: 45574.

113 Chen, S., Schulz, M., Lechner, B.-D. et al. (2015). *Polym. Chem.* 6: 7988–7994.

114 Xu, J., Tao, L., Boyer, C. et al. (2010). *Macromolecules* 43: 20–24.

115 Rosen, B.M., Lligadas, G., Hahn, C., and Percec, V. (2009). *J. Polym. Sci. Part A: Polym. Chem.* 47: 3931–3939.

116 Alvarez-Paino, M., Marcelo, G., Munoz-Bonilla, A. et al. (2013). *Polym. Chem.* 4: 986–995.

117 Babiuch, K., Pretzel, D., Tolstik, T. et al. (2012). *Macromol. Biosci.* 12: 1190–1199.

118 Alvarez-Paino, M., Munoz-Bonilla, A., Marcelo, G. et al. (2012). *Polym. Chem.* 3: 3282–3288.

119 Zhang, Y., Chen, G., Lin, Y. et al. (2015). *Polym. Chem.* 6: 97–105.

120 Beyer, V.P., Cattoz, B., Strong, A. et al. (2019). *Polym. Chem.* 10: 4259–4270.

121 Wild, A., Winter, A., Hager, M.D. et al. (2012). *Macromol. Rapid Commun.* 33: 517–521.

5

Synthesis of Polythioesters

Li-Yang Wang and Wei-Min Ren

Dalian University of Technology, State Key Laboratory of Fine Chemicals, 2 Linggong Road, Dalian 116024, China

5.1 Introduction

Sulfur, the sixth abundant nonmetal element in the earth, widely exists in a variety of food ingredients, agrochemicals, and natural medicines. To date, using in-depth studies of sulfur chemistry and diversified chemical transformation strategies have been developed to synthesize a variety of sulfur-containing compounds, such as urea, episulfide, and a large category of sulfur-containing medicinal compounds [1–3]. Also, the incorporation of sulfur atoms into the polymer backbone has captured the attention of polymeric chemistry scientists, as the newly induced sulfur atoms endow the polymer with accessional attractive properties [4], e.g. enhanced thermodynamic properties and excellent optical and electronic properties. These excellent properties result in their wide applications. For instance, polymers with high sulfur contents are a class of electroactive cathode materials for Li–S batteries, which exhibit high specific capacity and energy density [5]. The copolymers of dithiocarbonates show excellent metal-ion removal ability and can extract Pd and Au salts from their aqueous solution in an easy and efficient way [6]. Sulfur-containing polymers with disulfide groups have also been used as self-healing materials [7–9], which is a research hotspot in the field of polymers. The synthesis of hydrogels with sulfur-containing functional groups is also attractive, as these groups exhibit cross-link properties [10].

On the other hand, in recent years, degradable polymers have attracted much attention for the following reasons: the global consumption of plastics is high and increasing annually, and waste plastics are generally landfilled due to their non-degradability and poor recyclability, resulting in "white pollution". Furthermore, modern plastics are mainly non-renewable petroleum-based polymer material. [11] Among the polymer families, polyesters are attractive degradable polymer materials because the ester bonds in their main chain are susceptible to nucleophilic attack. When the oxygen atoms of ester units are partially or totally substituted by sulfur atoms, the resultant polythioesters also exhibit good

Sulfur-Containing Polymers: From Synthesis to Functional Materials,
First Edition. Edited by Xing-Hong Zhang and Patrick Theato.

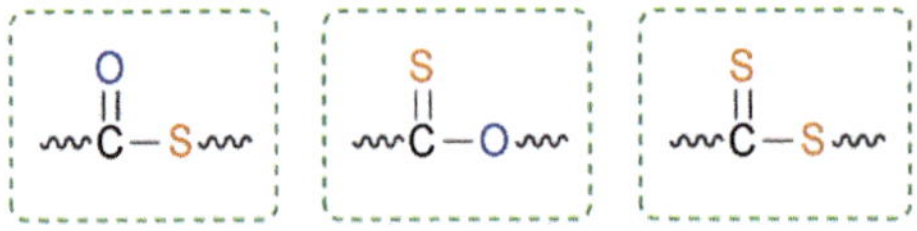

Figure 5.1 Schematic representation of the functional groups that are classified as polythioesters.

degradability and reactivity. Compared with esters, thioesters are more electrophilic and therefore easier to react with nucleophiles, making them excellent acyl transfer reagents [12]. As a consequence, polymers containing thioester in the main chain are a kind of good degradable materials. In addition, the recent advances in polymerization methodology and the elegant design of monomers have enabled the synthesis of well-defined sulfur-containing polymers [5, 13–22]. This book chapter mainly focuses on the progress concerning on the synthesis of polythioesters. It should be specifically stated that the term polythioesters are used to denote polymers containing either of the thioester groups in the backbone (Figure 5.1).

The term "polythioester" is not an accurate definition for a specific chemical structure, owing to the diversity of the thioester functional group. The thioester group is not as singular as the ester group, where there are two oxygen atoms that can be replaced by sulfur atoms. The concept of polythioesters was described in detail in a review in 2007 [23]. In the past few decades, polyesters have attracted great interest from both academic researchers and chemical companies. Polyesters have a wide range of applications because of their functional diversity and bioreactivity [24], and several polyesters have been commercialized. For example, poly(ethylene terephthalate) (PET) is widely used in fibers, blow-molded bottles, and other packaging applications due to its good mechanical strength and barrier properties, with an annual output of billions of pounds [25]. Compared with polyesters, polythioesters received less attention in the past. Due to the attractive properties of sulfur-containing polymers, as well as the flourishing development of degradable materials and biocompatible materials, the research on polythioesters shows a booming trend.

5.2 Synthesis of Aromatic Polythioesters

In aromatic polythioesters, each structural unit in the polymer chain contains aromatic groups. Since aromatic polythioesters generally have enhanced thermodynamic properties and strong mechanical and electrical properties [26, 27], they show potential for packaging material applications. Polycondensation is a common method for preparing aromatic polythioesters. Podkoscielny et al. [26, 28–30] prepared aromatic polythioesters via the interfacial polycondensation of dithiols and dicarboxylic acid dichlorides (Scheme 5.1). The obtained polymers generally showed good thermal, mechanical, and electrical properties. Hirano et al. [27] prepared aromatic polythioesters from bis(4,4-mercaptophenyl)sulfide (MPS) and aromatic acid dichlorides via solution polycondensation (Scheme 5.2). The polythioester from isophthaloyl chloride was a typical crystalline polymer with a melting temperature (T_m) of 249 °C, whereas that made from MPS and

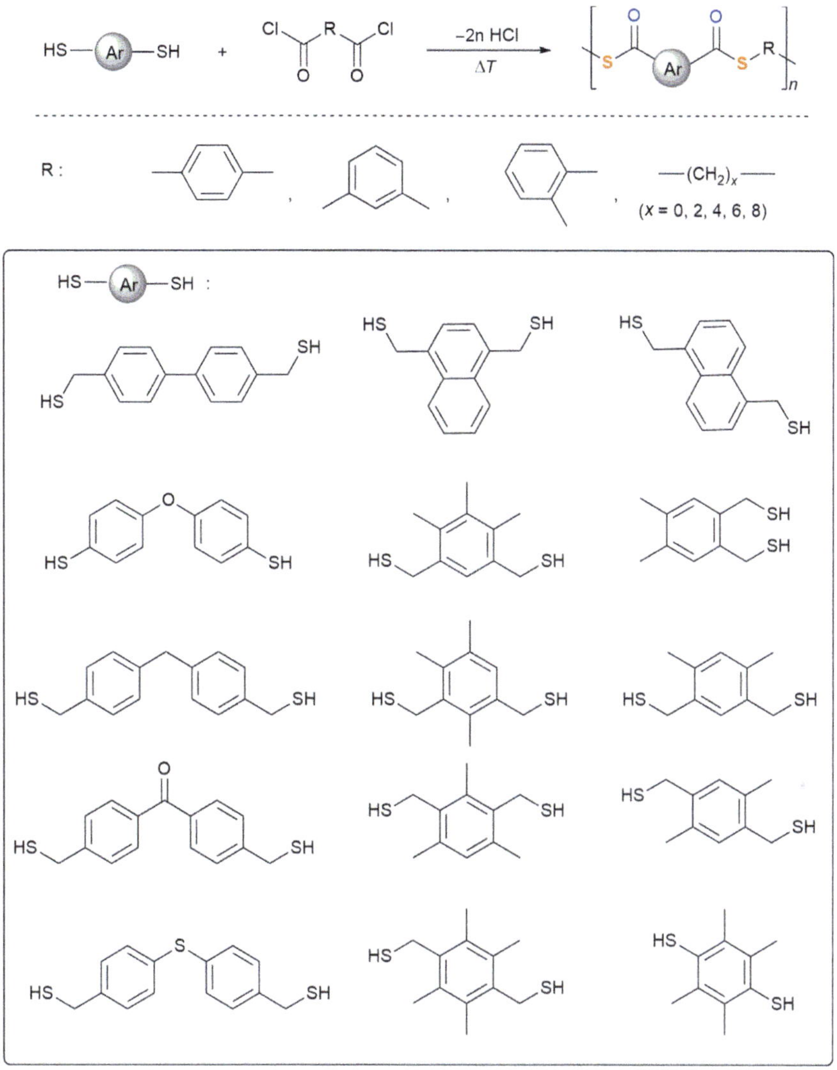

Scheme 5.1 Interfacial polycondensation of dithiols and diacyl halides. Source: Adapted from Refs. [26, 28–30].

t-butyl-isophthaloyl chloride was amorphous. Both exhibited high thermal stability (5% mass loss temperatures were 389 and 379 °C, respectively). Figueroa's group [31] also prepared aromatic poly(dithioester)s from the bis-piperidimum salt of tetrathioterephthalic acid and bis-bromoacetyl aromatic compounds at room temperature. When the poly(dithioester)s were cyclized in sulfuric acid, the resulting polymers containing dithiolium rings exhibited electrical conductivity.

Additionally, in a seminal work, Nishikubo and cowokers [32] synthesized a multi-sequence-ordered polythioester by inserting thiiranes into a polythioester backbone, with quaternary ammonium salts as the catalysts. Subsequently, they

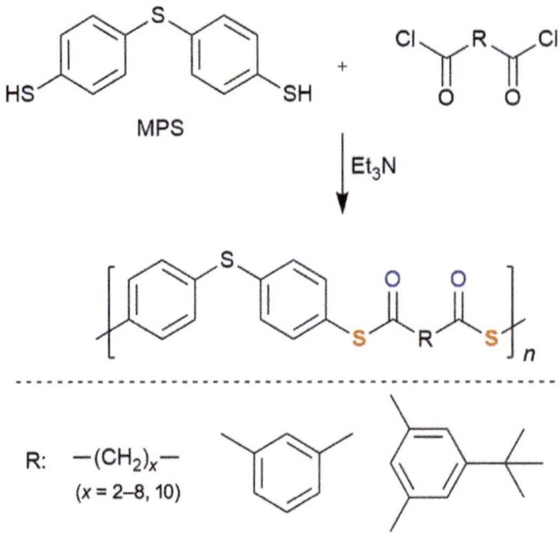

Scheme 5.2 Solution polycondensation of MPS and aromatic acid dichlorides. Source: Adapted from Ref. [27].

developed a novel approach for obtaining aromatic polythioesters via the polyaddition of bis(thiirane)s with diacyl chlorides using quaternary onium salts as the catalysts (Scheme 5.3) [33]. The study showed that aromatic solvents are more favorable for obtaining polymers with high molecular weights than aprotic polar solvents. The catalysts with chloride as the counteranion had a higher activity than those with bromide or iodide. With 5 mol% of tetrabutylammonium chloride or tetrabutylphosphonium chloride as catalyst and anisole as solvent, bisphenol A bis(2,3-epithiopropyl) ether and terephthaloyl chloride or isophthaloyl chloride reacted at 90 °C for 24 hours. The aromatic polythioesters with an average molecular weight of 40 000 were synthesized.

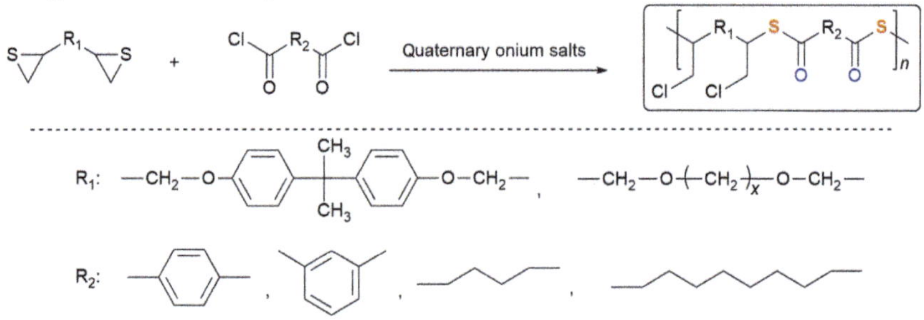

Scheme 5.3 Aromatic poly(S-thioester)s from polyaddition of bis(thiirane)s and diacyl chlorides using quaternary onium salts as catalysts. Source: Adapted from Ref. [33].

5.3 Synthesis of Semi-aromatic Polythioesters

Semi-aromatic polythioesters are a kind of polythioesters containing both aromatic groups and aliphatic groups in repeating units. Compared with aromatic

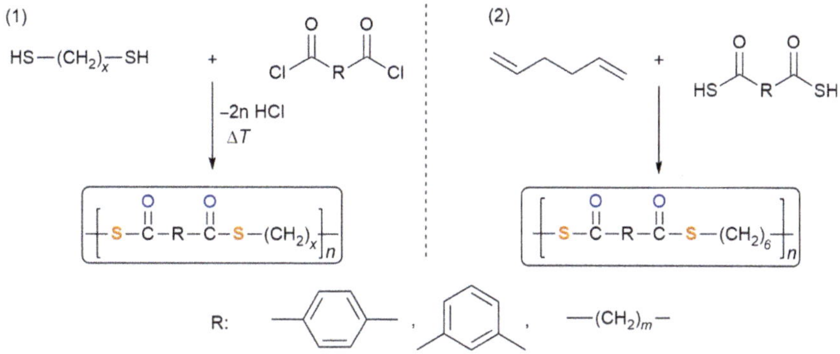

Scheme 5.4 Polythioesters reported by Marvel et al.: (1) polythioesters from polycondensation, and (2) polythioesters from radical polyaddition. Source: Adapted from Refs.[36, 37].

polythioesters, semi-aromatic polythioesters have thermotropic liquid crystal characteristics in addition to good thermodynamic and mechanical properties, and exhibit potential for use as liquid crystal materials [34, 35]. The first semi-aromatic polythioesters (Scheme 5.4, **1**) were reported by Marvel and Kotch in 1951 [36]. They described a different approach to prepare semi-aromatic polythioesters (Scheme 5.4, **2**) via the radical polyaddition of dibasic thio acids and the non-conjugated diolefin biallyl [36, 37] induced by ultraviolet light. Similar polythioesters synthesized using different methods had almost the same melting temperature, which was higher than that of their oxygen analogs. For instance, the T_m range of polyethylene dithiolterephthalate is 335–340 °C, which is much higher than that of polyethylene terephthalate (255–260 °C).

Similar to aromatic polythioesters, semi-aromatic polythioesters can also be prepared via polycondensation, i.e. through the polycondensation of dibasic acid chlorides and dithiols. Podkoscielny *et al.* [28–30] synthesized semi-aromatic polythioesters by interfacial polymerization from aromatic dithiol and aliphatic dicarboxylic acid chloride (Scheme 5.1). Subsequently, Hirano et al. [34] and Isogai et al. [35] used polycondensation to synthesize a series of semi-aromatic polythioesters based on the synthetic routes of Podkoscielny's group; they were named thermotropic liquid crystal polymers (TLCPs) containing sulfur in the main chain. These TLCPs exhibit a mesophase above their melting points and are capable of flowing in a structurally regular manner, similar to a crystal. In their study, MPS semi-aromatic polythioesters (Scheme 5.2) containing sulfur were investigated as epoxy resin modifiers, and they proved to be effective in enhancing the heterojunction of copper and epoxy resins [38].

Recently, Sasanuma and coworkers reported semi-aromatic poly(dithioester)s (Figure 5.2, **A**) prepared via an ionic polycondensation between tetrathioterephthalic acid complexed with piperidinium and an α,ω-dibromoalkane. Polythioesters **B** were also prepared for comparison [39, 40]. The glass transition temperature (T_g) of **A** increases with a reduction in y (from 26 °C ($y = 5$) to 71 °C ($y = 2$)), whereas the T_g of **B** is near room temperature. The decomposition temperature of **A** is in

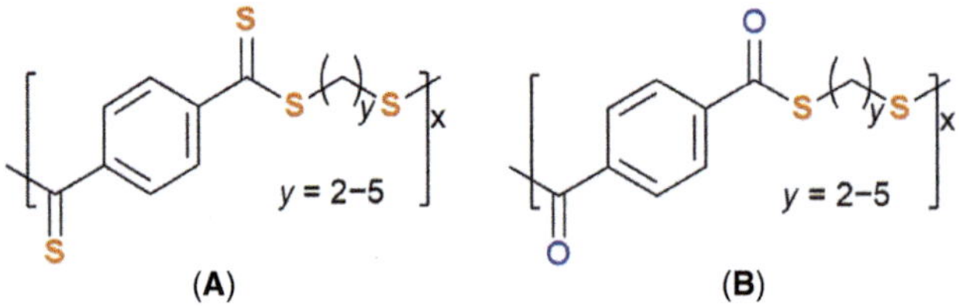

Figure 5.2 Chemical structures of semi-aromatic polythioesters: (**A**) semi-aromatic poly(dithioester)s; (**B**) semi-aromatic polythioesters. Source: Adapted from Ref. [39].

the range 220–260 °C. Compounds **B** decompose at around 350 °C, which is above the melting point. This indicates that the thermal resistance of **A** is lower than that of **B**. Compounds **B** are semi-crystalline, whereas compounds **A** are generally amorphous.

Very recently, Gutekunst and coworkers successfully incorporated the thioester into radical polymers, which is an all-carbon backbone polymer, by copolymerizing the thionolactone with acrylates in a reversible-activation fragmentation-transfer radical (RAFT) polymerization manner for the purpose of diversifying the structure of radical polymer, which is not easy to be degraded into small molecular (Scheme 5.5) [16]. As described, two significant challenges exist in current radical ring-opening systems featuring cyclic ketene acetal (CKA) monomers: (i) minimal control over the rate of addition to the CKA due to the high energy radical intermediate; (ii) the competition between the propagation rate of the intermediate radical and the ring-opening rate of the monomer. One elegant solution to resolve these issues is to judiciously select monomers. The fact that thiocarbonyl compounds are

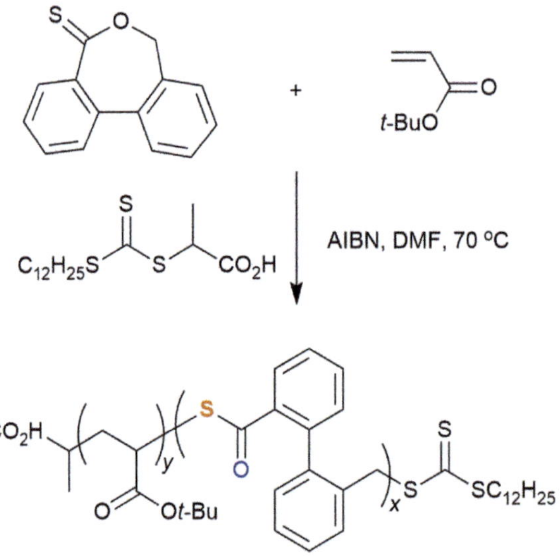

Scheme 5.5 Reversible addition–fragmentation chain transfer polymerization of thionolactone and acrylate. Source: Adapted from Ref. [16].

often used as radical initiators and thionoester often acts as the chain transfer agent inspired Gutekunst to design a new radical ring-opening polymerization (ROP) by cyclizing the known transfer agent to synthesize thionoester monomer. The newly synthesized monomer was applied for the copolymerization with *tert*-butyl acrylate (95 : 5) under free radical conditions. The decrease polymerization rate and visible color change suggest the incorporation of thionoester, which is further confirmed by the ^{1}H NMR spectroscopy. The polymer with variety of molecular weights and low dispersities can be obtained by varying the monomer to initiator ratio from 25 to 250 with a feed of 5% of thionoester in high degrees of thioester conversion. The feed ratio of thionoester can be increased up to 50% with a lower conversion, affording the copolymer with different thioester content, of which the T_g values vary from 46 to 81 °C. The further increase in feed ratio of thionoester leads to the entirely failed of the copolymerization, which could be attributed to the slow initial rates of polymerization as the intermediates radical can be quenched by each comonomer ring opening. The chain-extension experiments revealed the "livingness" nature of the copolymer, as the poly(*tert*-butyl acrylate) copolymer (M_n = 3 900) can act as a macroinitiator of RAFT polymerization with *n*-butyl acrylate and thionoester, affording a higher molecular weight diblock polymer (M_n = 11 200). The reversibly deactivating nature of thionoester leads to the usual kinetic behavior that the thionolactone monomer not only readily copolymerizes with *tert*-butyl acrylate but is consumed at an even faster rate, which is different from the case of the ROP of acrylates reported previously, of the copolymerization of thionoester and *tert*-butyl acrylate. The degradation experiments furtherly supported the kinetic study and verified the presence of thioester in the polymer chain. This work provides a novel synthetic strategy for obtaining polythioester with tunable microstructure and properties.

More recently, we developed a novel strategy for obtaining semi-aromatic polythioesters by copolymerization of propylene oxide (PO) and phthalic thioanhydride (PTA) using a binary system of the salenM(III) complex and [PPN]Cl ([PPN] = bis(triphenylphosphine)iminium) as the catalyst (Scheme 5.6) [17]. The catalytic system of the salenCr(III) complex and [PPN]Cl showed an outstanding activity, with a turnover frequency (TOF) up to 1420 h^{-1} at 100 °C. Transesterification occurred throughout the reaction, at both low (25 °C) and high temperatures (100 °C). A comprehensive study was devoted to elucidating the possible mechanism for the copolymerization of PO and PTA, which guides the synthesis of polythioesters with different thioester-containing units. As described, copolymers with a significant number of thioester bonds were produced at ambient temperature, whereas at a higher temperature, the polymerization selectivity is poor, yielding a copolymer with diverse structures. The structures of the polymer backbone, however, were different; the refractive index of such polythioesters are up to 1.6 (measured on an Abbe refractometer, cast film, 20 °C). NMR spectroscopic analysis showed that the resultant copolymers had multiple repeating units in their main chains due to transesterification, especially at higher reaction temperatures. The thiolate anion terminus originating from transesterification could undergo chain propagation by multiple routes (Scheme 5.7). We concluded that thioester

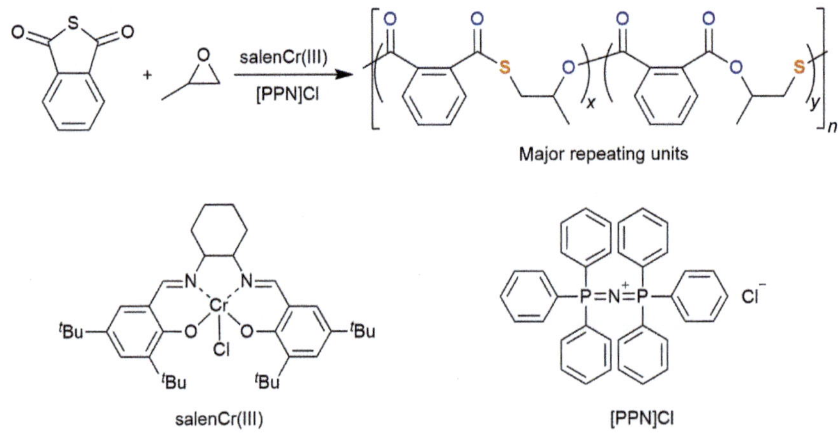

Scheme 5.6 Copolymerization of PO and PTA. Source: Adapted from Ref. [17].

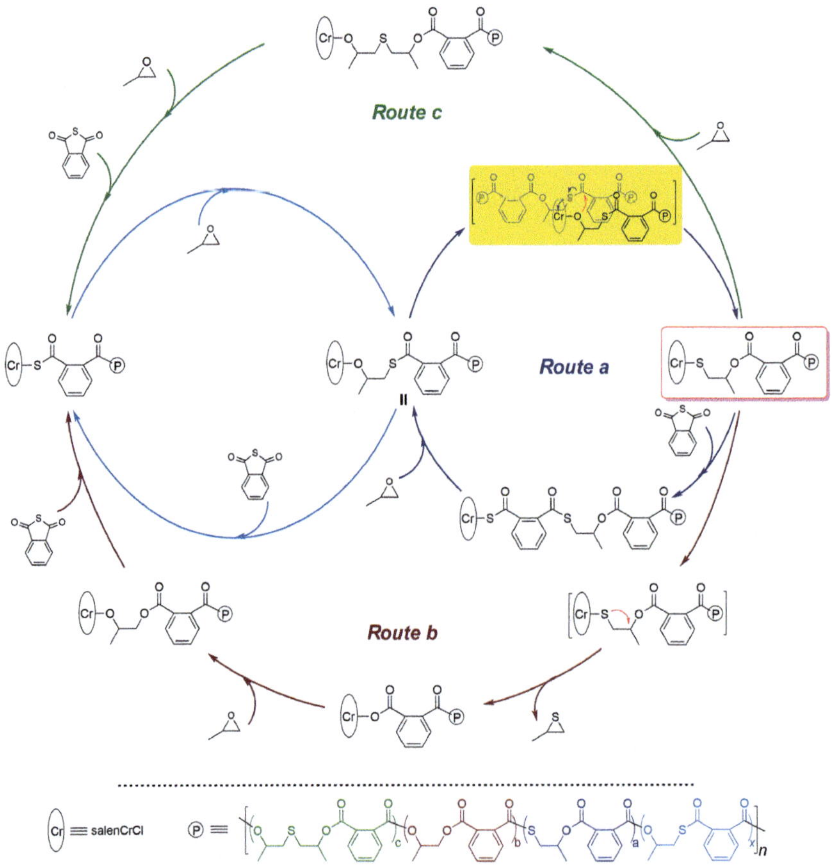

Scheme 5.7 Possible mechanism for the copolymerization of PTA and PO.

linkage (route a) was predominant at 25 °C, yielding a copolymer with a significant quantity of thioester linkages. Routes b and c were favored at higher reaction temperatures, yielding copolymers with diverse structures.

5.4 Synthesis of Aliphatic Polythioesters

Aliphatic polythioesters are a kind of potentially biocompatible polymers [41] and hold promise because analogous polyesters have a wide range of applications in drug delivery systems, regenerative medicine, and artificial tissue engineering [42–44]. Compared with aromatic and semi-aromatic polythioesters, the synthesis methods of aliphatic polythioesters are more diversified. Owing to the widespread distribution of thioester bonds in biological systems [12] and the potential biocompatibility of aliphatic polythioesters, the biological synthesis of these polythioesters is a more attractive method compared with polycondensation and polyaddition. ROP is another important method for the synthesis of aliphatic polythioesters.

Generally, the preparation of aliphatic polythioesters by polycondensation [36] and polyaddition [33, 37] are studied along with that of aromatic and semi-aromatic polythioesters (Schemes 5.3 and 5.4). Recently, new thermoplastic aliphatic polythioesters (Scheme 5.8) were synthesized via polycondensation [45]. The resultant polythioesters have good thermal stability, high refractive indices (1.6149–1.6271) and good resistance against coloration (event at 200 °C).

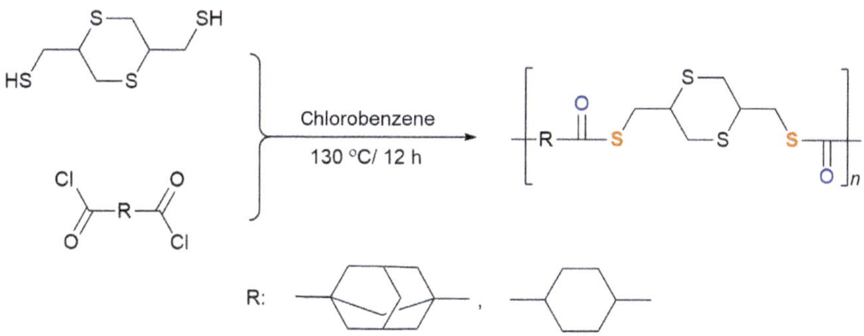

Scheme 5.8 Synthesis of aliphatic polythioesters. Source: Adapted from Ref. [45].

In addition, the biological synthesis of aliphatic polythioesters is an important method [46–49]. Steinbüchel demonstrated a biosynthesis procedure for obtaining novel thermoplastic polythioester by engineered *Escherichia coli* [50]. As described, there is a common sense that the sulfur moiety group often located in the side chain of the polyhydroxyalkanoates (PHAs) in the process of synthesis of biomaterials. The first example of thioester linkage in the polymer main chain was obtained from the isolation of the polythioester from the mixed PHA-accumulating bacterium *Ralstonia eutropha*, comprising 3-mercaptopropionate (3MP) or 3-mercaptobutyrate (3MB) in addition to 3-hydroxybutyrate (3HB). One longstanding existed issue of biosynthesis of polythioester, including poly(3-mercaptopropionate) (PMP) and poly(3-mercapto-butyrate) (PMB), is that it is unavailable to obtain the polythioester (PTEs) without 3HB, due to the unavoidable biosynthesis pathway of poly(3HB). To solve this issue, Steinbüchel used a recombinant strain of *E. coli*, harboring plasmid pBPP1, which encodes a butyrate kinase (Buk) and a phosphotransbutyrylase (Ptb) from Clostridium acetobutylicum, as well as a PHA synthase from Thiocapsa pfennigii. With this strategy, three novel biopolymers, including PMP; PMB; and poly(3-mercaptovalerate), PMV, were synthesized, of which the nascent morphology were examined by TEM. Subsequently, these three polythioesters were isolated by decomposition of the surrounding cell materials, affording corresponding purified PTEs, which are confirmed by the means of mass spectrum. The specific structure of these three polythioester was further identified by ^{13}C NMR after screening the most suitable solvent for these three PTEs, respectively, from a series of variety of solvents as their different solubility in various solvents. PMP shows a poorer solubility comparing with the rest two PTEs. This can be attributed to the higher crystalline degree order of PMP as the characterization of crystallinity of these PTEs in the means of X-ray powder-diffraction. Additionally, PTEs possess higher heat resistance than that of poly(3-hydroxybutyrate) (PHB) as confirmed by thermal gravimetric traces with a residual weight value of PMP of 74.6% at 295 °C, while the residual weight value of PHB is only 4.1% at 295 °C. Besides, T_m and T_g of PTEs were texted with the application of differential scanning calorimetry. The PMP sample melted at 170 °C, which is much higher than that reported for PHP, in keeping with the observed trend in the T_m results for aliphatic polyethers compared with the sulfur equivalents. However, PMB and PMV showed T_m endotherms at 100 and 84 °C, respectively; these values are lower than those of the oxygen analogues, PHB and PHV. If higher T_m for PTE compared with PHA is to be a rule, then the observed lower T_m for PMB and PMV are an exception, which could be explained by low molecular weight and poor crystalline organization. Observed T_g values at 8 °C for PMB and at −1 °C for PMV are higher than those of equivalent oxyanalogues. Polythioesters are interesting materials because of their different physical and chemical properties compared with polyoxyesters. The biosynthesis strategy extended the synthesis library of obtaining polythioesters with diversity structure.

An aliphatic polythioester was enzymatically synthesized by the direct polycondensation of mercaptoalkanoic acid using immobilized lipase of Candida Antarctica (lipase CA) in bulk, affording the poly(11-mercaptoundecanoate) with M_w of 34 kg mol^{-1} in a high yield at 110 °C [51]. It was found that the presence of 4 Å molecular sieves as water absorbents was beneficial for the reaction. The resulting polythioester is readily degraded into a cyclic oligomer by lipase, and can be re-polymerized by ring-opening using lipase, for a sustainable chemical recycling. Matsumura and coworkers prepared aliphatic polythioesters by direct polycondensation of hexane-1,6-dithiol and a range of diacid diesters, using immobilized lipase CA in the presence of 4 Å molecular sieves [52]. The melting and crystallization temperatures of the obtained polythioesters were higher than those of the corresponding polyoxyesters. Polythioesters prepared through a two-step procedure had a high molecular weight. Recently, lipase CA was used for the ROP of a cyclic thioester, which was prepared using the same lipase CA, to obtain a high-molecular-weight polythioester [53]. The polythioesters thus prepared had a high M_w of about 120 kg mol^{-1}, which was significantly higher than that of a polythioester obtained by direct polycondensation.

The ROP of lactones is an effective strategy for obtaining polyesters, and is also suitable for polythioesters. For instance, the anion-induced ROP of ε-thiocaprolactone (tCL) to synthesize poly(thiocaprolactone) with potassium *tert*-butoxide (*t*-BuOK) as the initiator was described for the first time by Overberger in 1964 [54]. Thereafter, poly(thiovalerolactone) was synthesized similarly [55]. It was confirmed that the 5-membered *γ*-thiolactone could not be ring-opened in an anionic or cationic manner (Scheme 5.9, 1) [55, 56]. However, the 5-membered γ-thionobutyrolactone can undergo ROP in the presence of cationic initiators [56]. The polymerization was performed at 100 °C for two hours with various cationic initiators (such as scandium trifluoromethanesulfonate). Interestingly, the resulting polymer was proven to be a polythioester rather than a poly(thionoester) (Scheme 5.9, 2). (*S*)-*α*-*p*-toluenesulfonamido-β-propiothiolactone (NTs-PTL) was used as the monomer for the ROP of 4-membered β-thiolactone [57]. Suzuki et al. [58] described the controlled ROP of the β-thiolactone derived from *N*-Boc cysteine using *N*-Boc-L-cysteine methyl ester as the initiator in NMP at room temperature. The polythioester was further functionalized via a thiol–ene click reaction. When the *N*-Boc groups were deprotected, the polythioester transformed into polycysteine via *S*-to-*N* acyl migration.

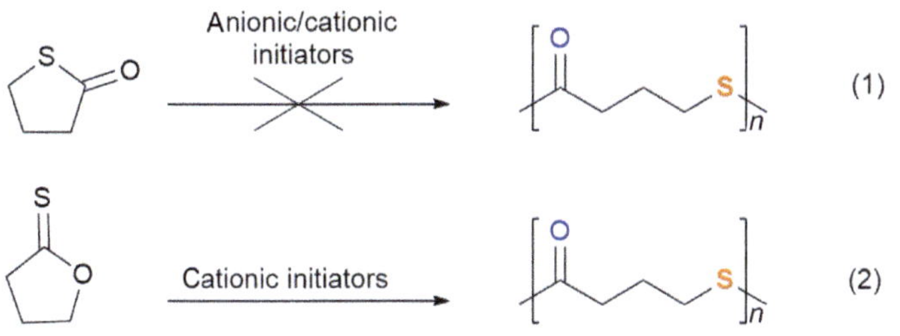

Scheme 5.9 (1) Conditions for ROP of 5-membered γ-thiolactone, and (2) ROP of 5-membered γ-thionolactone. Source: Adapted from Refs. [55, 56].

In 1999, Endo's group [59] reported the first anionic ring-opening of ε-thionocaprolactone (tnCL) (Scheme 5.10, 1). When the anionic ROP was carried out using organolithiums, Grignard reagents, and *t*-BuOLi as the initiators, the corresponding thiocarboxylic *O*-ester unit **A** formed. When the anionic ROP was carried out using *t*-BuOK and 1,8-diazabicyclo[5.4.0]undec-7-ene (DBU), a mixture of thiocarboxylic *O*-ester unit **A** and thiocarboxylic *S*-ester unit **B** were obtained; the content of **B** was higher. The polymerization rate was higher in tetrahydrofuran than in toluene, and led to the formation of a cyclic byproduct. However, the polythioesters obtained using this method had drawbacks: copolymerization, M_n control, and end group selection were not allowed. Soon thereafter, a cationic ROP strategy was put forward

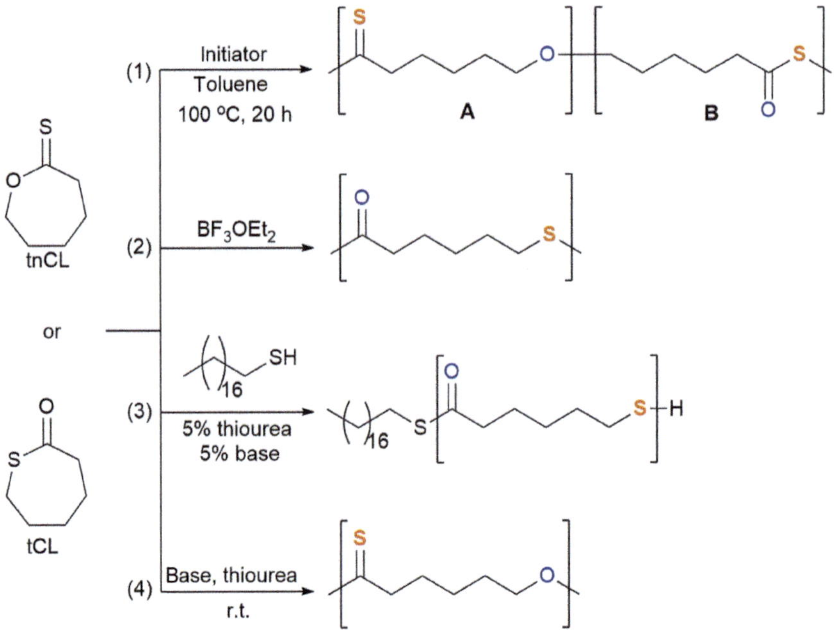

Scheme 5.10 ROP of ε-thionocaprolactone or ε-thiocaprolactone under different conditions. Source: Adapted from Refs. [59–62].

to realize the ROP of tnCL for the first time (Scheme 5.10, 2) [60]. When BF_3OEt_2 was used as the initiator, the poly(thiocaprolactone) reported by Overberger was obtained instead of poly(thionocaprolactone). This was due to the quantitative inversion of substitution at the thionoester. The monomer conversion and M_n of the resulting polymer increased with the temperature from −15 to 28 °C, and the M_n effectively improved upon increasing the initial monomer concentration.

Recently, Kiesewetter and coworker made an effort to realize the ROP of tCL via organocatalysis, which afforded poly(thiocaprolactone) instead of poly(thionocaprolactone) (Scheme 5.10) [61]. Initially, they measured the thermodynamics of polymerization by employing the rapid 1,5,7-triazabicyclododecene (TBD)-catalyzed ROP of tCL. The results suggest that tCL is energetically more similar to caprolactam (no ceiling temperature) than it is ε-caprolactone or δ-valerolactone (~534 K and Tc ~ 422 K, respectively). The screen of base catalysts revealed that only strong nucleophilic bases including DBU and 7-methyl-1,5,7-triazabicyclo[4.4.0]dec-5-ene (MTBD) are active for the ROP of tCL, demonstrating a high catalysis activity in a "living polymerization" manner, while the base with weaker nucleophilic like BEMP (2-tert-butylimino-2-diethylamino1, 3-dimethylperhydro-1,3,2-diazaphosphorine), a strong and bulky Brønsted base, is less effective for the polymerization of tCL, revealing a nucleophilic ROP mechanism that DBU and MTBD are not acting as general bases but rather are effecting ROP via nucleophilic attack at the thioester moiety, providing the corresponding polythiocapratone with moderate M_w/M_n, which gets bigger at the later polymerization stage due to the increased nucleophilicity of thiols. The decreased reaction time and lowered M_w/M_n were observed as the result of addition of thiourea into the catalyst system, which renders thiourea H-bond for the activation of tCL, regardless of the small binding constant between thiourea and tCL.

Subsequently, Kiesewetter and coworkers realized the ROP of tnCL in a controlled manner, affording the homopoly(ε-thionocaprolactone) (PtnCL) without carbonyl scrambling in the presence of organocatalyst system consisted by the thiourea and organic bases for the first time (Scheme 5.10, 4) [62]. The early approaches for obtaining PtnCL mainly rely on the cation or anion polymerization of tnCL with BF_3OEt_2 or alkyllithiums as initiator, leading to uncontrolled character of the ROP of tnCL. Another drawback of O/S scrambling appeals when the initiator with weaker nucleophilicity was employed to the polymerization. The application of organic bases with strong nucleophilicity such as DBU, MTBD as initiators in the presence of alcohol, or thiol for the homopolymerization of tnCL was successfully solved the problem described earlier, providing the polymer with the well-defined structure as no O/S scrambling reaction happens during the polymerization, and exhibited a linear evolution of M_n versus time and an initially narrow M_w/M_n that broadens throughout the ROP. The precise synthesis of PtnCL was realized by the addition of thiourea, which has much effect on the polymerization: linear evolution of M_n versus conversion, narrow M_w/M_n (=1.10), first-order evolution of [tnCL] versus time, and an M_n that is predictable from $[M]_0/[I]_0$. Additionally, the incorporation of δ-valerolactone in to the polythionocarprolactone was achieved in the presence of organocatalyst system consisted of thiourea, organic base, and alcohol or thiol, providing the method of construction of random (statistical), gradient, or block copolymers with traditional esters and offering the potential for tuning materials properties for custom tailored applications.

The polymers containing thioester functional group usually exhibit a dynamic or responsive behavior via the thiol–thioester exchange reaction. Benefit from this character, Bowman presented a straightforward template direct synthesis (TDS) strategy via the ROP of the thymidine or deoxycytidine appended thiolactone monomer, providing the corresponding polythioesters in a chain propagating mechanism manner for the purpose of meeting the considerable drawbacks, including broad dispersity, low molecular weights at moderate conversions, and unproductive cyclization of the monomers, generated in the step growth polymerizations (Scheme 5.11). To achieve this goal, two kinds of thiolactone appended thymidine and deoxycytidine were synthesized through multistep reaction, respectively [13]. The ROP of these two thiolactone monomer were performed with thiolate as initiator in the presence of a stoichiometric amount

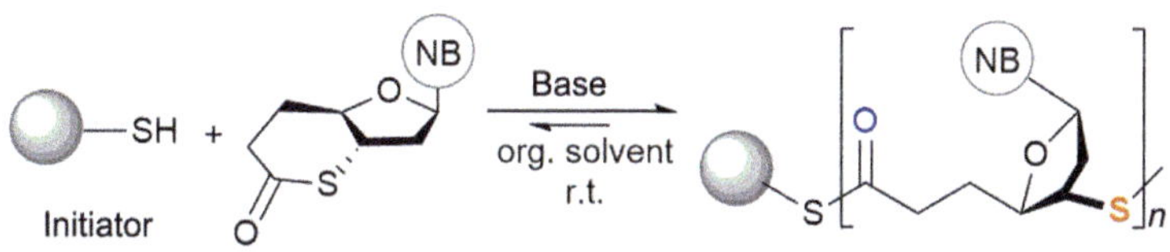

Scheme 5.11 Synthesis of DNA-like polymers. Source: Adapted from Ref. [13].

of organic base *N,N*-diisopropylethylamine (DIPEA), resulting the corresponding polythioesters with uncontrolled nature, as the molecular weight of the resultant polymer determined by the gel permeation chromatography (GPC) after the capping reaction of the polymer deviates from the calculated one. This can be attributed to the poor solubility of the polymer in THF, as at higher concentrations a solid slowly precipitated out of solution; while at lower concentrations the reaction stayed homogeneous. The addition of thiolate into a capped polymer resulting in a response by reduction of molecular weight revealed the dynamic exchange behavior of the thioester functional group via thio-thioester exchange, which provides a possibility of enabling complementary polymers of DNA via dynamic TDS by employing the thioester-based approach and the success in coalescing two homopolymers of different chemical compositions into a single copolymer validated the hypothesis. The application of thioester appended nucleic acid as dynamic and responsive group for the TDS constitutes the first step towards a platform that enables for the routine synthesis of sequence controlled polymers via dynamic template directed synthesis.

Recently, Lu presented a new paradigm of sustainable polymers from renewable resources by ROP of a newly designed bicyclic thiolactone monomer synthesized from an abundant biosourced feedstock through one-pot reaction, affording the polymer with recyclability in a fully controlled manner (Scheme 5.12) [14]. Owing to the intrinsically labile thioester bond and susceptibility to nucleophiles of thioester, they made an attempt to synthesize recyclable polymers by using the thioester unit as recycle driving force. Initially, the homopolymerization of the synthesized *N*-substituted *cis*-4-thio-L-proline thiolactone (NR-PTL) were performed in high polarity solvents like *N,N*-dimethylformamide, leading to either an entirely failure or the uncontrolled of the reaction, while the success with rapid and controlled polymerization was achieved when lower polarity solvents, including dichloromethane and chloroform, mediated by an organocatalyst system, consisted by benzyl mercaptan as initiator and triethylamine (TEA) as catalyst/base. It was confirmed that the kinetic behavior of the Boc-PTL ROP is a clear first-order kinetic character relative to [M], which cannot be affected by the concentration of the base, by the kinetic study. Increasing the monomer/initiator feed ratio results in the different molecular weight of the polymer, which is in well agree with the calculated molecular weight, with narrow dispersities, suggesting the well-controlled nature of the polymerization. The variety of the polymers, resulted from the variation of the substitutes on N atom, are easily decomposed into clear monomer in the presence of thiol and organic base with stronger nucleophilicity,

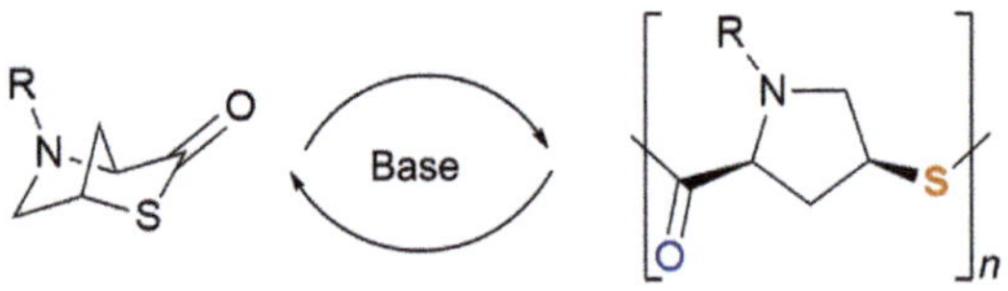

Scheme 5.12 ROP of 4-hydroxyproline-derived bicyclic thiolactone monomers and depolymerization of the corresponding polythioesters. Source: Adapted from Ref. [14].

like DBU, while the polymers is stable in the conditions without bases. Additionally, the depolymerization underwent smoothly in a milder condition, which is much different from the cases of polyester and polycarbonate, as the elevated temperature or strong catalysts is essential for the depolymerization, following a zero-order kinetic character, with a temperature-dependent rate constant. The new developed monomer for the access to obtain polythioester with excellent recyclability enlarged the family of the sustainable polymers, and provided a direction for the synthesis of more polymers in a "monomer–polymer–monomer" model with well-controlled behavior.

Although the methods described earlier can be used to synthesize polythioesters, there are still some shortcomings: (i) a large number of byproducts are produced; (ii) the cost is high; and (iii) a single type of monomer leads to a single structure of the polymer. Therefore, the development of a more efficient strategy for inexpensively obtaining polythioesters with diverse structures and without any byproduct is of much importance. Very recently, we made an effort to develop a novel approach for synthesizing aliphatic polythioesters with diverse structures through the ring-opening copolymerization of episulfides and cyclic thioanhydrides (Scheme 5.13) [15]. Utilization of readily available monomers opens up a new route to polythioesters with diverse structures and functions, offering the opportunity for post-polymerization modification. The copolymerization reaction readily proceeded using simple organic ammonium salts (such as [PPN]OAc) under mild conditions. This metal-free strategy can afford polythioesters that may be utilized as high-performance engineering plastics, as well as optical, optoelectronic, and photochemical materials. The controlled character was clearly evidenced by the linear increase in M_n of the resulting polythioesters, with cyclic thioanhydride conversion and narrow polydispersity (<1.3) even at a high monomer conversion. This advantage enables the successful synthesis of block copolymers via stepwise addition of different monomers. The resulting polymer from succinic thioanhydride and propylene sulfide is a typical semicrystalline material, with a high refractive index of up to 1.78 (measured on an Abbe refractometer, cast film, 20 °C), which is one of the largest values reported for sulfur-containing polymers.

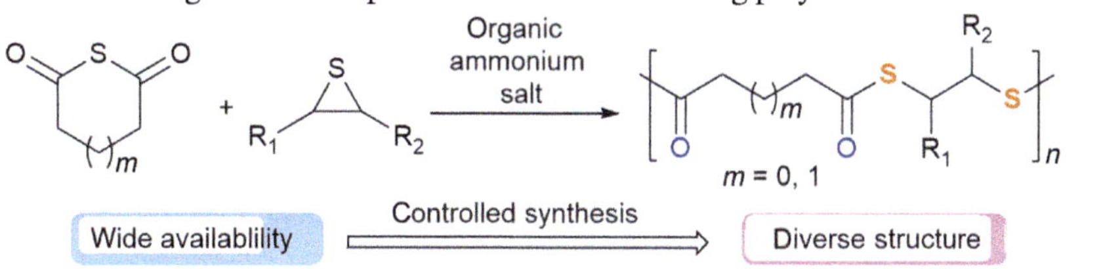

Scheme 5.13 Copolymerization of episulfides and cyclic thioanhydrides. Source: Adapted from Ref. [15].

5.5 Summary and Concluding Remarks

This review therefore aims to principally showcase the recent progress regarding the synthesis of polythioesters, including aromatic, semi-aromatic, and aliphatic polymers. Although these mentioned reactions suffered from problems, e.g. the lack of structurally diverse monomers, these investigations are important because they address the feasibility of using polythioesters as functional materials. To our delightsome, the discovery of the copolymerization of cyclic thioanhydride with epoxides or episulfides to afford the corresponding polymers with completely alternating structures, controlled molecular weights, and low dispersities makes polythioesters become truly competitive. Potential applications for this sulfur-containing copolymer are optical, optoelectronic, and photochemical materials, as well as drug-delivery systems, artificial tissues, high-performance engineering plastics. It is hoped that further advances in catalysis will allow for expanding the monomer scope and controlling the stereochemistry.

On the other hand, chemical recycling between polymers and the starting monomers in high efficiency is an ideal solution to reduce the need of finite raw materials, and minimize the negative impact on the environment. Due to their easy purification, the recycled small-molecule monomers could be repolymerized to form virgin-quality polymers. Thankfully, a recent study on highly efficient depolymerization of a kind of biosourced polythioesters into the corresponding monomers was reported by Li and Lu's group. The controllability of ROP and the ability of the polymer to depolymerize enable a sustainable way to create high-value materials such as degradable biomaterials and optical materials. Obviously, the process at the present level of technology is not commercially viable yet, but it reminds us that nature's ingenuity should not blind us to our own. It is highly desired to develop highly active catalysts and new catalytic processes for efficient degradation of polythioesters into useful chemicals, as well as even the starting monomers. Also, it urgently needs to develop novel polymerization processes without any catalyst and prepare recyclable smart sulfur-containing polymers, which exhibit pH, magnetic, or photo-response triggered depolymerization.

Acknowledgments

This work is supported by the National Natural Science Foundation of China (NSFC, Grant 21674015, 21722402, 21690073) and the Program for Changjiang Scholars and Innovative Research Team in University (IRT-17R14).

References

1 Feng, M., Tang, B., Liang, S.H., and Jiang, X. (2016). Sulfur containing scaffolds in drugs: synthesis and application in medicinal chemistry. *Curr. Top. Med. Chem.* 16: 1200–1216.

2 Wang, M. and Jiang, X. (2018). Sulfur–sulfur bond construction. *Top. Curr. Chem.* 376: 14.

3 Xiao, X., Xue, J., and Jiang, X. (2018). Polysulfurating reagent design for unsymmetrical polysulfide construction. *Nat. Commun.* 9: 2191.

4 Mutlu, H., Ceper, E.B., Li, X. et al. (2019). Sulfur chemistry in polymer and materials science. *Macromol. Rapid Commun.* 40: 1800650.

5 Chung, W.J., Griebel, J.J., Kim, E.T. et al. (2013). The use of elemental sulfur as an alternative feedstock for polymeric materials. *Nat. Chem.* 5: 518–524.

6 Ochiai, B., Ogihara, T., Mashiko, M., and Endo, T. (2009). Synthesis of rare-metal absorbing polymer by three-component polyaddition through combination of chemo-selective nucleophilic and radical additions. *J. Am. Chem. Soc.* 131: 1636–1637.

7 Hager, M.D., Greil, P., Leyens, C. et al. (2010). Self-healing materials. *Adv. Mater.* 22: 5424–5430.

8 Yoon, J.A., Kamada, J., Koynov, K. et al. (2011). Self-healing polymer films based on thiol–disulfide exchange reactions and self-healing kinetics measured using atomic force microscopy. *Macromolecules* 45: 142–149.

9 Bapat, A.P., Ray, J.G., Savin, D.A., and Sumerlin, B.S. (2013). Redox-responsive dynamic-covalent assemblies: stars and miktoarm stars. *Macromolecules* 46: 2188–2198.

10 Kharkar, P.M., Rehmann, M.S., Skeens, K.M. et al. (2016). Thiol-ene click hydrogels for therapeutic delivery. *ACS Biomater Sci. Eng.* 2: 165–179.

11 Zhang, X., Fevre, M., Jones, G.O., and Waymouth, R.M. (2018). Catalysis as an enabling science for sustainable polymers. *Chem. Rev.* 118: 839–885.

12 Aksakal, S., Aksakal, R., and Becer, C.R. (2018). Thioester functional polymers. *Polym. Chem.* 9: 4507–4516.

13 Mavila, S., Worrell, B.T., Culver, H.R. et al. (2018). Dynamic and responsive DNA-like polymers. *J. Am. Chem. Soc.* 140: 13594–13598.

14 Yuan, J., Xiong, W., Zhou, X. et al. (2019). 4-Hydroxyproline-derived sustainable polythioesters: controlled ring-opening polymerization, complete recyclability, and facile functionalization. *J. Am. Chem. Soc.* 141: 4928–4935.

15 Yue, T.-J., Zhang, M.-C., Gu, G.-G. et al. (2019). Precise synthesis of polythioesters with diverse structures by copolymerization of cyclic thioanhydrides and episulfides mediated by organic ammonium salts. *Angew. Chem. Int. Ed.* 58: 618–623.

16 Smith, R.A., Fu, G., McAteer, O. et al. (2019). Radical approach to thioester-containing polymers. *J. Am. Chem. Soc.* 141: 1446–1451.

17 Wang, L.-Y., Gu, G.-G., Yue, T.-J. et al. (2019). Semiaromatic polythioester from the copolymerization of phthalic thioanhydride and epoxide: synthesis, structure, and properties. *Macromolecules* 52: 2439–2445.

18 Yue, T.-J., Ren, W.-M., Chen, L. et al. (2018). Synthesis of chiral sulfur-containing polymers: asymmetric copolymerization of meso-epoxides and carbonyl sulfide. *Angew. Chem. Int. Ed.* 57: 12670–12674.

19 Luo, M., Zhang, X.-H., and Darensbourg, D.J. (2016). Poly(monothiocarbonate)s from the alternating and regioselective copolymerization of carbonyl sulfide with epoxides. *Acc. Chem. Res.* 49: 2209–2219.

20 Yang, J.-L., Wu, H.-L., Li, Y. et al. (2017). Perfectly alternating and regioselective copolymerization of carbonyl sulfide and epoxides by metal-free Lewis pairs. *Angew. Chem. Int. Ed.* 56: 5774–5779.

21 Kultys, A. (2010). *Sulfur-Containing Polymers in Encyclopedia of Polymer Science and Technology*, 4e. Hoboken: Wiley.

22 Kultys, A., Zulfiqar, S., and Sarwar, M.I. (2014). Recent developments in sulfur-containing polymers. *Polym. Rev.* 54: 185–267.

23 Kricheldorf, H.R. and Schwarz, G. (2007). Poly(thioester)s. *J. Macromol. Sci. Part A: Pure Appl. Chem.* 44: 625–649.

24 Longo, J.M., DiCiccio, A.M., and Coates, G.W. (2014). Poly(propylene succinate): a new polymer stereocomplex. *J. Am. Chem. Soc.* 136: 15897–15900.

25 Lecomte, P. and Jérôme, C. (2012). Recent developments in ring-opening polymerization of lactones. In: *Synthetic Biodegradable Polymers*, Advances in Polymer Science, vol. 245 (eds. B. Rieger, A. Künkel, G.W. Coates, et al.), 173–217. Berlin and Heidelberg, Germany: Springer-Verlag.

26 Podkoscielny, W. and Charmas, W. (1979). Linear polythioesters. II. Products of interfacial polycondensation of 1,4-di(mercaptomethyl)-naphthalene, 1,5-di(mercaptomethyl)naphthalene, and a mixture of 1,4- and 1,5-di(mercaptomethyl)-naphthalene with terephthaloyl and isophthaloyl chlorides. *J. Polym. Sci. Polym. Chem. Ed.* 17: 2429–2438.

27 Hirano, H., Watase, S., and Tanaka, M. (2005). Linear polymers with sulfur in the main chain III. Synthesis of polythioesters by polycondensation of bis(4,4′-mercaptophenyl)sulfide with various acid dichlorides and their properties. *J. Appl. Polym. Sci.* 96: 508–515.

28 Podkościelny, K.W. and Kowalewska, W. (1990). Linear polythioesters. XV. Products of interfacial polycondensation of 1,4-dimercapto-tetramethylbenzene with some aliphatic and aromatic acid dichlorides. *J. Appl. Polym. Sci.* 39: 1649–1656.

29 Podkoscielny, W. and Tarasiuk, B. (1993). Linear polythioesters, XX. Products of interfacial polycondensation of 4,4′-dis(mercaptomethyl)-biphenyl with some aliphatic and isophthaloyl dichlorides. *Angew. Makromol. Chem.* 204: 139–152.

30 Podkoscielny, W. and Rudz, W. (1994). Linear polythioesters. XXII. Products of interfacial polycondensation of bis(4-mercaptomethylphenyl)sulfide with some aliphatic and isomeric phthaloyl dichlorides. *J. Appl. Polym. Sci.* 53: 815–824.

31 Mulvaney, J.E. and Figueroa, F.R. (1986). Synthesis of polydithioesters and cyclization to 1,3-dithiolium(dithiafulvenium) salts. *J. Polym. Sci.* 24: 147–153.

32 Kameyama, A., Murakami, Y., and Nishikubo, T. (1996). Transformation of poly(*s*-aryl thioester) by insertion of thiiranes: a novel approach to achieve the synthesis of sequence-ordered polymers. *Macromolecules* 29: 6676–6678.

33 Kameyama, A., Kimura, Y., and Nishikubo, T. (1997). New synthesis of poly(S-thioester)s by regioselective addition reaction of bis(thiirane)s with diacyl chlorides using quaternary onium salts. *Macromolecules* 30: 6494–6497.

34 Hirano, H., Kadota, J., Agari, Y. et al. (2007). Linear polymers with sulfur in the main chain. IV. Synthesis of thermotropic liquid-crystalline polythioesters based on 4,4′-diphenyldithiol with excellent adhesive properties. *Polym. Eng. Sci.* 47: 262–269.

35 Isogai, H., Tadano, K., and Yano, S. (2008). Liquid crystallinity in polythioethers of 1,4-bis(mercaptomethyl)benzene (MB) and 4,4′-bis(mercaptomethyl)biphenyl

(MBP) with α, Ω-dibromo-N-alkanes and polythioesters of MB and MBP with N-alkanedioyl dichlorides. *Liq. Cryst.* 35: 689–694.
36 Marvel, C.S. and Kotch, A. (1951). Polythiolesters. *J. Am. Chem. Soc.* 73: 1100–1102.
37 Marvel, C.S. and Kraiman, E.A. (1953). Polythiolesters. II. Preparation in emulsions. *J. Org. Chem.* 18: 707–714.
38 Hirano, H., Okada, T., Nakamura, Y. et al. (2006). Enhancement of epoxy resin/copper heterojunction by introduction of sulfur-containing polymers. *Macromol. Mater. Eng.* 291: 205–209.
39 Abe, D., Fukuda, Y., and Sasanuma, Y. (2015). Chemistry of aromatic polythioesters and polydithioesters. *Polym. Chem.* 6: 3131–3142.
40 Abe, D. and Sasanuma, Y. (2012). Molecular design, synthesis and characterization of aromatic polythioester and polydithioester. *Polym. Chem.* 3: 1576–1578.
41 Duchiron, S.W., Pollet, E., Givry, S., and Avérous, L. (2017). Enzymatic synthesis of poly(ε-caprolactone-co-ε-thiocaprolactone). *Eur. Polym. J.* 87: 147–158.
42 Tian, H., Tang, Z., Zhuang, X. et al. (2012). Biodegradable synthetic polymers: preparation, functionalization and biomedical application. *Prog. Polym. Sci.* 37: 237–280.
43 Longo, J.M., Sanford, M.J., and Coates, G.W. (2016). Ring-opening copolymerization of epoxides and cyclic anhydrides with discrete metal complexes: structure–property relationships. *Chem. Rev.* 116: 15167–15197.
44 Ikada, Y. and Tsuji, H. (2000). Biodegradable polyesters for medical and ecological applications. *Macromol. Rapid Commun.* 21: 117–132.
45 You, N.-H., Higashihara, T., Yasuo, S. et al. (2010). Synthesis of sulfur-containing poly(thioester)s with high refractive indices and high abbe numbers. *Polym. Chem.* 1: 480–484.
46 Wübbeler, J.H. and Steinbüchel, A. (2014). New pathways for bacterial polythioesters. *Curr. Opin. Biotechnol.* 29: 85–92.
47 Lütke-Eversloh, T., Bergander, K., Luftmann, H., and Steinbüchel, A. (2001). Identification of a new class of biopolymer: bacterial synthesis of a sulfur-containing polymer with thioester linkages. *Microbiology* 147: 11–19.
48 Lütke-Eversloh, T., Bergander, K., Luftmann, H., and Steinbuchel, A. (2001). Biosynthesis of poly(3-hydroxybutyrate-co-3-mercaptobutyrate) as a sulfur analogue to poly(3-hydroxybutyrate) (PHB). *Biomacromolecules* 2: 1061–1065.
49 Lütke-Eversloh, T., Kawada, J., Marchessault, R.H., and Steinbuchel, A. (2002). Characterization of microbial polythioesters: physical properties of novel copolymers synthesized by *Ralstonia eutropha*. *Biomacromolecules* 3: 159–166.
50 Lütke-Eversloh, T., Fischer, A., Remminghorst, U. et al. (2002). Biosynthesis of novel thermoplastic polythioesters by engineered *Escherichia coli*. *Nat. Mater.* 1: 236–240.
51 Kato, M., Toshima, K., and Matsumura, S. (2005). Preparation of aliphatic poly(thioester) by the lipase-catalyzed direct polycondensation of 11-mercaptoundecanoic acid. *Biomacromolecules* 6: 2275–2280.

52 Kato, M., Toshima, K., and Matsumura, S. (2006). Enzyme-catalyzed preparation of aliphatic polythioester by direct polycondensation of diacid diester and dithiol. *Macromol. Rapid Commun.* 27: 605–610.

53 Kato, M., Toshima, K., and Matsumura, S. (2007). Enzymatic synthesis of polythioester by the ring-opening polymerization of cyclic thioester. *Biomacromolecules* 8: 3590–3596.

54 Overberger, C.G. and Weise, J. (1964). A polythioester by ring-opening polymerization. *Polym. Lett.* 2: 329–331.

55 Overberger, C.G. and Weise, J.K. (1968). Anionic ring-opening polymerization of thiolactones. *J. Am. Chem. Soc.* 90: 3533–3537.

56 Kikuchi, H., Tsubokawa, N., and Endo, T. (2005). First example of cationic ring-opening polymerization of γ-thionobutyrolactone. *Chem. Lett.* 34: 376–377.

57 Fles, D. and Tomasic, V. (1968). Preparation of poly-(S) (−)-[α-(p-toluenesulfonamido)-β-propiothiolactone]. *Polym. Lett.* 6: 809–813.

58 Suzuki, M., Makimura, K., and Matsuoka, S. (2016). Thiol-mediated controlled ring-opening polymerization of cysteine-derived β-thiolactone and unique features of product polythioester. *Biomacromolecules* 17: 1135–1141.

59 Sanda, F., Jirakanjana, D., Hitomi, M., and Endo, T. (1999). Anionic ring-opening polymerization of ε-thionocaprolactone. *Macromolecules* 32: 8010–8014.

60 Sanda, F., Jirakanjana, D., Hitomi, M. et al. (2000). Cationic ring-opening polymerization of ϵ-thionocaprolactone: selective formation of polythioester. *J. Polym. Sci. Part A* 38: 4057–4061.

61 Bannin, T.J. and Kiesewetter, M.K. (2015). Poly(thioester) by organocatalytic ring-opening polymerization. *Macromolecules* 48: 5481–5486.

62 Datta, P.P. and Kiesewetter, M.K. (2016). Controlled organocatalytic ring-opening polymerization of ε-thionocaprolactone. *Macromolecules* 49: 774–780.

6

Polymers with Sulfur-Nitrogen Bonds

Hatice Mutlu[1,2] *and Patrick Theato*[1,2]

[1] *Karlsruhe Institute of Technology (KIT), Institute for Biological Interfaces III, Soft Matter Synthesis Laboratory, Hermann-von-Helmholtz-Platz 1, Eggenstein-Leopoldshafen D-76344, Germany*
[2] *Karlsruhe Institute of Technology (KIT), Institute for Chemical Technology and Polymer Chemistry, Preparative Macromolecular Chemistry, Engesser Str. 18, Karlsruhe D-76131, Germany*

6.1 Introduction

Despite extensive knowledge of polymers with functional groups containing carbon–nitrogen double bonds (e.g. imines [1], hydrazones [2], and oximes [3]), surprisingly, little is known about the sulfur-containing analogues [4]. Specifically, knowing that due to its versatile reactivity, sulfur appears in many different functional groups allowing to address a widespread array of chemical and physical properties. On one hand, unique chemical properties of the sulfur-nitrogen bond (e.g. its polar character and the multiple valence states of sulfur) allow for a large diversity of sulfur-nitrogen bond-containing small organic compounds [5, 6], On the other hand, examples of macromolecules featuring sulfur-nitrogen bonds and its derivatives within the backbone are scarcely reported in comparison with the vast literature available on polymers decorated with versatile sulfur functional groups (such as thioethers, disulfides, oligosulfides, thioesters, thiocarbonates, thiourethanes, thioamides, and thioureas, amongst others) [7]. An exception is polythiazyl, also recognized as polymeric sulfur nitride $(SN)_x$, the first example of a polymeric metal (i.e. inorganic polymer) that does not include any carbon atoms neither in the backbone nor as pendant units, and is composed solely of non-metallic elements: sulfur and nitrogen, respectively [8]. Intriguingly, it was regarded that organic polymers consisting primarily of carbon and hydrogen atoms along with sulfur-nitrogen functional groups have not been known in the field of macromolecules until a decade ago when the very first synthesis of diverse sulfur-nitrogen containing polymers, such as poly(sulfenamide), poly(diaminosulfide), and poly(diaminodisulfide) was postulated [9, 10]. While the latter is partially correct, eventually, an in-depth literature survey demonstrated that the existence of some of these polymers is dating back to early 1990s, when Visco et al. developed various poly(diaminosulfide)s that were applicable as novel solid redox electrodes [11]. Critically, the lack of a comprehensive review on

Sulfur-Containing Polymers: From Synthesis to Functional Materials,
First Edition. Edited by Xing-Hong Zhang and Patrick Theato.

sulfur-nitrogen polymers, which presents this class of polymers in a coherent manner, has made it difficult to assess the first pioneering studies for each class of polymers in this field, thus leading sometimes to misconceptions. To address this problem, it may be instructive to review systematically the sporadically developed synthetic methods that allow for an efficient synthesis of polymers decorated with diverse sulfur-nitrogen bonds. Concurrently, it is also of fundamental importance to understand the distinctive chemical characteristics of these polymers, which ultimately dictate their performance in different fields of application ranging from energy storage to biomedicine. It shall be emphasized that the selection of examples was made in order to stimulate the reader in regard to what is synthetically possible, and hence makes no claim to be completed [12]. Therefore, in the following, we will first discuss the most common sulfur-nitrogen groups (Scheme 6.1) organized by structural class, and the respective monomers and polymers thereof, which have been investigated in past and recent years. Next, we will highlight selected examples of applications, which rely on the efficient use of these polymers.

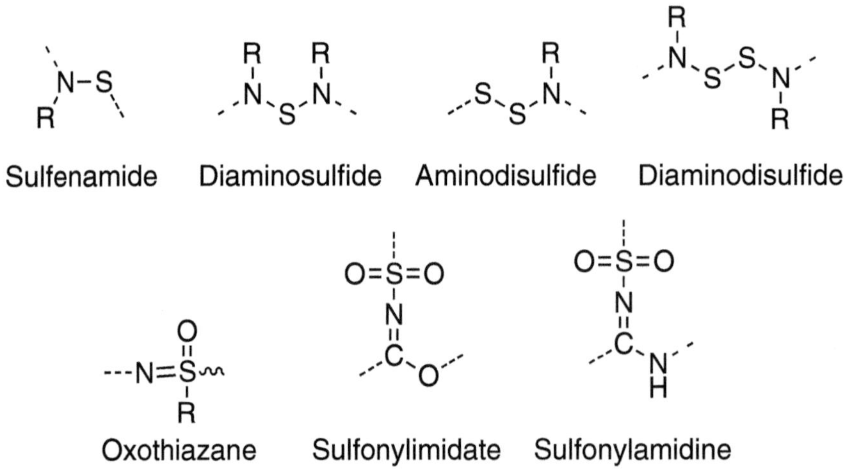

Scheme 6.1 Overview of the principal functional groups possessing diverse sulfur-nitrogen bonds that will be discussed thereafter.

It is important to note that approaches focusing on the equally important synthesis of the inorganic sulfur-nitrogen (i.e. polythiazyl) and sulfur-nitrogen-phosphorous polymers [i.e. poly(thiophosphazene)s, and poly(thionylphosphazene)s] are omitted. Indeed, the reader is encouraged to refer to the literature [13], which is written focusing on the synthesis and applications of these respective polymers.

6.2 Synthesis of Sulfur-Nitrogen Containing Polymers

6.2.1 Poly(sulfenamide)s

Sulfenamides (also named as thiohydroxylamines, Scheme 6.1) are compounds containing a bond of divalent sulfur and trivalent nitrogen (i.e. –S–N–), which by homolytical cleavage can generate aminyl and thiyl radicals [14]. Furthermore, due to the polarization of the S–N bond, sulfenamide derivatives can be subjected to attack by nucleophiles or electrophiles, at either sulfur or nitrogen atom, respectively. In addition, oxidation can take place at either the sulfur or the nitrogen atom. Last but not at least, the S–N bond is stable at neutral pH, but can be hydrolyzed in water under acidic conditions [15].

A critical drawback for this class of compounds is that their nomenclature has not yet been thoroughly systematized. Indeed, a variety of different functional groups are still identified as sulfenamides (e.g. $RSSNR_2$, RSN(R)SR, and R_2NSNR). For clarity reasons within this chapter, the sulfenamide functional group will be refereed, in particular, to the structure unit depicted in Scheme 6.1. Nevertheless, the first synthesis of this class of small molecules dates back to 1911 [16]. Since then, sulfenamides have been extensively studied as intermediates in organic and peptide synthesis [17]. Further, due to their aforementioned attractive chemical properties, sulfenamides were utilized in various industries including as flame retardants, pharmaceuticals, fungicides, and in rubber vulcanization [18–20]. In fact, some of these properties made sulfonamides very attractive functional groups to pursue the synthesis of macromolecules.

Literature survey reveals that the synthesis of main-chain poly(sulfenamide)s, which was reported by Bowden and coworkers in 2012, was initially attempted on two different methods, both of them relying on transamination reaction of diamines with activated dithiol derivatives ((**2**) and (**3**) in Scheme 6.2, respectively) [21]. The classical method of transamination (i.e. amine exchange) for activated thiol compounds involves the displacement of an amine of relatively weak basicity by another of stronger basicity. While the reaction is thermodynamically favorable, the equilibrium is displaced by removing the weaker amine from the reaction mixture. The efficiency of the reaction is dependent on the volatility of the replaced amine; in other words, the amine being displaced should be more volatile than the displacing amine. Alternatively, by using an excess of the displacing amine, amines of similar basicity could be also exchanged.

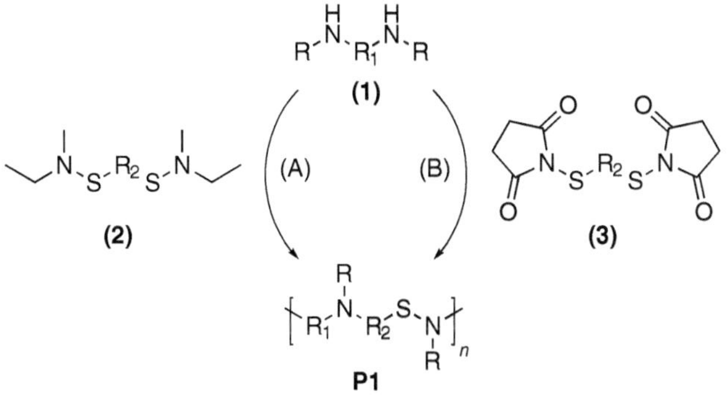

Scheme 6.2 The synthesis of main-chain poly(sulfenamide)s (**P1**) attempted on two different methods (A) and (B), both of them relying on transamination reaction of diamines with activated dithiol derivatives (**2**) and (**3**), respectively.

Respectively, in the first approach (i.e. (A)) disulfenamide derivatives (**2**) were utilized as the activated dithiol derivatives for the transamination reaction of diamine (**1**), while in the second approach (e.g. (B)), dithiosuccinimides (**3**) were the reaction counterparts of choice. The sulfenamides (**2**), which are employed in the former method, were obtained from the reaction of *N*-ethylmethylamine and disulfenylchloride derivatives. The latter compound (**3**), in turn, was generated *in situ* from the reaction of the respective dithiol with sulfuryl chloride, SO_2Cl_2. The main drawback of this approach is that the disulfenylchloride derivatives are considerably unstable and hard to be isolated. Once left overnight exposed to air, these molecules generate disulfides. Furthermore, polymerization with method (A) was performed at temperatures above the boiling point of the side product, i.e. *N*-ethylmethylamine (e.g. 37 °C), or even higher reflux temperatures (depending on the solvent used for the polymerization reaction) in order to drive the polymerization to completion. On the contrary, in the second method, dithiols were activated with a milder agent, *N*-chlorosuccinimide, to yield the more reactive dithiosuccinimide derivatives (**3**).

Whereas, the former method (e.g. method (A)) of polymerization was only effective in a polar protic solvent (i.e. methanol), the kinetics of the polymerization were hindered by the lack of solubility of the formed polymer. Hence, only oligomers were obtained in a low yield of a modest molecular weight (M_w = 2900 g mol^{-1}). Noteworthy, the second method, i.e. (B), based on the synthesis of poly(sulfenamide)s from the bench-stable dithiosuccinimides (**3**) could also proceed in polar aprotic solvent (e.g. methylene chloride or chloroform, 1.2 M) under catalyst-free and considerably mild conditions, i.e. at ambient temperature. By taking advantage of this approach, diverse aliphatic and cycloaliphatic dithiosuccinimides were reacted with three different secondary diamines for 24 hours to deliver well-defined poly(sulfenamide)s **P2–P6** with number-average molecular weights up to 5200 g mol^{-1} and yields of polymerization that ranged from 90 to 97%, respectively (Scheme 6.3), with dispersity (*Đ*) values in the range of 1.06–1.60. Crucially, secondary diamines were utilized, because the synthesis of polymers in the presence of primary diamines resulted in crosslinked, insoluble polymers. Mechanistically, it was postulated that the polymerization reaction proceeded through an S_N2 reaction pathway. Intriguingly, while these polymers were stable in most conventional organic solvents, they readily degraded in water with rate constants that were several orders of magnitude higher than the neutral and acidic degradation of esters [22]. For instance, slow rates of decomposition were found under neutral ($k = 4.7 \times 10^{-4}$ s^{-1}) and basic conditions ($k = 8.0 \times 10^{-5}$ s^{-1}), whereas the decomposition was excessively rapid under acidic conditions to measure a rate constant. Complementary nuclear magnetic resonance (NMR) studies indicated that ~75% of the degradation product was the thiosulfinate.

Alternatively to the main-chain poly(sulfenamide) synthesis, a couple of attempts were undertaken to synthesize polymers decorated with side-chain sulfenamide functional groups. For instance in 2016, Wilen and coworkers, reported on polymers bearing the sulfenamide functionality [23]. Accordingly, two commercially available polymers, i.e. Uvinul® 5050 H (**P7** in Scheme 6.4) and Sabo®stab UV 94,

	Structure	M_n (g·mol^{-1})[a]	Đ
(P2)	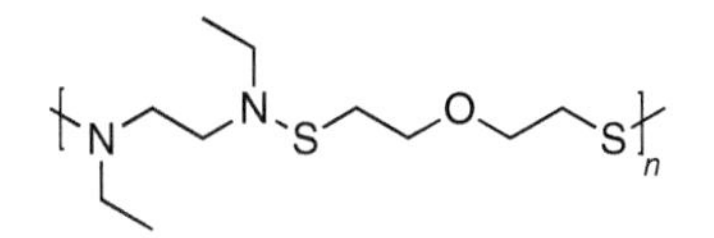	2 000	1.60
(P3)	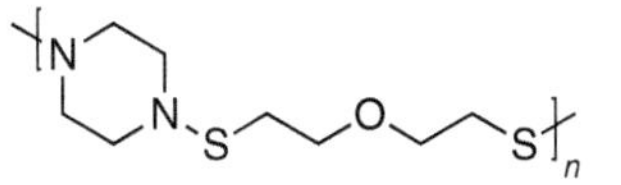	2 600	1.42
(P4)	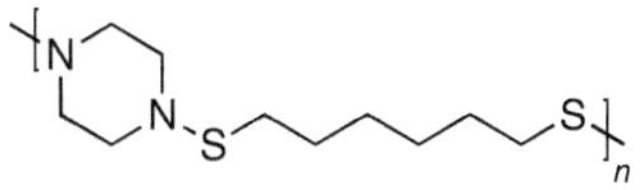	5 200	1.06
(P5[b])	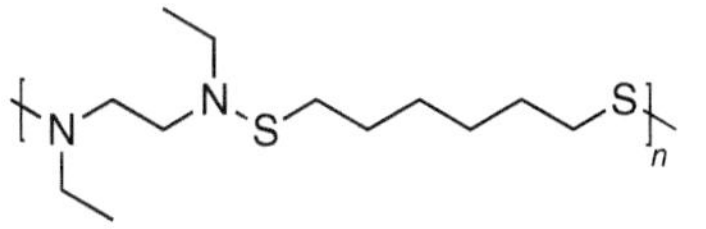	3 500	1.42
(P6[b])	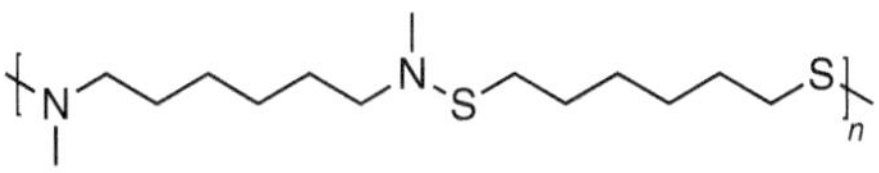	2 600	1.13

Scheme 6.3 Synthesis of poly(sulfenamide)s **P2–P6** from the bench-stable dithiosuccinimides (3) by employing method (B) (from Scheme 6.2); [a]The absolute M_n and $Đ$ were measured with multi-angle laser light scattering and refractive index detectors unless otherwise noted. [b]The M_n and $Đ$ were measured versus polystyrene standards using a calibration curve and the reflective index peaks of the polymer.

were post-modified with benzenesulfenyl chloride at ambient temperature to yield the sulfenamide substituted derivatives [18]. These materials demonstrated unique flame-resistant properties, which will be further discussed at Section 6.3.3. In particular, the increased number of substitutions of the nitrogen atom increased the thermal stability of these materials.

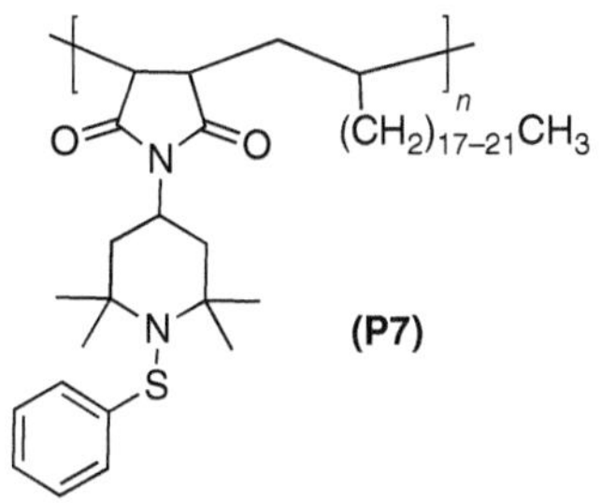

(P7)

Scheme 6.4 Tentative structures for a side-chain poly(sulfenamide) derivative, i.e. sulfenamide modified Uvinul® 5050 H.

In summary, due to the fact that poly(sulfenamide)s possess a functional group that was previously unknown in polymer chemistry, this class of polymers presents new opportunities in macromolecular chemistry. In fact, we can emphasize that the present literature survey clearly depicts the numerous methods, which can be adopted as alternative novel synthesis methodologies towards poly(sulfenamide)s. For instance, recently, the metal-free, iodine-catalyzed dehydrocoupling reaction in the presence of hydrogen peroxide has been described as an efficient method for the preparation of sulfenamides from anilines and various thiols in good to excellent yields (Scheme 6.5) [24]. This methodology has been characterized by its utilization of a non-toxic reaction medium (i.e. polyethyleneglycol, PEG_{400}), high atom economy and wide functional group tolerance. Hence, this reaction can be considered as a mild, clean, and efficient procedure for the synthesis of a variety of poly(sulfenamide)s. In fact, the employed aniline derivatives are also readily commercially available starting materials.

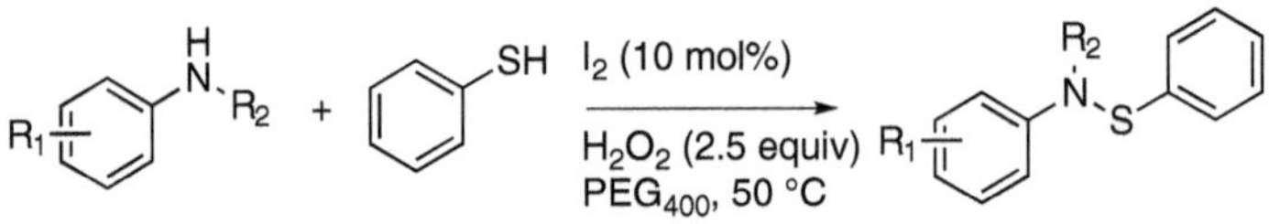

Scheme 6.5 The unprecedented iodine-catalyzed N–H/S–H dehydrocoupling reaction of thiophenol and various anilines in the presence of hydrogen peroxide has been described as an efficient method for the preparation of sulfenamides in good to excellent yields. Source: Based on Ref. [24].

6.2.2 Poly(diaminosulfide)s

As aforementioned, the adoption of unusual functional groups into polymers would not only enhance, but would also impart polymers with new and interesting physicochemical properties. In this context, similarly to sulfenamide, diaminosulfide (i.e. a sulfenamide substituted with a second amine) is still a rather unrecognized kind of functional group in the area of macromolecules, although a century has passed since the discovery of diaminosulfides [25]. Nevertheless, polymer chemists have shown interest in its utilization as monomers for the synthesis of diverse polymers. Particularly, diaminosulfides have been applied in the high-temperature vulcanization of diverse polymers (such as polyethylene) [26] and as sulfur transfer reagents [27]. Hence, motivated by the chemical properties of this functional group in small molecules, Bowden and coworkers were the first to perform pioneering studies on the synthesis of poly(diaminosulfide). Accordingly, as shown in Scheme 6.6, it was postulated that poly(diaminosulfide)s (**P8**) could be synthesized by reacting secondary diamines with small molecules that are highly successful sulfur transfer reagents, by using, respectively, methods (A) and (B) [28]. While, most of the employed diamine compounds were commercially available, the challenge in the polymerization was to develop the useful sulfur transfer agents, (**4**), (**5**), and (**6**). Indeed, initial polymerization attempts were considered to be performed in the presence of sulfur dichloride, i.e. SCl_2, which is quite often used in the synthesis of small molecule diaminosulfides. However, the use of SCl_2 has several drawbacks arising from the fact that this molecule has a low boiling point (59 °C) and must be handled under inert atmospheres. Further, SCl_2 is challenging to purify, reacts with several other functional groups such as alcohols and alkenes, and critically releases the corrosive gas HCl. Respectively, three different sulfur transfer agents were developed in order to pursue an efficient polymerization ((**4**), (**5**), and (**6**) in Scheme 6.6).

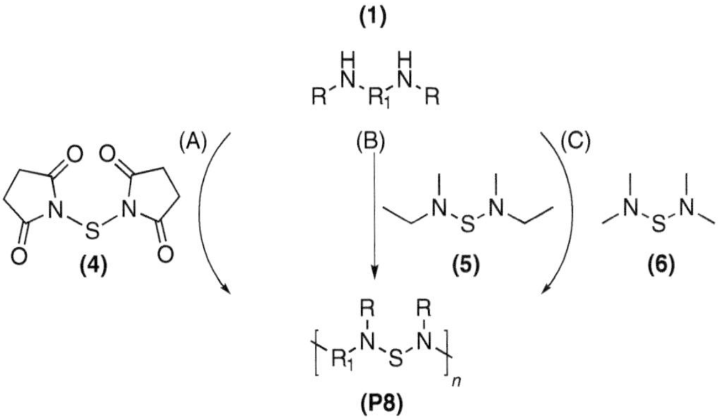

Scheme 6.6 The synthesis of main-chain poly(diaminosulfide)s (**P8**) attempted on two different methods (A), (B) and (C), all of them relying on the reaction of diamines with activated sulfur transfer agents, e.g. (**4**), (**5**), and (**6**), respectively.

Initially, the molecule utilized in approach (A) ((**4**), Scheme 6.6) was explored as a sulfur transfer reagent, since the reactions of *N*-thiosuccinimides with amines are well-recognized in organic synthesis [29]. Although the synthesis of molecule (**4**), depicted in Scheme 6.7a, was straightforward, its purification was challenging, because of its poor solubility in most conventional organic solvents (such as, benzene, chloroform, dimethyl sulfoxide (DMSO), and methylene chloride). In similar manner, using *N*-chlorophthalimide instead of *N*-chlorosuccinimide has not resolved the limited solubility of the sulfur transfer agents. Eventually, the sulfur transfer agents described in method (B) and (C), i.e. (**5**) and (**6**) depicted in Scheme 6.6, were applied in the synthesis of the targeted polymers (**P8**) that structurally resemble polythiazyl derivatives, which were first synthesized in 1953 by using S_2N_2 [8]. The sulfur transfer agent (**5**), as well (**6**), can be readily synthesized in two steps, feature significantly good solubility properties (see Scheme 6.7b) and can be directly utilized in the synthesis of polymers **P9–P12**, as shown in Scheme 6.8. The polymerizations were performed either in benzene or chloroform at the reflux temperatures of the respective solvents for up to 96 hours.

The isolated polymers were with molecular weight in the range of $1600 < M_n < 7200\ \mathrm{g\,mol^{-1}}$ ($1.6 < Đ < 3.7$). The polymerization with piperazine yielded an insoluble polymer **P9** in common organic solvents (Scheme 6.8), and the polymer synthesized with *N*-methyl-3,3′-bis(methylamino)dipropylamine, i.e. **P10**, had limited stability in polar protic solvents (such as methanol and water) during the isolation process via precipitation. It was postulated that the internal, tertiary amine unit within the polymer backbone underwent intramolecular reaction with the diaminosulfide repeating unit in the presence of the polar solvents, hence degrading into compounds with an analytically undefined structure. The latter finding was

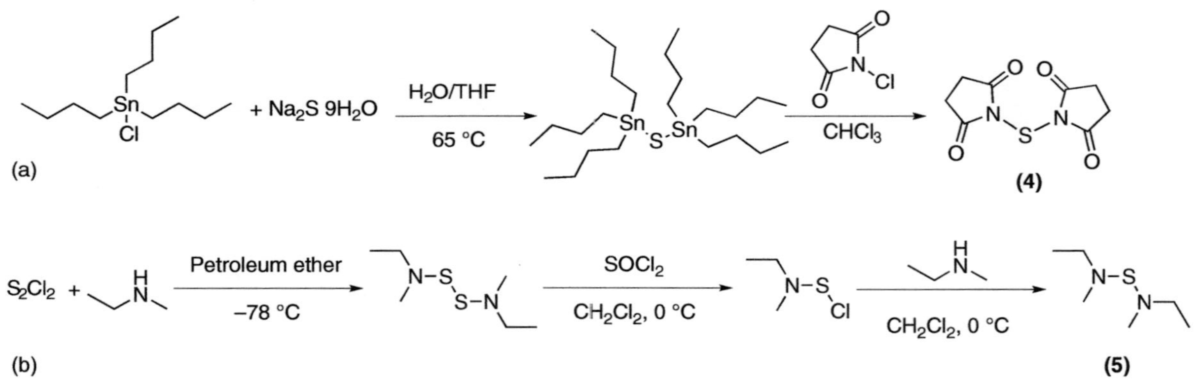

Scheme 6.7 The synthesis of two sets of sulfur transfer reagents. (a) The synthesis of a dithiosuccinimide (**4**). (b) representative synthesis of bisaminosulfide (**5**), which is also applicable for the synthesis of sulfur transfer agent (**6**).

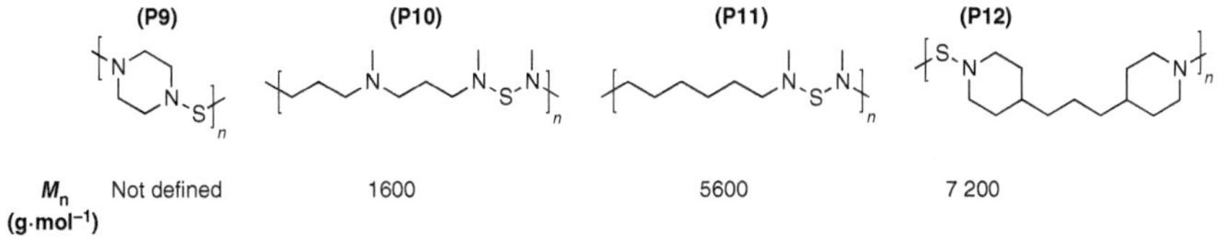

Scheme 6.8 Synthesis of N–S–N bond polymers, i.e. poly(diaminosulfide)s **P9–P12**.

also in agreement with a prior study demonstrating that diaminosulfides reacted in water to form sulfur monoxide, which in turn possessed a half-life of seconds and decomposed into SO_2 and elemental sulfur, respectively [30]. Nevertheless, these results clearly implied that the diaminosulfide functional group was stable in apolar aprotic solvents, and degraded slowly in polar protic solvents. In contrast, the polymerization reaction performed with 1,6-bis(methylamino)hexane yielded polymer **P11** with moderate molecular weights (M_n of 5600 g mol^{-1}, $Đ$ of 3.4). In similar manner, poly(diaminosulfide) **P12** was synthesized from 1,3-bis(4-piperidyl)propane, as diamine source and could be isolated with a M_n of 7200 g mol^{-1} ($Đ = 3.3$) in excellent yield (<96%). Its composition was characterized by elemental analysis to provide further evidence that it possessed the indicated structure shown in Scheme 6.8. Accordingly, the calculated weight composition of the repeating unit was C (64.95%), H (10.06%), N (11.65%), and S (13.34%), and the measured weight composition of the polymer was C (64.70%), H (9.97%), N (11.76%), and S (13.44%). Hence, this agreement between the calculated and measured elemental compositions for polymer **P12** confirmed that only one sulfur atom tethering the two nitrogen atoms is present in each repeating unit of the polymer. Unfortunately, the thermal and mechanical properties of these polymers were not investigated. Nonetheless, they were not prone to oxidation; in fact, no evidence of oxidation of the sulfur atoms was reported for the any of the aforementioned polymer samples.

6.2.3 Poly(aminodisulfide)s and Poly(diaminodisulfide)s

The disulfide bond, as one of the most intriguing dynamic covalent bonds featuring a simple redox chemistry, has a paramount importance not only in biochemistry [31], but also in polymer science [32]. Indeed, a rich literature is pointing out the advantages of the dynamic nature of the single disulfide bridges to explore adaptive self-sorting, self-healing, and degrading properties of various polymer materials in response to diverse stimuli (such as redox conditions, thiol nucleophiles, heat or light) [33]. Despite the widespread utilization of alkyl/aryl disulfide bonds in organic synthesis and polymer chemistry, it is relevant to note that amino- and diaminodisulfide functional groups (Scheme 6.1, –S–S–N– and –N–S–S–N, respectively) are very understudied in organic chemistry, and consequently only a scarce number of polymers that contain these functional groups are known. Nonetheless, there is continuing interest towards these novel polymers with either amino- or diaminodisulfide bond, respectively. Henceforth, it is essential to comment on some of the pioneering studies, particularly applicable for the synthesis of these polymers in separate sub-chapters.

6.2.3.1 Poly(aminodisulfide)s

A representative of a rare and extraordinary class nitrogen-sulfur polymers, i.e. poly(aminodisulfide)s, also known as poly(aza-disulfide)s, was synthesized unprecedently by the reaction of aromatic amines (i.e. aniline) with sulfur monochloride (i.e. S_2Cl_2) in early 1970s (Scheme 6.9, **P13**) [34]. On one hand, there was no evident report on the synthesis of small molecules possessing this unusual sulfur-nitrogen functional group up to 2018 [35], presumably due to the fact the construction of aza-disulfides is still a challenging transformation due to the high reactivity of S–S bond towards nucleophiles (such as amines). On the other hand, polymers synthesized at 30–50 °C until no more HCI was liberated turned out to be stable (air stability up to 300–350 °C) with melting temperatures (T_m) in the range of 182–189 °C, respectively, and were found to be paramagnetic. For instance, the concentration of paramagnetic particles was 10^{18}–10^{20} spin g^{-1} with a large electron displacement capacity, thus resulting in a polyvalent macroradical with additional aromatic stabilization effect. As a result, this novel class of polymers was utilized as thermal stabilizers of vulcanized polyethylene, due to their excellent physicochemical properties [26].

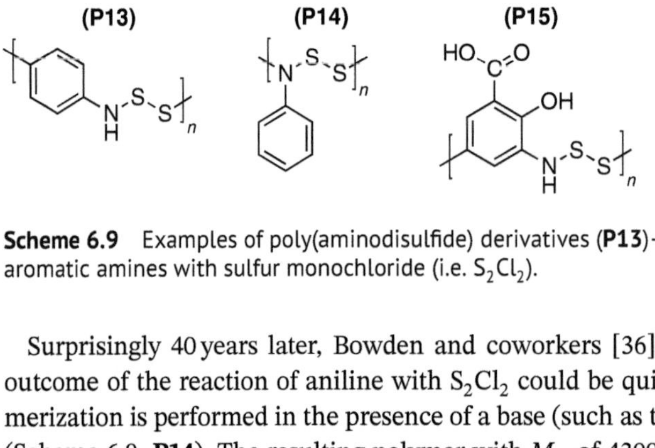

Scheme 6.9 Examples of poly(aminodisulfide) derivatives (**P13**)–(**P15**) synthesized from aromatic amines with sulfur monochloride (i.e. S_2Cl_2).

Surprisingly 40 years later, Bowden and coworkers [36] demonstrated that the outcome of the reaction of aniline with S_2Cl_2 could be quite different, if the polymerization is performed in the presence of a base (such as triethylamine) at −78 °C (Scheme 6.9, **P14**). The resulting polymer with M_w of 4300 g mol^{-1} ($Đ = 1.31$) was possessing a different chemical structure in comparison to the one reported by Losev and Paushkin [26]. In fact, the conjugation of the macromolecular chain was through the inorganic aminodisulfide functional group, whereas the aromatic ring bonded to the nitrogen would allow the conjugation to be altered by varying the presence of functional groups. Thus, this polymer combined conductivity through inorganic functional groups with the ability to alter the conductivity by the presence of organic functional groups.

Another seminal report focused on the synthesis of poly(aminodisulfide) is based on the polymerization of 5-aminosalicylic acid (5-ASA) in the presence of S_2Cl_2 [37]. The polymer (**P15** in Scheme 6.9) with a weight-average molecular weight of ~2200 g mol^{-1}, corresponding to on average of ten monomeric units, was characterized by UV absorbance spectrometry. Two absorption maximums at 241 and 286 nm with ε values of 49 200 and 94 200 M^{-1} cm^{-1}, respectively, were in accordance with the UV spectrum of 5-ASA in 10% DMSO, which was also characterized by absorption maxima at 240 and 300 nm with ε values of 3310 and 3470 M^{-1} cm^{-1}, respectively, thus implying the successful incorporation of 5-ASA within the poly(aminodisulfide).

6.2.3.2 Poly(diaminodisulfide)s

The simplest diaminodisulfide, e.g. NH_2–S–S–NH_2, possesses a low bond dissociation energy (180 kJ mol^{-1}) [37] in reference to the diaryl and dialkyl disulfide bonds (190–230 kJ mol^{-1} and 250–290 kJ mol^{-1}, respectively) [39]. The latter is result of the high stability of the respective thiyl radicals that are formed via the catalyst-free thermally (or possibly photochemically) induced reversible dissociation. Interestingly, in spite of the low bond dissociation energy for the disulfide bond in disulfidediamines, molecules possessing this functional group are stable and readily handled at ambient temperatures. Therefore, diverse disulfidediamine derivatives, with first syntheses dating back to 1895 [40], have been used as thermally active crosslinking agents in the vulcanization of rubber, due to the facile homolytic cleavage of its S–S bond at temperatures below 200 °C [41].

Attracted by these unique redox features and the promising potential as a dynamic scaffold, the first related polymeric diaminodisulfide derivatives (**P16**)–(**P18**), depicted in Scheme 6.10, were synthesized by Visco and coworkers in 1991 [11]. Accordingly, diverse primary and secondary amines (such as ethylenediamine, 1,2-dimethylethylenediamine, piperazine, poly(ethylene imine)) were reacted with S_2Cl_2 in chloroform (Scheme 6.10). In some cases, 4-dimethylaminopyridine (DMAP) was utilized to scavenge the HCl generated during the reaction. While, the polymer (**P17**) obtained from ethylenediamine decomposed under vacuum, elemental analysis clearly indicated the efficient synthesis of poly(diaminodisulfide)s (**P16**) starting from piperazine. However, this early study is not revealing any complementary characterization of the obtained poly(diaminodisulfide)s (**P16**)–(**P18**). The crucial information disclosed in this study is that in fact for the efficient synthesis of poly(diaminodisulfide)s, it is important wherever the diamine monomer is primary or secondary.

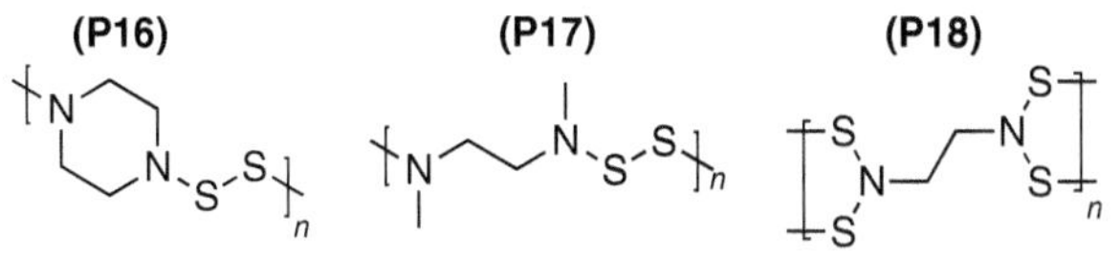

Scheme 6.10 Poly(diaminodisulfide) derivatives (**P16**)–(**P18**) synthesized from aliphatic diamines with sulfur monochloride (i.e. S_2Cl_2).

It took more than 20 years, until when the synthesis and in depth characterization of polymers possessing the disulfidediamine moiety was explicitly reported [36]. The polymerization attempts were based on the utilization of three different disulfidediamine transfer agents (e.g. dithiobissuccinimide, dithiobisphthalimide, and dithiobiscyclohexanedicarboximide, respectively, (**7**), (**8**), and (**9**) in Scheme 6.11) for the synthesis of poly(disulfidediamine)s. On one hand, the synthesis of dithiobissuccinimide (**7**) was straightforward; nevertheless, its purification was tedious, due to its limited solubility in common organic solvents, with an upper limit for solubility of 50 mg mL^{-1} in dichloromethane. Furthermore, the synthesis was accompanied by side products that were challenging to remove. On the other hand, dithiobisphthalimide (**8**) was synthesized in good yield and high purity, however, its solubility was more limited than dithiobissuccinimide (**7**), which in turn would hinder its usefulness in step polymerizations. Eventually, dithiobiscyclohexanedicarboximide (**9**), a novel disulfur monomer, was developed, as shown in Scheme 6.11, and readily obtained by crystallization up to >20 g without any limitation. The only drawback of this compound was that 18% of it decomposed in CD_3OD within five hours.

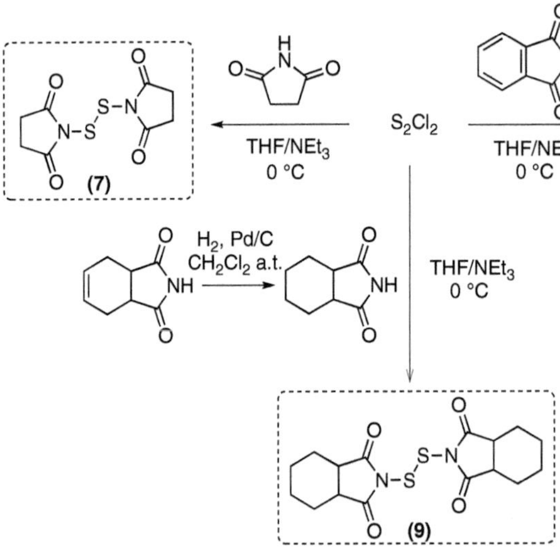

Scheme 6.11 The synthesis of three different disulfidediamine transfer agents (e.g. dithiobissuccinimide, dithiobisphthalimide, and dithiobiscyclohexanedicarboximide, respectively (**7**), (**8**), and (**9**)), further applied for the synthesis of poly(disulfidediamine).

Accordingly, a library of different poly(diaminodisulfide)s (**P16**), (**P19**)–(**P24**) in Scheme 6.12, respectively, with moderate molecular weights ($3900 < M_w < 11\,400$ g mol^{-1}, $1.2 < Đ < 1.8$) were synthesized in dichloromethane at ambient atmospheric conditions for 24 hours with dithiobiscyclohexanedicarboximide (**9**), as the disulfide transfer agent monomer, and diverse secondary diamine derivatives. In similar manner to poly(sulfenamide)s, the dispersity $Đ$ values were lower than expected for step polymerizations for several possible reasons. For instance, the polymers may have been fractionated in the isolation steps resulting in a loss of some low molecular weight polymer. In addition, the authors postulated that the choice of columns for the size exclusion chromatography (SEC) analysis was not optimized for polymers with these molecular weights. To provide further evidence for the presence of disulfidediamine functional groups, the polymer derived from

N,N′-dimethyl-1,6-hexanediamine, (**P20**) was studied by elemental analysis. The theoretical mass percentages (C, 57.3%; N, 10.3%; S, 23.5%; H, 8.8%) for the repeat unit closely matched the values that were measured (C, 55.3%; N, 9.7%; S, 23.1%; H, 8.5%). Thus, the analyzed elements accounted for 97% of the initial mass of the sample. The thermogravimetric analysis of (**P20**) revealed thermal stability at elevated temperatures, but with a sudden weight loss at 175 °C.

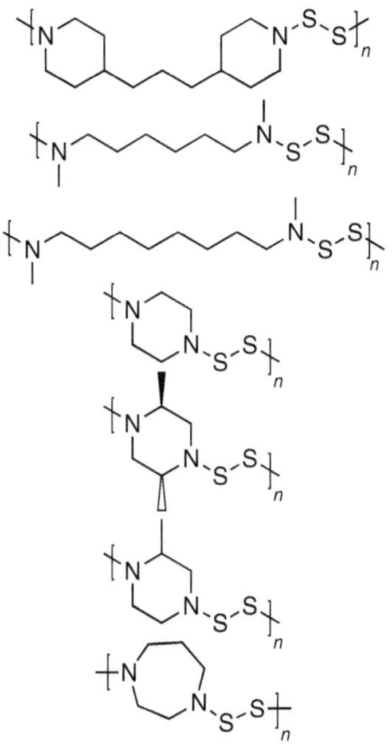

	M_w (g mol^{-1})	Đ
(P19)	11400	1.4
(P20)	5000	1.4
(P21)	4100	1.2
(P16)	- Not determined -	
(P22)	3900	1.2
(P23)	5300	1.2
(P24)	8700	1.8

Scheme 6.12 Poly(diaminodisulfide) derivatives (**P16**), (**P19**)–(**P24**) synthesized from secondary aliphatic diamines with dithiobiscyclohexanedicarboximide (**9**), as the disulfide transfer agent monomer.

Quite recently, Otsuka and coworkers [42] enriched the toolbox of the monomers applicable for the synthesis of poly(diaminodisulfide)s by developing 2,2,6,6-tetramethylpiperidine-1-sulfanyl (TEMPS) dimers as a new class of dynamic disulfide linkages and demonstrating its utility as a functional component in linear and network polymer systems (Scheme 6.13). Indeed, the dimer derivatives of TEMPS were first reported by Bennett et al. in 1967 [43] to be thermally and reversibly dissociable under moderate heating. Despite its unique features and the promising potential as a dynamic constituent in various molecular systems, Otsuka and coworkers are the first to explore TEMPS and its dimers as monomeric unit within a macromolecular system.

Accordingly, bis(4-hydroxy-2,2,6,6-tetramethylpiperidin-1-yl)disulfide (BiTEMPS-OH, (**10**) in Scheme 6.14) was designed as a versatile unit that can be easily modified at its OH groups to create a broad library of derivatives. Interestingly, the starting

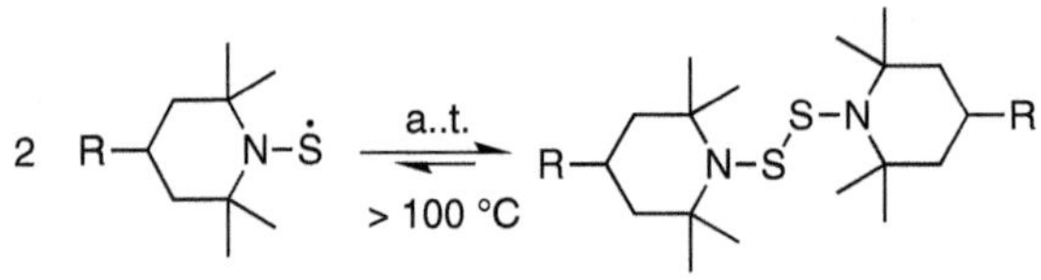

Scheme 6.13 Characteristic thermal and reversible dissociation of BiTEMPS dimers to TEMPS under moderate heating.

BiTEMPS-OH compound, which was successfully synthesized in three steps from commercially available compounds on a multi-gram scale, possessed sufficient air stability that in turn permits its facile use in atmospheric conditions.

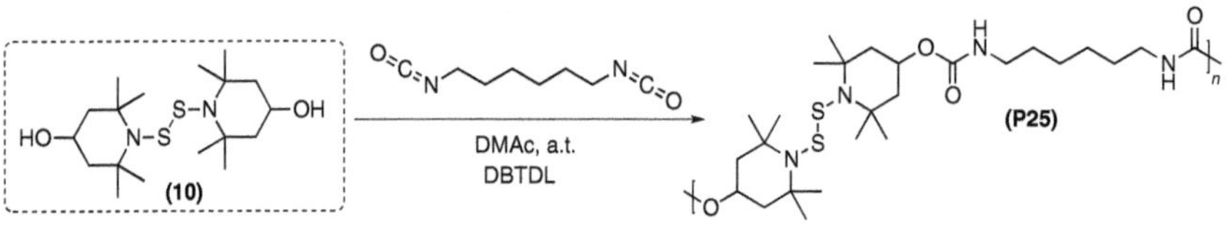

Scheme 6.14 Synthesis of BiTEMPS-incorporated linear polyurethane derivative (**P25**) from BiTEMPS-OH and hexamethylene diisocyanate (HDI) using di-*n*-butyltin dilaurate (DBTDL) as the catalyst.

Respectively, polymer (**P25**) with M_n of 14 900 g mol^{-1} and unimodal molecular weight distribution ($Đ = 1.8$) was first synthesized from BiTEMPS-OH and hexamethylene diisocyanate (HDI) using di-*n*-butyltin dilaurate (DBTDL) as the catalyst (Scheme 6.14). Since, each repeating unit was tethered by the dynamic BiTEMPS linkage, and the molecular weight of the polymer was thermally tunable through the exchange reaction according to the solution concentration without special care. Indeed, evaluation of the SEC data of a 0.5 wt% 1,4-dioxane solution of the polymer at 100 °C demonstrated a rapid shift of the original peak to the lower molecular weight region in 10 minutes, followed by gradual fragmentation, and subsequent degradation. SEC measurements showed a restoration of the polymer with the peak-top molecular weight $M_{\text{peak top}}$ of 25 200 g mol^{-1} (original: 27 700 g mol^{-1}), once the degraded solution was concentrated to 10 wt% and then heated at 100 °C.

In an additional experiment, the authors showed the dynamicity of BiTEMPS in a network polymer system using the network systems prepared according to Scheme 6.15 [44]. The BiTEMPS-containing poly(hexyl methacrylate) P(HMA-*co*-BTA) networks (**PN1** in Scheme 6.15) were synthesized via the facile free-radical polymerizations using a BiTEMPS-containing crosslinker (**11**). The polymerization was carried out at <30 °C in the presence of the low-temperature radical initiator V-70 to prevent the homolytic dissociation of BiTEMPS, which afforded a gel that is insoluble in common organic solvents. The chemical structures of these colorless, insoluble polymer network films were easily characterized by de-crosslinking experiments via bond-exchange reactions with external low molecular-weight BiTEMPS compounds. Furthermore, a thermogravimetric analysis revealed comparable thermal decomposition temperatures ($275 < T_{d,5\%} < 315\,°C$) for all networks, despite the differences in bond dissociation energy, which indicates that the incorporation of BiTEMPS moieties did not affect the thermal stability of the obtained materials in the temperature range examined. Swelling experiments and stress-relaxation measurements confirmed the readily inducible chain-exchanging properties based on the dissociation of BiTEMPS upon moderate heating (120–140 °C). External catalysts and special care to exclude atmospheric moisture or oxygen were not required, due to the high stability of the TEMPS radical mediators, which endows these network polymers with on-demand healability and reprocessability properties in the bulk.

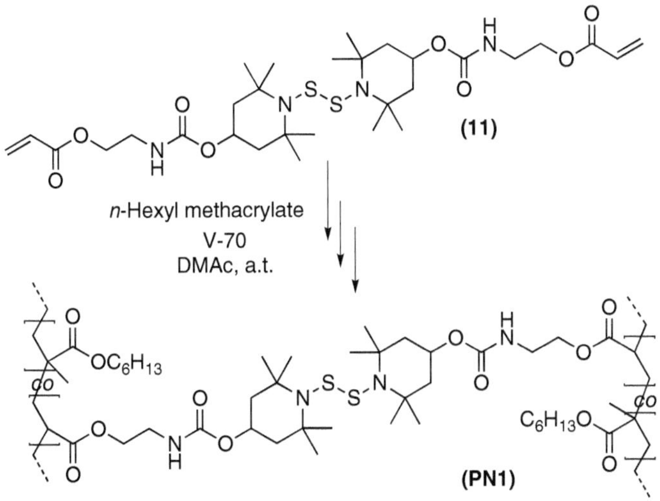

Scheme 6.15 Synthesis of BiTEMPS-crosslinked poly(methacrylate) network (**PN1**) via the facile free-radical polymerization of hexyl methacrylate using a BiTEMPS-containing crosslinker (**11**), which endows these network polymers with on-demand healability and reprocessability properties in the bulk. Source: Based on Ref. [44].

6.2.4 Poly(oxothiazene)s

Poly(oxothiazene)s, polymers with an alternating sulfur–nitrogen backbone in which both an oxygen atom and an alkyl or aryl group are attached to the sulfur ((**P26**)–(**P29**) in Scheme 6.16), represent another less explored class of polymers decorated with sulfur-nitrogen bond within the main chain. Intriguingly, their chemical structures are presumably analogous to the industrially important polyphosphazenes [45]. However, only a very limited number of well-characterized poly(oxothiazenes) have been described in the literature [46, 47]. In the 1960s, poly(fluorooxothiazenes) and poly(aminooxothiazenes) had been reported as the products formed by the condensation of $O{=}SF_4$ with ammonia [48, 49]. Similarly, poly(phenyloxothiazene) has been reported from the condensation of either phenyl benzenesulfonimidate or *N,N*-dimethylarenesulfonimidamides with arenesulfinyl chlorides [50, 51]. Unfortunately, little characterization data were provided to support the suggested polymer architectures in either of these cases. It was only reported that some of the polymers were soluble in solvents of high polarity, such as DMSO or *N,N′*-dimethylformamide (DMF). Complementary, thermal characterization has revealed that the glass transition temperature (T_g) decreased as alkyl substituents increased in size; although considered unusual, it was postulated that this trend was analogous to the alkoxy phosphazenes.

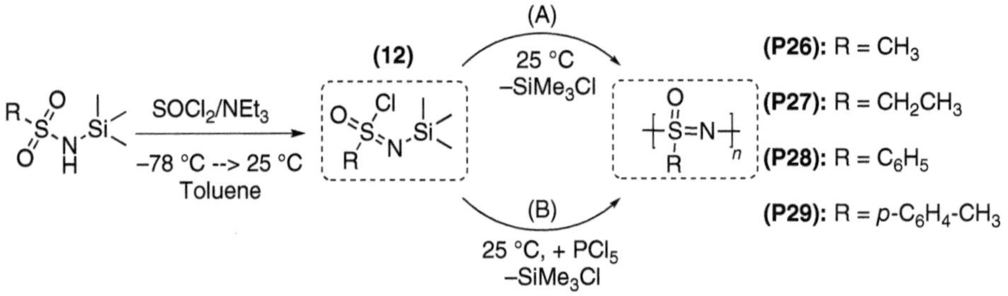

Scheme 6.16 Preparation of *N*-silylsulfonimidoyl chloride derivatives (**12**) and their *in situ* polymerization to deliver polymers (**P26**)–(**P29**). Source: Based on Ref. [53].

In fact, the first well-characterized examples were synthesized by the thermal condensation (e.g. 120–170 °C) of *N*-silylsulfonimidates ((**12**) in Scheme 6.16) in the early 1990s, in analogy to the synthesis of poly(phosphazene)s that is based on the phosphoranimine route [52]. The driving force for this polymerization, which was performed either under catalyst-free conditions or catalyzed by Lewis acids and bases (e.g. BF_3Et_2O, $AlCl_3$, fluoride ion, and phenoxide ion), was the elimination of trimethyl silyl ether. In an alternative method, in which "silyl-free" sulfonimidates were utilized, the reaction was performed at temperatures 20–40 °C lower (i.e. 85–120 °C), since the elimination products were simple alcohol derivatives (such as methanol) with considerably low boiling temperatures, in comparison with the trimethyl silyl ethers. Furthermore, this analogues polymerization approach was at rate approximately two orders of magnitude faster, and resulting in polymers with relatively high molecular weight ($6000 < M_n < 230\,000\,g\,mol^{-1}$). The purity of the obtained products was proven by showing complementary spectroscopic characterization (e.g. SEC or NMR data). Noteworthy, thermal characterization via differential scanning calorimetry (DSC) demonstrated that no clear melting transition has been observed with any of the polymers, which is in line with the polymers being essentially atactic. Nevertheless, there were several limitations to these pathways. These limitations include the necessity for long reactions times (>96 hours) at high reaction temperatures (up to 160 °C) at catalyst-free conditions and ~ 120 °C in the presence of Lewis acids catalysts. Furthermore, a lack of molecular weight control was another limitation; hence occasionally polymers with bimodal molecular weight distribution and high dispersity, i.e. $Đ$ up to 21, were isolated.

Quite recently, in order to address the previously reported limitations, an *in situ* Lewis acid (e.g. PCl_5) catalyzed polycondensation of *N*-silylsulfonimidoyl chlorides (**12**) was postulated as a potential novel reaction pathway (Scheme 6.16, methods (A) and (B)) [53]. This method was conducted at ambient temperature and afforded macromolecules (**P26**)–(**P29**) of high molecular weight and low *Đ* values ($10\,000 < M_w < 30\,000\,g\,mol^{-1}$, and $1.2 < Đ < 1.6$), accompanied by the formation of the condensation byproduct, Me_3SiCl. Interestingly, the properties of the poly(oxothiazene)s prepared by this method differed markedly from the materials prepared from the polycondensation reactions based on *N*-silylsulfonimidates or silyl-free analogues. In other words, these polymers prepared by the *in-situ* route demonstrated reasonably good solubility in both polar and apolar solvents, and possessed unique thermal behavior. Explicitly, DSC analysis of the resulting materials revealed differences in the properties of the polymers formed at ambient temperature in comparison with those synthesized previously at elevated temperatures. For instance, on one hand, in the case of methyl substituted poly(oxythiazene), (**P26**), the observed T_g (41 °C) was significantly lower than the corresponding material formed at reactions performed at elevated temperatures, i.e. T_g = 55–65 °C. In similar manner, replacing the methyl group with aromatic side chains resulted (e.g. (**P28**)) also in a decrease of 20–30 °C in the T_g. It was possible that this difference reflected a variance in the tacticity of the polymers formed by the two different methods, which would result in different packing effects in the two materials. Alternatively, the differences in T_g could be attributed to the differences in the molecular weight distributions that arise from the different synthetic approaches. On the other hand, in stark contrast, T_g values for the ethyl substituted poly(oxothiazene) derivative (**P27**), formed by the two different methods, were in the similar temperature range, i.e. 29–33 °C.

6.2.5 Poly(sulfonylimidate)s

With an interest in synthesizing potentially bioactive compounds possessing sulfur-nitrogen bonds, synthetic chemists have explored the synthesis of sulfonylimidates [54], i.e. sulfur analogues of imidates [55], which are known to be important pharmacophores [56]. On one hand sulfonylimidates have the potential to be useful as prodrug [57] for carboxylic acid esters derived from aliphatic alcohols as well as for drugs containing a primary sulfonamide group. On the other hand, sulfonylimidates have been also recognized as useful synthetic building blocks by virtue of their potent nucleophilicity, in similar way as esters [58]. Thus, sulfonylimidate formation has been often viewed as a way to activate the otherwise relatively inert or unreactive amide equivalents [59]. Given the wide existence of sulfonylimidate derivatives in synthetic chemistry, in addition to their interesting characteristics, recently, polymer chemists have devoted great efforts on developing methods to prepare poly(sulfonylimidate)s. While, the synthesis of conventional poly(imidate)s is limited to two-component step-growth polymerizations using diols and moisture-sensitive imidoyl chlorides [60], since the imidate formation itself is quite challenging, the pioneering studies on the preparation of poly(sulfonylimidate)s are based on the copper (I) catalyzed multicomponent reaction (MCR) of diynes (**13**), sulfonyl azides (**14**), and diols (**15**) (Scheme 6.17) [61].

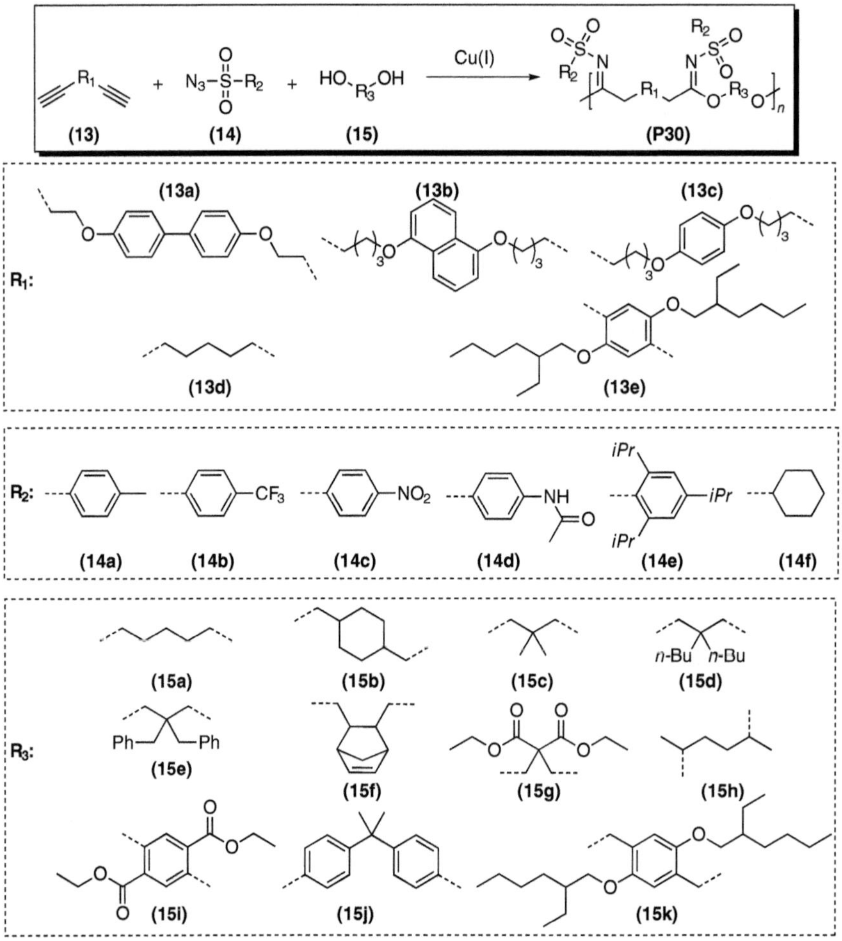

Scheme 6.17 Facile synthesis of poly(sulfonylimidate)s, e.g. (**P30**), based on the copper (I) catalyzed multicomponent reaction of diynes (**13**), sulfonyl azides (**14**), and diols (**15**). Source: Based on Ref. [61].

In fact, in the recent decade, multicomponent polymerizations (MCPs) [62] based on MCRs have been recognized as a highly efficient process for making complex polymer structures by using orthogonally reactive monomers. Particularly, MCPs display a series of advantages such as high efficiency, mild reaction conditions, atom economy, and operational simplicity and thus are representing a powerful new approach for the preparation of functional polymers. Accordingly, Choi and Kim [61] synthesized a library of 24 different poly(sulfonylimidates) (depicted with the general chemical structure of **P30**) with moderate to high molecular weights ($9500 < M_w < 33\,500\,\mathrm{g\,mol^{-1}}$, $1.43 < Đ < 2.98$) via the Cu-catalyzed (10.0 mol%) MCP of readily available and bench stable monomers (e.g. diynes, sulfonyl azides, and diols) (Scheme 6.17). The polymerizations were performed at ambient temperature in the presence of additive, e.g. triethylamine, to accelerate the polymerization. Nevertheless, the reactions took almost 48 hours to reach high polymerization yields. Minimizing water contamination was key to achieving the successful MCP of poly(sulfonylimidate)s. Therefore, hygroscopic polar aprotic solvents (such as DMSO and DMF) were omitted, and dry chlorinated solvents (e.g. chloroform and dichloromethane) were found to be the best solvents for the synthesis of the targeted polymers. Employing rigid diynes or diols produced polymers with higher molecular weight than when flexible linkers are used, presumably because of the suppression of intramolecular cyclization; whereas, the structures of the arylsulfonyl azides (**14**) did not have a significant effect on the MCP. Interestingly, phenolic diol monomers (e.g. (**15i**)–(**15k**)) with weak nucleophilicity were also suitable monomers for the polymerization. Last but not at least, it is important to emphasize that the obtained polymers generally had high thermal stability with decomposition temperatures ranging from 233 to 307 °C.

Attracted by this pioneering example, Tang and coworkers [63] explored the feasibility of employing the aforementioned MCP reaction for polymer synthesis in the presence of water as co-reactant (Scheme 6.18).

Even though it was reported that even a small amount of water contamination is inevitably limiting the efficiency of the reaction, since water reacts with the electrophilic keteneimine intermediate, which in turn terminates the polycondensation by forming *N*-sulfonylamide (Scheme 6.19).

Respectively, the one-pot three-component polymerizations using diyne (**13f**), disulfonyl azide (**16**), and water was performed in the presence of Cu(I) and NEt_3 under inert gas flow at ambient temperature. Nevertheless, formation of insoluble gels was detected within the first two hours of the polymerization. The gelation was attributed to the hydrogen bonding between the polymer chains, which physically crosslinked the polymer. In order to prevent the latter, mono-functionalized alcohol (i.e. ethanol) was utilized as the forth reactant in the MCP. In fact, a polymeric product (**P31**) in a yield of 87% with a satisfactory M_w value (13 700 g mol^{-1}) and reasonable dispersity ($Đ = 3.3$) was isolated.

Since, the Cu(I)-catalyzed cycloaddition reaction between alkynes and sulfonyl azides produces ketenimine species via nitrogen release from the *in situ* generated 1,2,3-triazole derivatives [64], the authors postulated that the polymerization can be potentially carried out in air. Satisfactorily, the polymerization reaction performed in air delivered polymer in comparable yield (85%), molecular weight (M_w of 16 500 g mol^{-1}), and well-defined structure ($Đ = 3.3$), which demonstrated the high oxygen and moisture tolerance of the polymerization. Systemically, the impact of catalyst loading and choice of solvent on the efficiency of polymerization was also explored. While, the optimum catalyst concentration was not playing a crucial role, the solvent exerted a strong effect on the reaction. In the former case, for instance, polymers with M_w value of 16 00 and 13 900 g mol^{-1} were obtained in the presence of 0.02 and 0.03 M of catalyst, respectively. While for the latter case, precipitation took place during the polymerization in tetrahydrofuran (THF), and

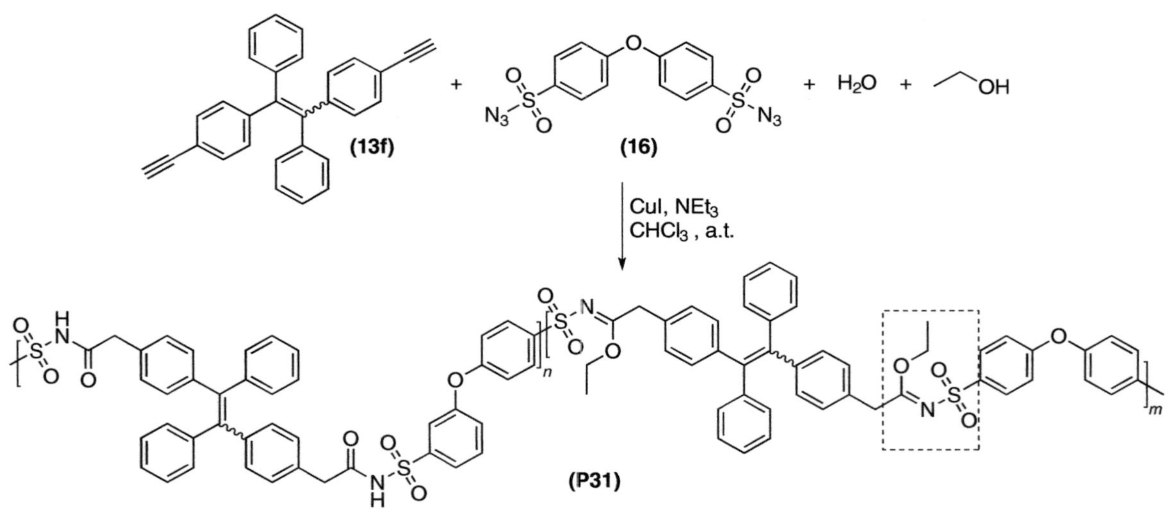

Scheme 6.18 One-pot four-component polymerizations using diyne (**13f**), disulfonyl azide (**16**), water, and ethanol as reactants to synthesize poly(sulfonylimidate) (**P31**).

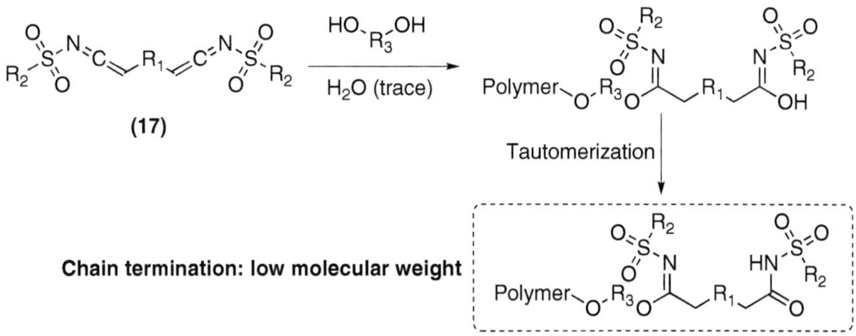

Scheme 6.19 Cu(I)-catalyzed multicomponent polymerization in the presence of trace amounts of water resulting in side reaction, i.e. chain termination, due to the fact that water reacts with the electrophilic keteneimine intermediate (**17**) to terminate the polycondensation by forming *N*-sulfonyl amide derivative.

gels were formed in 1,4-dioxane. Eventually, the polymerization proceeded with high efficiency in chlorinated solvents, such as chloroform and dichloromethane. Furthermore, it was also revealed that by controlling the feed ratio of water to ethanol, the ratio of amide to sulfonylimidate units in the polymer backbone can be readily modulated, which was proven by detailed characterization via X-ray crystallography, fluorescence microscopy along standard spectroscopic techniques. Moreover, the obtained polymers showed a feature of aggregation-induced emission (AIE), since they emitted more intensely in the aggregate state than in the solution state due to presence of AIE-active moieties in the polymer chain. Importantly, this MCP approach is meeting the requirements of "green chemistry," since it fulfils some the crucial requirements of "green chemistry," such as high efficiency, good atom economy, simple operation, cheap monomers, and high oxygen and moisture tolerance.

Encouraged by the aforementioned fruitful achievements, Tang and coworkers [65, 66] expanded the toolbox of Cu(I)-catalyzed cycloaddition reactions between alkynes and sulfonyl azides towards the synthesis of fused heterocyclic polymers with advanced functionalities. Following the seminal works, commercially available salicylic aldehyde (**18a**) or *ortho*-hydroxyacetophenone (**18b**) was reacted with diynes (**13**) and disulfonyl azide (**16**) in the presence of CuI and triethylamine (NEt_3) under nitrogen environment at ambient temperature to generate poly(sulfonylimidate)s (**P32**) integrated with fused heterocyclic unit of iminocoumarins (Scheme 6.20) [65].

On the basis of the preliminary polymerization attempts with salicylic aldehyde (**18a**), DMAc was the most suitable solvent for efficient polymerization among different common organic solvents, including THF, chloroform ($CHCl_3$), and DMSO, thus polymers with the highest M_w and yield were obtained, i.e. 42 400 g mol^{-1} and 91%, respectively. In similar way, the best polymerization result with M_w of 64 600 g mol^{-1} in 91% yield was achieved with monomer concentration of [(**13**)] = [(**16**)] = 0.40 M and optimal feed ratio of [(**13**)]:[(**16**)]:[(**18a**)] = 1 : 1 : 2.2, respectively. On one hand, remarkably, a polymer with M_w of 49 400 g mol^{-1} was

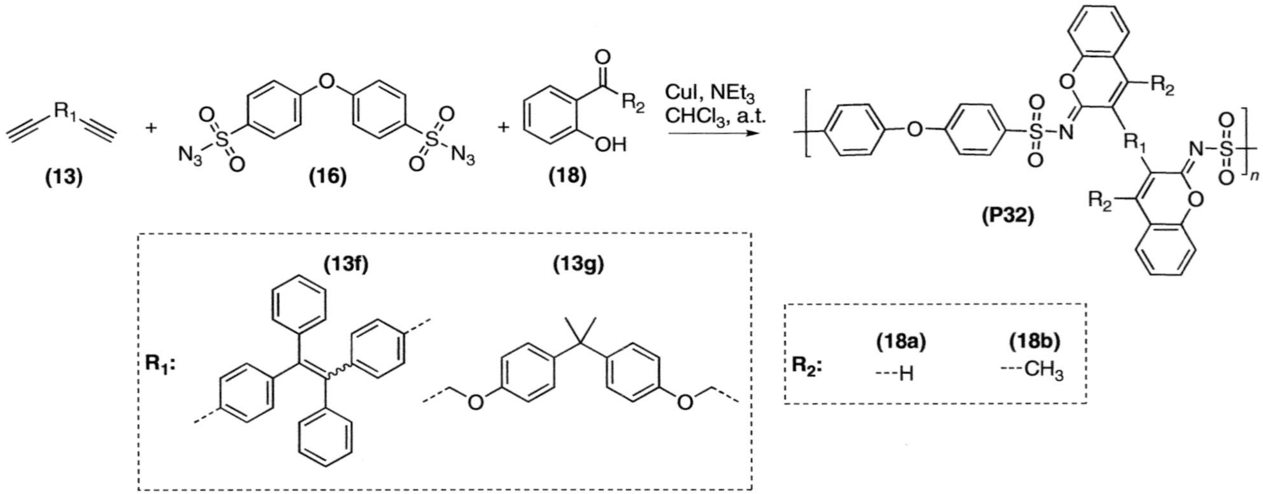

Scheme 6.20 MCPs of diynes (**13**), disulfonyl azides (**16**), with commercially available salicylic aldehyde (**18a**) or *ortho*-hydroxyacetophenone (**18b**) towards functional poly(sulfonylimidate) derivatives (**P32**). Source: Modified from Ref. [65].

obtained within 0.5 hour, whereas prolonging the reaction time to one hour resulted in M_w of 64 600 g mol^{-1}. On the other hand, further extending the reaction time (i.e. more than one hour) didn't have any crucial impact neither on the M_w nor the yield of the polymerization.

When salicylic aldehyde was replaced by *ortho*-hydroxyacetophenone (**18b**) in Scheme 6.20, while keeping other conditions intact, rapid gel formation was observed. In order to prevent the gel formation and yield soluble macromolecules, the concentration of [(**13**)] = [(**16**)] was adjusted to 0.05 M with the constant feed ratio of monomers, i.e. [(**13**)]:[(**16**)]:[(**18b**)] = 1 : 1 : 2.2, respectively. Thus, the general applicability of this powerful and efficient polymerization was manifested by obtaining polymer with M_w of 50 200 g mol^{-1}.

Remarkably, the resulting polymers exhibited an outstanding film-forming ability, high thermal stability ($286 < T_{d,5\%} < 341$ °C and $135 < T_g < 230$ °C), and good morphological stability. Furthermore, due to the highly polarized conjugated structures, the polymer films possessed outstanding refractive indices (1.9284–1.7734) in the visible and near-IR regions (400–893 nm). With abundant photosensitive groups embedded in the polymer chain, the refractive index (n) values of polymer films were also easily modulated by UV treatment. For instance, the n values dropped to 1.6604–1.5849 from initial 1.9284–1.7734 with a decrease of 0.1282 at 632.8 nm after 15 minutes treatment with UV light, thus demonstrating proficient refractivity tunability.

In another effort, 2-hydroxybenzonitrile (**19**) [66] was reacted with diynes (**13**) and disulfonyl azides (**16**) in the presence of CuCl and NEt_3, as catalysts, thus also generating iminocoumarin-containing poly(sulfonylimidate)s ((**P33**) in Scheme 6.21) with weight-average molecular weight (M_w) varying from 13 900 to 37 700 g mol^{-1} with unimodal distribution ($1.47 < Đ < 2.85$), and high yields (up to 96%).

However, in contrast to the previous work, the best polymerization results were achieved with monomer concentration of [(**13**)] = [(**16**)] = 0.05 M and optimal feed ratio of [(**13**)]:[(**16**)]:[(**19**)] = 1 : 1 : 2.5, respectively, at 40 °C in dichloromethane within 90 minutes. Critically, due to the presence of abundant and multiple types of intermolecular noncovalent interactions including hydrogen bonds and π–π stacking interactions in the macromolecules with high values of M_w, insoluble gel formation was detected when higher monomer concentration (<0.05 M), longer polymerization time (<90 min), or higher polymerization temperature (<40 °C) was utilized. Nonetheless, despite of the rigid conjugated fragments and hydrogen bonds, the poly(sulfonyimidate) (**P33**) exhibited satisfactory solubility in polar organic solvents such as DMSO and DMF. Furthermore, thermogravimetric analysis implied the high thermal stability of this class of polymers with decomposition temperatures (T_d) at 5% weight loss ranging from 272 to 344 °C.

6.2.6 Poly(sulfonylamidine)s

In similar way to imidates, amidines [67], which are known as useful synthons to assemble complex molecules, are also one of the most widely exploited moieties in medicinal chemistry. Intriguingly, recent research focus tends to favor the

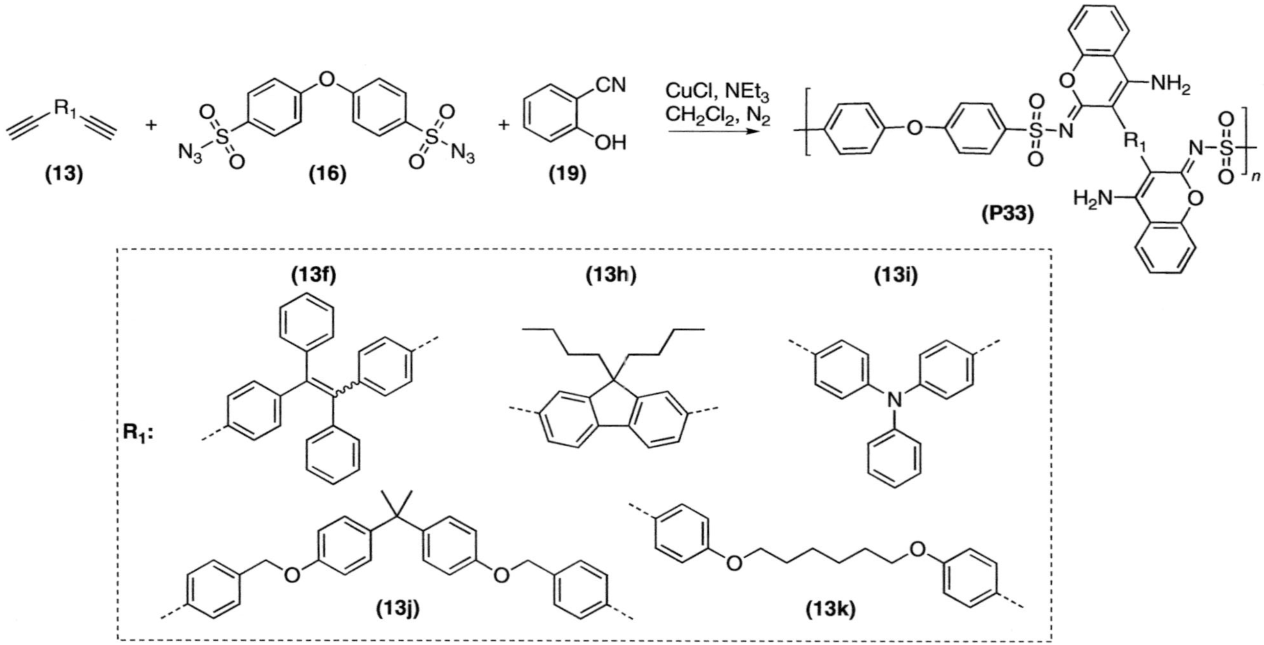

Scheme 6.21 Synthesis of fused heterocyclic-, i.e. iminocoumarin-, containing poly(sulfonylimidate) derivatives (**P33**) from 2-hydroxybenzonitrile (**19**) that was reacted with diynes (**13**) and disulfonyl azides (**16**) in the presence of CuCl and NEt_3.

emergence of sulfonylamidines as bioactive pharmacophores, since numerous new synthetic methods have been developed to access this polar, structurally rigid, and hydrogen-bond-rich moiety. In fact, with the endeavor to develop chemically diverse polymers possessing sulfur-nitrogen bonds, the synthesis of small molecule sulfonylamidines was tailored towards the development of poly(sulfonylamidine)s, the sulfur analogue of poly(amidine)s [68] that have attracted great attention by virtue of their potential role as biomaterials, metallic conductors, and photosensitive semiconductors. Having this in mind, Choi and coworkers [69] explored the three-component polymerization of diynes (**13**), sulfonyl azides (**14**), and diamines (**20**) to deliver 26 well-defined, defect-free high-molecular-weight poly(sulfonylamidine) derivatives ($4700 < M_n < 23\,500\,g\,mol^{-1}$ with $1.50 < Đ < 5.92$) with the generic chemical structure (**P34**) prepared from 26 different monomers [possessing diverse moieties such as alkyl, aryl, diene, and poly(ethylene glycol)], thus demonstrating the broad substrate scope of this methodology (Scheme 6.22).

Through a series of optimizations, it was implied that the addition of excess triethylamine (5 equiv) in the presence of CuCl as catalyst, and the use of DMF as a solvent were key factors to ensure efficient MCP. In addition, the stoichiometric balance between the diyne (**13**) and the diamine (**20**) was also crucial, with monomer concentration of 1.0 M, since both reactants were taking part in the chain-growth process of the step-growth polymerization. Furthermore, excess of sulfonyl azide derivative (**14**) (3 equiv) was essential in order to increase the conversion without limiting the molecular weight of the resulting poly(sulfonylamidine). Regardless of the structure, in addition to alkylsulfonyl azides, various aromatic sulfonyl azides containing electron-donating groups (−OMe, −NHCOMe), electron-withdrawing groups ($-CF_3$, $-NO_2$), or a sterically hindered group (−iPr) at various positions on the aromatic ring were successfully employed in the MCP. Ultimately, the conversion of the polymerization reaction was also significantly enhanced by high reaction temperatures, i.e. 70 °C. Critically, formation of cyclic poly(amidine)s was a side reaction during the polymerization, which was reduced by incorporating long or rigid moieties into the diyne (**13**) or diamine (**20**) monomer.

Inspired by the successful preparation of a library of poly(sulfonylamidine)s from diynes, sulfonyl azides, and diamines through Cu(I) catalyzed MCP, Tang and coworkers [70] extended the MCP from diamines to monomer with asymmetric center (i.e. amino ester) as depicted in Scheme 6.23, and reported a versatile method for the preparation of chiral poly(sulfonylamidines)s (**P35**) by Cu(I)-catalyzed MCP from various diynes (**13**), disulfonyl azides (**16**), and amino esters (**21**) (such as L-leucine methyl ester hydrochloride, D-leucine methyl ester hydrochloride, L-valine methyl ester hydrochloride, and L-alanine methyl ester hydrochloride).

$CHCl_3$ was identified as the most suitable solvent for the polymerization, since a soluble polymer with the highest $M_n = 15\,400\,g\,mol^{-1}$ and in the highest yield (79%) was isolated under the following conditions: monomer concentration of [(**13**)] = [(**16**)] = 0.20 M and optimal feed ratio of [(**13**)]:[(**16**)]:[(**21**)] = 1 : 1 : 2.4, respectively, in the presence of 20 mol% CuI and 0.40 M NEt_3 at ambient temperature within 24 hours. While, further diluting the concentration of diyne to 0.10 M or even 0.05 M at a constant monomer ratio, produced polymers with good

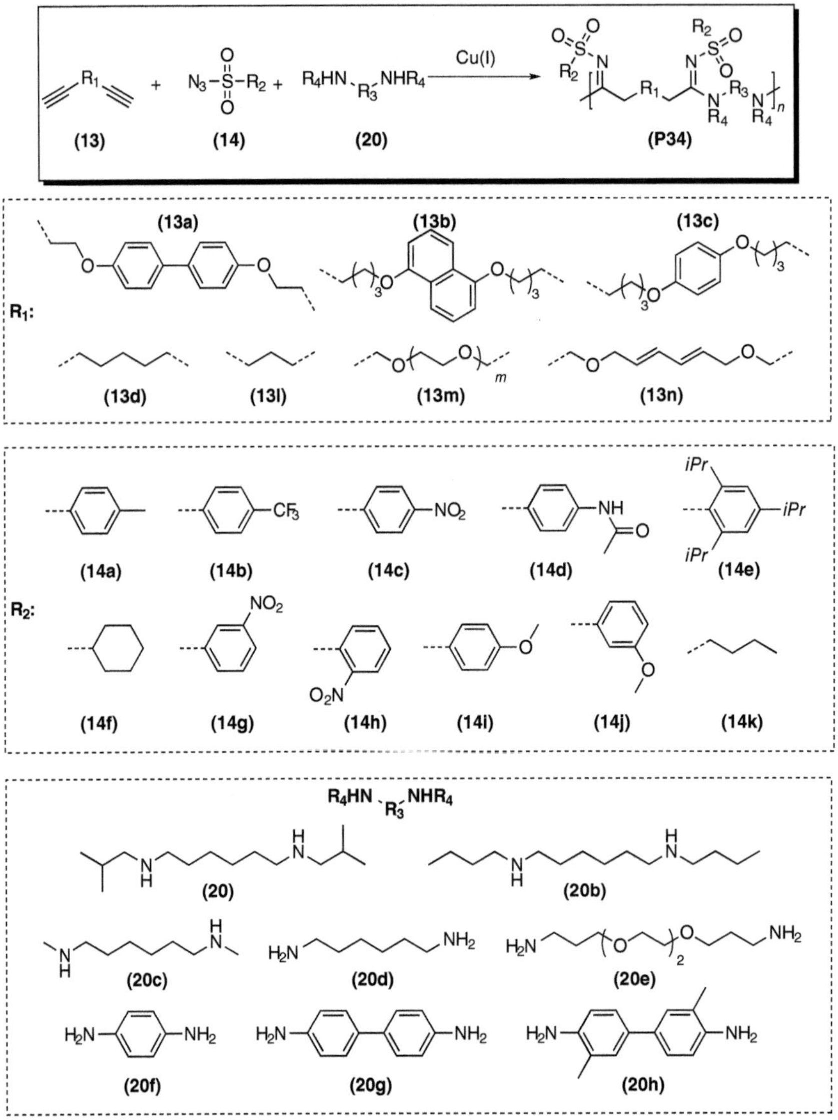

Scheme 6.22 One-pot facile synthesis of poly(sulfonylamidine) derivatives, e.g. (**P34**), based on the copper (I) catalyzed multicomponent reaction of diynes (**13**), sulfonyl azides (**14**), and diamines (**20**).

solubility, albeit in diminished isolated yields (61–70%) and molecular weights (M_n of ~8000 g mol^{-1}), insoluble gels were formed quantitatively when the polymerization was carried out at a relatively high monomer concentration of 0.60 M for diyne, i.e. (**13**). In general, the polymerization was very efficient with loadings of 20 mol% CuI and could be performed in short reaction time (e.g. six hours) to result in high molecular weight polymers (M_n of ~13 000 g mol^{-1}) with unimodal distribution ($Đ$ of 1.8) in favorable yields. Nevertheless, reducing the CuI amount

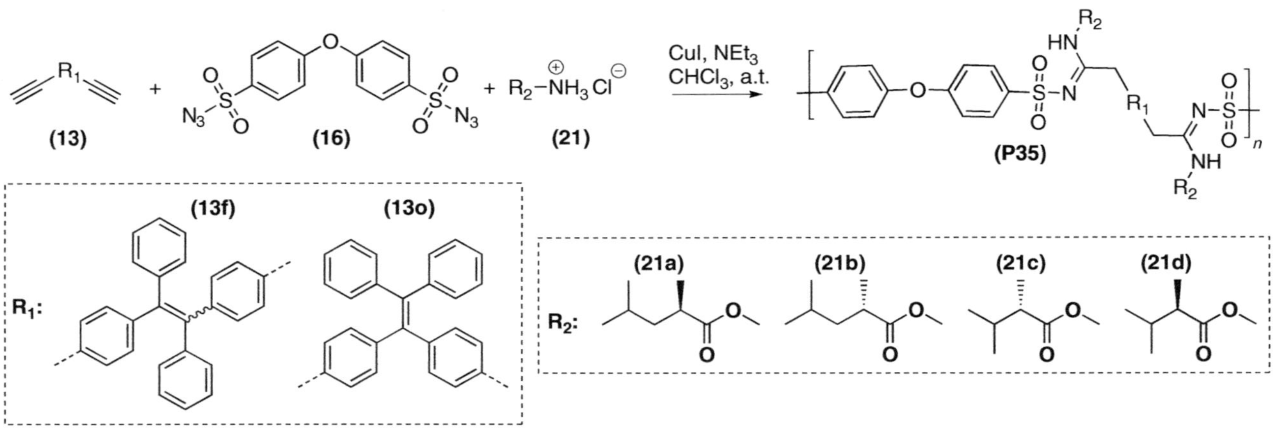

Scheme 6.23 Multicomponent polymerization of various diynes (**13**), disulfonyl azides (**16**), and amino esters (**21**) to obtain helical and luminescent polymer derivatives (**P35**).

dramatically decreased the efficiency of the polymerization reaction. For instance, in the presence of 10 mol% CuI, a polymer with M_n of 6100 g mol^{-1} (Đ of 1.7) was obtained.

Interestingly, all the polymers possessed excellent solubility in common organic solvents, despite the highly aromatic backbone. Indeed, due to the good solubility properties, most of the polymers were readily fabricated into uniform thin films by the spin-coating technique.

Compared with their aforementioned oxygen counter partners, i.e. poly(sulfonylimidate)s, the poly(sulfonylamidine) derivatives demonstrated lower thermal stability with 5% weight losses at temperature (T_d) ranging from 244 to 293 °C under nitrogen. Besides, circular dichroism (CD) signals associated with the positive Cotton effect, centered at ~287 nm, were observed in the absorption region of the polymer backbones, indicating the chirality of the amino ester pendant was transferred to the polymer backbone. The thin films of these polymers possess high refractive indices (RI = 1.7327–1.6419) in a wide wavelength region of 400–893 nm and low chromic dispersions (CD down to 0.0247). Complementary, the sulfonyl and ester groups incorporated into the skeleton of the poly(sulfonylamidine)s endowed these polymers with photosensitivity, which in turn enabled them to be fabricated into fluorescent photopatterns through the photolithography process. In fact, two-dimensional photoresist patterns with high resolution and sharp edges were observed under a fluorescence microscope.

Along with the successful three-component reaction between alkynes, sulfonyl azides, and amine derivatives, the group of Tang intensively explored new reactant partners for the ketenimines (the intermediate product formed from the reaction of alkynes and sulfonyl azides) and has recently proposed a new reactant for the Cu(I) MCP, i.e. the commercially available iminophosphorane, e.g. (**22**) [71]. In fact, iminophosphorane is a reactant with bipolar nature whose MCR with alkyne and sulfonyl azide can generate unique poly(sulfonylamidine)s (**P36**), decorated with phosphorous moieties, under mild reaction conditions (Scheme 6.24). Accordingly, the best polymerization results were obtained when the MCP was carried out under nitrogen atmosphere in polar solvent (i.e. THF) at ambient temperature for 12 hours in the presence of CuI (0.02 M) and NEt_3 (0.20 M). The monomer feeding ratio was fixed at [(**13**)]:[(**16**)]:[(**22**)] = 1 : 1 : 2.5 and monomer concentration of 0.1 M. In general, under these optimized polymerization conditions, a wide monomer scope including aromatic and aliphatic alkynes could be employed in the MCP together with aromatic sulfonyl azides and iminophosphorane, in order to generate polymers (with the generic chemical structure (**P36**)) with good to excellent yields of up to 92% and $14\,400 < M_w < 85\,600$ g mol^{-1} ($1.58 < Đ < 3.16$).

While the polymer structures were characterized through the comparison of the conventional spectroscopic techniques such as IR as well as ^{1}H, ^{13}C, and ^{31}P NMR spectra of the monomers, model compounds, and polymers, thermogravimetric analysis of them implied satisfactory thermal stability (i.e. $T_{d,5\%}$ ranging from 219 to 237 °C under nitrogen). The abundant heteroatoms and polarizable aromatic rings of these poly(sulfonylamidine)s endowed them not only with high n value in the

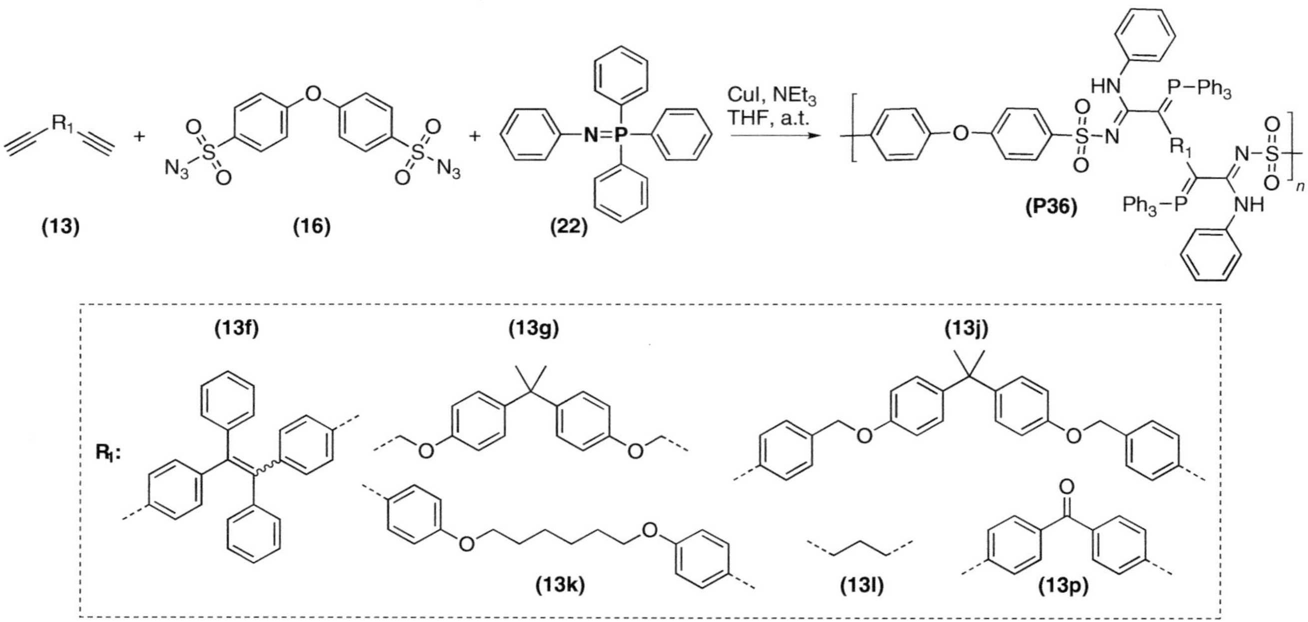

Scheme 6.24 Multicomponent polymerizations of alkynes, sulfonyl azides, and iminophosphorane (**22**) towards heteroatom-rich multifunctional poly(sulfonylamidine)s (**P36**).

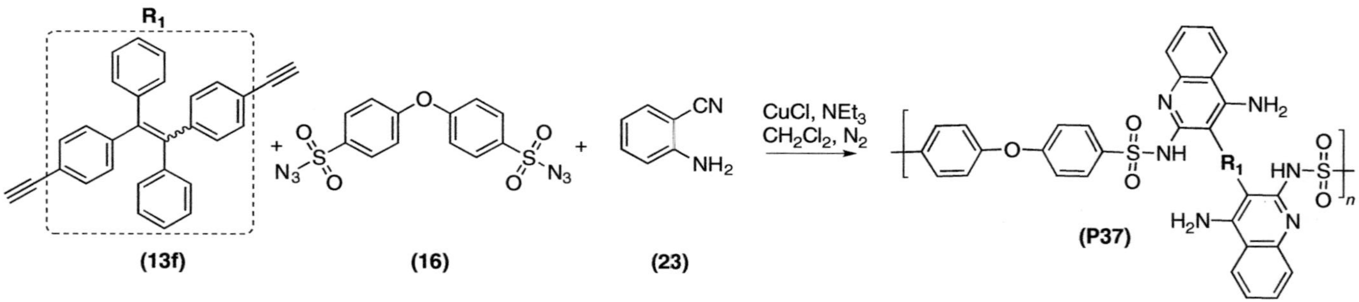

Scheme 6.25 A facile and efficient multicomponent click polymerization route mediated by Cu(I) to construct aminoquinoline-containing poly(sulfonylamidine) derivative (**P37**) with advanced functionalities. Source: Modified from Ref. [66].

range of 1.6287–1.8145, but also with a series of properties such as good solubility in polar solvents and good film-forming ability.

Attracted by the straightforward construction of quinoline structures via the MCRs of alkynes, sulfonyl azides, and 2-aminobenzonitrile (**23**) [72], Tang and coworkers developed the efficient MCP of diynes, disulfonyl azides, and 2-aminobenzonitrile (**22**) to furnish aminoquinoline-containing poly(sulfonylamidine) derivative (**P37**) with high molecular weight (M_w of 32 800 g mol^{-1}, $Đ < 2.13$) and reasonable high polymerization yield (82%) in 30 minutes (Scheme 6.25) [66].

Last but not least, it should be emphasized that the group of Theato [73] validated the synthetic utility of Cu(I)-catalyzed MCR for post-polymerization modification of polymers featuring acetylene moieties (e.g. (**P38**)) with sulfonyl azides (**14**) and amines (**24**) in the presence of a catalytic amount of $Cu(PPh_3)_3Br$ (Scheme 6.26) to yield polymeric *N*-sulfonylamidine derivatives (**P39**).

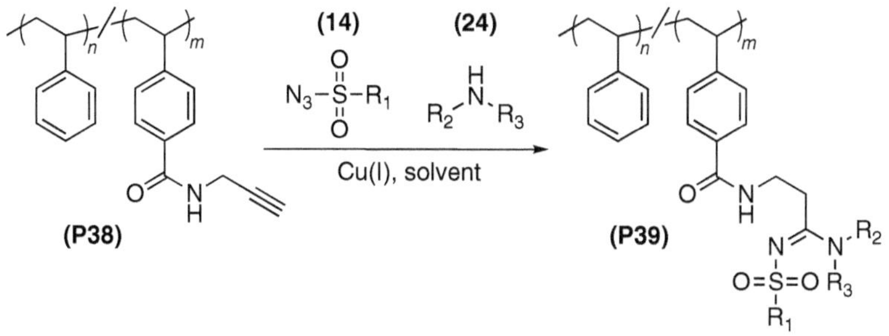

Scheme 6.26 Schematic representation of the post-modification process based on a Cu-catalyzed three-component reaction to furnish polymeric *N*-sulfonylamidine derivatives (**P39**).

6.3 Applications of Polymers with Sulfur-Nitrogen Bond

On one hand, sulfur polymers and materials thereof are well established, and endow broad range of application from biomedical to energy storage, which have been frequently covered in diverse general reviews [7, 74, 75]. On the other hand, the applications of polymers with sulfur-nitrogen bonds have not been previously in-depth summarized. Therefore, the exciting applications of current research involving biomedical applications, energy storage materials, metal-ion detection, and flame retardancy are outlined.

6.3.1 Biomedical Applications

Notably, sulfur-nitrogen bonds are rarely found in biochemistry, and there are only about 100 naturally abundant compounds that contain N–S bonds [6]. In general, these compounds are most commonly utilized in pharmaceuticals (both as active components and as intermediates in the synthesis of bioactive compounds) and antimicrobial agents (e.g. antibiotics or industrial antimicrobial finishes for textiles) [76].

Yet, biomaterials, which are recognized as significant technology, are promising candidate adjuvants (a pharmacological or immunological agent that modifies the effect of other agent) to enhance vaccine efficacy [77]. Through adjuvant design, the use of vaccines to diseases (such as AIDS, malaria, and cancer) is broadened. In this context, delivery systems based on polymeric particles have been considered promising candidates for vaccine delivery because they possess pathogen-mimicking properties in terms of shape, size, and morphology. Indeed, there are varieties of polymers that can be employed for the fabrication of these particles; however, those polymers that offer desirable characteristics are considered for use as cancer vaccine delivery systems. Specifically, polymers should be biodegradable and biocompatible and critically, the degradation byproducts of these polymers should be readily metabolized and non-toxic. Having this in mind, Salem and coworkers developed polymer that possessed unique sulfur-nitrogen bonds ((**P4**) in Scheme 6.3), i.e. sulfenamide [78], and explored its utilization for the fabrication of microparticles, which in turn were applied as cancer vaccine delivery system. Intriguingly, in comparison with the most synthetic polymers (e.g. polyesters, polyanhydrides, or polyamides among others) that are used to deliver pharmaceutical drugs in the body, polymers decorated with sulfur-nitrogen bonds are stable in polar protic and aprotic solvents, but degrade rapidly when exposed to acidic conditions.

A water-in-oil-in-water single [9] and double [78] emulsion solvent-evaporation methods were the method of choice for the fabrication of the microparticles. Importantly, in order to produce drug-loaded particles, fluorescein isothiocyanate (FITC)-dextran was used as a model hydrophilic compound. At the optimal formulation conditions, the external morphology of the poly(sulfenamide) based microparticles was examined by scanning electron microscopy to show the formation of smooth-surfaced spherical particles with low polydispersity. While the microparticles had a net negative surface charge (−23 mV) as analyzed by a zetasizer, the drug encapsulation efficiency of the particles and the drug loading were found to be a function of the following parameters: (i) drug molecular weight, (ii) amount of FITC-dextran used in fabricating FITC-dextran–loaded microparticles, (iii) concentration of poly(sulfenamide) and surfactant, and (iv) volume of the internal and external water phases. FITC-dextran was found to be distributed throughout the poly(sulfenamide) microparticles, and its release profile the microparticles demonstrated gradual and continuous release over two weeks following an initial burst release. Last but not at least, confocal laser scanning microscopy was used to qualitatively observe the cellular uptake of poly(sulfenamide) microparticles and indicated localization of the particles mainly in the cell cytoplasm of human embryonic kidney-293 (HEK-293) cells, and to some extent, in the nucleus.

In concomitant study, microparticles based on another poly(sulfenamide) derivative, i.e. (**P3**), shown in Scheme 6.3 [21], were fabricated by using an oil-in-water single emulsion solvent evaporation methodology, to which drugs and fluorescent molecules, such as rhodamine, were incorporated into by mixing them in the polymer/chloroform mixture prior to emulsification (Figure 6.1). The average size of the microparticle was 2.6 μm with a polydispersity of 0.62 (Figure 6.1B). They were spherical in shape, had a smooth surface, and had a negative charge of −41 mV in PBS buffer at pH = 7.4. The negative charge was attributed to the slow degradation of poly(sulfenamide) to yield sulfinic acid that deprotonated and led to a negatively charged surface. Interestingly, while the degradation of the sulfenamide bond was rapid in water, these microparticles fabricated from poly(sulfenamide)s, exhibited different degradation profile based on the monomer used in its synthesis. For instance, in Figure 6.1C, microparticles of (**P3**) at various stages of degradation can be observed after incubation in acetate buffer (pH 4.0) for 14 days.

Nonetheless, these microparticles were readily internalized by two different cell lines (e.g. murine dendritic cell line (i.e. JAWSII) and HEK293 cells) and their uptake was studied using confocal microscopy, which in turn revealed that this system has potential in immunological applications, especially in vaccine delivery and potential for gene therapy applications (Figure 6.1D). Indeed, confocal images clearly outlined that the microparticles were found on the interior of the cells, i.e. near the cytoplasm, hence indicating the phagocytosis of the microparticles by cells.

Likewise poly(sulfenamide)s, macromolecules (e.g. (**P12**) in Scheme 6.12) possessing the unique diaminosulfide [28, 79] functional group found application in the fabrication of microparticles as drug delivery vesicles (Figure 6.2a). Particularly due to the fact that, as previously mentioned, the diaminosulfide functional group degrades several orders of magnitude faster than conventional ester bonds under acidic conditions, and they possess reasonable stabilities in water under neutral conditions [22]. Hence, in their seminal work, Bowden and coworkers prepared microparticles according to a water/oil/water double emulsion-solvent evaporation method using poly(vinyl alcohol) as a surfactant. The particles, which surface charge was determined to be −11.6 ± 0.8 mV, were spherical in shape and possessed a smooth, nonporous surface. In addition, the *z*-average particle size determined by dynamic light scattering was 660 nm and consistent with the SEM micrograph shown in Figure 6.2b that were absorbed into cells. While, being nontoxic towards HEK-293 cells, their successful internalization into the cells was demonstrated in Figure 6.2c. Indeed, these microparticles were capable of being co-loaded with antigen (ovalbumin) and adjuvant (cytosine–phosphate–guanosine motifs), which upon vaccination displayed the capacity to generate detectable cell-mediated immune responses, as defined by increased levels of ovalbumin-specific cytotoxic T cell lymphocytes, and promoted increased survival of tumor challenged mice [79]. Respectively, in this prophylactic setting the antitumor effect of these co-loaded microparticles was at least comparable with that achieved with co-loaded poly(lactic acid-*co*-glycolic acid) microparticles or adenovirus type 5 encoding ovalbumin, suggesting that vaccine delivery systems fabricated from poly(diaminosulfide)s may have promise in cancer therapy.

Iminocoumarin derivatives have been recognized as compounds with pharmacological properties; [80] therefore, Tang and coworkers [66] investigated the antibacterial property of 4-amino-iminocoumarin-containing poly(sulfonylimidate) fluorescent (**P33**) shown in Scheme 6.21, revealing that this unique class of polymer can

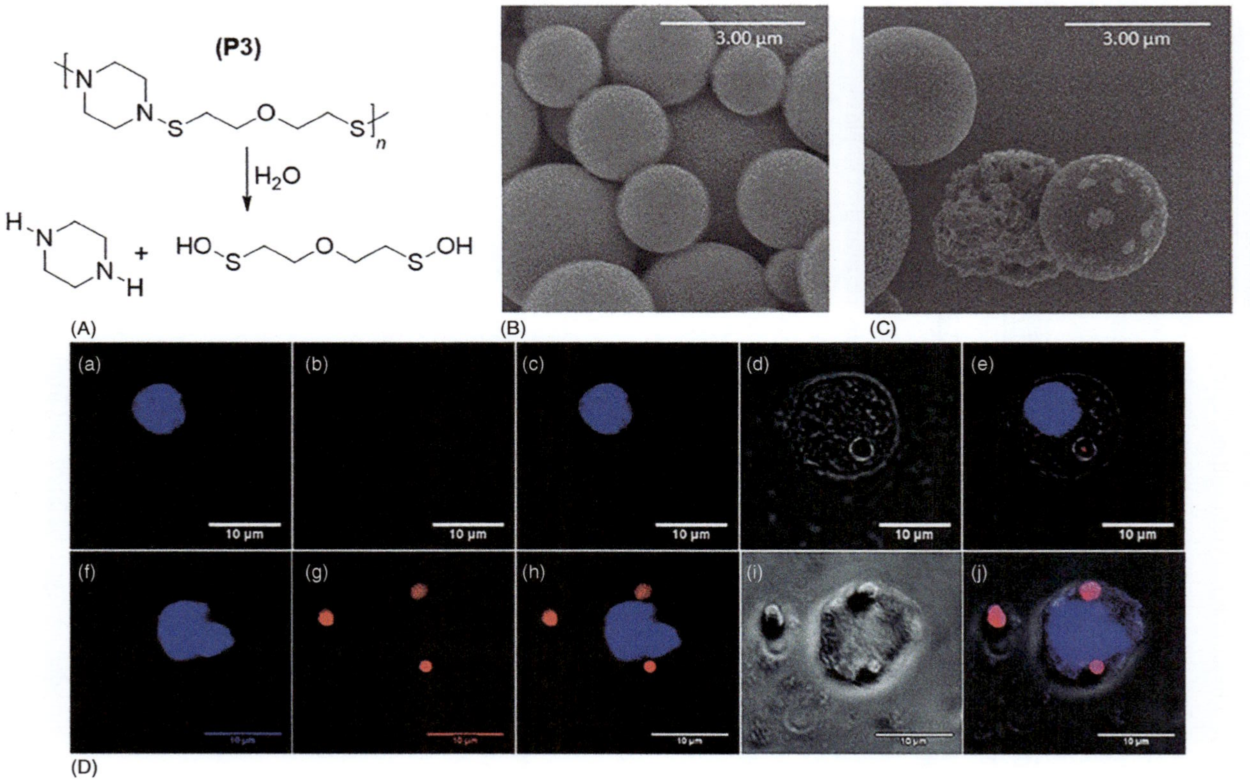

Figure 6.1 (A) Degradation of poly(sulfenamide) (**P3**) in water to release an amine and sulfenic acids. (B) Surface scanning electron microscopy image of microparticles fabricated at the optimal formulation conditions. (C) A confocal image is depicting the presence of green fluorescent FITC-dextran loaded microparticles mainly in the cell cytoplasm of (HEK-293) cells, and to some extent, in the nucleus. (D) Confocal microscopy images showing the uptake of microparticles fabricated from (**P3**) in Scheme 6.3, and loaded with rhodamine: (a–e) are JAWSII immature dendritic cells, and (f–j) are HEK293 cells; the nucleus of the cell was stained with 4′,6-diamidino-2-phenylindole to appear blue and the rhodamine appeared red; in (d) and (i) the transmitted light images indicated the periphery of the cells and the location of the microparticles, while in (e) and (j) are the merged images, which clearly outline the periphery of the cell, the location of the nucleus, and the microparticles to depict that the microparticles were found on the interior of the cells. Source: Reproduced from Ref. [21] with permission from the American Chemical Society.

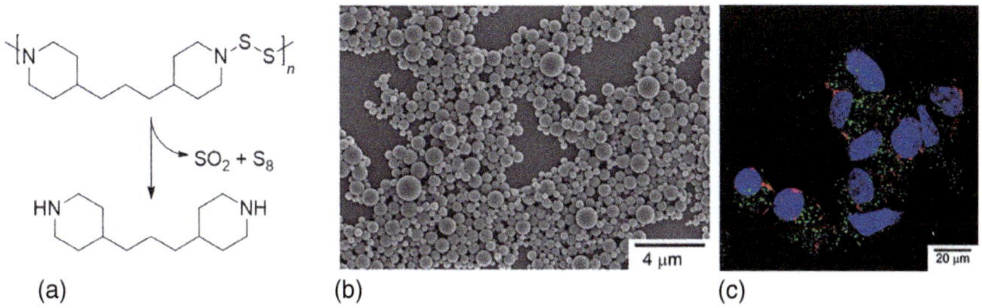

Figure 6.2 (a) Degradation of poly(diaminosulfide) (**P12**) in water to release an amine and sulfur monoxide, which in turn possessed a half-life of seconds and decomposed to release SO_2 and elemental sulfur. (b) SEM micrographs of microparticles fabricated from the polymer (**P12**) in Scheme 6.12. (c) A laser scanning microscopic image, which shows HEK-293 cells that were exposed to microparticles loaded with FITC-dextran (green) for 24 hours and then washed to remove microparticles that were not internalized into the cells. The nuclei of the cells are stained blue by 4′,6-diamidino-2-phenylindole, and the cytoplasm/cell membranes are stained red by phalloidin. Thus, this image revealed that the microparticles were internalized into the cells. Source: Reproduced from Ref. [28] with permission from the American Chemical Society.

be utilized in the inhibition of the growth of bacteria such as *Staphylococcus aureus*. Since, the structure of poly(sulfonylimidate) endowed it unique interaction with *S. aureus*, enabling its accumulation around *S. aureus* and further inhibition on the growth of the bacteria. In fact, the calculated inhibition ratio reached 58% when *S. aureus* was incubated with 32 μM of poly(sulfonylimidate) for 12 hours.

6.3.2 Metal-Ion Detection

Rapid and sensitive detection of chemical contaminants, particularly heavy metal ions, in environmental samples is significant and challenging. The contamination of heavy metal ions in the environment has received increasing attention because of its severe risks for human health, though metals are important to carry out cellular functions. Therefore, it is important to develop new materials for highly selective and sensitive detection of metal ions.

On one hand, AIE polymers are extensively studied as fluorescent chemosensors for metal ion detection, owing to their high sensitivity, selectivity, and reliability [81]. On the other hand, the heteroatom-rich nature of the polymers possessing diverse sulfur-nitrogen bonds attributes to these macromolecules' unique and specific properties (such as good solubility in polar solvents, good film-forming ability, satisfactory thermal stability, and high refractivity with small chromatic dispersions), thus they could serve as environmentally sensitive probes to visualize metal-ions. For example, the poly(sulfonylamidine), decorated with phosphorous moieties, which was synthesized by Tang and coworkers [71], was utilized as ligand for platinum group metal ions, particularly palladium. The nanoaggregates of the polymer (**P36**), in Scheme 6.24, based on (**13f**)/(**16**)/(**22**) [i.e. ($\mathbf{P36}_{(13f)/(16)/(22)}$)] in DMSO/$H_2O$ mixtures with 50 vol% water were studied as an example.

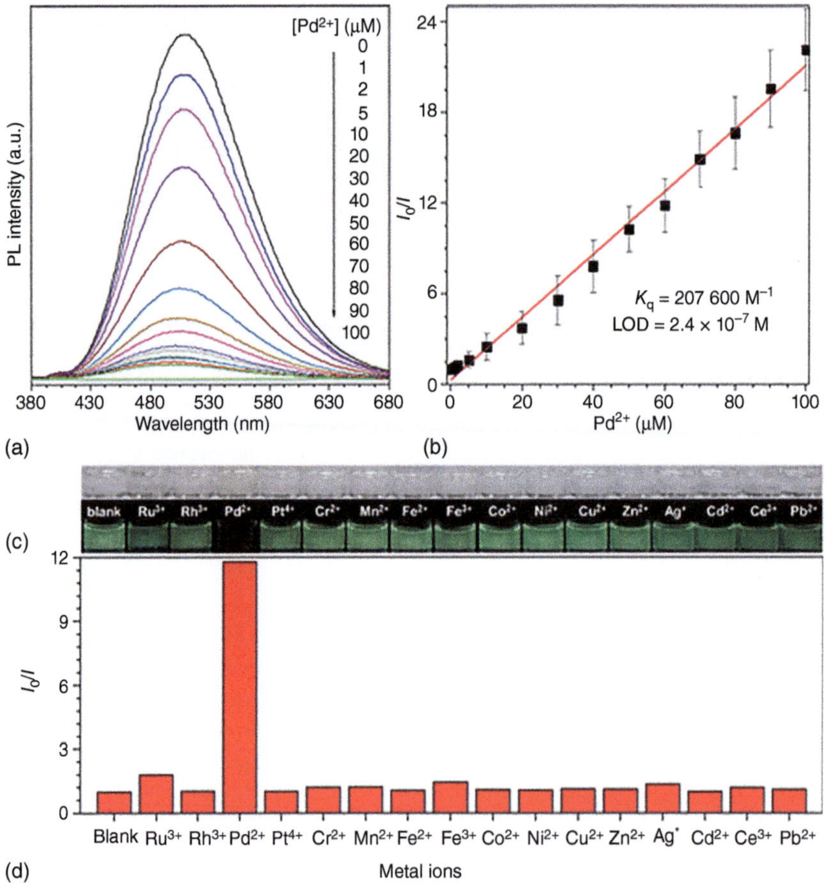

Figure 6.3 (a) PL spectra of (**P36**$_{(13f)/(16)/(22)}$) in DMSO/water mixture with 50 vol% water in the presence of different amounts of Pd^{2+}. (b) Stern–Volmer plot of relative intensity (I_0/I) versus the Pd^{2+} concentration. I_0 = PL intensity without Pd^{2+} (Concentration: 10 μM). (c) Images of (**P36**$_{(13f)/(16)/(22)}$) in DMSO/water mixtures with 50 vol% water containing different metal ions taken (upper) in daylight and (lower) under UV illumination. (d) Relative intensity (I_0/I) of aqueous solution of (**P36**$_{(13f)/(16)/(22)}$) versus different metal ions. I_0 = PL intensity without metal ions, at concentration of 10 and 50 μM, respectively, for polymer (**P36**$_{(13f)/(16)/(22)}$) and metal ion. Source: Reproduced from Ref. [71] with permission from the American Chemical Society.

Interestingly, when the concentration of $PdCl_2$ was increased from 0 to 100 μM, the photoluminescence (PL) intensity dramatically dropped (Figure 6.3a), respectively, the Stern–Volmer plot of relative PL intensity (I_0/I) versus the Pd^{2+} concentration appeared as a straight line with a large quenching constant of 207 600 M^{-1}, thus revealing a limit of detection (LOD), which was calculated to be 2.4×10^{-7} M (Figure 6.3b). Accordingly, suggesting a high sensitivity of this specific polymer (**P36**$_{(13f)/(16)/(22)}$) towards Pd^{2+} ions. Remarkably, when the PL spectra of (**P36**$_{(13f)/(16)/(22)}$) in 50 vol% aqueous mixture were studied in the presence of 16 different metal ions including Ru^{3+}, Rh^{3+}, Pd^{2+}, Pt^{4+}, Cr^{2+}, Mn^{2+},

Fe^{2+}, Fe^{3+}, Co^{2+}, Ni^{2+}, Cu^{2+}, Zn^{2+}, Ag^{+}, Cd^{2+}, Ce^{3+}, and Pb^{2+} under the same condition for comparison, only Pd^{2+} showed reasonable fluorescence response, indicating the high selectivity of this fluorescent chemosensor (Figure 6.3c and d, respectively).

In another representative example, fluorescence response of AIE-active poly(sulfonylimidate) (**P33**) derivative (i.e. ($\mathbf{P33}_{(13f)/(16)/(19)}$)) upon coordination with various metal ions was investigated, considering that the multiple heteroatoms and amino groups of poly(sulfonylimidate) may serve as metal ligands [66]. Into the DMSO/H_2O mixed solution of ($\mathbf{P33}_{(13f)/(16)/(19)}$) with 50 vol% H_2O, different metal ions such as Ru^{3+}, Rh^{3+}, Pd^{2+}, Pt^{4+}, Cr^{2+}, Mn^{2+}, Fe^{2+}, Fe^{3+}, Co^{2+}, Ni^{2+}, Cu^{2+}, Zn^{2+}, Ag^{+}, Cd^{2+}, Ce^{3+}, and Pb^{2+} were added. The fluorescence of ($\mathbf{P33}_{(13f)/(16)/(19)}$) was selectively quenched upon the addition of Ru^{3+}, while no significant change was observed with other metal ions. In similar manner, the control experiments with the mixture of different metal ions and Ru^{3+} suggested that the fluorescence quenching was only affected by Ru^{3+}, indicating high selectivity of the fluorescence detection of Ru^{3+}. It was also observed that, when increasing the concentration of $RuCl_3$ from 0 to 100 μM, the emission intensity of ($\mathbf{P33}_{(13f)/(16)/(19)}$) decreased, while keeping the spectral profile intact. Indeed, the emission of ($\mathbf{P33}_{(13f)/(16)/(19)}$) and the absorption of $RuCl_3$ in DMSO/H_2O mixture with 50 vol% water content overlapped at the wavelength range of 400–650 nm, indicating potential energy transfer from ($\mathbf{P33}_{(13f)/(16)/(19)}$)) to Ru^{3+}, which could be the reason for fluorescence quenching.

6.3.3 Flame Retardant Chemicals Based on Polymers with Sulfur-Nitrogen Bonds

Flame retardants have become a range of chemicals, which receive continuous scientific and public attention [82]. Generally, flame retardancy can be achieved by different physical and chemical mechanisms, thus, there is also a great variety in products, which can be used to achieve a lowering of the flammability of materials. Along with phosphorus, nitrogen, and silicon, sulfur is one of the known flame retardant elements, but compared with the others, sulfur-containing compounds have been much less extensively studied as flame retardants [82]. In line with this, it was discovered that sulfenamides and polymers, thereof, provide excellent flame retarding properties to polymer substrates [18, 23, 83]. Respectively, the excellent flame retarding action of sulfenamide based polymers (for instance (**P7**) in Scheme 6.4) on polypropylene, low density polyethylene, and polystyrene when exposed to small heat source was reported. Indeed, the small molecule and poly(sulfenamide) flame retardants enabled polypropylene films and fibers to self-extinguish even at a very low loading of 0.5–1.0 wt%. The nitrogen-sulfur bond of low basicity demonstrated stability at conventional normal polymer processing temperatures of polypropylene, but rapidly decomposed under fire conditions. The generated aminyl and sulfenyl radicals were capable of interrupting the radical processes otherwise leading to the fire propagation, hence exhibiting appropriate thermal properties for potentially being suitable flame retardants for polypropylene. In similar manner, the potential of poly(sulfenamide)s as synergists with conventional halogen free-flame retardants was investigated, implying the suitability of poly(sulfenamide) derivatives as standalone flame retardant or as synergists with conventional flame retardants [83].

6.3.4 Energy Storage Applications

In the context of the energy storage, particularly, in the field of organic batteries, numerous organosulfur compounds, which mainly cover disulfides, polysulfides, and thioethers, have been reported in the last three decades as possible cathode materials for organic rechargeable batteries [84]. The pioneering studies on organosulfur compounds in the field of energy storage date back to late 1990s, when Visco et al. [11] introduced different poly(diaminodisulfide)s (e.g. (**P16**) and (**P18**) in scheme 6.10) as alternative electrode materials in high temperature battery systems such as Li–FeS_2 and Na–Na_2S_x. One-electron-transfer electrochemistry based on the cleavage and formation of S–S bonds during discharge and charge, respectively, accounted for the rate-determining step, while the total number of transferred electrons in the overall reaction was two. Unfortunately, the cell configuration seemed sophisticated, and cyclic voltammetry results implied very poor kinetics of the redox reaction, due to the large separation between the cathodic and anodic peak potentials ($\Delta E\text{p} = 1.25–1.63$ V). As an additional intrinsic drawback, these organic materials operated at elevated temperatures (typically 80–130 °C). Nevertheless, these studies underpinned the plausibility of using poly(diaminodisulfide)s as cathodes in rechargeable lithium batteries, mainly due the fact that the highly heteroatom nature of the sulfur-nitrogen bond presented in each repeating unit of the polymer greatly accelerated the electrode kinetics.

It took more than 30 years until Bowden and coworkers [36] fabricated a poly(aminodisulfide) derivative (**P14**) (Scheme 6.9) into a layered device to measure its current–voltage curve in the solid state. The cross-sectional area was 16.9×10^{-3} mm^2 and the thickness of the poly(disulfidediamine) was 40 nm. According the current–voltage curve shown in Figure 6.4, the polymer demonstrated diode characteristics and was weakly electrically conducting. The conductivity between 9 and 10 V was estimated to be 4.73×10^{-8} S cm^{-1}. Whereas, this value was very low and similar to that of distilled water, it was comparable with values for other

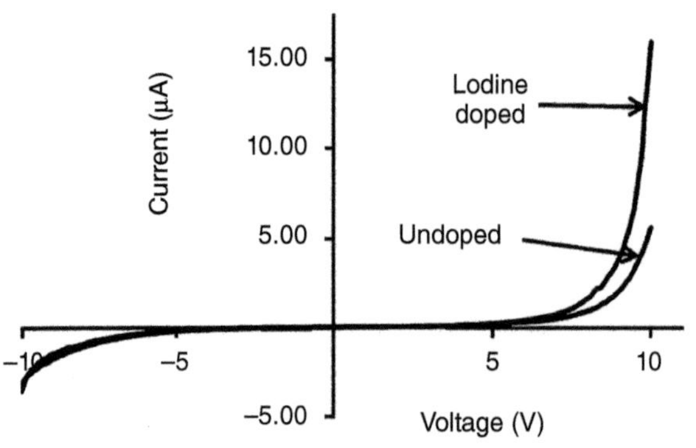

Figure 6.4 Current–voltage curve for polymer (**P14**), which reveals that (**P14**) possessed diode characteristics and was weakly electrical conducting. Source: Reproduced from Ref. [36] with permission from the American Chemical Society.

undoped polymers, such as polythiophene, polyaniline, and polyacetylene, which have conductivities of approximately 10^{-5}–10^{-9} S cm^{-1}. Besides, the impact of dopants on (**P14**) was briefly tested by exposing the device to iodine vapor for five minutes and then immediately measuring the conductivity. Indeed, the iodine doping enhanced the electrical conductivity up to 2.85×10^{-7} S cm^{-1}.

6.4 Conclusion and Outlook

With a series of synthetic breakthroughs in recent years, new chemistries suitable for the synthesis of various polymers containing sulfur-nitrogen functional groups have been developed, and thus novel materials with interesting properties have been created. Indeed, polymers with sulfur-nitrogen bond can be considered as a sleeping beauty and (re)appear to be promising to new materials that combine the advantages of sulfur and nitrogen solely. Nevertheless, to endeavor polymers with sulfur-nitrogen bonds more streamlined, it is essential to enable an interdisciplinary combination of conventional synthetic methods drawn from organic, inorganic, and polymer chemistry, as well as fabrication and processing techniques to create new materials.

References

1 Grigoras, M. and Catanescu, O. (2005). Imine oligomers and polymers. *J. Macromol. Sci. Polym. Rev.* 44 (2): 131–173.

2 Tatum, L.A., Su, X., and Aprahamian, I. (2014). Simple hydrazone building blocks for complicated functional materials. *Acc. Chem. Res.* 47 (7): 2141–2149.

3 Collins, J., Xiao, Z., Müllner, M., and Connal, L.A. (2016). The emergence of oxime click chemistry and its utility in polymer science. *Polym. Chem.* 7 (23): 3812–3826.

4 Davis, F.A. (2006). Adventures in sulfur–nitrogen chemistry. *J. Org. Chem.* 71 (24): 8993–9003.

5 Herberhold, M. (1988). Small reactive sulfur-nitrogen compounds and their transition metal complexes. *Comm. Inorg. Chem.* 7 (2): 53–72.

6 Petkowski, J.J., Bains, W., and Seager, S. (2018). Natural products containing a nitrogen–sulfur bond. *J. Nat. Prod.* 81 (2): 423–446.

7 Mutlu, H., Ceper, E.B., Li, X. et al. (2019). Sulfur chemistry in polymer and materials science. *Macromol. Rapid Commun.* 40 (1): 1800650.

8 Labes, M.M., Love, P., and Nichols, L.F. (1979). Polysulfur nitride—a metallic, superconducting polymer. *Chem. Rev.* 79 (1): 1–15.

9 Yoo, J. (2011). Synthesis of new biodegradable polysulfenamides for applications in medicine. PhD thesis. University of Iowa, Iowa City, IA, USA.

10 Graf, T.A. (2012). Poly(disulfidediamines); new biodegradable polymers. PhD thesis. University of Iowa, Iowa City, IA, USA.

11 Liu, M., Visco, S.J., and Jonghe, C.D. (1991). Novel solid redox polymerization electrodes. *J. Electrochem. Soc.* 138 (7): 1891–1895.

12 Mutlu, H. and Theato, P. (2020). Making the Best of Polymers with Sulfur-Nitrogen Bonds: From Sources to Innovative Materials. *Theato Macromolecular Rapid Communications*, 1800650.

13 Chandrasekhar, V. (2005). Other inorganic polymers that contain phosphorus, boron and sulfur. In: *Inorganic and Organometallic Polymers*, 189–200. Berlin, Heidelberg: Springer.

14 Craine, L. and Raban, M. (1989). The chemistry of sulfenamides. *Chem. Rev.* 89 (4): 689–711.

15 Guarino, V.R., Karunaratne, V., and Stella, V.J. (2007). Sulfenamides as prodrugs of NH-acidic compounds: a new prodrug option for the amide bond. *Bioorg. Med. Chem. Lett.* 17 (17): 4910–4913.

16 Zincke, T. (1911). New series of aromatic sulfur compounds. *Ber. Dtsch. Chem. Ges.* 44 (1): 769–771.

17 Koval, I.V. (1996). Synthesis and application of sulfenamides. *Russ. Chem. Rev.* 65 (5): 421–440.

18 Wilen, C.E., Aubert, M., Tirri, T., and Pawelec, W. (2015). Sulfenamides as flame retardants. WO 2015067736A1.

19 Nti-Addae, K.W. and Stella, J.V. (2011). In vitro conversion of model sulfenamide prodrugs in the presence of small molecule thiols. *J. Pharm. Sci.* 100 (3): 1001–1008.

20 Cohen, M.P. and D'Sidocky, R.M. (2002). Rubber chemicals. In: Herman F. Mark *Encyclopedia of Polymer Science and Technology*, 577–612. Hoboken, NJ: Wiley.

21 Yoo, J., Kuruvilla, D.J., D'Mello, S.R. et al. (2012). New class of biodegradable polymers formed from reactions of an inorganic functional group. *Macromolecules* 45 (5): 2292–2300.

22 Wolfenden, R. and Yuan, Y. (2011). The "neutral" hydrolysis of simple carboxylic esters in water and the rate enhancements produced by acetylcholinesterase and other carboxylic acid esterases. *J. Am. Chem. Soc.* 133 (35): 13821–13823.

23 Tirri, T., Aubert, M., Pawelec, W. et al. (2016). Structure–property studies on a new family of halogen -free flame retardants based on sulfenamide and related structures. *Polymers* 8 (10): 360.

24 Yang, L., Feng, J., Qiao, M., and Zeng, Q. (2018). Synthesis of *N*-sulfenyl-sulfoximines and -sulfenamides through a metal-free N–H/S–H dehydrocoupling reaction. *Org. Chem. Front.* 5 (1): 24–28.

25 Michaelis, A. (1895). Ueber die einwirkung anorganischer chloride auf piperidin und aliphatische amine. *Ber. Dtsch. Chem. Ges.* 28 (1): 1012–1019.

26 Losev, Y.P. and Paushkin, Y.M. (1992). Stabilization of vulcanized polyethylene. *J. Appl. Polym. Sci.* 45 (9): 1517–1520.

27 Harpp, D.N., Steliou, K., and Chan, T.H. (1978). Synthesis and reactions of some new sulfur transfer reagents. *J. Am. Chem. Soc.* 100 (4): 1222–1228.

28 Yoo, J., D'Mello, S.R., Graf, T. et al. (2012). Synthesis of the first poly(diaminosulfide)s and an investigation of their applications as drug delivery vehicles. *Macromolecules* 45 (2): 688–697.

29 Harpp, D.N. and Back, T.G. (1971). A general synthesis of sulfonamides. *Tetrahedron Lett.* 12 (52): 4953–4956.

30 Schenk, P.W. and Steudel, R. (1968). *Inorganic Sulfur Chemistry*. Amsterdam: Elsevier.

31 Góngora-Benítez, M., Tulla-Puche, J., and Albericio, F. (2014). Multi-faceted roles of disulfide bonds. peptides as therapeutics. *Chem. Rev.* 114 (2): 901–925.

32 Martin, R., Rekondo, A., de Luzuriaga, A.R. et al. (2016). Dynamic sulfur chemistry as a key tool in the design of self-healing polymers. *Smart Mater. Struct.* 25 (8): 084017.

33 Bang, E.-K., Lista, M., Sforazzini, G. et al. (2012). Poly(disulfide)s. *Chem. Sci.* 3 (6): 1752–1763.

34 Losev, Y.P., Paushkin, Y.M., and Isakovich, V.N. (1974). Inhibition of thermal degradation of crosslinked polyethylene. *Polym. Sci. U.S.S.R* 16 (11): 2502–2505.

35 Xiao, X., Xue, J., and Jiang, X. (2018). Polysulfurating reagent design for unsymmetrical polysulfide construction. *Nat. Commun.* 9 (1): 2191.

36 Graf, T.A., Yoo, J., Brummett, A.B. et al. (2012). New polymers possessing a disulfide bond in a unique environment. *Macromolecules* 45 (20): 8193–8200.

37 Losev, Y.P., Fedulov, A.S., and Mezen, N.I. (1998). Belarus Author's certificate No. 2684, MKI VU (11) 2684 C 1 G 08 C 75/14, A 61 K 31/795.

38 Denk, M.K. (2009). The Variable strength of the sulfur–sulfur bond: 78 to 41 kcal – G3, CBS-Q, and DFT bond energies of sulfur (S_8) and disulfanes XSSX (X = H, F, Cl,CH_3,CN,NH_2, OH, SH). *Eur. J. Inorg. Chem.* 2009 (10): 1358–1368.

39 Miyashita, T., Matsuda, M., and Iino, M. (1975). Kinetics of the thermal decompositions of diaryl and dialky disulfides. *Bull. Chem. Soc. Jpn.* 48 (11): 3230–3235.

40 Michaelis, A. and Luxembourg, K. (1895). Ueber eine nue Reihe schwefelhaltiger Derivate der aliphatischen Amine. *Ber. Dtsch. Chem. Ges.* 28 (1): 165–167.

41 Datta, R.N., Das, P.K., and Basu, D.K. (1986). Studies on the reactions between thiocarbamyl sulfenamide and 2-(iminodithio)benzothiazole accelerator system in the early stage of vulcanization of NR. *J. Appl. Polym. Sci.* 32 (7): 5849–5864.

42 Takahashi, A., Goseki, R., and Otsuka, H. (2017). Thermally adjustable dynamic disulfide linkages mediated by highly air-stable 2,2,6,6-tetramethyl-piperidine-1-sulfanyl (TEMPS) radicals. *Angew. Chem. Int. Ed.* 56 (8): 2016–2021.

43 Bennett, J.E., Sieper, H., and Tavs, P. (1967). 2,2,6,6-Tetramethylpiperidyl-1-thiyl: a stable new radical. *Tetrahedron* 23 (4): 1697–1699.

44 Takahashi, A., Goseki, R., Ito, K., and Otsuka, H. (2017). Thermally healable and reprocessable bis(hindered amino)disulfide-cross-linked polymethacrylate networks. *ACS Macro Lett.* 6 (11): 1280–1284.

45 Rothemund, S. and Teasdale, I. (2016). Preparation of polyphosphazenes: a tutorial review. *Chem. Soc. Rev.* 45 (19): 5200–5215.

46 Roy, A.K. (1992). The first, unambiguous synthesis of poly(alkyl/aryl oxothiazenes). A novel route to precursors, and synthesis of polymers. *J. Am. Chem. Soc.* 114 (4): 1530–1531.

47 Roy, A.K., Burns, G.T., George, C.L., and Grigoras, S. (1993). *J. Am. Chem. Soc.* 115 (7): 2604–2612.

48 Seel, F. and Simon, G. (1960). Die umsetzung des thionyltetrafluorides mit ammoniak. polymeres sulfanurfluorid. *Angew. Chem.* 72 (18): 709.

49 Parshall, G.W., Cramer, R., and Foster, R.E. (1962). Iminosulfur oxydifluoride and poly-(oxofluorosulfur nitride). *Inorg. Chem.* 1 (3): 677–679.

50 Levchenko, E.S., Kozlov, E.S., and Kirsanov, A.V. (1962). Phenylesters of areniminosulfonic acids. *Zhurnal Obshchei Khimii* 32 (8): 2535–2592.

51 Levchenko, E.S., Kozlov, E.S., and Kirsanov, A.V. (1963). Amides of areneiminosulfonic acids. *Zhurnal Obshchei Khimii* 33: 565–571.

52 Neilson, R.H. and Wisian-Neilson, P. (1988). Poly(alkyl/arylphosphazenes) and their precursors. *Chem. Rev.* 88 (3): 541–562.

53 Aldeva, V., Gezahegn, S., Panahi, B. et al. (2013). Synthesis, characterization, and theoretical analysis of soluble poly(oxothiazenes): the ambient temperature Lewis acid catalyzed in situ polymerization of n-silylsulfonimidoyl chlorides. *Macromolecules* 46 (7): 2562–2568.

54 Aly, A.A., Bräse, S., and Goma, M.A.-M. (2018). Amidines: their synthesis, reactivity, and applications in heterocyclesynthesis. *ARKIVOC* 2018 (vi): 85–138.

55 Roger, R. and Neilson, D.G. (1961). The chemistry of imidates. *Chem. Rev.* 61 (2): 179–211.

56 Volkova, Y.A., Kozlov, A.S., Kolokolova, M.K. et al. (2019). Steroidal *N*-sulfonylimidates: synthesis and biological evaluation in breast cancer cells. *Eur. J. Med. Chem.* 179 (1): 694–706.

57 Bundgaard, H. and Larsen, J.D. (1988). *N*-Sulfonyl imidates as a novel prodrug form for an ester function or a sulfonamide group. *J. Med. Chem.* 31 (11): 2069–2071.

58 Matsubara, R., Berthiol, F., and Kobayashi, S. (2008). *J. Am. Chem. Soc.* 130 (6): 1804–1805.

59 Bartrum, H.E., Viceriat, A., Carret, S., and Poisson, J.-F. (2014). Sulfinylimidates as chiral amide equivalents for irreversible, asymmetric aldol reactions. *Org. Lett.* 16 (7): 1972–1975.

60 Sanzhizhapov, D.B., Tonevitskii, Y.V., Mognonov, D.M., and Doroshenko, Y.E. (2002). New heat-resistant polyimidates derived from bisphenolsand mono- and dicarboxylic acid imidoyl chlorides. *Russ. J. Appl. Chem.* 76 (4): 619–622.

61 Kim, H. and Choi, T.-L. (2014). Preparation of a library of poly(*N*-sulfonylimidates) by Cu-catalyzed multicomponent polymerization. *ACS Macro Lett.* 3 (8): 791–794.

62 Theato, P. (2015). *Multi-Component and Sequential Reactions in Polymer Synthesis*, 269 pp. Heidelberg: Springer International Publishing.

63 Deng, H., Han, T., Zhao, E. et al. (2016). Multicomponent polymerization: development of a one-pot synthetic route to functional polymers using diyne, N-sulfonyl azide and water/ethanol as reactants. *Polym. Chem.* 7 (36): 5646–5654.

64 Bae, I., Han, H., and Chang, S. (2005). Highly efficient one-pot synthesis of *N*-sulfonylamidines by Cu-catalyzed three-component coupling of sulfonyl azide, alkyne, and amine. *J. Am. Chem. Soc.* 127 (7): 2038–2039.

65 Deng, H., Han, T., Zhao, E. et al. (2016). Multicomponent click polymerization: a facile strategy toward fused heterocyclic polymers. *Macromolecules* 49 (15): 5475–5483.

66 Xu, L., Zhau, T., Min, H. et al. (2019). Multicomponent polymerizations of alkynes, sulfonyl azides, and 2-hydroxybenzonitrile/2-aminobenzonitrile toward multifunctional iminocoumarin/quinoline-containing poly(*N*-sulfonylimine)s. *ACS Macro Lett.* 8 (2): 101–106.

67 Aly, A.A. and Nour-El-Din, A.M. (2008). Functionality of amidines and amidrazones. *ARKIVOC* 2008 (i): 153–194.

68 Burdukovskii, V.F. and Kholkhoev, B.C. (2015). Synthesis, properties, and transformations of polyamidines. *Polym. Sci. Ser. B* 57 (2): 61–70.

69 Lee, I.H., Kim, H., and Choi, T.-L. (2013). Cu-catalyzed multicomponent polymerization to synthesize a library of poly(*N*-sulfonylamidines). *J. Am. Chem. Soc.* 135 (10): 3760–3763.

70 Deng, H., Zhao, E., Li, H. et al. (2015). Multifunctional poly(*N*-sulfonylamidine)s constructed by Cu-catalyzed three-component polycouplings of diynes, disulfonyl azide, and amino esters. *Macromolecules* 48 (10): 3180–3189.

71 Xu, L., Hu, R., and Tang, B.Z. (2017). Room temperature multicomponent polymerizations of alkynes, sulfonyl azides, and iminophosphorane toward heteroatom-rich multifunctional poly(phosphorus amidine)s. *Macromolecules* 50 (16): 6043–6053.

72 Yi, F., Zhang, S., Huang, Y. et al. (2017). An efficient one-pot protocol for the synthesis of polysubstituted 4-amino-iminocoumarins and 4-aminoquinolines by a copper-catalyzed three-component reaction. *Eur. J. Org. Chem.* 2017 (1): 102–110.

73 Kakuchi, R. and Theato, P. (2013). Three-component reactions for post-polymerization modifications. *ACS Macro Lett.* 2 (5): 419–422.

74 Worthington, M.J.H., Kucera, R.L., and Chalker, J.M. (2017). Green chemistry and polymers made from sulfur. *Green Chem.* 19 (12): 2748–2761.

75 Vo, C.D., Kilcher, G., and Tirelli, N. (2009). Polymers and sulfur: what are organic polysulfides good for? Preparative strategies and biological applications. *Macromol. Rapid Commun.* 30 (4–5): 299–315.

76 Lesch, J.E. (2006). *The First Miracle Drugs: How the Sulfa Drugs Transformed Medicine.* New York: Oxford University Press.

77 Jones, K.S. (2008). Biomaterials as vaccine adjuvant. *Biotechnol. Prog.* 24 (4): 807–814.

78 D'Mello, S.R., Yoo, J., Bowden, N.B., and Salem, A.K. (2014). Microparticles prepared from sulfenamide-based polymers. *J. Microencapsul.* 31 (2): 137–146.

79 Geary, S.M., Hu, Q., Joshi, V.B. et al. (2015). Diaminosulfide based polymer microparticles as cancer vaccine delivery systems. *J. Control. Release* 220 (Pt B): 682–690.

80 Srikrishna, D., Godugu, C., and Dubey, P.K. (2018). A review on pharmacological properties of coumarins. *Mini-Rev. Med. Chem.* 18 (2): 113–141.

81 Mei, J., Leung, N.L., Kwok, R.T. et al. (2015). Aggregation-induced emission: together we shine, united we soar! *Chem. Rev.* 115 (21): 11718–11940.

82 Hirschler, M.M. (2015). Flame retardants and heat release: review of data on individual polymers. *Fire Mater.* 39 (3): 232–258.

83 Tirri, T., Aubert, M., Azizi, H. et al. (2019). Sulfenamides in synergistic combination with halogen free flame retardants in polypropylene. *Polym. Degrad. Stab.* 164 (1): 75–89.

84 Liu, J., Wang, M.F., Xu, N. et al. (2018). Progress and perspective of organosulfur polymers as cathode materials for advanced lithium-sulfur batteries. *Energy Storage Mater.* 15 (1): 53–64.

7

Thioester Functional Polymers

Suzan Aksakal[1], Resat Aksakal[1], and C. Remzi Becer[2]

[1] *Queen Mary University of London, School of Engineering and Materials Science, Mile End Road, London E1 4NS, UK*
[2] *University of Warwick, Department of Chemistry, Gibbet Hill, Coventry CV4 7AL, UK*

7.1 Introduction

Sulfur-containing polymers represent an attractive tool for the next generation of functional materials. These materials have already attracted significant interest and led to multifaceted applications, including the construction of thiolated polymers with high mucoadhesive ability [1], optical materials with higher refractive indices [2], and robust self-assembled monolayers [3–5] (SAMs) including gold nanoparticles (AuNP) [6] for imaging and bioconjugates of polymers and thiol-containing biomolecules [7]. However, the use of free thiols in polymerization reactions requires protected thiols due to their low stability and high reactivity compared with esters [8–10], which can be named as their hydroxy counterparts [11]. Nevertheless, there are many straightforward ways to introduce thiols in a protected state into a polymer [12, 13] and numerous studies on various designs, consisting of a source of thiols in acrylic, vinyl or hetereocyclic systems (e.g. thiazole [14], thiophene [15, 16], or thiolactones (TLs) [17–19]) have been reported up to date. The latter one is still of great interest as TLs have been demonstrated to be a versatile and unique structural motif in the construction of sequence defined oligomeric and polymeric materials [17–19].

Inspired by the uniqueness and ubiquity of thioesters in nature, much attention has been paid to thioester functionalized materials, yielding applications ranging from responsive polymers to bioconjugates and (bio)degradable polymers.

In this chapter we review the progress in the monomer design and synthesis of thioester containing monomers and their polymerization. The generation of a thioester via polymerization processes or the presence of thioesters in chain ends, such as initiators or chain transfer agents (CTAs), is discussed. Several strategies of thioesters in polymer science are highlighted and their advances and promises are discussed in detail. In brief, recent examples of post-polymerization modifications with various compounds (e.g. thiols, azides, amines, and cysteine containing

Sulfur-Containing Polymers: From Synthesis to Functional Materials,
First Edition. Edited by Xing-Hong Zhang and Patrick Theato.

peptides) to enable modification via pathways such as ligation, amidation or exchange reactions are also presented.

7.2 Thioesters: Structural Features, Reactivities, and Reactions

The thioester bond is a common functional group in biology as it displays a wealth of attractive properties and plays a fundamental role in biosynthetic pathways, such as the coenzyme A (CoA). In general, CoA promotes the metabolism of cellular components (e.g. peptides, fatty acids, sterols, terpenes, lipids, and porphyrins) [20].

Esters are stabilized by a contribution from a resonance structure where the lone pair of the single bonded oxygen atom donates electron density into the antibonding orbital of the carbonyl structure. This is manifested in a partial double bond character and hindered rotation around the C–O single bond. In a similar structure with a thioester as in CoA, the carbonyl is not stabilized by resonance as the ester carbonyl. The poor resonance overlap of the 2*p* orbital from the carbon and the sulfur 3*p* orbital makes a thioester less resonance stabilized and thus more reactive [21]. Thioesters are mainly stabilized from a resonance structure where both electron pairs are localized on the oxygen, resulting in a full negative charge and the carbonyl carbon in a full positive charge and susceptible to nucleophilic attack [22]. Additionally, a thiolate anion and thiol presents a good leaving group because of its high polarizability and low degree of polarization [23]. Thus the aforementioned factors make the more reactive carbonyl in thioesters more favorable against oxoesters for the linkage in CoA.

There are numerous known methodologies for the synthesis of thioesters. Esterification reactions of an acyl compound (e.g. carboxylic acid [24–26], acid anhydride [27, 28], or acid chloride [29–32]) with a thiol or disulfide conducted in the presence of a base is a convenient and standard protocol for the synthesis of these organosulfur compounds. Aldehydes have also been widely explored for the synthesis of thioesters [28, 33–38].

An alternative methodology is the acylation of thiols using various catalysts (e.g. triflates, CsF, NBS, zeolites, rongalite, Lewis acids, zinc, ionic liquids) [39]. An inexpensive and efficient method for the synthesis of thioesters by an acid-catalyzed *S*-acetylation of a wide range of aromatic and aliphatic thiols with isopropenyl acetate has also been reported recently [40]. Additionally, one-step conversion of the thioester group into other functional groups is very attractive in organic synthesis. A particularly significant conversion is the transformation of thioesters into amides. Reaction of thioesters with primary and secondary amines, azides [41–43], or cysteine-containing structures [44, 45] results in an amide functionality in the molecule, whereas a reaction with a thiol in a thiol-thioester exchange reaction [46, 47] will form a new thioester bond. Numerous studies have elaborated their reactivity in contrast to oxoester analogues and proven a 100-fold faster reaction rate with amines [22]. The mechanism and progress of this amidation process with mono- and bifunctional amines on a dithioester has been described in 1990 [48], catalyzed with arylthiols [49], and outlined in more details by Castro for thioesters and thiocarbonates [50]. Thioesters can undergo both acid and base-catalyzed hydrolysis [20].

Treatment with triethylsilane and a catalytic amount of Pd/C can lead to the respective aldehyde [51, 52], whereas treatment with an organozinc compound furnishes ketones [51]. Very recently, the transformation of thioesters to their thioethers has been reported on arylthioesters via Pd- and Ni-catalyzed decarbonylative conversion, under base and thiol free conditions [53]. In an attempt to develop new rigid-rod polymers, the thia-Michael reaction was employed between molecular rods, bearing terminal thiols and bis maleimides [54]. Interestingly, cyclic oligomers were mainly obtained contrary to the expected linear polymers, which was attributed to the folded conformation of the building blocks.

7.3 Preparation of Thioester Containing Structures

In the past few years, the manifoldness of thioesters has become widely available and has drawn widespread attention in polymer science. For example, the aminolysis of dithiocarbonates, dithiocarbamates, and thioesters has been evaluated as thiol protection groups in atom transfer radical polymerization (ATRP) and reversible addition fragmentation chain transfer (RAFT) polymerization [55, 56]. Interestingly, dithiocarbamate group could not be cleaved under mild conditions and led to side reactions during the radical polymerization of styrene and (meth)acrylates. Thiocarbonyl and thioesters did not induce any side reactions and could be cleaved under mild conditions. Moreover, up to eight times higher reactivity towards amines of the dialkyl xanthate in comparison to an alkyl thioacetate was reported [11].

One approach to incorporate a thioester functionality into a polymer describes the direct utilization of thioester containing monomers within the polymerization process next to several other successful protocols that have been reported to date (Scheme 7.1).

Among these in particular are three main classes, each of which corresponds to the nature of the monomeric unit, namely acrylics, vinyls, and cyclic monomers. Type A can be classified as acrylics, for instance, thio(meth)acrylates, thioester containing (meth)acrylates, and (meth)acrylamides. Type B monomers include olefins, such as thioester functionalized cyclopentene and N-(allyloxy) carbonyl-homocysteine TL. Cyclic monomers, which are polymerized by diverse ring-opening polymerization techniques (e.g. TLs), fall into Type C. Additionally a 100-fold higher reactivity towards thiolate nucleophiles and at least a 2000-fold higher reactivity towards carbanion nucleophiles have been observed in experiments and computational studies, attributed mainly to the lower electron delocalization in thioesters compared to oxoester [22].

Polymers that contain thioester functionalities can be prepared via different routes as listed in Table 7.1. The range of polymerization techniques, concerning monomers with pendant thioester groups range from free radical polymerization (FRP) to various reversible deactivation radical polymerization (RDRP) techniques,

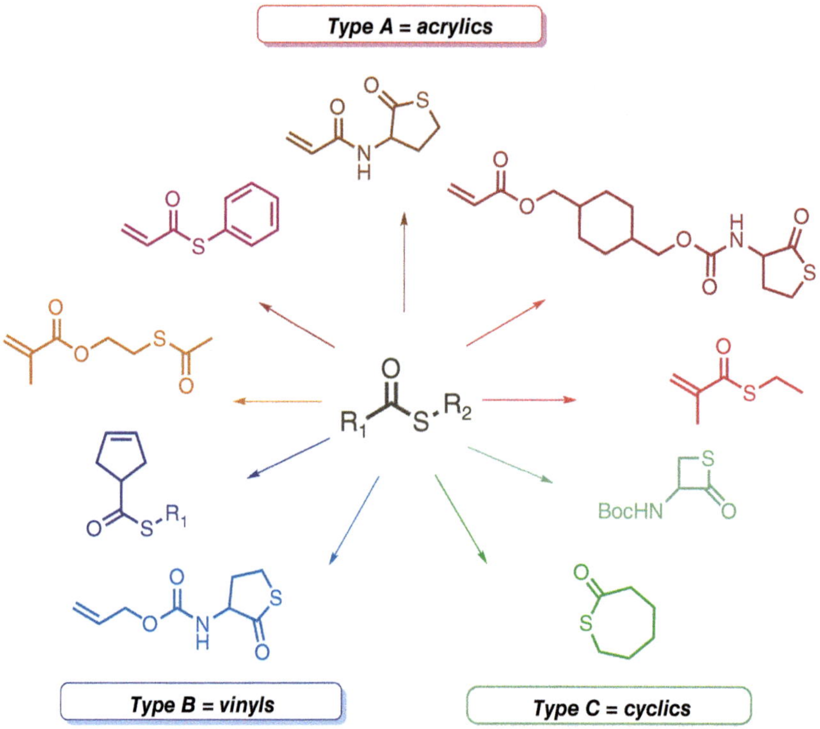

Scheme 7.1 Categorization of thioester containing monomers.

such as copper-mediated RDRP (Cu-RDRP), ATRP, and RAFT. Furthermore, thioesters can not only be incorporated in form of a thioester containing monomer, but also formed throughout a polymerization process such as β-thioesters via reactions of thiols with either an acrylate in a thiol-ene process or in an acid condensation reaction. Thioesters can also be found in the polymer chain end, integrated as thioester functionalized initiators or CTAs. Moreover, cyclic thioesters (e.g. TLs with different number of carbon atoms in the heterocycle) exhibit thioesters distributed along the polymer chain.

7.3.1 Access to Thioester Containing Polymers: Thioesters in the Side Chain

Based on reports in the literature, the introduction of thioester group into a polymerizable unit from Type A can be further divided into three groups, depending on the nature of the thioester. The organosulfur compound can be incorporated as a thioacrylic derivative in form of thio(meth)acrylates, thioester(meth)acrylate, or thioester(meth)acrylamide. One distinct factor is the local occurrence of the thioester, as it can be either somewhere in the side chain as in a (meth)acrylate or (meth)acrylamide based structure or it can be directly connected to the polymerizable unit as a thio(meth)acrylate. The latter being still of a lower level of

Table 7.1 Thioester containing monomer, initiator, CTAs used in various polymerization techniques.

Technique		Compound	Polymer structure	Refs.
FRP	Monomer	Thioacrylates (TA) Thio methacrylates (TMA)	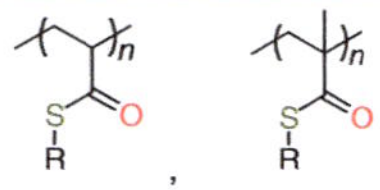	[57–59]
		Maleimide thiolactone (MITla)	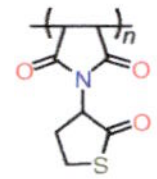	[60]
NMP	Monomer	Thioacrylates (TA)	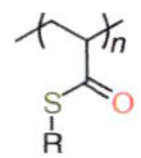	[61]
RAFT	Monomer	Thioacrylates (TA)	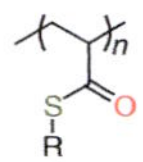	[62]
		2-(Acetylthio)ethyl methacrylate (AcSEMA)	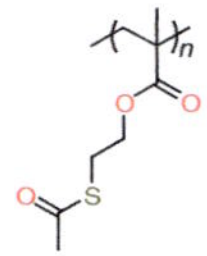	[63]
		N-Thiolactone acrylamide (TLAm)	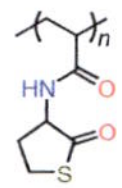	[64, 65]
		Maleimide thiolactone (MITL)	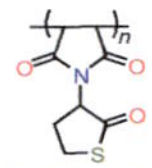	[60]

(Continued)

Table 7.1 *(Continued.)*

Technique		Compound	Polymer structure	Refs.
	CTA	Thioester trithiocarbonate	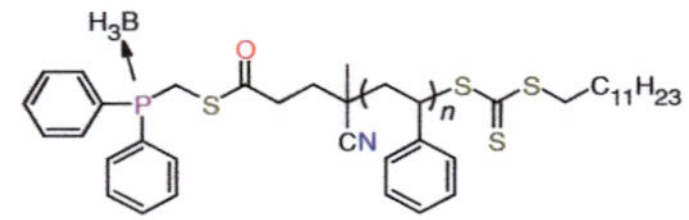	[43]
Cu-RDRP	Monomer	Thiolactone acrylate (TLA) Thioacrylates (TA)	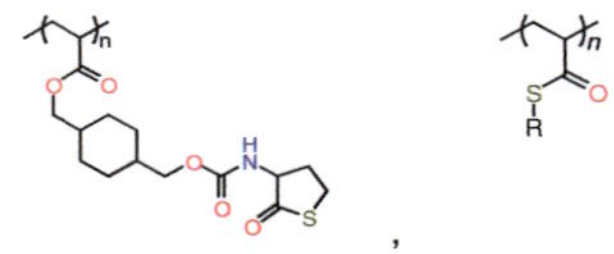	[66, 67]
	Initiator	Thioester ATRP initiator	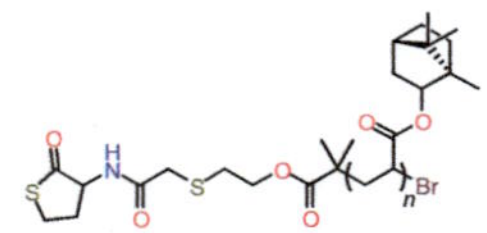	[68]
Cu-RDRP	Initiator	Linear thioester Initiator 4-arm thioester initiator	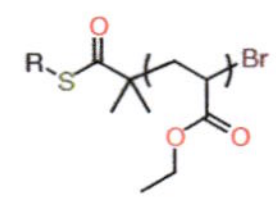 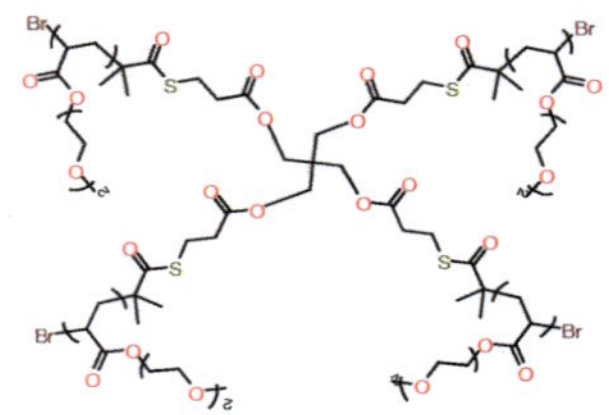	[69]
ROMP	Monomer	Cyclopentene thioesters		[70]
ROP	Monomer	β-Thiolactone	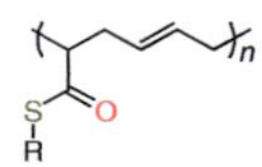	[71]

		γ-Thiolactone δ-Thiolactone	or 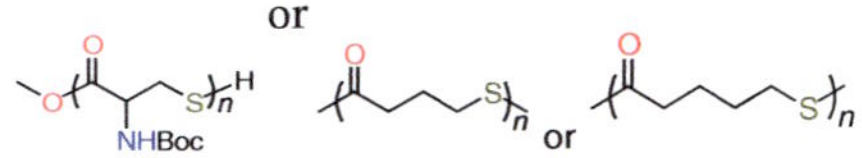	[72]
		ε-Thiolactone	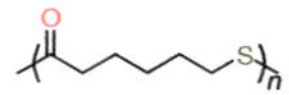	[72–76]
eROP	Monomer	ε-Thiolactone	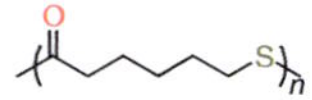	[77]
		Cyclic dithioester	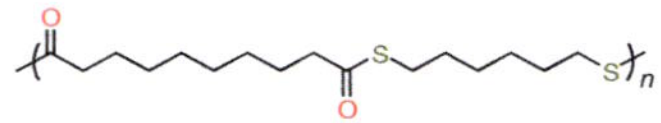	[78]
Thiol-ene step-growth	Monomer	Thiol + acrylate	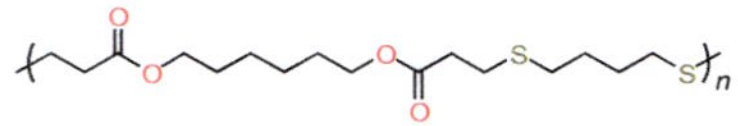	[79]
		Thiol + acrylate	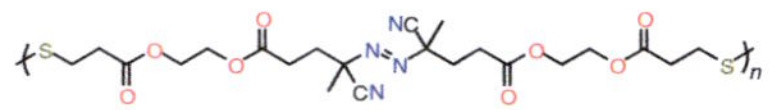	[80]
		Thiol + allyl	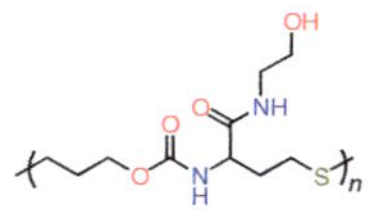	[17]

esteem than other acrylics, reports in the past showed promising features. For this class of monomers, sporadic entries to the literature concerning the (meth)acrylic thioester counterpart to (meth)acrylates have appeared in 1956. In these initial reports Marvel et al. showed the synthesis of a range of alkyl thioacrylates by the reaction of α,β-dibromopropionyl chloride with alkylthiols [57].

In 1977 Hadjichristidis et al. reported the preparation of thiomethacrylates by treating methacryloyl chloride with respective thiols in aqueous sodium hydroxide solution with successive distillation [58, 59]. More recently, Becer and coworker described a modified synthesis for thioacrylates to overcome the Michael addition of used thiols with a double bond of the (meth)acryloylchloride. For this purpose, a four-step protocol via thioesterfication of alkyl- and arylthiols with either bromoacetic acid or bromoacetyl bromide followed by the subsequent addition of triphenylphosphine to yield the phosphonium salt. Deprotonation with potassium carbonate gave the corresponding phosphonium ylide. Reacting the ylide with paraformaldehyde gave the respective thioacrylates via a Wittig reaction [62].

Another elegant and more recent example presents a cyclic monomer with a thioester as a pendant group, such as cyclopent-3-enecarbothioate (Scheme 7.2). In a one-step synthesis, monomers were obtained by simple thioesterfication of 3-cyclopentene-1-carboxylic acid with four alkythiols of different length. Differential scanning calorimetry (DSC) analysis highlighted higher glass transition temperatures (T_g) and melting temperature (T_m) for thioester containing polymers than for the structurally identical counterparts with ester pendants [70].

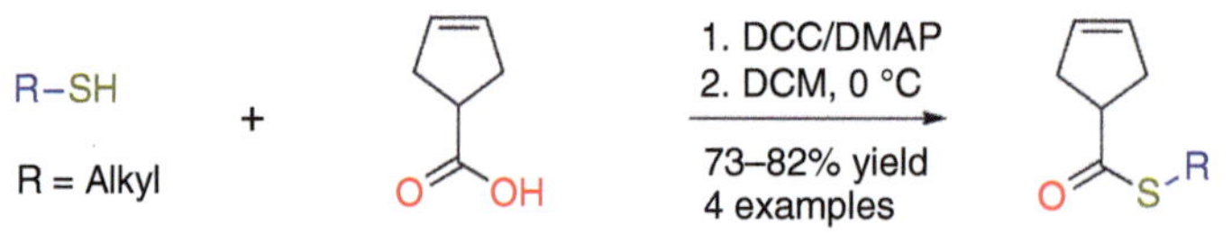

Scheme 7.2 Synthesis of thioester functionalized cyclopentene monomer. Source: Adapted from Ref. [70].

Thermogravimetric analysis (TGA) elucidated slightly lower thermal stability than their counterparts without sulfur [70, 81]. Additionally to thiols as starting materials, TLs have also been proven to be ideal precursors for thioester containing monomers. Among these in particular, vinylic TLs are the most established building blocks. A large set of TL based structures, derived from either itaconic acid or homocysteine TL, have been reported in multi-gram amounts (Scheme 7.3) [82].

A reaction of TL isocyanate with different alcohols yielded in TL carbamides. The same TL can be functionalized with allylamine and resulted in a TL urea (Route A in Scheme 7.3).

Homocysteine TL reacted with either an acid or cyclic anhydrides with different alcohols yielded in TL amides. Bisthiolactones derived from difunctional acid chlorides were obtained (Route B in Scheme 7.3) and the esterification of thioparaconic acid with alcohols lead to a set of TL esters. Additionally, a TL amide could be synthesized from the same starting material, but via a thioparaconic acid chloride and allylamine (Route C in Scheme 7.3).

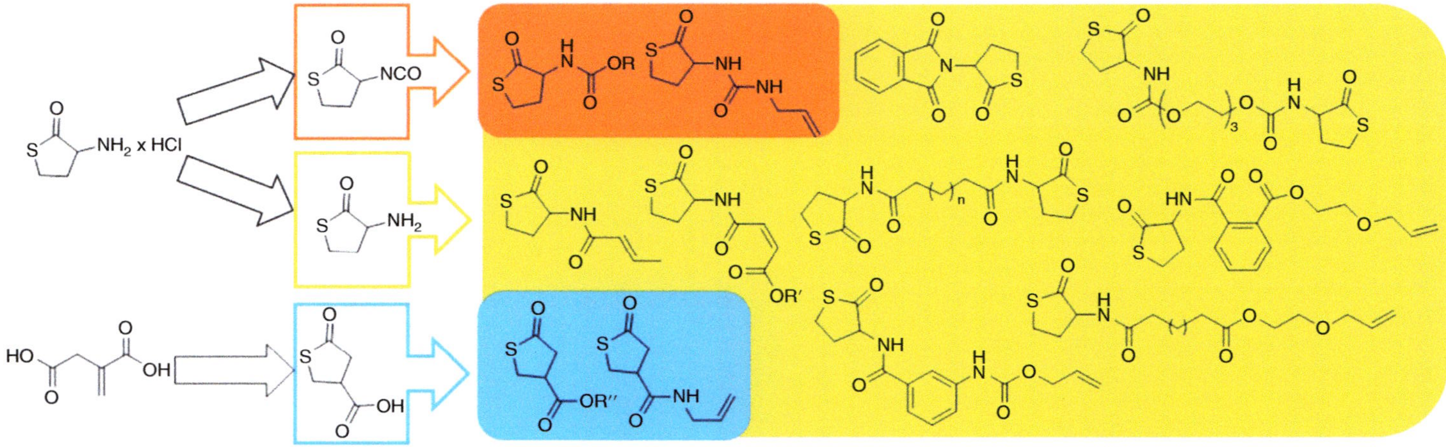

Scheme 7.3 Overview of the chemical synthesis of thiolactone derived vinyl monomers in Ref. [82].

On the other hand, a prominent example of a thioester bearing methacrylate is 2-(acetylthio)ethyl methacrylate (AcSEMA), resulting from a two-step synthesis [63, 83]. For this, methacryloyl chloride was reacted with 2-bromoethanol under basic conditions. The resulting 2-bromoethyl methacrylate was further reacted with potassium thioacetate to yield the thioester containing methacrylate (Scheme 7.4) [63].

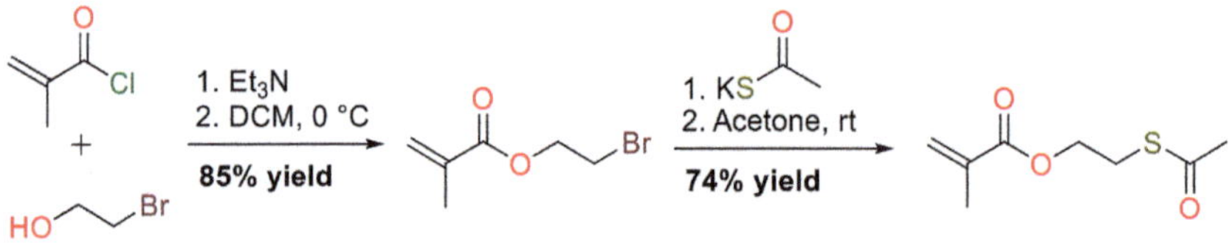

Scheme 7.4 Synthesis of the AcSEMA monomer. Source: Adapted from Ref. [63].

Möller and coworkers reported the synthesis of thiol functionalized poly(meth) acrylates, starting from modified methyl acrylate (MA) or methyl methacrylate (MMA) monomers by enzymatic transacylation with lipase (Novozym 435) with various alcohols at ambient temperatures [84]. After removal of the enzyme from the reaction mixture, polymerization were carried out in bulk, utilizing azobisisobutyronitrile (AIBN) as initiator in FRP gave polymers with weights up to 40 kDa.

The incorporation of the thioester group into a specific position on a polymer chain proves to be easy and useful when further functionalization is desired. In general, not only the thioester, but also any functional group, if present, is typically found along the backbone, chain end or in the monomeric repeating unit. By incorporating the thioester functionality in the monomer, it is possible to tailor the polymer by varying the targeted DP (degree of polymerization) or by introducing comonomers to obtain random or block copolymers. Hence, this methodology allows the further functionalization of thioesters at specific points along the polymer, which could yield a change in physical, thermal, and structural properties.

In the 1980s, Hadjichristidis et al. reported the synthesis of a new class of an acrylic monomer, namely, thiomethacrylate, in which the methacrylic ester oxygen was replaced with a sulfur atom. A series of phenylic poly methacrylates and poly thiomethacrylates were prepared at different chain lengths via free radical polymerization, using AIBN at 50 °C with relatively good control (PDI < 1.4). The chain flexibility of the polymers prepared was investigated from their thermal properties. Generally, the substituents in poly(phenyl thiomethacrylate) were found to be more flexible (higher flexibility factor σ) when compared with poly(phenyl methacrylate), which was attributed to the increased length of the less polar bond between the sulfur and carbon, caused by the substitution of the oxygen atom by a sulfur atom [58, 59].

Inspired by this, Becer and coworker recently reported the synthesis and controlled radical polymerization of a range of thioacrylates via RAFT polymerization to obtain excellent control over the polymerization (PDI 1.11–1.19) with quantitative conversions (Scheme 7.5) [62].

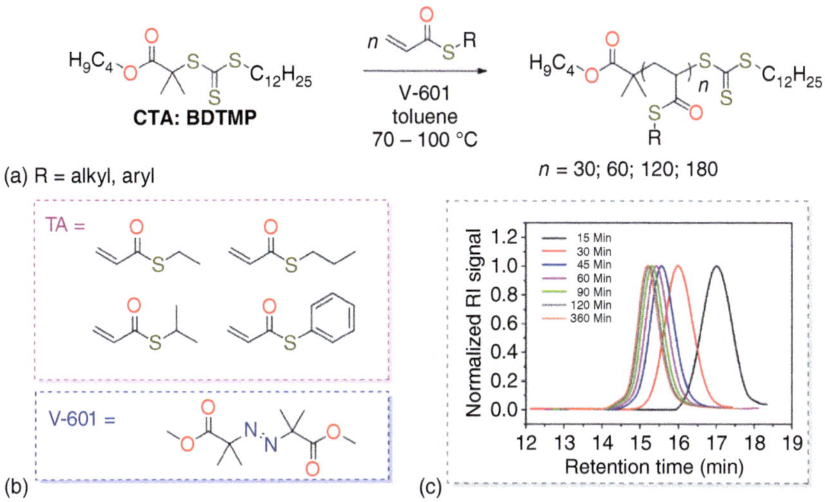

Scheme 7.5 (a) Conditions for homopolymerization of thioacrylates via RAFT, (b) structure of monomers used in this work, (c) GPC traces of poly(isopropyl thioacrylate) with DP = 60 in toluene at 70 °C and [I] = 0.1 mol%. Source: Adapted with permission from Aksakal and Becer [62]. Published by The Royal Society of Chemistry.

For example, ethyl thioacrylate (ETA) was found to polymerize at a similar rate compared with its oxoester acrylic counterpart ethyl acrylate (EA), whereas polydispersities remained low and observed molecular weight was close to the theoretical. It was shown that this new monomer class can be polymerized via more modern controlled radical polymerization techniques, allowing better control over the architecture. Finally, both the water contact angle and the T_g for P(ETA) were observed to be significantly higher than for P(EA), which provides evidence that thioacrylates give access to new materials with different properties, that could help to improve our understanding in material–property relationship [62].

Similarly, Song et al. have reported the polymerization of a range of thioester functionalized cyclopentenes via ring-opening metathesis polymerization (ROMP) with high conversions, while retaining control over dispersity [70]. A comparison of the thermal properties between the oxoester containing polymers and the reported thioester containing polymers revealed a similar trend as above, in which the latter displayed a higher T_g.

Another method of incorporating a thioester in to the polymer was shown by Boyer et al. in the form of a thioacetate monomer. In this case, AcSEMA was copolymerized with oligo(ethylene glycol) methyl ether methacrylate (OEGMA) via RAFT at different molar ratios and thoroughly characterised [63]. The obtained polymers were hydrolyzed to yield a free thiol, allowing the formation of gold nanoclusters. Interestingly, random copolymer stabilized gold nanoclusters displayed higher emission intensity in comparison to block copolymer stabilized gold nanoclusters.

Du Prez and coworkers generated a large library of TL containing monomers, which found application in a wide range of polymerization techniques. For example, a radical amine-thiol-ene polymerization has been performed in a one pot reaction via nucleophilic ring opening of a TL with an amine (aminolysis), followed by a radical thiol-ene conjugation [19, 85, 86]. Typically, a free thiol group was generated *in situ* after aminolysis, which reacted with a double bond already present in the same reactor. Using this protocol, a TL bearing a double bond (i.e. *N*-(allyloxy)carbonylhomocysteine TL) and various amines were employed to obtain linear polymers and networks in a radical photopolymerization process [17]. Radical thiol-ene polymerization has been performed on *N*-(allyloxy)carbonylhomocysteine TL and 10-undecenoyl thiolactonamide after aminolysis [17]. The same group used a similar strategy to polymerize a TL containing acrylate via Cu-RDRP LRP to form the backbone of a graft copolymer. Furthermore, propylamine and poly(ethylene glycol) acrylate were added to react with the TL ring, forming the brush structures [66].

7.3.2 Access to Thioester Containing Polymers: Thioester in the Chain End

Instead of having a number of thioesters incorporated, initiators for controlled radical polymerization (CRP) can be functionalized to provide thioester functionality at the chain end of a polymer. Cu(0)-mediated polymerization of isobornyl acrylate using a TL-based initiator has been evaluated as an efficient ATRP type initiator, while the presence of the thioester end group of the polymer has been confirmed by ^{1}H NMR (Scheme 7.6) [68].

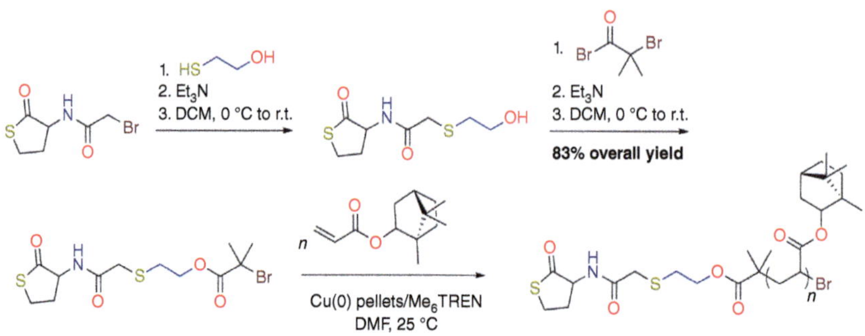

Scheme 7.6 One-pot, two-step synthesis of a thiolactone containing initiator and Cu(0)-mediated polymerization of isobornyl acrylate via a thiolactone-initiator. Source: Adapted from Ref. [68].

In a similar manner, the synthesis of functionalized cyclic polymers via RAFT utilizing a TL based CTA was shown. After the aminolysis on the thioester functionality, the free thiol was rearranged with the ω-chain end under high dilution, which afforded the targeted cyclic polymers [87].

In another approach, a CTA has been tethered to the polymer end group and have been used for the homopolymerization of styrene with further modification. Voit and coworkers reported a protocol for the esterification of a carboxylic-acid containing CTA with pentaflurophenyl and subsequently reacted with thiolphosphine. The obtained CTA was utilized for the synthesis of polystyrene via RAFT (Scheme 7.7) [43].

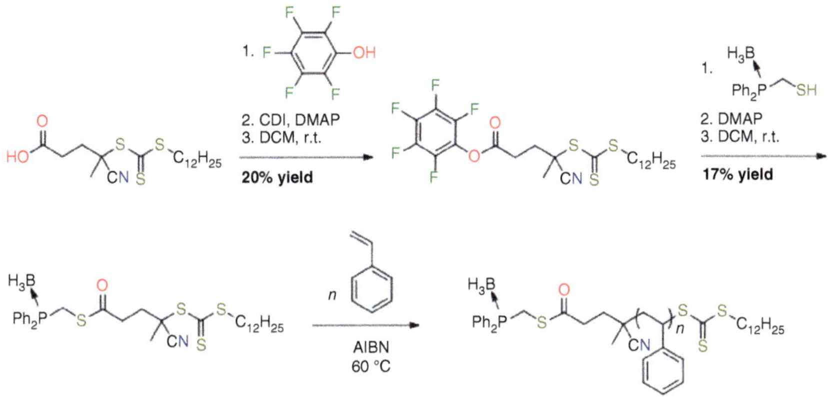

Scheme 7.7 Two-step preparation of thioester containing CTA and polymerization of styrene. Source: Adapted with permission from Pötzsch et al. [43]. Copyright 2011 American Chemical Society.

7.3.3 Access to Thioester Containing Polymers via Polymerization Process

As stated before, polythioesters can be described as polymers containing thioester moieties along the backbone. The synthesis of a polythioester was first reported by Kotch as early as 1951, where a range of dibasic acid chlorides were reacted with aliphatic dithiols, such as the reaction between adipoyl chloride and hexane-1,6-dithiol [88]. The obtained polythioesters were found to be low in molecular weight and displayed evidence of some crystallinity in their X-ray patterns, with their melting points being higher compared with their oxygen analogues. Similarly, polycondensations of methyl mercaptoacetate at different lengths were also described [89]. Later, the enzyme catalyzed polycondensation of hexane-1,6-dithiol with a range of diesters was demonstrated using lipase (Novozym 435), yielding low molecular weight polythioesters with 75–90% yield (M_n 3700–6000 g/mol, PDI 1.7–2.0) [90].

Higher molecular weights could also be obtained, when aromatic groups were included in the backbone. For example, Kowalewska and coworker showed the synthesis of an aromatic polythioester by interfacial polycondensation of 1,4-di(mercaptomethyl)-tetramethylbenzene phtaloyl, isophtaloyl, and terephthaloyl chloride, in which the aqueous to organic solvent ratio, as well as the type of organic phase and molar ratio of reagents was investigated, in order to determine the optimal reaction conditions to allow high conversion [91].

Another route to polythioesters is the ring-opening polymerization (ROP) [76]. Further experiments confirmed the same trend for higher melting points compared with corresponding polyesters. Polymers were prepared by anionic ROP of ε-thiocaprolactone for the first time, initiated with potassium *tert*-butoxide [92]. An extended study by the same group investigated the effect of the number of carbon atoms in the TL ring, and reactivity of the TLs towards polymerization dependent on ring size. Polymerization occurred for four-, six-, and seven-membered TLs, whereas the five-membered TL could not be polymerized. While this trend could also be observed for lactones [72], γ-TLs have been shown to copolymerize in the presence of trimethylene carbonate. Although NMR and MALDI-TOF-MS analysis has proven its incorporation, the percentage of TL found was limited to only 9 mol% [93].

Furthermore, when ε-thiocaprolactone has been also used as initiator for the anionic polymerization of ε-caprolactame, higher polymerization rate with the initiation by sulfur compound compared with the oxygen analogue was observed [94]. Moreover, different metal alkoxides (Sn, Cd, Mn, etc.) as catalysts and thiols or alcohols initiators have been used for the polymerization of cyclic thioesters [73, 74]. Lipase have also been employed as an alternative "greener" enzymatic-catalyzed ring-opening polymerization (eROP) and generated high molecular weight polymers [77, 78]. Polythioesters with higher molecular weight ($M_n > 50\,000$ g/mol, PDI 2.3) could be prepared under enzymatic ring-opening polymerization of cyclic polythioesters at 120 °C for 2 days [95].

Using various dithiols and diacrylates, Junkers and coworkers showed the design and synthesis of biodegradable poly(β-thioesters) via step-growth polymerization [79]. The obtained polymers were found to be semicrystalline materials with low T_g values due to the incorporation of thiols. Similarly, poly(β-thioesters) containing networks with broad molecular weight distributions were synthesized via thiol-ene Michael addition, when more than difunctional thiols and acrylates were used [96].

In the same manner, Michael addition was employed to obtain well-defined amphiphilic triblock copolymers of poly(β-thioesters)s, containing an azo linkage in the middle of the chain. TEM images of these polymers have shown their formation into micelles, which could be thermally degraded upon heating to 95 °C [80]. Long and coworkers on the other hand made use of the thia-Michael addition chemoselectivity towards an acrylate over a methacrylate to synthesize segmented poly(β-thioesters) in a one-pot approach [97]. Bis-thiol compounds were initially reacted with diacrylates to obtain thiol terminated soft segment oligomers, which were further reacted with oxamide containing dimethacrylates as the hard segments. Nishikubo and coworkers developed the synthesis of polythioethers by the acyl transfer polymerization of thiiranes with thioesters [98]. A similar approach was used when cyclic dithioesters were employed, in order to increase the ring size to yield cyclic(thioester-*alt*-thioether) [99].

Bowman and coworkers have recently shown that TL based nucleoside mimics can be polymerized via ring-opening polymerization to obtain well controlled DNA-like polymers resulting in high conversions (Scheme 7.8) [100]. The polymerization not only proceeded in an orderly sequence due to the addition of a DNA template, but also the poly(thioester) also depolymerized in the presence of a base, proving the dynamic and responsive character of the polymers. A similar example of polymer degradation was also shown by Bingham and Roth [101].

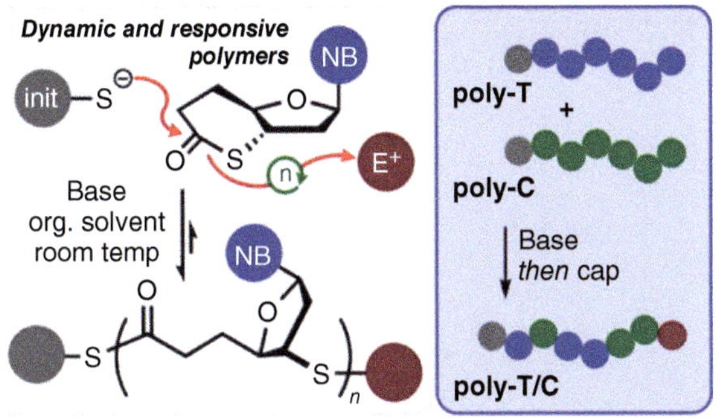

Scheme 7.8 Synthesis of dynamic and responsive "DNA-like" poly(thioester)s obtained via template directed polymerisation. Source: Adapted with permission from Mavila et al. [100]. Copyright 2018 American Chemical Society.

7.3.4 Access to Thioester-Bearing Structures via Post-modification Approach

It is possible to introduce thioesters to polymers that bear a functional group that is inert to the polymerization conditions and can be modified further after completion. One way to introduce thioesters into a polymer is by using thioacetic acid. For example, various dithiols and glycidyl propargyl ether were used in an attempt to obtain polymers via thiol-yne chemistry [102]. In further post-polymerization functionalization, the epoxide in the side chain was ring-opened using thioacetic acid in the presence of triethyl amine, to form thioesters in the side chain. Similarly, a highly efficient synthesis pathway to polysiloxanes containing thiols as end or side groups was demonstrated, when vinyl end-functionalized polydimethylsiloxane or polymethyl-vinylsiloxanes were reacted with thioacetic acid [103]. This allowed the formation of thioesters along the side chain or at the chain ends, which were reduced to the corresponding thiol. Although relatively stable to aminolysis, it is well known that thioesters can readily undergo transesterification with thiol groups to form new thioesters. An elegant example of a thioester incorporation via an exchange reaction was demonstrated by Liu et al. [104] (Scheme 7.9).

To construct dynamic nucleobase containing copolymers, dimethylacrylamide was copolymerized with pyridyldisulfide ethylacrylamide via RAFT. Thymine thioester reacted with the *in situ* generated pendant thiol group via a thiol thioester exchange reaction. This newly formed reversible thioester linkage was also shown to be glutathione responsive. Recently, Durmaz and coworkers have modified a polyester containing alkyne functionality in the polymer backbone, using an azide functional TL [105]. In an additional step, the TL group was ring-opened using various amines, displaying a fast and straight forward approach to obtain thioester functional polymers.

Grinstaff and coworkers have similarly obtained hydrogels when a poly(ethylene glycol)-based peptide dendron possessing four terminal thiols was crosslinked over thioesters with poly(ethylene glycol disuccinimidyl valerate) (SVA-PEG-SVA) (Scheme 7.10) [106].

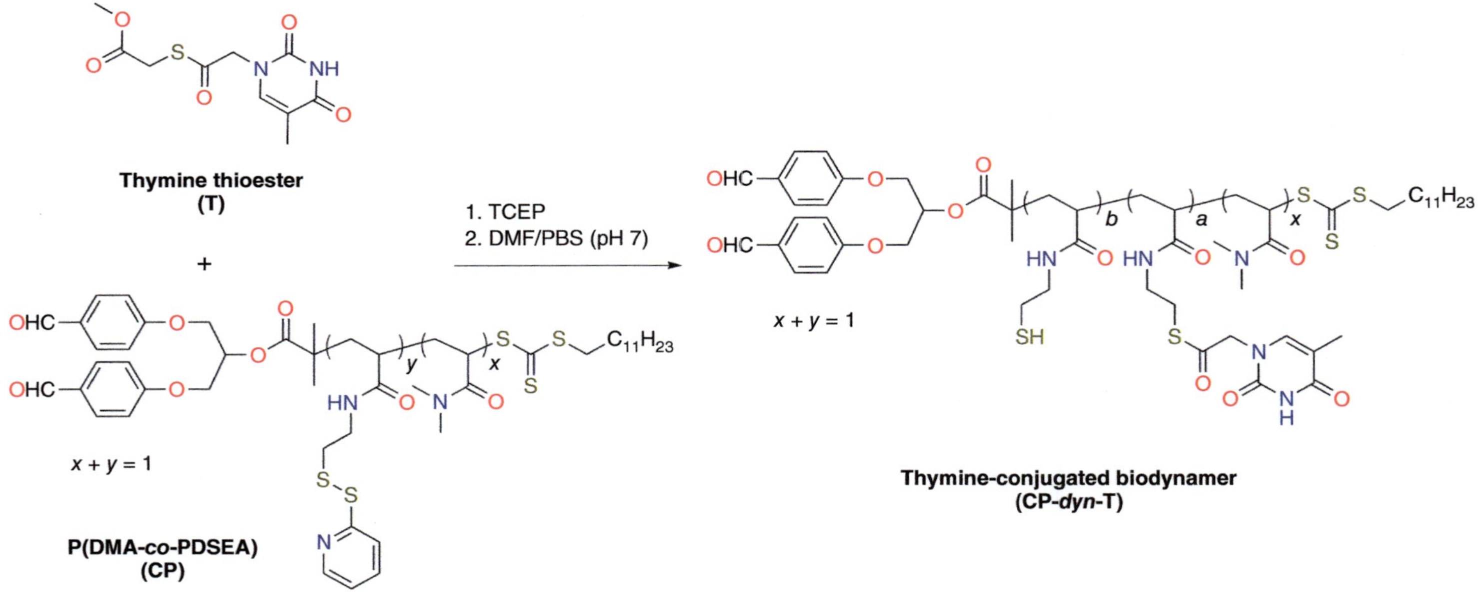

Scheme 7.9 Outline for the synthesis of thymine-conjugated biodynamer CP-dyn-T by RAFT polymerization via a thiol–thioester exchange reaction. Source: Adapted from Ref. [104].

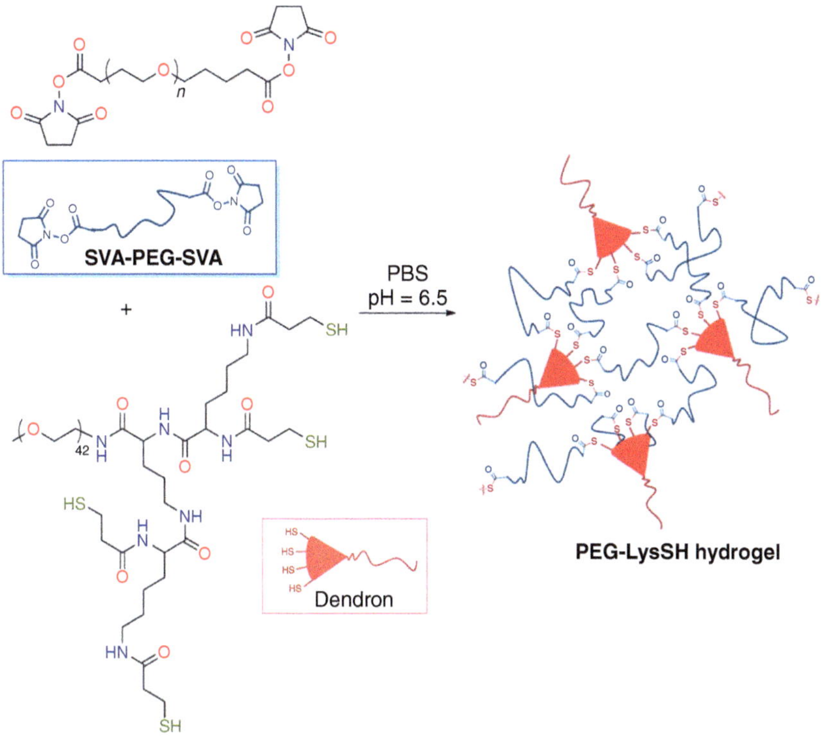

Scheme 7.10 A crosslinked PEG-LysSH hydrogel formed by the reaction of Dendron and SVA-PEG-SVA by thiol–thioester exchange. Source: Adapted from Ref. [106].

Unlike other examples, their dissolution by breaking the thioester bond was shown for the first time. Among others, L-cysteine methyl ester was used to fully promote the dissolution of the hydrogel via a thiol-thioester exchange reaction within 12 minutes.

7.4 Post-Polymerization Modification of Thioesters

Reactions, where new bonds are formed are the most used tools in synthesis and much progress have been made in the field of thioester containing reaction partners, yielding valuable products such as amides. For example, a thioester can lead to an *S*-acyl intermediate in the presence of cysteine, which can then spontaneously form amide bonds over an *S*- to *N*-acyl migration. This has been reported by Messersmith and coworkers, when initially a four arm poly(ethylene glycol) tetra amine was transformed into a tetra thioester [107]. The native chemical ligation (NCL) method was employed to covalently crosslink these macromonomers with a four-arm cysteine macromonomer into a hydrogel. It was also shown that no concurrent reactions to form hydrogels (i.e. disulfide bond, thioester exchange) took place during the ligation. Likewise, Kempe and coworkers have employed NCL using a cystein

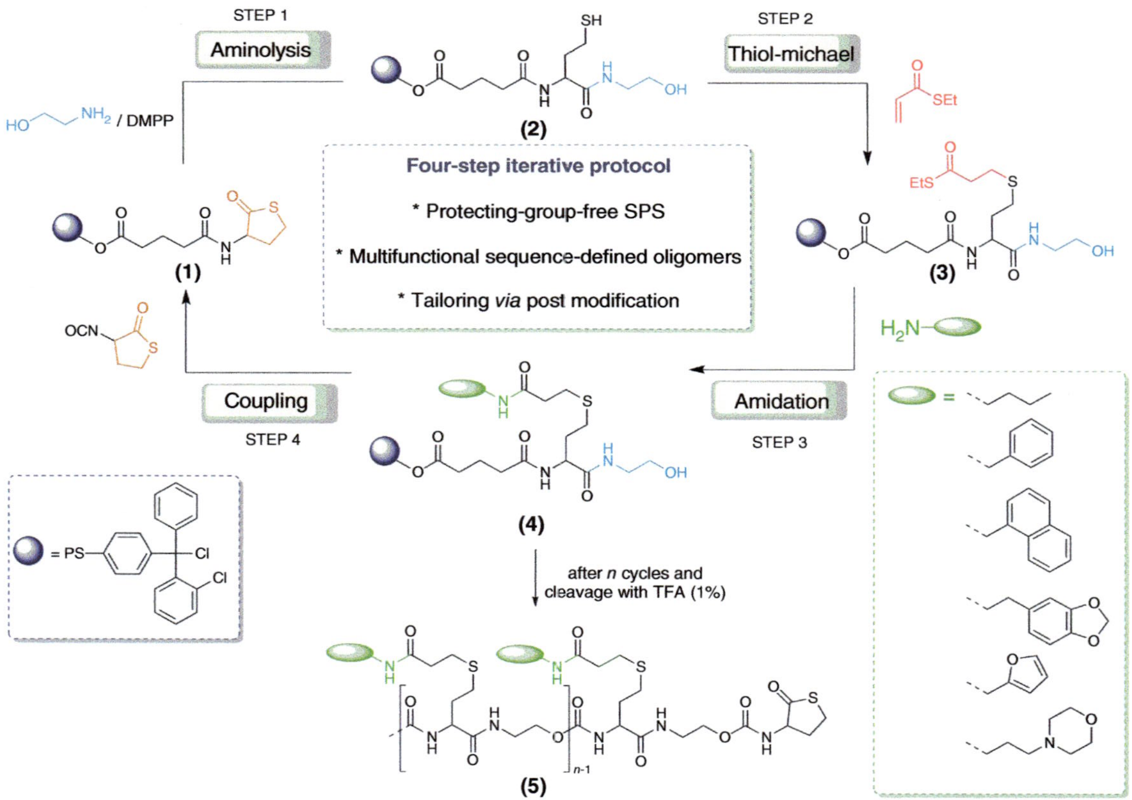

Scheme 7.11 Solid phase protocol with incorporated thioacrylates via thia-Michael reaction and subsequent amidation with various primary amines. Source: Adapted from Ref. [109].

end-functional PEG in a more controlled fashion [108]. Upon the amidation reaction of a thioester functional poly(oxazoline), the free thiol was reacted with acrylate in a further reaction resulting in well-defined 3-arm mikto star polymers.

In a collaborative work between the groups of Becer and coworkers, a new synthetic protocol for the synthesis of sequence-defined oligomers was developed (Scheme 7.11). Using an iterative protocol based on TL chemistry, a backbone was obtained. Test reactions between the free thiol and acrylamides to form a side chain were proven to be unsuccessful. By incorporating thioacrylates into the side chain over a thiol-ene chemistry however, it was possible to introduce different amines over an amidation reaction. This overcame the inefficiency of the reactions between the thiol and acrylamides, yielding in sequences of high purity [109].

As mentioned in Table 7.1, Voit and coworkers demonstrated the synthesis of poly(styrene) using a phosphine containing thioester based CTA, which was incorporated to allow further functionalization with an azide via Staudinger Ligation (Scheme 7.12). Hence, using 3,6,9-trioxodecyl azide, the end group modification was carried out on the thioester, yielding an amide bond on the chain end in almost quantitative manner [43].

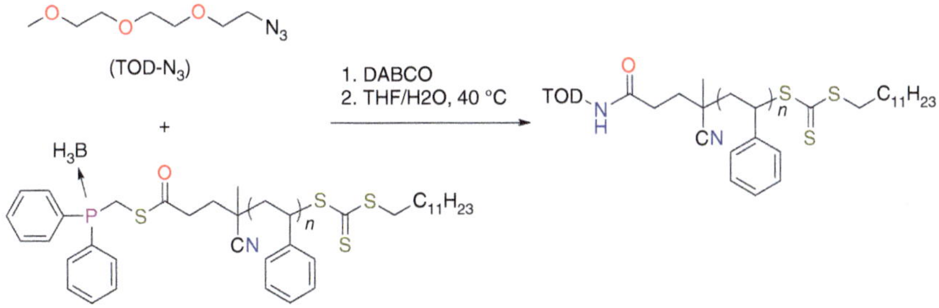

Scheme 7.12 Staudinger Ligation of thioester containing CTA with TOD-N_3. Source: Adapted from Ref. [43].

Furthermore, París and coworkers prepared thioester containing polymers via modification on the polymer end of a typical bromine functionalized ATRP initiator [110]. In a one-step protocol, various poly-methacrylates have been synthesized and the bromine end group has been transformed into corresponding thioester by substitution reaction with potassium thioacetate. After subsequent hydrolysis, a thiol terminated polymer was generated and could be further modified via a thiol-ene reaction with a fluorescent alkene (a synthetic alquene tethered to a 4,4-difluoro-4-bora-3a,4a-diaza-s-indacene (BODIPY) fragment). By this, quantification of the quantitative thiol-ene process could be calculated from UV–Vis measurements (Scheme 7.13).

The incorporation of a thioester to the chain end and its modification can also be achieved by utilizing a thioester based initiator (e.g. Cu-RDRP). To date, majority of the ATRP-type initiators were based on an ester neighboring group. While the use of amide based initiators is also to be found in the literature, the investigation of a thioester based initiator, however, was only reported recently for the

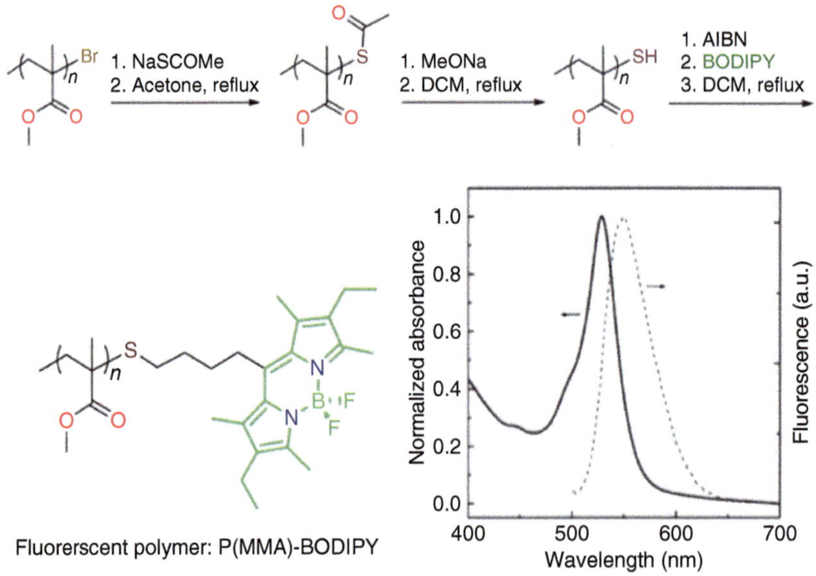

Scheme 7.13 Route for the formation of BODIPY end-functionalized P(MMA) via thio-ester end-functionalization, subsequent hydrolysis to a thiol end-functionalized polymer. "Click" reaction with BODIPY to yield a fluorescent polymer, which absorption spectrum is displayed on the bottom right (solid lines) and fluorescence (dashed lines). Source: Adapted with permission from Liras et al. [110]. Copyright 2011 American Chemical Society.

first time [69]. Becer and coworkers have described that thioester-based initiators polymerized just as good as ester based initiators, with the added benefit to allow for further modification under very mild reaction conditions (Scheme 7.14). For example, a tetra-thioester functional initiator was used to obtain an ethylene glycol based four-arm star polymer. Periodic sampling of the star polymer in the presence of thiophenol and L-cysteine methyl ester (Cys-ME) has revealed lower molecular weight signals in GPC spectra, indicating its transformation into a linear polymer. The end polymer was analyzed via MALDI-TOF-MS and NMR spectroscopy and was found to be in agreement with the predicted linear structure.

7.5 Conclusion and Outlook

In this chapter, a brief overview of the literature on polymers containing thioesters is provided on selected examples, to give the reader an initial understanding of interesting reports in their synthesis and their possible further functionalization routes. Thioester groups are easily included into polymers, which can be from a thioester containing monomer or that can be generated during the polymerization process. Moreover, reports display that the position of the thioester can be specific and easily incorporated in any region and position along a polymer, giving materials tailored thermal or structural properties.

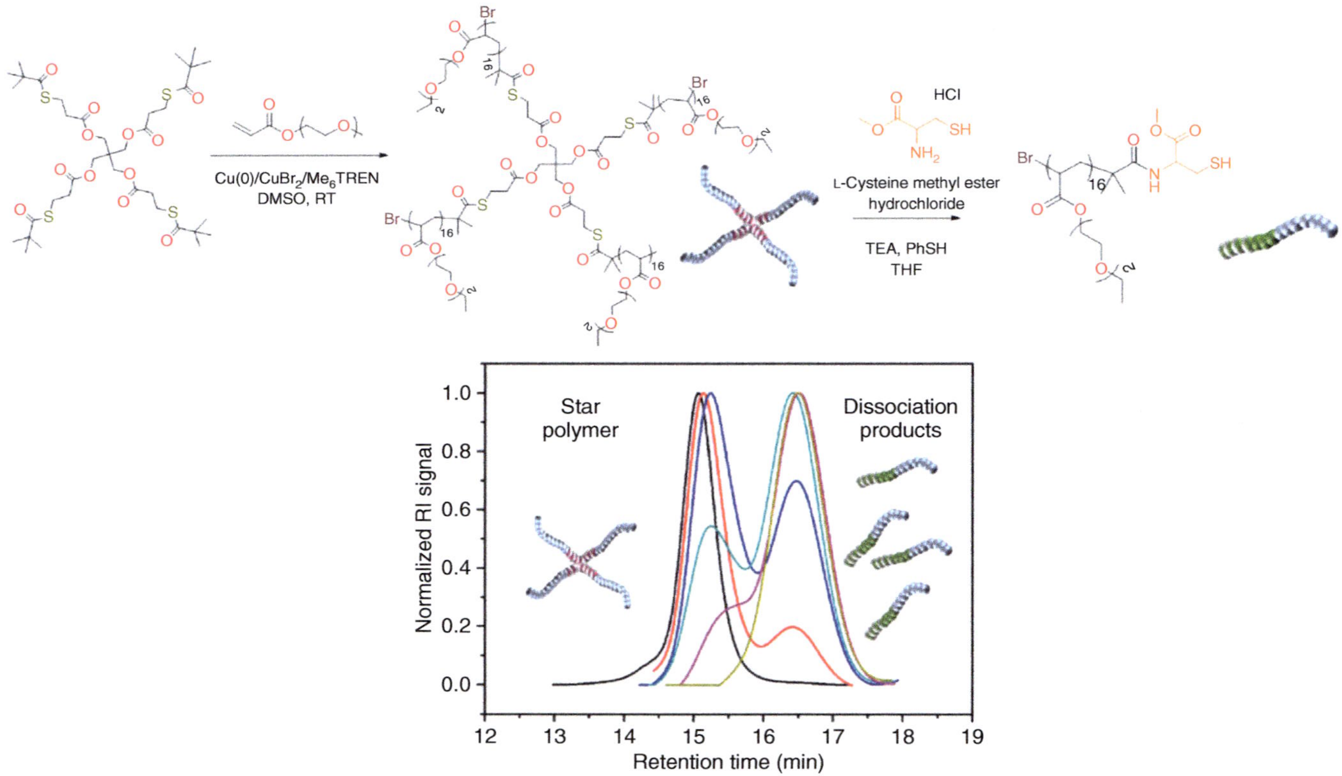

Scheme 7.14 Top: A tetrathioester initiator (left) is used to synthesize a four-arm star polymer (middle). Its transformation into a linear polymer (right) was shown using Cys-ME. Bottom: Overlay of the GPC traces obtained during the dissociation of the star polymer, displaying a shift towards lower retention times [69].

While there are reports on thioester-containing initiators for ATRP or CTAs for RAFT that allow the positioning thereof mainly in the chain end, their inclusion in the side chain is best achieved via RAFT polymerization. On the other hand, ring-opening, polycondensation, or thia-Michael polymerization techniques allow the incorporation of (α-, β-) thioesters along the backbone. In addition, thioesters allow reactions, which would otherwise lead to side reactions, when carried out, for example, in the presence of free thiols, which are rather reactive in radical and nucleophilic reactions (e.g. early termination in ATRP due to thio-bromo "click" reaction). The free thiol can then be generated via post polymerization modification. The introduction of thioesters to change the flexibility of the backbone or the sidechain in contrast to its ester or amide analogues allows for modification of the thermal properties, while retaining similarities in the core structure (i.e. control of T_g in poly(ethyl acrylate) versus poly(ethyl thioacrylate)). Exchange reactions, amidation on the thioester, and ligation reactions are main functionalization pathways of thioesters and are well established in incorporating further functionalities. Many reports also contain the synthesis of complex structures, whereas the thioester bond can be transformed into different linkages. Contrary, current limitations and focus are mainly on finding novel uses that can be transformed into direct applications. Although thioesters on polymers are well understood, future research on the implementation of thioester chemistry in living systems to polymers could improve the understanding of polymer interactions with organisms (as biomacromolecule mimics, in polymer-peptide conjugates or drug delivery etc.) that could allow new strategies and chemistries to be developed.

References

1 Kast, C.E. and Bernkop-Schnürch, A. (2001). Thiolated polymers—thiomers: development and in vitro evaluation of chitosan–thioglycolic acid conjugates. *Biomaterials* 22: 2345–2352.
2 Tatsuhito, M., Yasuaki, F., Masahiro, Y. et al. (2000). Optical material of high refractive index resin composed of sulfur-containing aliphatic and alicyclic methacrylates. *J. Appl. Polym. Sci.* 76: 45–49.
3 Xue, Y., Li, X., Li, H., and Zhang, W. (2014). Quantifying thiol–gold interactions towards the efficient strength control. *Nat. Commun.* 5: 4348.
4 Häkkinen, H. (2012). The gold–sulfur interface at the nanoscale. *Nat. Chem.* 4: 443.
5 Love, J.C., Estroff, L.A., Kriebel, J.K. et al. (2005). Self-assembled monolayers of thiolates on metals as a form of nanotechnology. *Chem. Rev.* 105: 1103–1170.
6 Boisselier, E. and Astruc, D. (2009). Gold nanoparticles in nanomedicine: preparations, imaging, diagnostics, therapies and toxicity. *Chem. Soc. Rev.* 38: 1759–1782.
7 Stenzel, M.H. (2013). Bioconjugation using thiols: old chemistry rediscovered to connect polymers with nature's building blocks. *ACS Macro Lett.* 2: 14–18.

8 Verbraeken, B., Monnery, B.D., Lava, K., and Hoogenboom, R. (2017). The chemistry of poly(2-oxazoline)s. *Eur. Polym. J.* 88: 451–469.

9 Delaittre, G. and Barner, L. (2018). The para-fluoro-thiol reaction as an efficient tool in polymer chemistry. *Polym. Chem.* 9: 2679–2684.

10 Agar, S., Baysak, E., Hizal, G. et al. (2018). An emerging post-polymerization modification technique: the promise of thiol-para-fluoro click reaction. *J. Polym. Sci. A Polym. Chem.* 56: 1181–1198.

11 Le Neindre, M. and Nicolay, R. (2014). Polythiol copolymers with precise architectures: a platform for functional materials. *Polym. Chem.* 5: 4601–4611.

12 Yu, S.H., Ercole, F., Veldhuis, N.A. et al. (2017). Polymers with acyl-protected perthiol chain termini as convenient building blocks for doubly responsive H2S-donating nanoparticles. *Polym. Chem.* 8: 6362–6367.

13 Iimura, S., Manabe, K., and Kobayashi, S. (2003). Hydrophobic polymer-supported catalyst for organic reactions in water: acid-catalyzed hydrolysis of thioesters and transprotection of thiols. *Org. Lett.* 5: 101–103.

14 Nakabayashi, K., Matsumura, A., Abiko, Y., and Mori, H. (2016). Controlled synthesis of thiazole-based polymers and block copolymers by RAFT polymerization of azolyl S-vinyl sulfides and metal complexation. *Macromolecules* 49: 1616–1629.

15 Qiao, Y., Yin, X., and Tang, C. (2015). Progress in side-chain thiophene-containing polymers: synthesis, properties and applications. *Sci. China Chem.* 58: 1641–1650.

16 Zhao, C., Zhang, Y., Pan, S. et al. (2007). Synthesis, characterization, and properties of homopolymers functionalized with oligothiophene derivatives in the side chain. *Macromolecules* 40: 1816–1823.

17 Espeel, P., Goethals, F., and Du Prez, F.E. (2011). One-pot multistep reactions based on thiolactones: extending the realm of thiol–ene chemistry in polymer synthesis. *J. Am. Chem. Soc.* 133: 1678–1681.

18 Pieter, E., Carrette, L.L.G., Katarzyna, B. et al. (2013). Multifunctionalized sequence-defined oligomers from a single building block. *Angew. Chem. Int. Ed.* 52: 13261–13264.

19 Martens, S., Van den Begin, J., Madder, A. et al. (2016). Automated synthesis of monodisperse oligomers, featuring sequence control and tailored functionalization. *J. Am. Chem. Soc.* 138: 14182–14185.

20 Bracher, P., Snyder, P., Bohall, B., and Whitesides, G. (2011). The relative rates of thiol–thioester exchange and hydrolysis for alkyl and aryl thioalkanoates in water. *Orig. Life Evol. Biosph.* 41: 399–412.

21 Marlier, J.F., Fogle, E.J., Redman, R.L. et al. (2015). A mechanistic study of thioester hydrolysis with heavy atom kinetic isotope effects. *J. Org. Chem.* 80: 1905–1908.

22 Yang, W. and Drueckhammer, D.G. (2001). Understanding the relative acyl-transfer reactivity of oxoesters and thioesters: computational analysis of transition state delocalization effects. *J. Am. Chem. Soc.* 123: 11004–11009.

23 Koval, I.V. (2005). Synthesis, structure, and physicochemical characteristics of thiols. *Russ. J. Org. Chem.* 41: 631–648.

24 Neises, B. and Steglich, W. (1978). Simple method for the esterification of carboxylic acids. *Angew. Chem. Int. Ed. Engl.* 17: 522–524.

25 Iranpoor, N., Firouzabadi, H., Khalili, D., and Motevalli, S. (2008). Easily prepared azopyridines as potent and recyclable reagents for facile esterification reactions. An efficient modified mitsunobu reaction. *J. Org. Chem.* 73: 4882–4887.

26 Chou, Y.-L., Jhong, Y., Swain, S.P., and Hou, D.-R. (2017). Microwave-assisted direct thioesterification of carboxylic acids. *J. Org. Chem.* 82: 10201–10208.

27 Temperini, A., Annesi, D., Testaferri, L., and Tiecco, M. (2010). A simple acylation of thiols with anhydrides. *Tetrahedron Lett.* 51 (41): 5368–5371.

28 Khan, A.T., Choudhury, L.H., and Subrata, G. (2005). Acetonyltriphenylphosphonium bromide (ATPB): a versatile reagent for the acylation of alcohols, phenols, thiols and amines and for 1,1-diacylation of aldehydes under solvent-free conditions. *Eur. J. Org. Chem.* 2005: 2782–2787.

29 Lakouraj, M., Movassagh, B., and Fadaei, Z. (2002). Convenient synthesis of thiol esters from acyl chlorides and disulfides using Zn-$AlCl_3$. *Monatshefte fuer Chemie* 133: 1085–1088.

30 Ranu, B.C., Dey, S.S., and Hajra, A. (2003). Highly efficient acylation of alcohols, amines and thiols under solvent-free and catalyst-free conditions. *Green Chem.* 5: 44–46.

31 Basu, B., Paul, S., and Nanda, A.K. (2010). Silica-promoted facile synthesis of thioesters and thioethers: a highly efficient, reusable and environmentally safe solid support. *Green Chem.* 12: 767–771.

32 Singh, P. and Peddinti, R.K. (2017). Harnessing the catalytic behaviour of 1,1,1,3,3,3-hexafluoro-2-propanol (HFIP): an expeditious synthesis of thioesters. *Tetrahedron Lett.* 58 (19): 1875–1878.

33 Ogawa, K.A. and Boydston, A.J. (2014). Organocatalyzed anodic oxidation of aldehydes to thioesters. *Org. Lett.* 16: 1928–1931.

34 Yost, J.M., Zhou, G., and Coltart, D.M. (2006). A facile and efficient direct aldol addition of simple thioesters. *Org. Lett.* 8: 1503–1506.

35 Junyong, C., Ryung, S.U., Supill, C., and Keun, C.Y. (2016). Poly(3,4-dimethyl-5-vinylthiazolium)/DBU-catalyzed thioesterification of aldehydes with thiols. *ChemCatChem* 8: 318–321.

36 He, C., Qian, X., and Sun, P. (2014). Syntheses of thiol and selenol esters by oxidative coupling reaction of aldehydes with RYYR (Y = S, Se) under metal-free conditions. *Org. Biomol. Chem.* 12: 6072–6075.

37 Huang, Y.-T., Lu, S.-Y., Yi, C.-L., and Lee, C.-F. (2014). Iron-catalyzed synthesis of thioesters from thiols and aldehydes in water. *J. Org. Chem.* 79: 4561–4568.

38 Yi, C.-L., Huang, Y.-T., and Lee, C.-F. (2013). Synthesis of thioesters through copper-catalyzed coupling of aldehydes with thiols in water. *Green Chem.* 15: 2476–2484.

39 Kazemi, M. and Shiri, L. (2015). Thioesters synthesis: Recent adventures in the esterification of thiols. *J. Sulfur Chem.* 36 (6): 613–623.

40 Kuciński, K. and Hreczycho, G. (2018). S-acetylation of thiols mediated by triflic acid: a novel route to thioesters. *Org. Process Res. Dev.* 22: 489–493.

41 Staudinger, H. and Jules, M. (1919). Über neue organische Phosphorverbindungen III. Phosphinmethylenderivate und Phosphinimine. *Helv. Chim. Acta* 2: 635–646.

42 Nilsson, B.L., Kiessling, L.L., and Raines, R.T. (2000). Staudinger ligation: a peptide from a thioester and azide. *Org. Lett.* 2: 1939–1941.

43 Pötzsch, R., Fleischmann, S., Tock, C. et al. (2011). Combining RAFT and staudinger ligation: a potentially new synthetic tool for bioconjugate formation. *Macromolecules* 44: 3260–3269.

44 Johnson, E.C.B. and Kent, S.B.H. (2006). Insights into the mechanism and catalysis of the native chemical ligation reaction. *J. Am. Chem. Soc.* 128: 6640–6646.

45 Chen, W., Qing-Xiang, G., and Yao, F. (2011). Theoretical analysis of the detailed mechanism of native chemical ligation reactions. *Chem. Asian J.* 6: 1241–1251.

46 Woll, M.G. and Gellman, S.H. (2004). Backbone thioester exchange: a new approach to evaluating higher order structural stability in polypeptides. *J. Am. Chem. Soc.* 126: 11172–11174.

47 Sun, X.-H., Yu, H.-Z., Pei, S.-Q., and Dang, Z.-M. (2015). Theoretical investigations on the thiol–thioester exchange steps of different thioesters. *Chin. Chem. Lett.* 26: 1259–1264.

48 Deletre, M. and Levesque, G. (1990). Kinetics and mechanism of polythioamidation in solution. 1. Reaction of mono- and bis(dithioester)s with excess amine. *Macromolecules* 23: 4733–4741.

49 Stuhr-Hansen, N., Bork, N., and Stromgaard, K. (2014). Mechanistic insight into benzenethiol catalyzed amide bond formations from thioesters and primary amines. *Org. Biomol. Chem.* 12: 5745–5751.

50 Castro, E. (2009). ChemInform abstract: kinetics and mechanism of the aminolysis of thioesters and thiocarbonates in solution. *ChemInform* 81 (4): 685–696.

51 Tokuyama, H., Yokoshima, S., Yamashita, T. et al. (1998). Facile palladium-mediated conversion of ethanethiol esters to aldehydes and ketones. *J. Braz. Chem. Soc.* 9: 381–387.

52 Fukuyama, T., Lin, S.C., and Li, L. (1990). Facile reduction of ethyl thiol esters to aldehydes: application to a total synthesis of (+)-neothramycin A methyl ether. *J. Am. Chem. Soc.* 112: 7050–7051.

53 Ichiishi, N., Malapit, C.A., Woźniak, Ł., and Sanford, M.S. (2018). Palladium- and nickel-catalyzed decarbonylative C–S coupling to convert thioesters to thioethers. *Org. Lett.* 20: 44–47.

54 Wessig, P., Schulze, T., Pfennig, A. et al. (2017). Thiol–ene polymerization of oligospiroketal rods. *Polym. Chem.* 8: 6879–6885.

55 Willcock, H. and O'Reilly, R.K. (2010). End group removal and modification of RAFT polymers. *Polym. Chem.* 1: 149–157.

56 Xu, J., He, J., Fan, D. et al. (2006). Aminolysis of polymers with thiocarbonylthio termini prepared by RAFT polymerization: the difference between polystyrene and polymethacrylates. *Macromolecules* 39: 8616–8624.

57 Marvel, C.S., Jacobs, S.L., Taft, W.K., and Labbe, B.G. (1956). Alkyl thiolacrylates: their preparation and polymerization. *J. Polym. Sci.* 19: 59–72.
58 Hadjichristidis, N. (1981). Solution properties and chain flexibility of polythiolmethacrylates: 1. Poly(phenyl thiolmethacrylate) and poly(O-methyl phenyl thiolmethacrylate). *Polymer* 22: 63–66.
59 Hadjichristidis, N. (1981). Solution properties and chain flexibility of poly(thiolmethacrylates). 2. Poly(cyclohexyl thiolmethacrylate). *Macromolecules* 14: 128–130.
60 Rudolph, T., Espeel, P., Du Prez, F.E., and Schacher, F.H. (2015). Poly(thiolactone) homo- and copolymers from maleimide thiolactone: synthesis and functionalization. *Polym. Chem.* 6: 4240–4251.
61 Aksakal, S., Liu, R., Aksakal, R., and Becer, C.R. (2020). Nitroxide-mediated polymerisation of thioacrylates and their transformation into poly(acrylamide)s. *Polym. Chem.* 11: 982–989.
62 Aksakal, S. and Becer, C.R. (2016). Poly(thioacrylate)s: expanding the monomer toolbox of functional polymers. *Polym. Chem.* 7: 7011–7018.
63 Adnan, N.N.M., Ahmad, S., Kuchel, R.P., and Boyer, C. (2017). Exploring the potential of linear polymer structures for the synthesis of fluorescent gold nanoclusters. *Mater. Chem. Front.* 1: 80–90.
64 Reinicke, S., Espeel, P., Stamenovic, M.M., and Du Prez, F.E. (2014). Synthesis of multi-functionalized hydrogels by a thiolactone-based synthetic protocol. *Polym. Chem.* 5: 5461–5470.
65 Montolio, S., Zagorodko, O., Porcar, R. et al. (2017). Poly(acrylamide-homocysteine thiolactone) as a synthetic platform for the preparation of polymeric ionic liquids by post ring-opening-orthogonal modifications. *Polym. Chem.* 8: 4789–4797.
66 Driessen, F., Herckens, R., Espeel, P., and Du Prez, F.E. (2016). Thiolactone chemistry and copper-mediated CRP for the development of well-defined amphiphilic dispersing agents. *Polym. Chem.* 7: 1632–1641.
67 Aksakal, S., Beyer, V.P., Aksakal, R., and Becer, C.R. (2019). Copper mediated RDRP of thioacrylates and their combination with acrylates and acrylamides. *Polym. Chem.* 10: 6622–6629.
68 Driessen, F., Du Prez, F.E., and Espeel, P. (2015). Precision multisegmented macromolecular lineups: a display of unique control over backbone structure and functionality. *ACS Macro Lett.* 4: 616–619.
69 Aksakal, S., Aksakal, R., and Becer, C.R. (2019). Transformation of thioester-initiated star polymers into linear arms via native chemical ligation. *Macromol. Rapid Commun.* 40: 1900247.
70 Song, S., Zhang, Z., Liu, X. et al. (2017). Synthesis and characterization of functional polyethylene with regularly distributed thioester pendants via ring-opening metathesis polymerization. *J. Polym. Sci. A Polym. Chem.* 55: 4027–4036.
71 Suzuki, M., Makimura, K., and Matsuoka, S.-i. (2016). Thiol-mediated controlled ring-opening polymerization of cysteine-derived β-thiolactone and unique features of product polythioester. *Biomacromolecules* 17: 1135–1141.

72 Overberger, C.G. and Weise, J.K. (1968). Anionic ring-opening polymerization of thiolactones. *J. Am. Chem. Soc.* 90: 3533–3537.

73 Fritze, P.E. Union Carbide; USA: thiolactone polymerization and catalysts. US Patent 3, 755,268, 16 June 1972.

74 Seefried, C.G. and Kodeske, J.V. (1976). Lactone polymers VII. Activation energies for glass transitions of poly-ε-caprolactone and poly-ε-thiocaprolactone. *Polym. Eng. Sci.* 16: 526–528.

75 Overberger, C.G. and Weise, J.K. (1968). Optical rotatory dispersion studies of asymmetric poly(thiol esters). *J. Am. Chem. Soc.* 90: 3538–3543.

76 Bannin, T.J. and Kiesewetter, M.K. (2015). Poly(thioester) by organocatalytic ring-opening polymerization. *Macromolecules* 48: 5481–5486.

77 Duchiron, S.W., Pollet, E., Givry, S., and Avérous, L. (2017). Enzymatic synthesis of poly(ε-caprolactone-co-ε-thiocaprolactone). *Eur. Polym. J.* 87: 147–158.

78 Kato, M., Toshima, K., and Matsumura, S. (2007). Enzymatic synthesis of polythioester by the ring-opening polymerization of cyclic thioester. *Biomacromolecules* 8: 3590–3596.

79 Zaquen, N., Wenn, B., Ranieri, K. et al. (2014). Facile design of degradable poly(β-thioester)s with tunable structure and functionality. *J. Polym. Sci. A Polym. Chem.* 52: 178–187.

80 Sun, H., Dobbins, D.J., Dai, Y. et al. (2016). Radical departure: thermally-triggered degradation of azo-containing poly(β-thioester)s. *ACS Macro Lett.* 5: 688–693.

81 Song, S., Fu, Z., Xu, J., and Fan, Z. (2017). Synthesis of functional polyolefins via ring-opening metathesis polymerization of ester-functionalized cyclopentene and its copolymerization with cyclic comonomers. *Polym. Chem.* 8: 5924–5933.

82 Frank, D., Espeel, P., Claessens, S. et al. (2016). Synthesis of thiolactone building blocks as potential precursors for sustainable functional materials. *Tetrahedron* 72: 6616–6625.

83 Liras, M., Quijada-Garrido, I., Palacios-Cuesta, M. et al. (2014). Acetyl protected thiol methacrylic polymers as effective ligands to keep quantum dots in luminescent standby mode. *Polym. Chem.* 5: 433–442.

84 Hrsic, E., Keul, H., and Möller, M. (2012). Synthesis of thiol functionalized poly(meth)acrylates through enzymatic catalysis and a subsequent one pot reaction process. *Eur. Polym. J.* 48: 761–768.

85 Martens, S., Landuyt, A., Espeel, P. et al. (2018). Multifunctional sequence-defined macromolecules for chemical data storage. *Nat. Commun.* 9: 4451.

86 Martens, S., Holloway, J.O., and Du Prez, F.E. (2017). Click and click-inspired chemistry for the design of sequence-controlled polymers. *Macromol. Rapid Commun.* 38: 1700469.

87 Stamenovic, M.M., Espeel, P., Baba, E. et al. (2013). Straightforward synthesis of functionalized cyclic polymers in high yield via RAFT and thiolactone-disulfide chemistry. *Polym. Chem.* 4: 184–193.

88 Marvel, C.S. and Kotch, A. (1951). Polythiolesters. *J. Am. Chem. Soc.* 73: 1100–1102.
89 Kricheldorf, H.R. and Bösinger, K. (1973). Über die polymerisation von 2,5-dioxo-1,3-oxathiolan und 2,4-dioxo-1,3-dithiolan. *Die Makromolekulare Chemie* 173: 67–80.
90 Kato, M., Toshima, K., and Matsumura, S. (2006). Enzyme-catalyzed preparation of aliphatic polythioester by direct polycondensation of diacid diester and dithiol. *Macromol. Rapid Commun.* 27: 605–610.
91 Podkoscielny, W. and Kowalewska, W. (1984). Linear polythioesters. VII. Products of interfacial polycondensation of 1,4-di(mercaptomethyl)-2,3,5,6-tetramethylbenzene with some aliphatic acid dichlorides. *J. Polym. Sci. Polym. Chem. Ed.* 22: 1025–1033.
92 Overberger, C.G. and Weise, J. (1964). A polythioester by ring-opening polymerization. *J. Polym. Sci. Part B: Polym. Lett.* 2: 329–331.
93 Langlais, M., Coutelier, O., Moins, S. et al. (2018). Scope and limitations of ring-opening copolymerization of trimethylene carbonate with substituted γ-thiolactones. *Polym. Chem.* 9: 2769–2774.
94 Matzner, M., McGrath, J.E., Chow, S.W. et al. (1973). Anionic polymerization of ε-caprolactam. II. Initiation by ε-thiocaprolactone. *J. Appl. Polym. Sci.* 17: 983–986.
95 Kato, M., Toshima, K., and Matsumura, S. (2005). Preparation of aliphatic poly(thioester) by the lipase-catalyzed direct polycondensation of 11-mercaptoundecanoic acid. *Biomacromolecules* 6: 2275–2280.
96 Vandenbergh, J., Peeters, M., Kretschmer, T. et al. (2014). Cross-linked degradable poly(β-thioester) networks via amine-catalyzed thiol-ene click polymerization. *Polymer* 55 (16): 3525–3532.
97 Moon, N.G., Mondschein, R.J., and Long, T.E. (2017). Poly(β-thioesters) containing monodisperse oxamide hard segments using a chemoselective thiol-Michael addition reaction. *Polym. Chem.* 8: 2598–2608.
98 Kameyama, A., Kiyota, M., and Nishikubo, T. (1994). New ring-opening reaction of thiiranes with carboxylic acid derivatives catalyzed by quaternary onium salts. *Tetrahedron Lett.* 35: 4571–4574.
99 Kudo, H., Makino, S., Kameyama, A., and Nishikubo, T. (2005). Synthesis of cyclic polymers: ring-expansion reaction of cyclic S-dithioester with thiiranes. *Macromolecules* 38: 5964–5969.
100 Mavila, S., Worrell, B.T., Culver, H.R. et al. (2018). Dynamic and responsive DNA-like polymers. *J. Am. Chem. Soc.* 140: 13594–13598.
101 Bingham, N.M. and Roth, P.J. (2019). Degradable vinyl copolymers through thiocarbonyl addition–ring-opening (TARO) polymerization. *Chem. Commun.* 55: 55–58.
102 Zheng, Y., Cai, S., Peng, L. et al. (2016). Group interval-controlled polymers: an example of epoxy functional polymers via step-growth thiol-yne polymerization. *Polym. Chem.* 7: 6202–6210.
103 Perju, E., Dünki, S.J., and Opris, D.M. (2016). A versatile synthetic path to thiol containing polysiloxanes. *J. Polym. Sci. A Polym. Chem.* 54: 2940–2948.

104 Liu, L., Wu, L., Tan, J. et al. (2015). "Reduction" responsive thymine-conjugated biodynamers: synthesis and solution properties. *Polym. Chem.* 6: 3934–3941.

105 Baysak, E., Gunay, U.S., Daglar, O., and Durmaz, H. (2019). Synthesis and post-polymerization modification of polyester containing pendant thiolactone units. *Eur. Polym. J.* 112: 241–247.

106 Ghobril, C., Charoen, K., Rodriguez, E.K. et al. (2013). A dendritic thioester hydrogel based on thiol–thioester exchange as a dissolvable sealant system for wound closure. *Angew. Chem. Int. Ed.* 52: 14070–14074.

107 Hu, B.-H., Su, J., and Messersmith, P.B. (2009). Hydrogels cross-linked by native chemical ligation. *Biomacromolecules* 10: 2194–2200.

108 Rajakanthan, A., De Jongh, P.A.J.M., Town, J.S. et al. (2019). A sequential native chemical ligation-thiol-Michael addition strategy for polymer–polymer ligation. *Polym. Chem.* https://doi.org/10.1039/C9PY01198F.

109 Holloway, J.O., Aksakal, S., Du Prez, F.E., and Becer, C.R. (2017). Tailored modification of thioacrylates in a versatile, sequence-defined procedure. *Macromol. Rapid Commun.* 38: 1700500.

110 Liras, M., García, O., Quijada-Garrido, I., and París, R. (2011). Transformation of the bromine end group into thiol in (meth)acrylic polymers synthesized by atom transfer radical polymerization. *Macromolecules* 44: 1335–1339.

8

Thiophene-Based Polymers: Synthesis and Applications

Haifeng Ji and Xiaojie Zhang

Hebei University of Technology, Hebei Key Laboratory of Functional Polymers, Department of Polymer Materials and Engineering, School of Chemical Engineering and Technology, 8 Guangrong Street, Tianjin 300130, P. R. China

8.1 Introduction

Since the first report in 1977, numerous π-conjugated polymers have been synthesized and reported as shown in Figure 8.1 [1]. Most of them are composed of common aromatic structural blocks, such as *p*-phenylenevinylene, thiophene, fluorene, carbazole, and fused heterocyclic rings [2–5]. Unlike traditional insulating polymers, conjugated polymers have conductive properties, which are derived from their conjugated linkages of aromatic or heteroaromatic units. Conjugated polymers have garnered significant research attention owing to their wide applications in optoelectronic devices [6–9] such as organic light-emitting diodes (OLEDs) [10], organic thin-film transistors (OTFTs) [11, 12], and organic photovoltaics (OPVs) [13–15]. Compared with conventional silicon-based electronic materials, conjugated polymer materials have predictable advantages in terms of processability in solution, low energy consumption, as well as reducing environmental impact [16, 17].

Today 40 years on, many conjugated polymers have been developed, and among these, polythiophene and its derivatives, all collectively referred to as thiophene-based polymers, represent one of the largest classes of conjugated polymer and utilized in organic electronic fields such as conducting materials, OTFTs and OPVs due to their excellent optical and electrical properties, environmental stability, ease of synthesis, and the ability to connect various groups on thiophene's β-positions or cross-couple with other aromatic/heteroaromatic units (Figure 8.1) to make their properties tunable [18–22]. In particular, poly(3-hexylthiophene) (P3HT) has been the model polymer for organic semiconductor physics and devices studies relating to charge transport and film morphology due to its precision of synthesis and good π–π stacking performance. In this chapter, we focus our discussion on the recent synthetic developments of thiophene-based polymers, followed by a short introduction on their applications as organic electronic materials.

Sulfur-Containing Polymers: From Synthesis to Functional Materials,
First Edition. Edited by Xing-Hong Zhang and Patrick Theato.

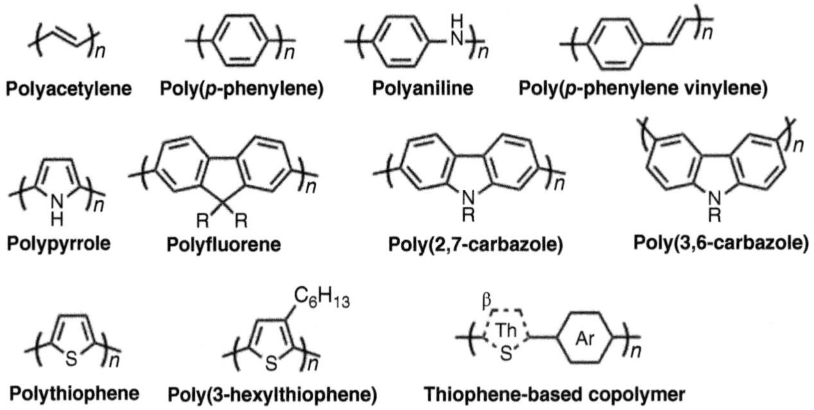

Figure 8.1 Chemical structures of a few conjugated polymers. Source: Based on Ref. [1].

8.2 Development of Synthetic Methods

Many methods for the synthesis of thiophene-based polymers have been developed. As shown in Figure 8.2a, thiophene-based polymers have been synthesized, in general, by traditional chemical oxidation polymerization (using oxidant) [23], electrochemical polymerization [24, 25], transmetalation polymerization [22, 26], photoinduced polymerization [27, 28], acid-assisted polymerization [29], and solid-state polymerization (SSP) [30]. Initial synthetic methods of non-substituted polythiophene involved chemical oxidation or electrochemical polymerization. However, the obtained product is intractable with poorly soluble in common organic solvents and difficult to process [31, 32]. The general and effective practices to improve the solubility and property of polythiophene are: (i) Introduction alkyl pendant groups onto thiophene's β-position; (ii) copolymerization with other soluble conjugated monomers to construct soluble alternating copolymers with unique properties (Figure 8.1). Since the difficulties to apply non-substituted polythiophene as organic electronic materials, this view will mainly focus on soluble thiophene-based polymers.

8.2.1 Oxidative Polymerization and Electrochemical Polymerization

In terms of molecular structure, the transformation from thiophene-based monomers to polymers lies in the removal of α-protons (at 2- and 5-positions, Figure 8.2b) from thiophene rings, so oxidative coupling is the most direct way to connect thiophene rings. The key step of electrochemical polymerization is the oxidative dehydrogenation of thiophene and its derivatives by applying bias voltage. Therefore, in essence, the principles of chemical oxidation polymerization (using oxidants) and electrochemical polymerization are basically the same [18].

Early synthesis of the soluble poly(3-alkylthiophene) (P3AT) by chemical oxidation or electrochemical polymerization was suffered from the gross defect that

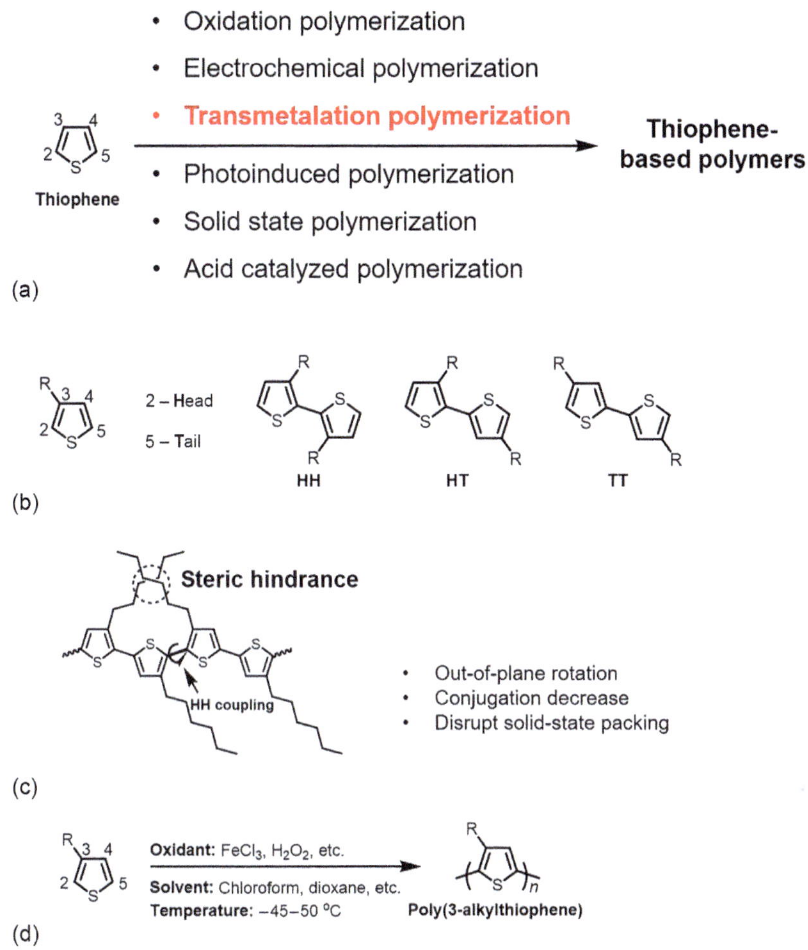

Figure 8.2 (a) Different synthetic method for preparation of thiophene-based polymers. (b) Different linkages in the 3-alkylthiophene dimers. (c) Effect of the regioirregularity: steric hindrance among side-chains cause the deplanarization of polythiophene backbone, that decreased the conjugation length and disrupt the solid-state packing. (d) Oxidative polymerization of 3-alkylthiophene.

the structure of the polymer is undefined and regioirregular [33]. The introduction of alkyl pendant chain at 3-position of thiophene ring is an efficient method to improve the processability of polythiophene. Conventionally, in the asymmetrical 3-substituted thiophene, the α-carbon (2-position) between the S atom and the alkyl chain is labeled as the "head" (H) and the other α-carbon (5-position) is labeled as the "tail" (T). As shown in Figure 8.2b, three different linkages can be formed during coupling: head-to-head (HH), head-to-tail (HT), and tail-to-tail (TT). Containing a mixture of those possible linkages results in a regioirregular poly(3-alkylthiophene) (*rr*-P3AT) in which the steric hindrances between alkyl pendant groups in HH-coupled units promote twisted, non-planar backbone conformations, as shown in Figure 8.2c, leading to loss of π-conjugation and disruption of solid-state packing. On the other hand, a P3AT containing only HT-linkages, can be defined as strictly regioregular (*rr*-P3AT), which shows a coplanar conformation, leads to better conjugation and more efficient π-stacking interactions between adjacent polymer chains. The percentage of HT-linkages in thiophene-based polymers is called regioregularity (*rr*) value [18, 25, 34, 35]. A comparison on P3ATs prepared by oxidative polymerization ($FeCl_3$ as oxidant) or electronic polymerization results in similar *rr* value (~80%) and molecular weight [25].

The synthesis of polythiophenes by chemical oxidation is usually achieved by dissolving monomers in aqueous or organic solutions, followed by adding appropriate oxidants under acidic conditions such as iron trichloride ($FeCl_3$), hydrogen peroxide (H_2O_2), ammonium persulfate ($(NH_4)_2S_2O_8$), vanadium(III) acetylacetonate, or potassium dichromate ($K_2Cr_2O_7$), as shown in Figure 8.2d [23, 36–38]. The types and concentrations of oxidants and acidic partner, reaction time and temperature, and concentration of monomers will all influence the results of oxidation reaction. In addition, the oxidation potential of monomer should be taken into account for selecting oxidant. The rate and mode of reactant addition, with the proper choice of reaction temperature, seem to be the crucial step in oxidative polymerization. The initial method of chemical oxidative polymerization performed the addition of the thiophene monomer to the oxidant suspension in chloroform or dioxane, providing *rr* value of 58%, while reversing the addition order, the regioregularity rose up to 70% [23]. Andersson et al. confirmed that the reverse addition of reactants could reduce Fe^{3+}/monomer ratio, Fe^{3+}/Fe^{2+} ratio and oxidation potential in solution. Though this study, they produced a poly(3-octylphenylthiophene) with high *rr* value of 94%, in a yield of 80% [39]. The effects of reaction temperature and time, oxidant/monomer molar ratio, solvent (chloroform, hexane, dichloromethane, etc.), and oxidant concentration were also fully studied by Sugimoto and Hai [40] increasing temperature, from room temperature to 50 °C, can slightly lower the polymer molecular weight and *rr* value. Meanwhile, lower reaction temperature (−45 °C) can significantly rise the regioregularity (91%) in chloroform. Balogh's group reported that both longer addition time and lower temperature in hexane or acetone can lead to further increase on the molecular weight (M_w up to 318 kDa and M_n up to 68 kDa, 81% regioregularity) [41]. $FeCl_3$/monomer ratio also shows critical effect on molecular weight of polythiophenes. Standard ratio of 1 : 1 gives high average molecular weight around 100 kDa; Lowering the ratio to 1 : 2 provided a molecular weight of 52 kDa, while a 1 : 8 ratio only provided 28 kDa [40, 41]. Okabe and coworkers have investigated the effect of monomer concentration [42]. The number-average molecular weight (M_n) increased with an increase of 3-hexylthiophene concentrations in reaction solution, which indicate that M_n of P3HT could be fuzzy-controlled in a wide range of 3.7 to 70 kDa with low polydispersity (PDI, below 2) by varying 3-hexylthiophene concentrations.

The first reported synthesis of polythiophene by electrochemical polymerization appeared in 1982 [43]. This reaction was performed under a three-electrode system, i.e. working electrode, counter electrode, and reference electrode. Electrochemical polymerization can be divided into cathodic reduction mode and anodic oxidation mode [44]. At present, anodic oxidation is widely used, including constant current method, potentiostatic method, pulse polarization method, and potentiodynamic scanning method. Recently, an improvement of electrochemical polymerization has been made by using microflow reaction, in which the molecular weight of the obtained polymer has been well controlled [45]. The electrochemical polymerization of thiophene is influenced by many factors, such as solvent, concentration, temperature, electrode properties, shape, supporting electrolyte and electric field conditions [24, 46, 47]. Furthermore, some strong acidic anions (AsF_6^-, BF_4^-, and PF_6^-) and cations (Li^+ and quaternary ammonium cation) are usually chosen as the supporting electrolyte [48–50]. The obtained polythiophene usually grows on the surface of metal electrodes for example Au and Pt, or directly adheres to conductive glass. Electrochemical polymerization is considered to be one of the best methods to prepare membrane materials due to the purity and conductivity of the products are relatively high, and the reaction conditions are simple and easy to control. By using this method, Shi and coworkers have developed a series of high-impact conductive polythiophene membrane with strength higher than that of metal aluminum [51, 52]. Although the electrochemical method is one of the most elegant in preparation of polythiophene, it also has many limitations, such as difficulty in large-scale preparation and lower regioregularity value. The optimal synthetic conditions still need to be further explored.

8.2.2 Transmetalation Polymerization

8.2.2.1 Polymerization with Ni Catalysis

In order to prepare P3ATs with high electronic performance, synthetic developments have been mainly aimed at the preparation of fully regioregular *rr*-P3ATs. The first breakthrough of synthesis of highly regioregular P3HT was reported by McCollough et al. in 1990s, as shown in Figure 8.3 [53, 54]. The initial intermediate (5-bromo-4-hexylthiophen-2-yl)magnesium bromide was prepared from 2-bromo-3-hexylthiophene by first deprotonation with LDA, followed by treatment with $MgBr_2$. Highly *rr*-P3HT (*rr* > 98%) with controlled molecular weight and narrow PDI was obtained by using 1,3-bis(diphenylphosphino)propane nickel(II) chloride ($Ni(dppp)Cl_2$) catalyzed Kumada-type homopolymerization, later termed as Kumada catalyst transfer polymerization (KCTP) [55]. Further evolution by using high reactive Rieke Zn (Zn*) instead of LDA/$MgBr_2$ system to prepare *rr*-P3HT is reported by Rieke and coworkers in 1992 [56, 57]. As shown in Figure 8.3, 2,5-dibromo-3-hexylthiophene was reacted with Zn* to yield two possible intermediates (A and B) in a ratio of 95 : 5. In the present of 1,3-bis(diphenylphosphino) ethane nickel(II) chloride ($Ni(dppe)Cl_2$), the organozinc intermediates can polymerize into *rr*-P3HT, via Negishi coupling conditions, with 75% yield and excellent *rr* value more than 97%. In last 20 years, many other attempts for producing highly *rr*-P3AT come to the fore [58–61]. For instance, a very recent paper prosed a KCTP of novel substrates as 2-phenylsulfonylthiophene [62]. Initial monomers

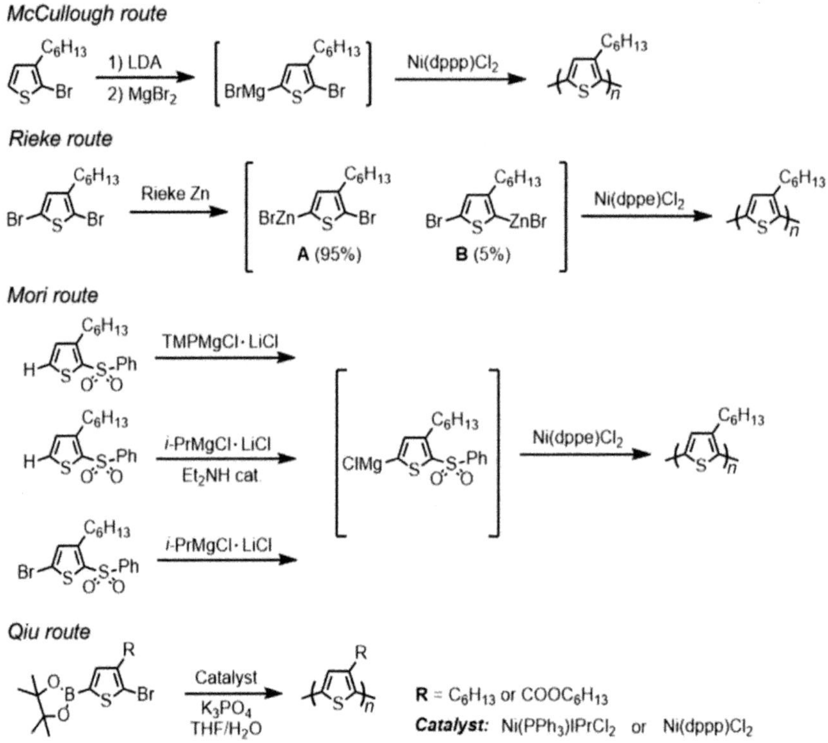

Figure 8.3 McCollough route for the synthesis of regioregular P3HT: dppp = 1,3-bis(diphenylphosphino)propane. Rieke route for the synthesis of regioregular P3HT: using Ni-catalyzed Negishi conditions, two zincate intermediates were involved with **A** (95%) and **B** (5%), respectively, dppe = 1,3-bis(diphenylphosphino)ethane. Mori route for the synthesis of regioregular P3HT: homopolymerization of thiophene derivatives via C–S bond cleavage. Qiu route: using nickel-catalyzed SCTP.

(4-hexyl-5-(phenylsulfonyl)thiophen-2-yl)magnesium chloride can be obtained through three different ways: (i) Knochel–Hauser base (TMPMgCl·LiCl) mediated deprotonation of 3-hexyl-2-(phenylsulfonyl)thiophene; (ii) Et_2NH-catalyzed deprotonation of 3-hexyl-2-(phenylsulfonyl)thiophene; (iii) Grignard metathesis (GRIM) metalation of 3-hexyl-5-bromo-2-(phenylsulfonyl)thiophene. The polymerization was then performed via C—S bond cleavage in the presence of $Ni(dppe)Cl_2$ as catalyst, generating *rr*-P3HT (M_n = 1.6 kDa, PDI = 1.68) with high *rr* value of almost 99%. In addition to the Kumada- and Negishi-type polymerizations, *rr*-P3HT also can be synthesized by using Ni-catalyzed Suzuki catalyst-transfer polymerization (SCTP), reported by Qiu et al. [63], in which an NHC-based catalyst ($Ni(PPh_3)IPrCl_2$) offered the high molecular weight *rr*-P3HT (up to 74 kDa). Furthermore, this protocol also provided an approach for the polymerization of the ester-substituted monomer, as shown in Figure 8.3.

Among all these Ni-catalyzed methods, GRIM method, based on KCTP, is the most important evolution for producing not only *rr*-P3AT but also some other

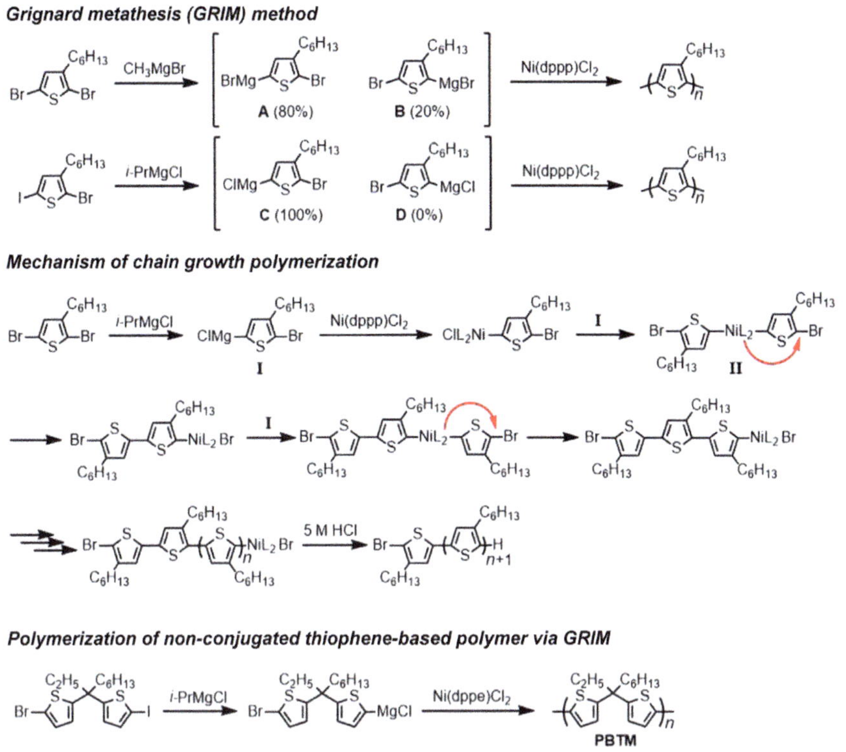

Figure 8.4 Synthesis of regioregular P3HT by Grignard Metathesis method. Mechanism of chain growth polymerization proposed by Yokozawa. Synthesis of non-conjugated thiophene-based polymer by the GRIM polymerization.

conjugated polymers (i.e. polyphenylenes, polyfluorenes, etc.) with predetermined molecular weight and narrow PDI [64]. In the 1990s, McCullough and Yokozawa independently developed a simple and convenient route for the synthesis of *rr*-P3HT by using cheap Grignard reagent (i.e. CH_3MgBr or *i*-PrMgCl) for metathesis with aryl-halide monomer followed by a Ni-catalyzed coupling to polymerize the organomagnesium intermediate [58, 60, 65–67]. As shown in Figure 8.4, 2,5-dibromo-3-hexylthiophene first reacted with CH_3MgBr in refluxing THF to form a mixture of 2-bromo-5-bromomagnesio-3-hexylthiophene (A) and 5-bromo-2- bromomagnesio-3-hexylthiophene (B) in a ratio of 4 to 1, referred to as GRIM. The organomagnesium intermediates were then polymerized *in situ* in present of Ni(dppp)Cl_2 to afford *rr*-P3HT with M_n of 20–35 kDa, high *rr* value (>99%), and low PDI (1.2–1.5). Further optimization of GRIM method was reported by Yokozawa group [58, 67], which involved changing the monomer to 2-bromo-5-iodo-3-hexylthiophene to afford nearly 100% metathesis selectivity.

Detailed investigations for monomers, catalysts, phosphorus ligands, and ancillary additives on GRIM method have been examined in the last two decades [68–76]. For example, Shi et al. have proposed a controlled synthesis of *rr*-P3HT with predictable molecular weight using NHC-based Ni catalyst (Ni(IPr)(acac)$_2$), which undergo a "quasi-living" chain growth polymerization [77]. Particularly, the M_n of resulting P3HT is well controlled in a very wide range, the M_n increased proportionally from 9.9 to 353 kDa according to the decrease of the catalyst content from 4.00 to 0.10 mol%. Notably the highest M_n (up to 416 kDa) also makes a record so far.

As the second example, external initiators have also been studied for synthesis of *rr*-P3HT via GRIM method [55, 70, 73, 78–82]. Initiating the polymerization from an external moiety makes an introducing of another aromatic building block, which can be served for preparing thiophene-based polymers with more geometric architectures such as brushes, multiarmed stars, and V-shape polymers. Kiriy's group has first reported an external initiator assisted polymerization for the synthesis of phenyl end-capped *rr*-P3HT [78], in which Ar-Ni(PPh$_3$)$_2$-Br was chosen as initiator. Based on this, Luscombe group and Koeckelberghs group have extended the work on various initiators, such as phosphonate aryl halide initiators and *cis*-chloro(aryl)(dppp)nickel complex [55, 79]. Moreover, Kiriy and coworkers also prepared P3HT brushes by surface-initiated polymerization of 2-bromo-5-chloromagnesio-3-hexylthiophene with nickel macroinitiator [78].

To develop a general synthetic method and extend the method to wider scope, understanding the reaction mechanism is very important and urgent. In recent years, extensive investigations have been carried out by many scientists in order to understand the mechanism of GRIM/KCTP process. The GRIM polymerizations follow a chain growth manner, which is realized through an intramolecular catalyst transfer, termed "ring walking" later [61, 68, 76]. As shown in Figure 8.4, the initial step produced a TT-connection thiophene dimer: (i) Chloro-atoms in Ni(dppp)Cl$_2$ were replaced by two Grignard monomers **I** via two oxidative additions, generating the intermediate **II**, in which Ni(0) is bound to two thiophene rings; (ii) A coupling reaction between these two thiophene units is accompanied by an intramolecular transfer of the Ni active complex (NiL$_2$) elimination and insertion into the terminal C-Br bond, see red arrows in Figure 8.4, which should be considered as the key of the intramolecular catalyst transfer for GRIM polymerization. Model reaction of 2-bromo-5-chloromagnesio-3-hexylthiophene and 50 mol% of Ni(dppp)Cl$_2$ was performed in THF at room temperature, followed by quenching with acid. The obtained product was characterized as 4,4′-dihexyl-2-bromo-5,5′-bithiophene by gas chromatography–mass spectrometry (GC-MS) and nuclear magnetic resonance (NMR), which confirmed intramolecular transfer [76]. Polymer chain growth occurred via consecutive cross-coupling between the growing fragments end-capped with NiL$_2$Br and Grignard monomer **I**. Finally quenching the reaction with 5 M HCl, rr-P3HT with exclusive presence of Br/H termini was obtained. Noted that, the absence of Br/Br or H/H termini also shows a "ring walking" phenomenon, in other words the Ni active complex does not fall off from the polymer chain during the chain growing.

In 2005, the "quasi-living" property of GRIM polymerization method was demonstrated by successful achievement of poly(3-hexylthiophene)-*co*-poly(3-dodecylthiophene) (P3HT-*co*-P3DT) copolymer, reported by McCullough group [61]. Utilizing the living property, various polythiophenes, polyphenylenes, polypyrroles, and related blocking copolymers with predictable molecular weight and narrow PDI have been synthesized. However, although GRIM polymerization of the monomers with longer molecular length, i.e. fluorenes and oligothiophenes, still follows the chain growth manner, it has no "living" or "quasi-living" characteristics, which may due to the longer migration distances lead to inefficient catalyst transfer [81, 83, 84]. In addition to the conjugated polymers, GRIM method can also be used for the synthesis of non-conjugated polymers. Wu et al. have first reported that bithienylmethane derivative, in which two thiophene rings are connected via a saturated carbon, can be polymerized by using GRIM method [85]. In present of $Ni(dppe)Cl_2$ as catalyst, non-conjugated poly(bithienylmethylene)s (PBTM) and related multiblock copolymers with an M_n close to the theoretical value and a narrow PDI (<1.4) have been successful synthesized. This result shows the "quasi-living" feature as discussed above, which may due to the close distance of the carbon-linked two thiophene rings in space, resulting in the efficient intrachain catalyst transfer.

8.2.2.2 Polymerization with Pd Catalysis

As a rapidly evolving method of synthesizing regioregular polythiophenes, Ni-catalyzed Kumada- and Negishi-type polymerization still has many limitations in the scope of application, such as the instability of the organometal reagents and the difficulty in cross-coupling reactions. Furthermore, many functional groups (i.e. carboxyl group, carbonyl group, and ester) are not compatible with these reactions due to the strong nucleophilic nature of organomagnesium and organozinc intermediates. The formation of C—C bonds between two aromatic units is considered to be the essence of the preparation of conjugated polymers. Among all C—C bonds formations reactions, palladium-catalyzed coupling reactions are indeed the most common synthetic strategy [26, 86–90], particularly in cross-coupling two symmetric bis-functionalized monomers to obtain alternating conjugated copolymers. The well acceptable mechanism for Pd-catalyzed cross-coupling reactions of electrophiles R_1X with organometals R_2M is described in Figure 8.5 [22]. It consists of three consecutive steps: (i) the oxidative addition of organic halide R_1X onto $Pd(0)L_n$ catalytic species, where L_n represents as coordinating ligands, and the catalytic Pd center changes from Pd(0) to Pd(II) during this process; (ii) transmetalation between $R_1Pd(II)L_nX$ and R_2M, which could be considered to be driven thermodynamically by the formation of MX; (iii) reductive elimination of $R_1R_2Pd(II)L_n$ to give R_1–R_2 via C—C bond formation, with the regeneration of $Pd(0)L_n$ for the next catalytic cycle.

So far, a library of thiophene-based polymers have been synthesized efficiently via Pd-catalyzed cross-coupling polycondensations using organotins (Stille coupling) [86] or organoborons (Suzuki coupling) [87] as monomers. The higher stability and ease of purification of these organometallic monomers have enabled thiophene-based polymers to be achieved with various functional groups and widely different performance. Due to the structural variety, the synthetic aspect of thiophene-based copolymers will be introduced later in conjunction with applications (see Section 8.3). The following discussion will mainly focus on another

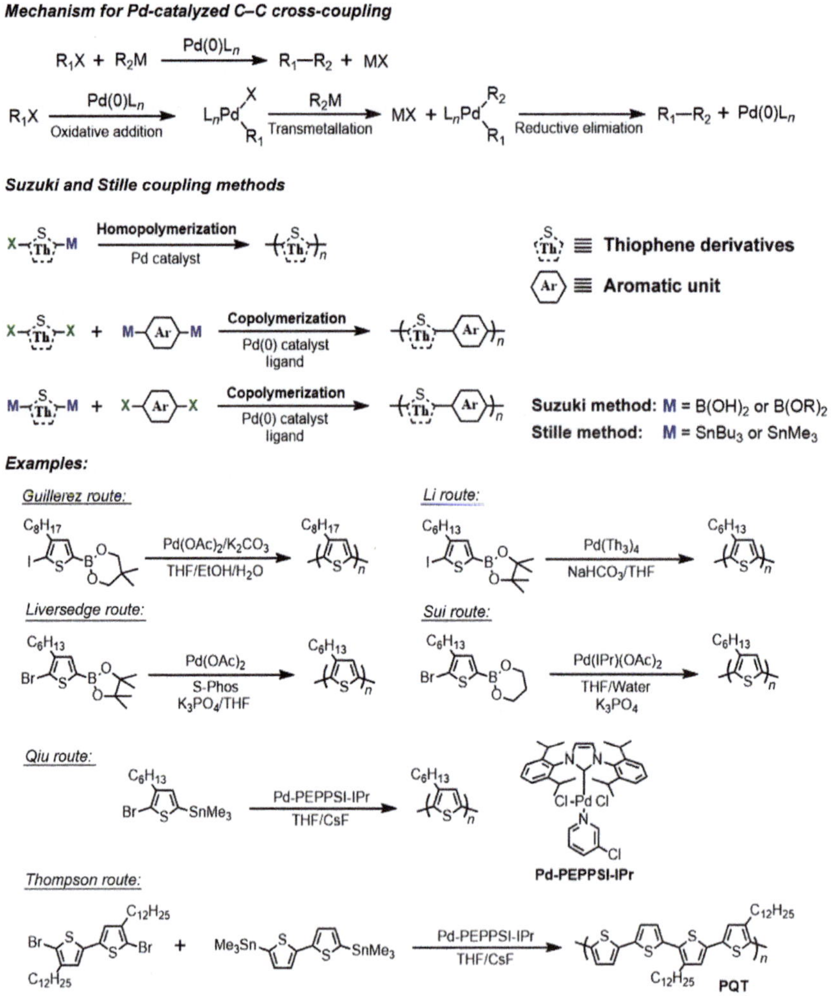

Figure 8.5 Mechanism for Pd-catalyzed cross-coupling reactions. Overview and examples of Pd-catalyzed Suzuki and Stille homo- or cross-coupling techniques for the preparation of thiophene-based polymers. Source: Based on Ref. [22].

synthetic Pd-catalyzed approach utilizing single monomer bearing both halogen group and organoboron/organotin group to produce regioregular homopolymer.

8.2.2.3 Suzuki Coupling Method

In addition to Ni-catalyzed Kumada- and Negishi-type polymerization, *rr*-P3AT also can be synthesized by Pd-catalyzed Suzuki coupling method, which requires the coupling of aryl halides with aryl boronic acids or esters, and the first report appeared in 1998. The first Suzuki polymerization was published by Guillierez and Bidan [91], by reacting [1′,3-(2,2′-dimethylpropylene)]-2-iodo-3-octylthienylboronate with $Pd(OAc)_2$ and K_2CO_3 in a solvent mixture of $THF/EtOH/H_2O$ at refluxing condition, yielding an *rr*-P3OT in 55% with high molecular weight ($M_n = 27$ kDa, PDI = 1.5) and 96–97% *rr* value. Obviously, compared with Negishi- and Kumada-type conditions, the synthesis of polythiophene by means of Suzuki method requires higher reaction temperature and more reaction time and tends to produce P3AT with low controllability (higher PDI and lower *rr* value). After this initial work, the Suzuki method was applied with more coordinative ligands, investigating their electron effect and steric hindrance effect. For instance, Liversedge et al. [92] have reported the use of electron-rich and steric demanding phosphine ligands (P^tBu_3 and S-Phos) for the Suzuki synthesis of highly regioregular P3HT (*rr* up to 98%, M_n around 9–1.7 kDa and 20% yield) from 2-Bpin-5-bromo-3-hexylthiophene (Bpin = pinacol borate). Other examples for P3HT prepared by Pd-catalyzed Suzuki reaction were reported by Bo's group [93]. They polymerized the 2-Bpin-5-iodo-3-hexylthiophene with palladium tetrakis(tris-(2-thienyl)phosphine) ($Pd(PTh_3)_4$) as catalyst to afford *rr*-P3HT with improved reaction yield (up to 72%), high molecular weights (M_n = 5–26 kDa), and high *rr* value (96–98%).

Contrary to GRIM chain-growth polymerization, most Suzuki polymerization follows the step-growth mechanism, thereby generally need more than days and result in producing polymers with uncontrollable molecular weight as well as broad PDI. In recent years, Ni-catalyzed SCTP has been demonstrated to follow the chain-growth mechanism, see Section 8.2.2.1. Pioneered by this work, the first P3HT prepared by Pd-catalyzed SCTP reaction was reported in 2014 by Geng's group [94]. In present of NHC-based $Pd(IPr)(OAc)_2$, 2-(4-hexyl-5-bromanyl-2-thienyl)-1,3,2-dioxaborinane can be polymerized into P3HT at room temperature in an almost quantitative yield with *rr* value of 99%. M_n of the obtained P3HT was linear to the conversion of monomer with PDIs of ~1.40 and could be controlled in the range of 9.5–64 kDa by tuning the feed ratios, which indicated that the polymerization proceeded in a "quasi-living" chain-growth manner. Moreover, the mainly H/Br end groups of the P3HT also shows that this $Pd(IPr)(OAc)_2$ catalyzed polymerization involves an intramolecular catalyst-transfer mechanism.

8.2.2.4 Stille Coupling Method

Stille homopolymerization requires the coupling of aryl halides with aryl tins and has also been used to make *rr*-P3HT [86, 95]. Reported at the first time in 1998 by Iraqi [26], 2-iodo-3-hexyl-5-(tri-*n*-butylstannyl)thiophene can be polymerized into regioregular HT P3HT in present of $Pd(PPh_3)_4$ as catalyst. The polymerization was undertaken in three different solvent, i.e. toluene, THF, and 1,2-dichloroethane, while toluene gives the best results (M_n = 16 kDa, PDI = 1.22, $rr > 96\%$) at refluxing condition. Interestingly, treatment of the producing P3HT with methanol at refluxing temperature leads to the cleavage of tin end groups, which is very important for increasing lifetime of its application in electronic devices. Obviously, this reaction requires much more time and higher temperature than that of Ni-catalyzed Kumada- and Negishi-type ones. Recently, Qiu et al. have reported an effective reaction system for Stille catalyst-transfer polymerization of thiophene monomers using a commercially available Pd-NHC catalyst (Pd-PEPPSI-IPr, see Figure 8.5) [96]. The polymerization proceeds in THF at 40 °C (reaction time less than 2.5 hours). The M_n of the resultant P3HT can be modulated (7–73 kDa, PDI = 1.14–1.53) by the ratio of conversion monomer and catalyst concentration, which indicated a chain-growth manner. The Pd catalyst has been shown to "ring-walk" based on the exclusive presence of H/Br polymeric termini.

The Stille protocol is known to be quite robust since the thienotins are normally much more stable than thienoborons, making it advantageous over the Suzuki reaction in preparation of thiophene-based alternating polymers [95]. For instance, a series of thiophene-based polymers containing TT bithiophene unit, such as poly(quaterthiophene) (PQT), and poly(bithiophene-*co*-dithieno[3,2-*b*:2′,3′-*d*]pyrrole) (PBT-DTP) can be easily synthesized by using Stille copolymerization of 5,5′-dibromo-4,4′-didodecyl-2,2′-bithiophene with related dithienotin partners, the synthesis of PQT was shown in Figure 8.5 as an example [97, 98].

8.2.2.5 Direct Arylation Method

Development of conjugated alternating copolymers, particularly donor–acceptor (D–A) copolymers has promoted the dramatic improvement of the performance of OTFTs and polymer solar cells (PSCs) [8, 99–103]. Currently, these kinds of polymers are mainly synthesized by polycondensation using Pd-catalyzed transmetalation cross-coupling reactions such as Suzuki and Stille couplings, as we described earlier [64, 104–106]. Although they did very well, these two protocols required the preparation of high purity bifunctional organoboron or organotin compounds as monomers. Meanwhile, the polycondensations also produce a stoichiometric amount of toxic byproducts such as organotin compounds. These will become key issues in the mass production of conjugated polymers. Recently, cross-coupling reactions between non-preactivated (hetero) arenes and aryl halides though C—H bond cleavage, so-called direct arylation as shown in Figure 8.6, have been developed [107–115]. As the prior preparation of organometallic monomers and the treatment of toxic byproducts are no longer necessary, this method represents an economically and ecologically benign alternative to the conventional transmetalation cross-couplings. The mechanism for Pd-catalyzed C–H activation cross-coupling reactions has been studied experimentally and computationally, and the possible mechanism includes electrophilic aromatic substitution, Heck-type coupling and concerted metalation-deprotonation (CMD) [116]. Most aromatic units, such as phenyls and thiophenes, are considered to follow an acid assisted CMD pathway [114]. Figure 8.6 depicts the carboxylate-mediated CMD process. In the presence of inorganic base (such as potassium carbonate) and carboxylate additive (such as pivalic acid, acetic acid), the oxidative addition of R_1X on $Pd(0)L_n$ species is followed by exchange of the halogen for the carboxylate anion to form a square-shaped Pd(II) complex. With the assistance from the carboxylate ligand, this complex then deprotonates the R_2H while simultaneously forming a metal-carbon bond (R_2-Pd). Finally, reductive elimination renders the aryl coupled product and regenerates the Pd catalyst for continuation of the catalytic cycle.

An embryonic form of direct arylation polymerization (DArP) was reported by Lemaire and coworkers in 1999 [117, 118]. In presence of $Pd(OAc)_2$ as catalyst, potassium carbonate as base and tetrabutylammonium salts, 2-iodo-3-octylthiophene can be polymerized into oligo-P3OT (M_n around 3000 Da, PDI around 2, which

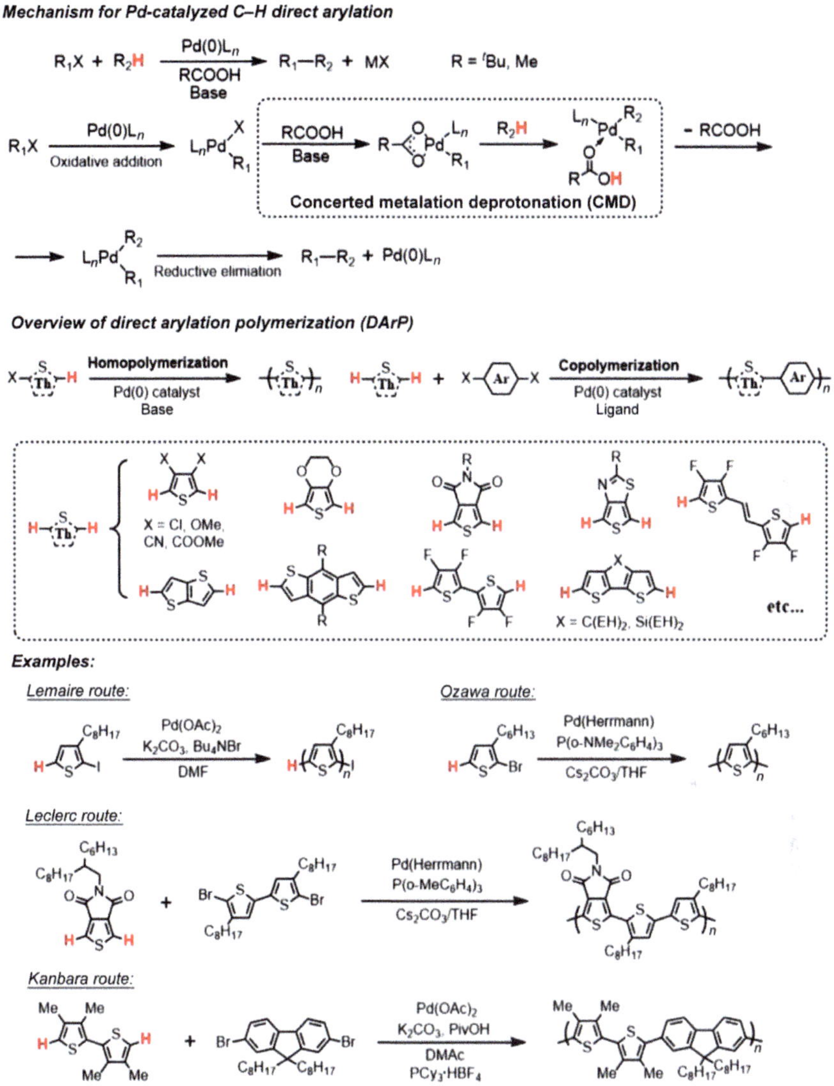

Figure 8.6 Mechanism for Pd-catalyzed C–H direct arylation. Overview and examples of Pd-catalyzed C–H direct arylation homo- or cross-coupling techniques for the preparation of thiophene-based polymers.

corresponding to 15–16 repeating units) at 80 °C in a high yield of 93%. Over 10 years later, Ozawa group optimized this chemistry and reported a milestone work for a DArP towards *rr*-P3HT with increased molecular weight up to 30.6 kDa and regioregularity up to 98% [119]. In this homopolymerization, the Herrmann's catalyst (trans-bis-(acetato)bis[*o*-(di-*o*-tolylphosphino)benzyl] dipalladium(II)) (Pd(Herrmann)) was exploited as catalyst due to the high reaction temperatures (130 °C). Ligands studies also manifested that electron-rich aryl phosphines (P(*o*-$NMe_2(C_6H_4))_3$ and P(*o*-$MeOC_6H_4)_3$ gave much better results than that of bulky alkyl phosphines (PCy_3 and P^tBu_3) [96, 120, 121]. Pioneered by this work, Thompson et al. also prepared a sample of P3HT by DArP method using palladium-pivalic acid cocatalyst combination ($Pd(OAc)_2$-PivOH), developed by Fagnou and coworkers [113, 114], in *N*,*N*-dimethylacetamide (DMAc), generating P3HT (M_n = 15.5 kDa, PDI = 2.9) with lower *rr* value of 88% [122]. This unsatisfactory result perhaps due to the less solubility of P3HT in DMAc and branched or crosslinked side-reactions happened on β-protons of thiophene backbones.

To date, there have been more and more studies addressing the utilization of DArP for synthesizing π-conjugated copolymers [123–126]. The first DArP copolymerization involving the thiophene-based monomer (3,4-dimethylthiophene or 3,3′,4,4′-tetramethylbithiophene) with a 2,7-dibromo-9,9-dioctylfluorene was carried out in presence of $Pd(OAc)_2$ as catalyst, $PCy_3 \cdot HBF_4$ as ligand and DMAc as solvent, respectively [127, 128]. In this case, the introduced methyl on 3- and 4-positions of thiophene rings were acting as protection group to avoid branched or crosslinked side-reactions. First example for copolymerization of electron-deficient monomer (5-alkyl[3,4-*c*]thienopyrrole-4,6-dione) and electron-rich monomer (5,5′-dibromo-4,4′-dioctyl-2,2′-bithiophene), which could be incorporated into D–A alternating copolymer, was reported by Leclerc et al. by using similar polymerization conditions described by Ozawa for the synthesis of P3HT [129]. Owing to the real high reactivity of α-thieno-proton, various thiophene-based monomers have been designed and synthesized for DArP, such as 5-alkyl[3,4-*c*]thienopyrrole-4,6-dione [129–131], 3,4-ethylenedioxythiophene [132–135], 3,4-substituted thiophenes, etc. [124, 136–138], listed in Figure 8.6. Among these monomers, most β-position of thiophene rings were protected since the β-proton of thiophene rings lead to the unwanted possible C–H activation. Recently, Geng and coworkers assessed the reactivity of a series of fluorine incorporated monomers [139–141]. The introduction of F-atoms in β-positions of thiophene rings not only evades the unwanted side-reaction, but also can remarkably enhance the reactivity of the thiophene moieties towards direct arylation due to the strong electron-withdrawing nature of F-atom. Under the optimized condition (Pd(Herrmann)/$P(o\text{-}MeOC_6H_4)_3$/PivOH/Cs_2CO_3/toluene), a series of D–A copolymer with high molecular weight have been achieved. Noted that fluorine substituents in thiophene-based polymer does not cause an increase of the dihedral angle of polymer backbones, which suggested that this F-atom enhanced DArP methodology could be a promising candidate for synthesizing high performance polymeric semiconductors.

8.2.3 Other Polymerization Methods

8.2.3.1 Photoinitiated Polymerization

During the past two decades, photoinitiated polymerizations, as called photoinduced polymerizations, that proceed under UV-light and visible-light have been the subject of intense scientific and industrial activity because of their real-life applications, such as adhesives, photoresists, photopolymerizable inks, highly pigmented coatings, dental filling materials, and laser-induced 3D curing [142–144].

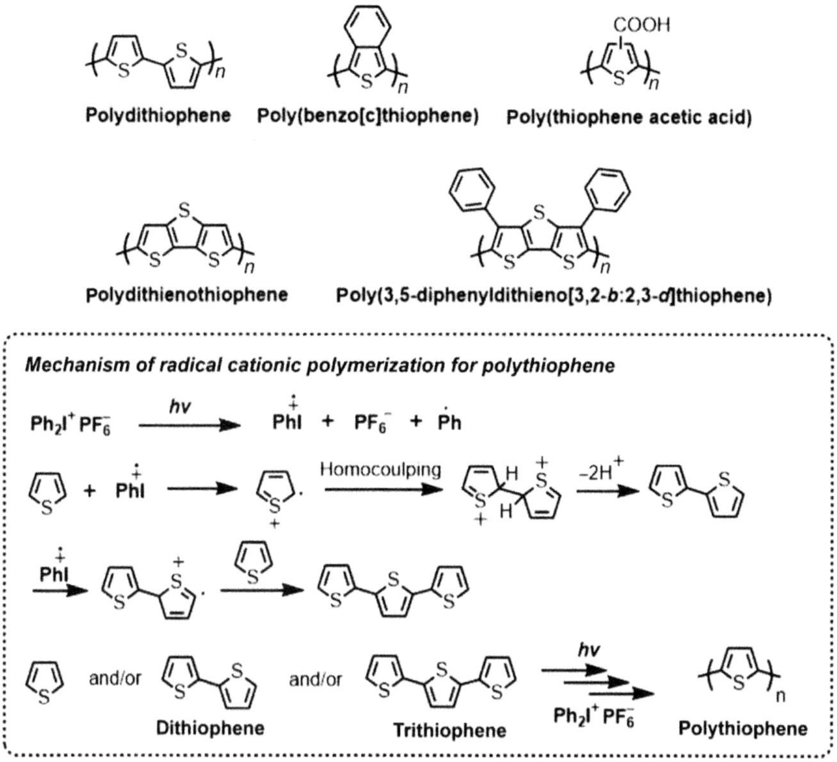

Figure 8.7 Some examples of thiophene-based polymers prepared by photoinitiated polymerization. And the mechanism of photoinitiated polymerization of thiophene using onium salts.

In order to expand the application scope of photoinitiated polymerizations, numerous reaction substrates have been hired for attempt, including thiophene unit and its derivatives. In 1991, the first thiophene-based (benzo[c]thiophene) photoinitiated polymerizations [145], using 500 W Xe-lamp as light source, acetonitrile as solvent, carbon tetrachloride, and tetrabutylammonium bromide mediated, were introduced by Shimidzu et al. to produce conducting poly(benzo[c]thiophene) (Figure 8.7) thin films with micron-scale thickness. Subsequently, various strategies have been utilized to make polymerizations of thiophene and its derivatives photochemically. For example, Shimidzu's group has developed other successful approach that dithienothiophene and oligothiophene (dimer or tetramer) has been photopolymerized in the presence of electron acceptors such as dinitrobenzene and carbon tetrachloride [146–148]. Piletsky et al. have reported the photoinduced polymerization of thiophene acetic acid in aqueous solution with potassium dichromate as the electron acceptor [149]. It was supposed that the polymerization proceeds undergo a charge-transfer reaction: a complex between the thiophene derivative and Cr(VI) formed in the excited state, leading after a charge separation to Cr(V) and an organic radical. This method provides a feasible technology for coating 0.5–2 mm thick conductive polymer layer on non-conductive surface and has potential applications in microelectronics.

It was found that, in aforementioned case, the photoinitiated polymerizations are carried out by the means of cationic radical process [150, 151]. The radical cation of the monomer involving in polymerization was efficiently formed by photoinduced charge-transfer from the excited thiophene ring to the electron acceptor. In recent decades, photoinitiated polymerization has developed vigorously since a variety of seminal cationic photoinitiators have been reported [146–153]. Among them, the onium-type photoinitiators such as iodonium, sulfonium, and alkoxypyridinium salts have become frequently used due to their thermal stability, good solubility in common organic solvent, and high photolysis quantum yields when irradiated by short- to mid-wavelength UV light (230–300 nm) [152, 154].

By using of onium salt photoinitiator, Yagci et al. produced the polythiophene by means of UV irradiation of thiophene in the presence a diphenyliodonium salt [152]. The polymerization is carried on the surface of the reaction vessel with the formation of the conducting polymer film. The polymerization mechanism involving photolysis of onium salts and electronic charge transfer from phenyliodinium radical cations to thiophenes, generating thiophene-radical-cations, was proposed in Figure 8.7. Subsequent laser flash photolysis and electron paramagnetic resonance (EPR) studies verified that polymerization mechanism theoretically. Radical cations produced by photolysis of diphenyliodonium salt rapidly oxidize thiophene to form thiophene radical-cations, followed by homocoupling and proton release to form thiophene dimer. Repeat of these similar reactions of monomers, dimers, and trimers eventually leads to polythiophene. Alkoxylpyridinium and triphenylsulfonium salts have also been provided to promote the photoinitiated polymerization of thiophene. Besides that, highly fused thiophene-based polymer, poly(3,5-diphenyldithieno[3,2-*b*:2,3-*d*]thiophene), can be achieved by onium-salts assisted photochemical means [27, 155]. In contrast to electrochemical polymerization, the described photoinitiated polymerization offers distinct advantages in membrane science. Under appropriate conditions this process will be potentially utilized for the *in situ* preparation of conductive coatings consisting of conjugated polymers, particularly, when the polymers are desired to be coated on various non-conductive substrates, such as glasses, since the non-conductive substrate cannot act as an electrode where electrochemical polymerizations typically occur.

8.2.3.2 Solid-State Polymerization

Of all thiophene-based polymers, one of the most successfully commercialized is poly(3,4-ethylenedioxythiophene) (PEDOT) [156–158]. As shown in Figure 8.8a, PEDOT chemical structural originates from polythiophene, by connecting ethylenedioxy group on thiophene's β- and β′-positions to endow them with small steric hindrance, low band gap, high stability in neutral and doped state, good solution processability, and excellent electronic properties (electrical conductivity up to 550 $S\,cm^{-1}$) [159]. The first conducting polymer product, poly(3,4-ethylenedioxythiophene)/poly(styrene sulfonate) (PEDOT/PSS), was initially commercialized as antistatic coating in 1990s. To date, PEDOT has represented

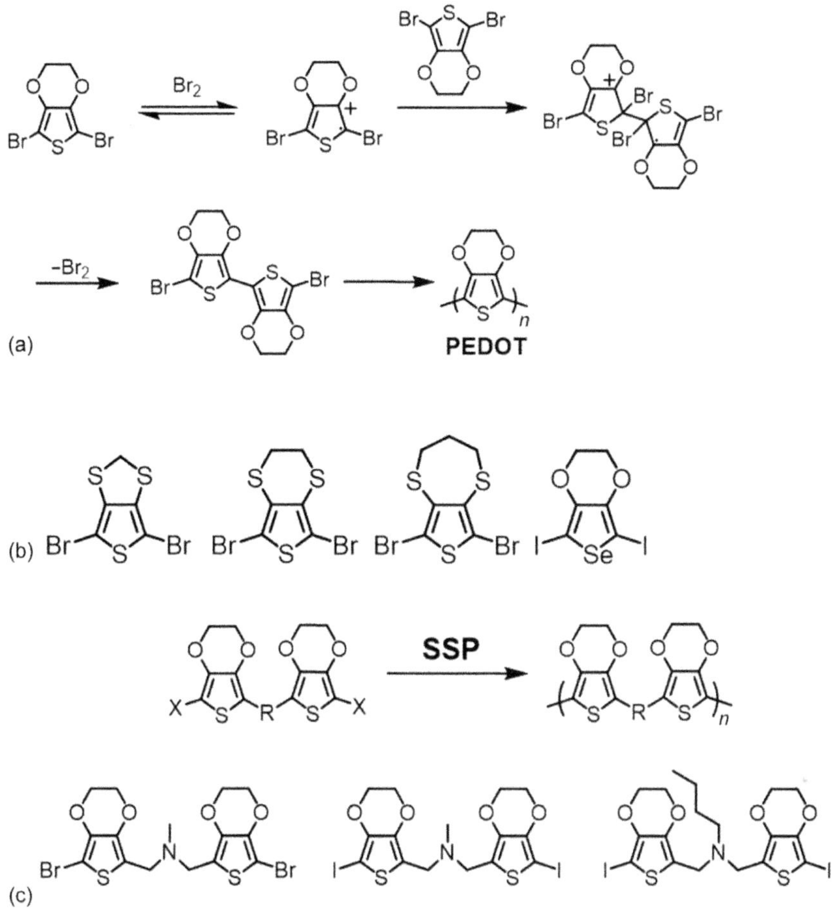

Figure 8.8 (a) Proposed mechanism of solid-state polymerization of 2,5-dibromo-3,4-ethylenedioxythiophene (DBEDOT). (b) The chemical structure of thiophene derivatives for solid-state polymerization. (c) The chemical structure of non-conjugated monomers and their respective polymers.

as one of the largest classes of conducting polymer and utilized in almost all organic electronic devices, for example, as transparent electrodes, as hole-injection layers (anode interfacial layer) in OLEDs, and as buffer layers in OPVs.

The conventional synthesis methods of PEDOT make rational use of chemical coupling reactions, involving chemical oxidants or transition metal catalysts [160–163], for example, a Ni(0) catalyzed polymerization of 2,5-dihalo-3,4-ethylenedioxythiophenes combined with magnesium. Although these coupling methods can be highly efficient, the purification of polymer products often requires cumbersome procedures, since the removal of these metal impurities away from these highly electron-rich and conjugated polythiophenes is quite difficult and complex. Recently, a novel method of SSP has been attracting much more attention due to its solvent and reagent free features. More importantly, SSP is not only a cheap and convenient approach for the synthesis of conducting polymer but also a feasible way to achieve a nearly defect-free, highly ordered polymers or even macroscopic "single-crystalline" conjugated polymer, which are necessary requirements in organic electronics, especially in OTFT [164–166].

The first crystalline polymers, polydiacetylenes and $(SN)_x$, were successfully realized by SSP in the 1960s and 1970s, respectively [167–171]. Based on these pioneering works, Wudl and coworkers have reported the first solid-state synthesis of conducting PEDOT by facile polymerization of 2,5-dibromo-3,4-ethylenedioxythiophene (DBEDOT) in the solid state [30]. The white crystalline DBEDOT can be obviously transformed to blue-black "crystals" with metallic luster after gentle heating for eight hours. The blue-black "crystals" was characterized and determined as PEDOT by elemental analysis, solid-state CNMR, and FT-IR spectroscopies. The reaction is not affected by air and light and can be performed not only in its crystalline state but also in a film state deposited on a plastic substrate or a glass slide. The proposed mechanism of the solid-state coupling is shown in Figure 8.8a. This polymerization is autocatalytic and initialized by the released bromine (formed from the bromide leaving groups) [159]. They proposed that the SSP of DBEDOT could be promoted by moderate intermolecular Br· · ·Br contacts, which offered a suitable intermolecular interaction and hold the thiophene units in proximity to each other. In their following work, Wudl and coworkers have studied the reactivity of different halogen substituent on SSP, and found that, the polymerization required a higher temperature (~140 °C) and longer time (two to three days) for 2,5-diiodo-3,4-ethylenedioxythiophene monomer, while cannot be performed for 2,5-dichloro-3,4-ethylenedioxythiophene [159]. The electron-donating group 3,4-ethylenedioxy pendant, which can further lower the oxidation potential of the thiophene unit, also plays an important role in this SSP. With this in mind, various dihalogen-substituted monomers, modified with other donor substituents on the vertical direction of the ethylenedioxythiophene (EDOT) units, have been synthesized and investigated on SSP, for instance, the replacement of the oxygen atom of 3,4-ethylenedioxy pendant with sulfur or replacement the sulfur atom of thiophene ring with selenium (Figure 8.8b) [172–175]. Spencer et al. have provided three 2,5-dibromothiophene-based monomers bearing methylenedithio-, ethylenedithio-, and propylenedithio-groups [172], respectively, which can undergo SSP to achieve relating polymers. Among them, poly(3,4-ethylenedithiathiophene), the sulfur analogue of PEDOT, has attracted significant attention due to its unique properties, driving from the stronger interchain interaction between the proximity of polymer chains through intermolecular S· · ·S attractions [174]. Besides conjugated PEDOT, non-conjugated polythiophene derivatives consisted of 3,4-ethylenedioxythiophene can also be achieved by SSP. As shown in Figure 8.8c, Xia and coworkers have developed a series of symmetrical dihalogen-substituted monomers by introducing kinds of flexible linkers or heteroatom (Si or P) between two EDOT units, and obtained the corresponding thiophene-based polymer by SSP [176, 177].

So far, although great breakthroughs have been made in SSP, the mechanism of the SSP of monomer crystals still cannot be drawn a full picture. Furthermore, exploring new monomers of this case still needs certain attention and studies.

8.2.3.3 Acid-Catalyzed Polymerization

The acid-catalyzed polymerization of thiophene can be divided into auto-acid-catalyzed polymerization and exo-acid-catalyzed polymerization. The method of auto-acid-catalyzed polymerization is almost the same as that of SSP. Noteworthy, acid-catalyzed polymeric monomers are not limited to solid state, i.e. bulk auto-polymerization of liquid monomers can also occur. Compared with exo-acid-catalyzed polymerization, usually proceed in a solution system, the auto-acid-catalyzed polymerization method adopted by Officer, Xia, and coworkers can be polymerized without any additional reagent and solvent and show more environmental characteristics [29, 178].

In 2011, Officer and his coworkers found that some 2-bromo-3-alkoxythiophene are unstable at room temperature [29]. They synthesized a series of alkoxythiophene derivatives and studied its spontaneous polymerization process. For example, oil-liked 2-dibromo-3,4-ethylenedioxythiophene (BEDOT) was synthesized and left to stand in a sealed container at −20 °C. Over a period of two weeks, black needle-like crystals formed in pure liquid BEDOT with the evolution of hydrogen bromide gas, which were characterized by infrared and Raman spectroscopy and considered to be PEDOT. The proposed mechanism of the auto-acid-catalyzed polymerization is shown in Figure 8.9a [178]. This mechanism obviously relies on the presence of a proton (H^+) to initiate the first step, and in this case, α-hydrogen of thiophene unit might be involved as the source of the proton, and activated via an intermolecular interaction. More importantly, only those molecules with an alkoxy and a bromo group ortho to each other can undergo auto-acid-catalyzed polymerization, which is consistent with the polymerization results of other alkoxythiophene monomers. Xia and coworkers reported the preparation of PEDOT with optimized auto-acid-catalyzed polymerization condition, heating for four hours at 60 °C. Under the exposure to air, the obtained PEDOT is in doped state to some extent, showing poor conductivity of lower at 10^{-6} S cm^{-1} [178].

One challenge in polymer synthesis is to achieve controlled/"living" polymerization. As we described earlier, some polythiophenes, such as P3HT, could be

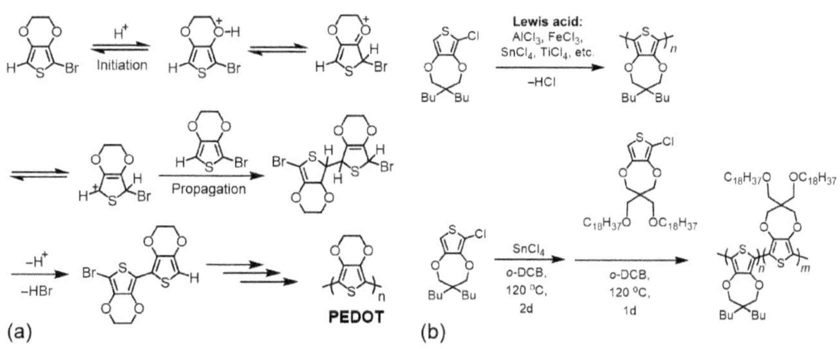

Figure 8.9 (a) Proposed mechanism of the auto-acid-catalyzed polymerization of 2-bromo-3,4-ethylenedioxythiophene (BEDOT). (b) Polymerization of 2-chloro-3,4-alkylenedioxythiophene and synthesis of the all conjugated thiophene-thiophene copolymers.

synthesized in a "living" or "quasi-living" manner using the combined reagents of magnesium metal with Ni(0) or Pd(0) as catalyst. However, the polythiophenes prepared by transmetalation polymerization methods have encountered a common problem, that is, it is contaminated by metal impurities which may have harmful effects on the lifetime of its application devices. The removal of these strongly coordinative transition metal ions away from these highly electron-rich and conjugated polythiophenes is quite difficult and complex and would be troublesome in the future commercialization of polythiophenes. Thus, it is of great practical significance to develop a controlled metal-free polymerization method for preparing polythiophenes. Recently, Swager and Bonillo have reported a cationic chain-growth polymerization to create polythiophenes from the reaction of 2-chloro-3,4-alkylenedioxythiophene and various Lewis acid (Figure 8.9b), such as tin tetrachloride ($SnCl_4$) that shows the best results [179]. The molecular weight of producing polythiophene is controllable with narrow PDI. The chain-growth and "living" character of this polymerization also enabled the synthesis of thiophene-based block copolymers by addition different thiophene based monomer step by step. Balasubramanian et al. also have studied the metal-free cationic chain-growth polymerization of 2-halogenated-3-substituted-thiophenes as promoted by Brønsted acids, e.g. trifluoroacetic acid (TFA) [180]. Compared with Lewis acid, Brønsted acids are considered to be more effective for initiating the chain-growth polymerization of 2-halogenated-3-substituted-thiophenes. Excitedly, Brønsted acids are not only suitable for highly active alkoxythiophenes, but also for the low active alkylthiothiophenes and alkylthiophenes. Their results also indicated that bromo-substituted thiophene monomer showed more reactivities and readiness undergo the chain-growth polymerization than that of chloro-substituted monomer, probably due to the easier elimination of HBr versus HCl.

As a developing method of synthesizing polythiophenes, acid-catalyzed polymerization still has many limitations in the scope of application, but it has demonstrated the feasibility to polymerize an electron rich aromatic monomer through a classical cationic methodology, which is a good start.

8.3 Applications of Polythiophene and Its Derivatives

For a half century, polythiophene and its derivatives represent one of the largest classes of all kinds of conjugated polymers. They were widely utilized in optoelectronic fields [6–9], such as OTFTs [11], OPVs, [13] and OLEDs [10], due to their excellent optical and electrical properties (suitable and adjustable energy gap, HOMO, and LUMO level), well stacking performance, environmental stability, and ease of synthesis and structural derivation. Besides that, polythiophenes also can be used in further fields such as biological applications, conductive coating, and anti-corrosion material, since they can easily be integrated and processed with other materials. In this section, we will make a short brief on their optoelectronic and biological applications.

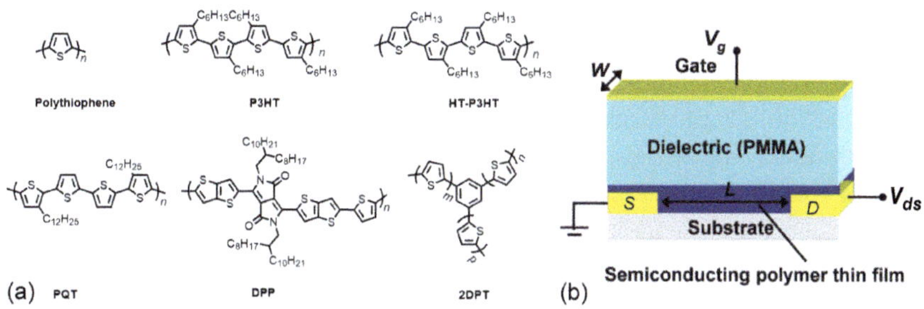

Figure 8.10 (a) The chemical structure of thiophene-based polymers for OTFTs. (b) Schematic of the top-gate/bottom-contact transistor structure. Source: (b) From Ref. [182]. Copyright © 2011 American Chemical Society.

8.3.1 Organic Thin-Film Transistors

Since the demonstration of the first OTFTs in the early 1980s [181], OTFTs have been widely used in microelectronics and optical displays. Thiophene-based polymers are the most important materials for OTFTs (Figure 8.10b). To date, enormous efforts have been made to enhance the performance of OTFTs made from thiophene-Based Polymers [8, 11, 12]. The chemical structures of some typical thiophene-based polymers for OTFTs are presented in Figure 8.10a.

In 1986, Tsumura and his coworkers first used a electrochemically polymerized thiophene in OTFT; however, this OTFT show very low mobility of 10^{-5} cm^2 V^{-1} s^{-1} [181]. Shortly after, soluble P3HT, prepared via $FeCl_3$-mediated oxidative coupling, was utilized and again with low mobility due to lack of crystalline orders in amorphous thin films [183]. With newly developed polymerization methods for polythiophenes [57, 184], regioregular head-to-tail poly (3-alkylthiophenes) (HT-P3AT), with high regioregularity (≥95%), were successfully synthesized. Regioregular HT-P3ATs such as poly(3-hexylthiophene) HT-P3HT exhibited solid-state properties (e.g. band gaps, crystallinity, and conductivity) significantly different from those of corresponding regiorandom counterparts. The high crystallinity of HT-P3HT semiconductor led to greatly improved performance in OTFTs. Bao et al. find that the film quality and field-effect mobility are strongly dependent on the choice of solvents. In addition, treating a film with ammonia or heating to 100 °C under N_2 can increase the on/off ratio without decreasing the mobility [185]. Processing method has a great influence on the structure of P3HT in OTFT. In 1999, Sirringhaus et al. use thin-film, field-effect transistor structures to probe the transport properties of the ordered microcrystalline domains in the conjugated polymer P3HT. Self-organization in P3HT results in a lamella structure with two-dimensional conjugated sheets formed by interchain stacking. They find that, depending on processing conditions, the lamellae can adopt two different orientations parallel and normal to the substrate the mobilities of which differ by more than a factor of 100, and can reach values as high as 0.1 cm^2 V^{-1} s^{-1} [12]. Stability is also an important indicator of OTFT. One way to achieve the balance between mobility and stability is by decreasing the alkyl chains, because a decrease in alkyl chains should not only lower the HOMO energy levels but also induce a twist of the thiophenes to shorten the extended π-conjugation. In 1999, Ong and his coworkers developed a class of solution-processable regioregular polythiophenes PQT (see Section 8.2.2.2, Figure 8.5), in which only half of the thiophene rings were substituted by alkyl chains. The HOMO energy level of PQT was only 0.1 eV lower than that of P3HT, but the mobility of dodecyl substituted PQT reached $0.14\,cm^2\,V^{-1}\,s^{-1}$. Slight degradation was observed after storing its devices in air for one month [98, 186].

Later, various thiophene-based polymers with new structure were prepared for OTFT. In 2011, Bronstein and his coworkers reported the synthesis and polymerization of a novel thieno[3,2-*b*]thiophene-diketopyrrolopyrrole based monomer. Copolymerization with thiophene afforded a polymer (DPP) with a maximum hole mobility of $1.95\,cm^2\,V^{-1}\,s^{-1}$ [182]. Solution-processed polythiophene thin films suffer from low efficient intermolecular carrier hopping and undermined conjugation arising from chain twist and random entanglement. In 2019, Bronstein and his coworkers synthesized a decimeter-sized 2D crosslinked polythiophene network (2DPT). The 2D crosslinked network improves charge carrier transport, resulting in a carrier mobility up to $4.6\,cm^2\,V^{-1}\,s^{-1}$ and conductance of $206\,S\,cm^{-1}$, much higher than that of normal solution processed conjugated polymer films [187].

8.3.2 Organic Photovoltaics

Solar photovoltaic energy is a clean, stable renewable energy source. The conductive polymer can be used as a photoelectric conversion material of a solar cell to convert solar energy into electrical energy. Though they are still behind their inorganic counterparts in terms of efficiency, these solar cells have several advantages, such as scalability, printability, flexibility, low cost, etc. Thiophene-based polymers are the most studied as donors and hole transporting materials for OPVs [9, 13–15, 188, 189].

Although pure polythiophenes are difficult to process, they can be improved by adding functional moieties. The chemical structures of thiophene-based polymers containing three types groups for OPVs are shown in Figure 8.11. In 1988, Hotta and his coworkers first synthesized P3HT that have good solubility and stretch ability. The stretched P3HT films exhibit a remarkably high conductivity, as high as $200\,S\,cm^{-1}$, after being doped with iodine [190]. However, P3HT is not conducive to obtaining a higher open circuit voltage because the HOMO of P3HT is too high. In 2009, a new P3AT derivative (P3HDTTT) was designed and prepared by reducing the number of electron donating groups, the alkyl side chains, the HOMO of P3ATs were successfully reduced from −4.9 eV as for P3HT to −5.3 eV as for P3HDTTT, without changing their optical band gap [191]. In 2011, Zhang and his coworkers synthesized three polythiophene derivatives PT-C1, PT-C2, and PT-C3 with electron withdrawing group, the carboxylate substituent. HOMO energy level of the polymers is successfully decreased from ca. −4.8 eV for P3HT to −5.15, −5.11, and −5.10 eV for PT-C1, PT-C2, and PT-C3, respectively [192]. It is well known that alkylthio chains have steric effects, which affect the photoelectric properties

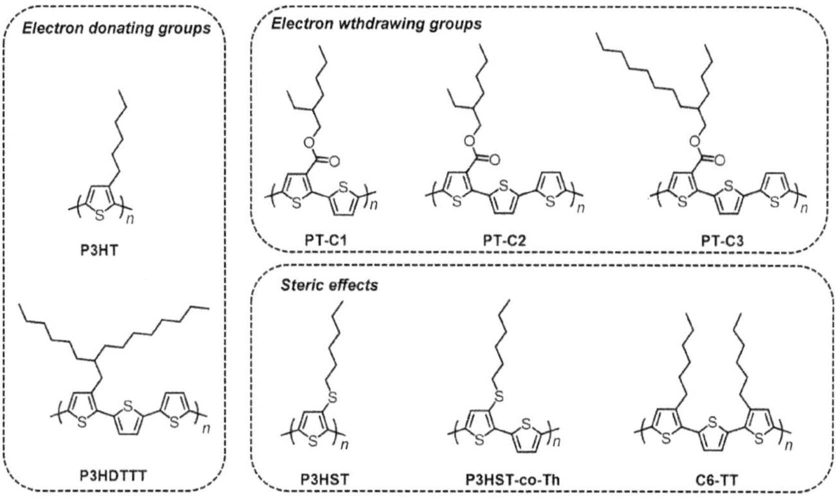

Figure 8.11 The chemical structure of thiophene-based polymers for OPVs.

of conjugated polymers. Two polythiophene derivatives P3HST and P3HST-*co*-Th were synthesized and characterized. Compared with alkyl-substituted P3HT, the introduction of alkyl sulfide side chains in P3HST and P3HST-*co*-Th redshifted the absorption spectra and the HOMO energy level moved downward [193]. Hexyl substituted polyterthiophenes are another classic polythiophene derivative for OPVs. It effectively lowers the HOMO by ca. 0.15 eV compared with the HOMO of P3HT, because the unsubstituted thiophene twists slightly out of plane, perturbing the planarity of the backbone compared with P3HT [194].

8.3.3 Organic Light-Emitting Diodes

OLEDs are solid-state light-emitting devices. They have become a mainstream display technology in consumer electronics due to self-emitting ability, transparency, true dark tone, and capability of being made flexible [195]. Among various electrochromic materials, thiophene-based polymers have obvious advantages over other electrochromic materials, such as stable color-changing performance, rich color change and high color contrast, high electrical conductivity, and fast response time. As electrochromic materials used in different fields, different thiophene-based polymers must have different optical properties.

Dyer et al. obtained the full color palette of thiophene-based polymers by through synthetic modifications and color tuning utilizing electron rich and donor acceptor repeat units, electron-donating substituents, and steric interactions with 3,4-alkylenedioxythiophene family of polymers [196]. In addition to the change from a single color, some polythiophene derivatives can complete the color conversion of tow or even multiple colors. Wang et al. obtained an electrochromic device from red to blue by synthesizing poly(3,4-ethylenedioxythiophene-didodecyloxybenzenes) [197]. Argun et al. prepared electrochromic devices that can complete the color conversion of multiple colors by synthesizing some new thiophene-based polymers

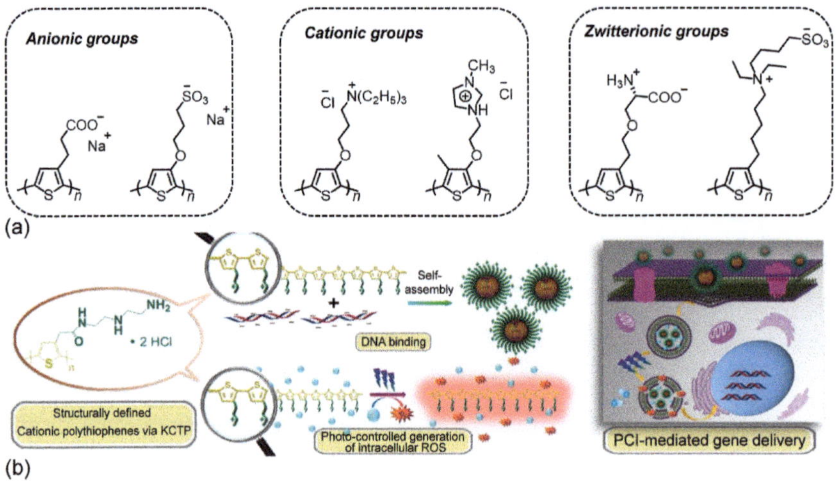

Figure 8.12 (a) The chemical structure of thiophene-based polymers for water-soluble polythiophenes. (b) Schematic illustration of the cationic polythiophenes for DNA binding and gene delivery. Source: (b) From Ref. [204] Copyright © 2018 American Chemical Society.

such as PBEDOT-NMeCz, PBEDOTPyrPyr, and PBEDOT-Pyr [198]. Moreover, the characteristics of the light-emitting diodes of thiophene-based polymers can be improved by incorporated polyethylene glycol [199], carbon nanotube [200], or multi-shell quantum dots [201].

8.3.4 Biological Applications

Water-soluble conjugated polymers provide a choice of fabricating sensor for the biologically relevant targets and polythiophene derivatives are excellent candidates due to tunable fluorescence property, forster resonance energy transfer (FRET), tunable electrical conductivity, superior biocompatibility, and environmental stability [202, 203]. Pristine polythiophene chain is hydrophobic. However, it exhibits aqueous solubility after attachment/grafting of ionic pendent groups or hydrophilic polymer chains on its backbone [21]. Growing applications of water-soluble polythiophene in the field of biological applications would be discussed here. They are classified as anionic, cationic, and zwitterionic conjugated polyelectrolytes (CPEs). Some examples of water-soluble polythiophenes are presented in Figure 8.12a.

In 1998, Faid and Leclerc reported the new synthesis of water-soluble biotin-functionalized polythiophene containing sulfonic acid group. And the polymer can perceive the binding of avidin (biotin binding protein) monitored from UV–vis spectroscopy, and the polymer solution exhibits a sharp color change (violet to yellow) after addition of avidin due to the strong interaction of avidin and biotin [205]. The *in vivo* deposition of amyloid fibrils is a hallmark of many devastating diseases known as the amyloidoses. Amyloid formation *in vitro* may also complicate production of proteins in the biotechnology industry. In 2005, Nilssion et al. developed a negatively charged polythiophene, which exhibits optical changes perceived either visually or by absorption and emission spectra. The optical property of the polyelectrolyte changed depending on the negative state or amyloid fibril conformation of the protein with which it formed the complex. The method is fast and simple and is based on noncovalent assembly between the anionic polyelectrolyte and the protein [206].

Ho et al. developed a novel methodology that allows simple optical (colorimetric or fluorometric) detection of nucleic acids depending the conformational change of the polythiophene containing cationic groups. The conformational change of the polymer is due to the electrostatic interaction with single-stranded oligonucleotides and double-stranded nucleic acids [207]. However, the use of water-soluble polythiophenes with controllability and regioregularity for biological applications have scarcely been reported. In 2018, Zhang et al. synthesized polythiophenes containing ester side chains (P3ET). In addition, they obtained cationic regioregular polythiophenes (cPTs) by ammonolysis of P3ET with varied chain lengths and studied DNA binding capability and gene delivery performance. Benefiting from photo-controlled generation of intracellular reactive oxygen species (ROS), the cationic polythiophenes successfully delivered DNA into tumor cells without additional polymer species. Schematic illustration of the cationic polythiophenes for DNA binding and gene delivery is shown in Figure 8.12b [204]. Water-soluble cationic polythiophene derivatives also can produce interpolymer complex with polysaccharides [208].

In 2004, Nilsson and his coworkers synthesized a kind of water-soluble, zwitterionic, and photoactive polythiophene derivative containing carboxyl and amino groups. Upon addition of Ca^{2+} the intensity of the emitted light is increased and blue-shifted. The geometrical alteration of the polymer chains can further be utilized for recording of the binding of calcineurin to the calcium-activated POWT-calmodulin complex. This offers a novel way to create microarrays without using covalent attachment of the receptor or labeling of the analyte [209]. The selective engineering of CPE-phospholipid interfaces is poised to play a key role in the design of advanced biomedical and biotechnological devices. In 2016, Houston and his coworkers performed a strategic study of conjugated poly(thiophene)s bearing anionic, cationic, and zwitterionic pendant groups to successfully demonstrate that the CPE penetration depth within a zwitterionic phospholipid bilayer can be controlled via charge mediation. With an overall neutral charge, the zwitterionic side group of P3Zwit, in combination with its hydrophobic poly(thiophene) backbone, means that P3Zwit can readily insert in between the dipalmitoyl phosphatidylcholine zwitterionic headgroups as well as, uniquely, the hydrophobic phospholipid tails. Cationic, anionic, and zwitterionic conjugated polyelectrolytes have different applications in the biological field [210].

Like most conjugated polymers, polythiophenes that combine the electronic properties of inorganic semiconductors and physical properties of organic polymers can be easily processed to form thin, uniform, and lightweight films. Thus, besides the application areas we discussed earlier, the application of thiophene-based conjugated polymer has been expanded to more and more areas, such as memory devices [6, 211, 212] and sensors [7, 213]. As a photoelectric display material, polythiophene polymer has a main advantage in that its luminescent color can be regulated by adjusting the length of conjugate chain to replace the type of base and polymer regulation to regulate recently. The purpose is to find and obtain new polythiophene materials with good stability, processing and luminescence properties, and some progress has been made [214–216]. Thiophene-based polymers have high Seebeck coefficient and low thermal conductivity, so they have good application prospects in thermoelectric materials [217, 218]. Thiophene-based polymers has excellent properties such as high electrical conductivity, good light transmittance, etc. and can be polymerized with other materials to prepare a conductive coating for anti-static electricity [219]. Thiophene-based polymers also are used for anti-corrosion materials through electrochemical protection due to their high electrical conductivity [220–222].

8.4 Conclusions and Future Scope

Polythiophene and its derivatives, we mention earlier, have good application prospects in many organic electronic fields. To date, more and more methods for the synthesis of thiophene-based polymers have been developed. In this chapter, several classical or innovative synthetic methods are introduced, along with their results in the structural characteristics, which leads to different properties towards different applications. Recently, significant progress has been made in various methods. Some methods, like Ni-catalyzed catalyst transfer homopolymerization, allow for a precise control of the regioregularity and the molecular weight. Particularly for that of using NHC ligands, generating true "living" polymers that enables a well controllable and ultra-high molecular weight. Transmetalation copolymerization, such as Stille and Suzuki couplings, is widely applicable to all kinds of thiophene-based polymers and provides remarkable control on the polymer properties. The simpler and cheaper coupling methods (e.g. SSP) produced high crystalline material, which are appreciated in the electronics devices. In our opinion, all kinds of polymerization methods have their unique advantages as well as the limited scope at present. How to use these well-designed polymerizations to meet the application requirements is the goal for contemporary scientists and technicians.

References

1 Chiang, C.K., Fincher, C.R., Park, Y.W. et al. (1977). Electrical conductivity in doped polyacetylene. *Phys. Rev. Lett.* 39: 1098–1101.

2 Yamamoto, T., Hayashi, Y., and Yamamoto, A. (1978). A novel type of polycondensation utilizing transition metal-catalyzed C-C coupling. I. Preparation of thermostable polyphenylene type polymers. *Bull. Chem. Soc. Jpn.* 51: 2091–2097.

3 Sekine, C., Tsubata, Y., Yamada, T. et al. (2014). Recent progress of high performance polymer OLED and OPV materials for organic printed electronics. *Sci. Technol. Adv. Mater.* 15: 034203.
4 Ballard, D.G.H., Courtis, A., Shirley, I.M., and Taylor, S.C. (1988). Synthesis of polyphenylene from a *cis*-dihydrocatechol, a biologically produced monomer. *Macromolecules* 21: 294–304.
5 Kovacic, P. and Jones, M.B. (1987). Dehydro coupling of aromatic nuclei by catalyst-oxidant systems: poly(p-phenylene). *Chem. Rev.* 87: 357–379.
6 Chou, Y.H., Chang, H.C., Liu, C.L., and Chen, W.C. (2015). Polymeric charge storage electrets for non-volatile organic field effect transistor memory devices. *Polym. Chem.* 6: 341–352.
7 Baeg, K.J., Binda, M., Natali, D. et al. (2013). Organic light detectors: photodiodes and phototransistors. *Adv. Mater.* 25: 4267–4295.
8 Wang, C.L., Dong, H.L., Hu, W.P. et al. (2012). Semiconducting π-conjugated systems in field-effect transistors: a material odyssey of organic electronics. *Chem. Rev.* 112: 2208–2267.
9 Guo, X., Baumgarten, M., and Müllen, K. (2013). Designing π-conjugated polymers for organic electronics. *Prog. Polym. Sci.* 38: 1832–1908.
10 Xu, F., Kim, H.U., Kim, J.H. et al. (2015). Progress and perspective of iridium-containing phosphorescent polymers for light-emitting diodes. *Prog. Polym. Sci.* 47: 92–121.
11 Holliday, S., Donaghey, J.E., and McCulloch, I. (2014). Advances in charge carrier mobilities of semiconducting polymers used in organic transistors. *Chem. Mater.* 26: 647–663.
12 Sirringhaus, H., Brown, P.J., Friend, R.H. et al. (1999). Two-dimensional charge transport in self-organized, high-mobility conjugated polymers. *Nature* 401: 685–688.
13 Holliday, S., Ashraf, R.S., Wadsworth, A. et al. (2016). High-efficiency and air-stable P3HT-based polymer solar cells with a new non-fullerene acceptor. *Nat. Commun.* 7: 11585.
14 Schmidt-Mende, L., Fechtenkötter, A., Müllen, K. et al. (2001). Self-organized discotic liquid crystals for high-efficiency organic photovoltaics. *Science* 293: 1119–1122.
15 Mazzio, K.A. and Luscombe, C.K. (2015). The future of organic photovoltaics. *Chem. Soc. Rev.* 44: 78–90.
16 Hösel, M., Dam, H.F., and Krebs, F.C. (2015). Development of lab-to-fab production equipment across several length scales for printed energy technologies, including solar cells. *Eng. Technol.* 3: 293–304.
17 Günes, S., Neugebauer, H., and Sariciftci, N.S. (2007). Conjugated polymer-based organic solar cells. *Chem. Rev.* 107: 1324–1338.
18 Quagliotto, P. and Fin, A. (2018). Advances in synthetic methods for the preparation of poly(3-hexylthiophene) (P3HT). *Lett. Org. Chem.* 15: 991–1006.
19 Li, C. and Shi, G.Q. (2013). Polythiophene-based optical sensors for small molecules. *ACS Appl. Mater. Inter.* 5: 4503–4510.

20 Kaloni, T.P., Giesbrecht, P.K., Schreckenbach, G., and Freund, M.S. (2017). Polythiophene: from fundamental perspectives to applications. *Chem. Mater.* 29: 10248–10283.

21 Das, S., Chatterjee, D.P.G., Ghosh, R., and Nandi, A.K. (2015). Water soluble polythiophenes: preparation and applications. *RSC Adv.* 5: 20160–20177.

22 Xu, S.Q., Kim, E.H., Wei, A., and Negishi, E.I. (2014). Pd- and Ni-catalyzed cross-coupling reactions in the synthesis of organic electronic materials. *Sci. Technol. Adv. Mater.* 15: 044201.

23 Yoshino, K., Hayashi, S., and Sugimoto, R.I. (1984). Preparation and properties of conducting heterocyclic polymer films by chemical method. *Jpn. J. Appl. Phys.* 23: L899–L900.

24 Abdulla, H.S. (2013). Electrochemical synthesis and vibrational mode analysis of poly (3-methelthiophene). *Int. J. Electrochem. Sci.* 8: 11782–11790.

25 Leclerc, M., Diaz, F.M., and Wegner, G. (1989). Structural analysis of poly(3-alkylthiophene). *Makromol, Chem.* 190: 3105–3114.

26 Iraqi, A. and Barker, G.W. (1998). Synthesis and characterisation of telechelic regioregular head-to-tail poly(3-alkylthiophenes). *J. Mater. Chem.* 8: 25–29.

27 Aydogan, B., Gunbas, G.E., Durmus, A. et al. (2010). Highly conjugated thiophene derivatives as new visible light sensitive photoinitiators for cationic polymerization. *Macromolecules* 43: 101–106.

28 Xia, J.B., Masaki, N., Lira-Cantu, M. et al. (2008). Influence of doped anions on poly (3,4-ethylenedioxythiophene) as hole conductors for iodine-free solid-state dye-sensitized solar cells. *J. Am. Chem. Soc.* 130: 1258–1263.

29 Wagner, P., Jolley, K.W., and Officer, D.L. (2011). Why do some alkoxybromothiophenes spontaneously polymerize? *Aust. J. Chem.* 64: 335–338.

30 Meng, H., Perepichka, D.F., and Wudl, F. (2003). Facile solid-state synthesis of highly conducting poly (ethylenedioxythiophene). *Angew. Chem. Int. Ed.* 42: 658–661.

31 Roncali, J. (1992). Conjugated poly (thiophenes): synthesis, functionalization, and applications. *Chem. Rev.* 92: 711–738.

32 Wei, Y., Chan, C.C., Tian, J. et al. (1991). Electrochemical polymerization of thiophenes in the presence of bithiophene or terthiophene: kinetics and mechanism of the polymerization. *Chem. Mater.* 3: 888–897.

33 Sato, M. and Morii, H. (1991). Nuclear magnetic resonance studies on electrochemically prepared poly (3-dodecylthiophene). *Macromolecules* 24: 1196–1200.

34 Cheng, Y.J., Yang, S.H., and Hsu, C.S. (2009). Synthesis of conjugated polymers for organic solar cell applications. *Chem. Rev.* 109: 5868–5923.

35 Pratt, C.M., Foot, P.J.S., and Davis, R. (2016). Effects of $CoCl_2$ and other additives on the oxidative chemical synthesis and properties of poly (3-hexylthiophene). *Polym. Polym. Compos.* 24: 185–190.

36 Takagi, K., Torii, C., and Yamashita, Y. (2009). Polymerization of branched thiophene monomers and optoelectronic properties of materials. *J. Polym. Sci. Pol. Chem.* 47: 3034–3044.

37 Liu, R.C. and Liu, Z.P. (2009). Polythiophene: synthesis in aqueous medium and controllable morphology. *Chinese. Sci. Bull.* 54: 2028–2032.

38 Tsuchiya, K. and Ogino, K. (2013). Catalytic oxidative polymerization of thiophene derivatives. *Polym. J.* 45: 281–286.

39 Andersson, M.R., Selse, D., Berggren, M. et al. (1994). Regioselective polymerization of 3-(4-octylphenyl)thiophene with $FeCl_3$. *Macromolecules.* 27: 6503–6506.

40 Hai, T.A.P. and Sugimoto, R. (2017). Effect of molar ratio of oxidizer/3-hexylthiophene monomer in chemical oxidative polymerization of poly(3-hexylthiophene). *J. Mol. Struct.* 1146: 660–668.

41 Facchinatto, W.M., Torres, B.B.M., and Balogh, D.T. (2016). One-pot synthesis of poly-(3-hexylthiophene) with variable degrees of molar mass and regioregularity. *J. Polym. Res.* 23: 187.

42 Shimizu, H., Yamada, M., Wada, R., and Okabe, M. (2007). Preparation and characterization of water self-dispersible poly(3-hexylthiophene) particles. *Polym. J.* 40: 33–36.

43 Tourillon, G. and Garnier, F.J. (1982). New electrochemically generated organic conducting polymers. *J. Electroanal. Chem.* 135: 173–178.

44 Xu, Z., Horowitz, G., and Garnier, F. (1988). Cathodic electropolymerization of polythiophene on platinum and various semiconducting electrodes. *J. Electroanal. Chem.* 246: 467.

45 Mizuno, M., Tateno, H., Matsumura, Y., and Atobe, M. (2017). Synthesis and molecular weight control of poly(3-hexylthiophene) using electrochemical polymerization in a flow microreactor. *React. Chem. Eng.* 2: 642–645.

46 Hillman, A.R. and Mallen, E.F. (1988). A spectroelectrochemical study of the electrodeposition of polythiophene films. *J. Electroanal. Chem.* 243: 403–417.

47 Waltman, R.J., Bargon, J., and Dlaz, A.F. (1983). Electrochemical studies of some conducting polythiophene films. *J. Phys. Chem.* 87: 1459.

48 Roncail, J., Yassar, A., and Garnier, F. (1988). Electrosynthesis of highly conducting poly(3-methylthiophene) thin films. *J. Chem. Soc. Chem. Commun.* 9: 581–582.

49 Yassar, A., Roncali, J., and Garnier, F. (1989). Conductivity and conjugation length in poly(3-methylthiophene) thin films. *Macromolecules* 22: 804.

50 Tanaka, S., Sato, M., and Kaeriyama, K. (1984). Electrochemical preparation and characterization of poly (2, 5-thiophenediyl). *Makromol. Chem.* 185: 1295–1306.

51 Fu, M., Zhu, Y., Tan, R., and Shi, G. (2001). Aligned polythiophene micro- and nanotubules. *Adv. Mater.* 13: 1874–1877.

52 Li, C., Bai, H., and Shi, G. (2009). Conducting polymer nanomaterials: electrosynthesis and applications. *Chem. Soc. Rev.* 38: 2397–2409.

53 McCullough, R.D. and Lowe, R.D. (1992). Enhanced electrical conductivity in regioselectively synthesized poly(3-alkylthiophenes). *J. Chem. Soc. Chem. Commun.* 1: 70–72.

54 McCullough, R.D., Lowe, R.D., Jayaraman, M., and Anderson, D.L. (1993). Design, synthesis, and control of conducting polymer architectures: structurally homogeneous poly (3-alkylthiophenes). *J. Org. Chem.* 58: 904–912.

55 Boyd, S.D., Jen, A.K.-Y., and Luscombe, C.K. (2009). Steric stabilization effects in nickle-catalyzed regioregular poly(3-hexylthiophene) synthesis. *Macromolecules* 42: 9387–9389.
56 Chen, T.A. and Rieke, R.D. (1992). The first regioregular head-to-tail poly(3-hexylthiophene-2,5-diyl) and a regiorandom isopolymer: nickel versus palladium catalysis of 2(5)-bromo-5(2)-(bromozincio)-3-hexylthiophene polymerization. *J. Am. Chem. Soc.* 114: 10087–10088.
57 Chen, T.A., Wu, X., and Rieke, R.D. (1995). Regiocontrolled synthesis of poly(3-alkylthiophenes) mediated by rieke zinc: their characterization and solid-state properties. *J. Am. Chem. Soc.* 117: 233–244.
58 Yokoyama, A., Miyakoshi, R., and Yokozawa, T. (2004). Chain-growth polymerization for poly(3-hexylthiophene) with a defined molecular weight and a low polydispersity. *Macromolecules* 37: 1169–1171.
59 McCullough, R.D. (1998). The chemistry of conducting polythiophenes. *Adv. Mater.* 10: 93–116.
60 Loewe, R.S., Ewbank, P.C., Liu, J.S. et al. (2001). Regioregular, head-to-tail coupled poly(3-alkylthiophenes) made easy by the GRIM method: investigation of the reaction and the origin of regioselectivity. *Macromolecules* 34: 4324–4333.
61 Iovu, M.C., Sheina, E.E., Gil, R.R., and McCullough, R.D. (2005). Experimental evidence for the quasi-"living" nature of the grignard metathesis method for the synthesis of regioregular poly(3-alkylthiophenes). *Macromolecules* 38: 8649–8656.
62 Tamba, S., Fuji, K., Nakamura, K., and Mori, A. (2014). Nickel (II)-catalyzed cross-coupling polycondensation of thiophenes via C-S bond cleavage. *Organometallics* 33: 12–15.
63 Qiu, Y., Worch, J.C., Fortney, A. et al. (2016). Nickel-catalyzed Suzuki polycondensation for controlled synthesis of ester-functionalized conjugated polymers. *Macromolecules* 49: 4757–4762.
64 Okamoto, K. and Luscombe, C.K. (2011). Controlled polymerizations for the synthesis of semiconducting conjugated polymers. *Polym. Chem.* 2: 2424–2434.
65 Loewe, R.S., Khersonsky, S.M., and McCullough, R.D. (1999). A simple method to prepare head-to-tail coupled, regioregular poly(3-alkylthiophenes) using grignard metathesis. *Adv. Mater.* 3: 250–253.
66 Sheina, E.E., Liu, J., Iovu, M.C. et al. (2004). Chain growth mechanism for regioregular nickel-initiated cross-coupling polymerizations. *Macromolecules* 37: 3526–3528.
67 Miyakoshi, R., Yokoyama, A., and Yokozawa, T. (2005). Catalyst-transfer polycondensation. Mechanism of Ni-catalyzed chain-growth polymerization leading to well-defined poly (3-hexylthiophene). *J. Am. Chem. Soc.* 127: 17542–17547.
68 Wu, S.P., Huang, L., Tian, H.K. et al. (2011). LiCl-promoted chain growth Kumada catalyst-transfer polycondensation of the "reversed" thiophene monomer. *Macromolecules* 44: 7558–7567.
69 Jeffries-EL, M., Sauvé, G., and McCullough, R.D. (2004). In-situ end-group functionalization of regioregular poly (3-alkylthiophene) using the Grignard metathesis polymerization method. *Adv. Mater.* 16: 1017–1019.

70 Doubina, N., Ho, A., Jen, A.K., and Luscombe, C.K. (2009). Effect of initiators on the Kumada catalyst-transfer polycondensation reaction. *Macromolecules* 42: 7670–7677.

71 Bronstein, H.A. and Luscombe, C.K. (2009). Externally initiated regioregular P3HT with controlled molecular weight and narrow polydispersity. *J. Am. Chem. Soc.* 131: 12894–12895.

72 Lanni, E.L., Locke, J.R., Gleave, C.M., and McNeil, A.J. (2011). Ligand-based steric effects in Ni-catalyzed chain-growth polymerizations using bis (dialkylphosphino) ethanes. *Macromolecules* 44: 5136–5145.

73 Tkachov, R., Senkovskyy, V., Komber, H., and Kiriy, A. (2011). Influence of alkyl substitution pattern on reactivity of thiophene-based monomers in Kumada catalyst-transfer polycondensation. *Macromolecules* 44: 2006–2015.

74 Lohwasser, R.H. and Thelakkat, M. (2011). Toward perfect control of end groups and polydispersity in poly (3-hexylthiophene) via catalyst transfer polymerization. *Macromolecules* 44: 3388–3397.

75 Tkachov, R., Senkovskyy, V., Komber, H. et al. (2010). Random catalyst walking along polymerized poly (3-hexylthiophene) chains in Kumada catalyst-transfer polycondensation. *J. Am. Chem. Soc.* 132: 7803–7810.

76 Achord, B.C. and Rawlins, J.W. (2009). Evidence of Ni(0) complex diffusion during Grignard metathesis polymerization of 2,5-dibromo-3-hexylthiophene. *Macromolecules* 42: 8634–8639.

77 Shi, X.C., Sui, A.G., Wang, Y.X. et al. (2015). Controlled synthesis of high molecular weight poly (3-hexylthiophene) s via Kumada catalyst transfer polycondensation with Ni (IPr)(acac)$_2$ as the catalyst. *Chem. Commun.* 51: 2138–2140.

78 Senkovskyy, V., Khanduyeva, N., Komber, H. et al. (2007). Conductive polymer brushes of regioregular head-to-tail poly (3-alkylthiophenes) via catalyst-transfer surface-initiated polycondensation. *J. Am. Chem. Soc.* 129: 6626–6632.

79 Senkoskyy, V., Sommer, M., Tkachov, R. et al. (2010). Convenient route to initiate Kumada catalyst-transfer polycondensation using Ni(dppe)Cl$_2$ or Ni(dppp)Cl$_2$ and sterically hindered Grignard compounds. *Macromolecules* 43: 10157–10161.

80 Doubina, N., Paniagua, S.A., Soldatova, A.V. et al. (2011). Steric effects of the initiator substituent position on the externally initiated polymerization of 2-bromo-5-iodo-3-hexylthiophene. *Macromolecules* 44: 512–520.

81 Beryozkina, T., Senkovskyy, V., Kaul, E., and Kiriy, A. (2008). Kumada catalyst-transfer polycondensation of thiophene-based oligomers: robustness of a chain-growth mechanism. *Macromolecules* 41: 7817–7823.

82 Senkovskyy, V., Tkachov, R., Beryozkina, T. et al. (2009). "Hairy" poly(3-hexylthiophene) particles prepared via surface-initiated kumada catalyst-transfer polycondensation. *J. Am. Chem.Soc.* 131: 16445–16453.

83 Stefan, M.C., Javier, A.E., Osaka, I., and McCullough, R.D. (2009). Grignard metathesis method (GRIM): toward a universal method for the synthesis of conjugated polymers. *Macromolecules* 42: 30–32.

84 Huang, L., Wu, S.P., Qu, Y. et al. (2008). Grignard metathesis chain-growth polymerization for polyfluorenes. *Macromolecules* 41: 8944–8947.

85 Wu, S.P., Sun, Y.Q., Huang, L. et al. (2010). Grignard metathesis chain-growth polymerization for poly(bithienylmethylene)s: Ni catalyst can transfer across the nonconjugated monomer. *Macromolecules* 43: 4438–4440.

86 Stille, J.K. (1986). The palladium-catalyzed cross-coupling reactions of organotin reagents with organic electrophiles. *Angew. Chem. Int. Ed.* 25: 508–524.

87 Suzuki, A. (1999). Recent advances in the cross-coupling reactions of organoboron derivatives with organic electrophiles, 1995–1998. *J. Organomet. Chem.* 576: 147–168.

88 Pomerantz, M., Yang, H., and Cheng, Y. (1995). Poly(alkyl thiophene-3-carboxylates). Synthesis and characterization of polythiophenes with a carbonyl group directly attached to the ring. *Macromolecules* 28: 5706–5708.

89 Pomerantz, M., Yang, C., Kasim, R.K., and Elsenbaumer, R.L. (1999). Poly (alkyl thiophene-3-carboxylates). Synthesis, properties and electroluminescence studies of polythiophenes containing a carbonyl group directly attached to the ring. *J. Mater. Chem.* 9: 2155–2163.

90 Coombs, B.A., Rutter, S.R., Goeta, A.E. et al. (2012). 2,5-bis (Arylethynyl) thienyl systems: preparation and photophysical properties. Part II. *RSC Adv.* 2: 1870–1876.

91 Guillerez, S. and Bidan, G. (1998). New convenient synthesis of highly regioregular poly (3-octylthiophene) based on the Suzuki coupling reaction. *Syn. Metal.* 93: 123–126.

92 Liversedge, I.A., Higgins, S.J., Giles, M. et al. (2006). Suzuki route to regioregular polyalkylthiophenes using Ir-catalysed borylation to make the monomer, and Pd complexes of bulky phosphanes as coupling catalysts for polymerisation. *Tetrahedron. lett.* 47: 5143–5146.

93 Li, W.W., Yang, H., Li, B.I. et al. (2008). Tris [tri (2-thienyl) phosphine] palladium as the catalyst precursor for thiophene-based Suzuki-Miyaura crosscoupling and polycondensation. *J. Polym. Sci. Part A: Polym. Chem.* 46: 4556–4563.

94 Sui, A.G., Shi, X.C., Tian, H.K. et al. (2014). Suzuki–Miyaura catalyst-transfer polycondensation with Pd (IPr)$(OAc)_2$ as the catalyst for the controlled synthesis of polyfluorenes and polythiophenes. *Polym. Chem.* 5: 7072–7080.

95 Kinzel, T., Zhang, Y., and Buchwald, S.L. (2010). A new palladium precatalyst allows for the fast Suzuki-Miyaura coupling reactions of unstable polyfluorophenyl and 2-heteroaryl boronic acids. *J. Am. Chem. Soc.* 132: 14073–14075.

96 Qiu, Y.Y., Mohin, J., Tsai, C.H. et al. (2015). Stille catalyst-transfer polycondensation using Pd-PEPPSI-IPr for high-molecular-weight regioregular poly (3-hexylthiophene). *Macromol. Rapid Commun.* 36: 840–844.

97 Thompson, B.C., Kim, B.J., Kavulak, D.F. et al. (2007). Influence of alkyl substitution pattern in thiophene copolymers on composite fullerene solar cell performance. *Macromolecules* 40: 7425–7428.

98 McCulloch, I., Heeney, M., Bailey, C. et al. (2006). Liquid-crystalline semiconducting polymers with high charge-carrier mobility. *Nat. Mater.* 5: 328–333.

99 You, J.B., Dou, L.T., Yoshimura, K. et al. (2013). A polymer tandem solar cell with 10.6% power conversion efficiency. *Nat. Commun.* 4: 1446–1455.

100 Li, J.L. and Grimsdale, A.C. (2010). Carbazole-based polymers for organic photovoltaic devices. *Chem. Soc. Rev.* 39: 2399–2410.

101 Grimsdale, A.C., Chan, K.L., Martin, R.E. et al. (2009). Synthesis of light-emitting conjugated polymers for applications in electroluminescent devices. *Chem. Rev.* 109: 897–1091.

102 Takeda, Y., Andrew, T.L., Lobez, J.M. et al. (2012). An air-stable low-bandgap n-type organic polymer semiconductor exhibiting selective solubility in perfluorinated solvents. *Angew. Chem. Int. Ed.* 51: 9042–9046.

103 Tsao, H.N. and Müllen, K. (2010). Improving polymer transistor performance via morphology control. *Chem. Soc. Rev.* 39: 2372–2386.

104 Carsten, B., He, F., Son, H.J. et al. (2011). Stille polycondensation for synthesis of functional materials. *Chem. Rev.* 111: 1493–1528.

105 Sakamoto, J., Rehahn, M., Wegner, G., and Schlüter, A.D. (2009). Suzuki polycondensation: Polyarylenes à la Carte. *Macromol. Rapid Commun.* 30: 653–687.

106 Amatore, C., Jutand, A., and Le Duc, G. (2012). The triple role of fluoride ions in palladium-catalyzed Suzuki-Miyaura reactions: unprecedented transmetalation from [$ArPdFL_2$] complexes. *Angew. Chem. Int. Ed.* 51: 1379–1382.

107 Schipper, D.J. and Fagnou, K. (2011). Direct arylation as a synthetic tool for the synthesis of thiophene-based organic electronic materials. *Chem. Mater.* 23: 1594–1600.

108 Alberico, D., Scott, M.E., and Lautens, M. (2007). Aryl-Aryl bond formation by transition-metal-catalyzed direct arylation. *Chem. Rev.* 107: 174–238.

109 Chiusoli, G.P., Catellani, M., Costa, M. et al. (2010). Catalytic C-C coupling through C-H arylation of arenes or heteroarenes. *Coord. Chem. Rev.* 254: 456–469.

110 Lia, B. and Shi, Z. (2012). From C(sp^2)-H to C(sp^3)-H: systematic studies on transition metal-catalyzed oxidative C-C formation. *Chem. Soc. Rev.* 41: 5588–5598.

111 Okamoto, K., Zhang, J.X., Housekeeper, J.B. et al. (2013). C-H arylation reaction: atom efficient and greener syntheses of π-conjugated small molecules and macromolecules for organic electronic materials. *Macromolecules* 46: 8059–8078.

112 Sun, H., Gorelsky, S.I., Stuart, D.R. et al. (2010). Mechanistic analysis of azine N-oxide direct arylation: evidence for a critical role of acetate in the $Pd(OAc)_2$ precatalyst. *J. Org. Chem.* 75: 8180–8189.

113 Campeau, L., Parisien, M., Jean, A., and Fagnou, K. (2006). Catalytic direct arylation with aryl chlorides, bromides, and iodides: intramolecular studies leading to new intermolecular reactions. *J. Am. Chem. Soc.* 128: 581–590.

114 Lafrance, M. and Fagnou, K. (2006). Palladium-catalyzed benzene arylation: incorporation of catalytic pivalic acid as a proton shuttle and a key element in catalyst design. *J. Am. Chem. Soc.* 128: 16496–16497.

115 Wang, Y.N., Guo, X.Q., Zhu, X.H. et al. (2012). $Pd(OAc)_2$ catalyzed direct arylation of electron-deficient arenes without ligands or with monoprotected amino acid assistance. *Chem. Commun.* 48: 10437–10439.

116 Gorelsky, S.I. (2013). Origins of regioselectivity of the palladium-catalyzed (aromatic) CH bond metalation-deprotonation. *Coord. Chem. Rev.* 257: 153–164.

117 Sévignon, M., Papillon, J., Schulz, E., and Lemaire, M. (1999). New synthetic method for the polymerization of alkylthiophenes. *Tetrahydron Letter* 40: 5873–5876.

118 Hassan, J., Schulz, E., Gozzi, C., and Lemaire, M. (2003). Palladium-catalyzed synthesis of oligo (alkylthiophenes). *J. Mol. Catal. A: Chem.* 195: 125–131.

119 Wang, Q., Takita, R., Kikuzaki, Y., and Ozawa, F. (2010). Palladium-catalyzed dehydrohalogenative polycondensation of 2-bromo-3-hexylthiophene: an efficient approach to head-to-tail poly (3-hexylthiophene). *J. Am. Chem. Soc* 132: 11420–11421.

120 Dou, J.Y., Chen, Z.Q., and Ma, C.Q. (2014). Regioregularity and properties of the poly (3-hexylthiophene) synthesized by palladium catalyzed direct CH arylation polycondensation under different reaction conditions. *Synth. Met.* 196: 117–124.

121 Pouliot, J.R., Wakioka, M., Ozawa, F. et al. (2016). Structural analysis of poly (3-hexylthiophene) prepared via direct heteroarylation polymerization. *Macromol. Chem. Phys.* 217: 1493–1500.

122 Rudenko, A.E., Latif, A.A., and Thompson, B.C. (2013). Influence of β-linkages on the morphology and performance of DArP P3HT-PC61BM solar cells. *Nanotechnology.* 25: 014005.

123 Zhang, X.J., Gao, Y., Li, S.D. et al. (2014). Synthesis of poly(5,6-difluoro-2,1,3-benzothiadiazole-alt-9,9-dioctyl-fluorene) via direct arylation polycondensation. *J. Polym. Sci., Part A: Polym. Chem.* 52: 2367–2374.

124 Pouliot, J.R., Grenier, F., Blaskovits, J.T. et al. (2016). Direct (hetero) arylation polymerization: simplicity for conjugated polymer synthesis. *Chem. Rev.* 116: 14225–14274.

125 Bura, T., Blaskovits, J.T., and Leclerc, M. (2016). Direct (hetero) arylation polymerization: trends and perspectives. *J. Am. Chem. Soc.* 138: 10056–10071.

126 Mercier, L.G. and Leclerc, M. (2013). Direct (hetero) arylation: a new tool for polymer chemists. *Accounts. Chem. Res.* 46: 1597–1605.

127 Fujinami, Y., Kuwabara, J., Lu, W. et al. (2011). Synthesis of thiophene-and bithiophene-based alternating copolymers via Pd-catalyzed direct C-H arylation. *ACS Macro. Lett.* 1: 67–70.

128 Kuwabara, J., Nohara, Y., Choi, S.J. et al. (2013). Direct arylation polycondensation for the synthesis of bithiophene-based alternating copolymers. *Polym. Chem.* 4: 947–953.

129 Berrouard, P., Najari, A., Pron, A. et al. (2012). Synthesis of 5-Alkyl[3,4-*c*] thienopyrrole-4,6-dione-based polymers by direct heteroarylation. *Angew Chem. Int. Ed.* 51: 2068–2071.

130 Grenier, F., Berrouard, P., Pouliot, J.R. et al. (2013). Synthesis of new n-type isoindigo copolymers. *Polym. Chem.* 4: 1836–1841.

131 Saito, H., Kuwabara, J., and Kanbara, T. (2015). Facile synthesis of fluorene-based π-conjugated polymers via sequential bromination/direct arylation polycondensation. *J. Polym. Sci., Part A: Polym. Chem.* 53: 2198–2201.

132 Hayashi, S. and Koizumi, T. (2015). Chloride-promoted Pd-catalyzed direct C-H arylation for highly efficient phosphine-free synthesis of π-conjugated polymers. *Polym. Chem.* 6: 5036–5039.

133 Li, W. and Michinobu, T. (2016). Structural effects of dibromocarbazoles on direct arylation polycondensation with 3,4-ethylenedioxythiophene. *Polym. Chem.* 7: 3165–3171.

134 Yamazaki, K., Kuwabara, J., and Kanbara, T. (2013). Detailed optimization of polycondensation reaction via direct C-H arylation of ethylenedioxythiophene. *Macromol. Rapid Commun.* 34: 69–73.

135 Kuwabara, J.P., Yamazaki, K., Yamagata, T. et al. (2014). The effect of solvent in direct arylation polycondensation of substituted thiophenes. *Polym. Chem.* 6: 891–895.

136 Sun, M.M., Wang, W., Lu, Y.L. et al. (2015). Substituent effects on direct arylation polycondensation and optical properties of alternating fluorene-thiophene copolymers. *Chin. J. Polym. Sci.* 33: 783–791.

137 Wang, X.C. and Wang, M.F. (2014). Synthesis of donor-acceptor conjugated polymers based on benzo[1,2-*b*:4,5-*b*′]dithiophene and 2,1,3-benzothiadiazole via direct arylation polycondensation: towards efficient C-H activation in nonpolar solvents. *Polym. Chem.* 5: 5784–5792.

138 Chang, S.W., Waters, H., Kettle, J. et al. (2012). Pd-catalysed direct arylation polymerisation for synthesis of low-bandgap conjugated polymers and photovoltaic performance. *Macromol. Rapid Commun.* 33: 1927–1932.

139 Gao, Y., Zhang, X., Tian, H. et al. (2015). High mobility ambipolar diketopyrrolopyrrole-based conjugated polymer synthesized via direct arylation polycondensation. *Adv. Mater.* 27: 6753–6759.

140 Gao, Y., Deng, Y.F., Tian, H.K. et al. (2017). Multifluorination toward high-mobility ambipolar and unipolar n-type donor-acceptor conjugated polymers based on isoindigo. *Adv. Mater.* 29: 1606217.

141 Gao, Y., Bai, J.H., Sui, Y. et al. (2018). High mobility ambipolar diketopyrrolopyrrole-based conjugated polymers synthesized via direct arylation polycondensation: influence of thiophene moieties and side chains. *Macromolecules* 51: 8752–8760.

142 Hutchison, J. and Ledwith, A. (1974). Photoinitiation of vinyl polymerization by aromatic carbonyl compounds. *Advances in Polym. Sci.* 14: 49–84.

143 Moszner, N. and Salz, U. (2001). New developments of polymeric dental composites. *Prog. Polym. Sci.* 26: 535–576.

144 Moszner, N. and Salz, U. (2007). Recent developments of new components for dental adhesives and composites. *Macromol. Mater. Eng.* 292: 245–271.

145 Iyoda, T., Kitano, M., and Shimidzu, T. (1991). New method for preparing poly (benzo[c]thiophene) thin films by photopolymerization. *J. Chem. Soc., Chem. Commun.* 15: 1618–1619.

146 Fujitsuka, M., Sato, T., Segawa, H., and Shimidzu, T. (1995). Photochemical polymerization of oligothiophenes. *Chem. Lett.* 69: 99–100.

147 Fujitsuka, M., Sato, T., Watanabe, A. et al. (1996). Photochemical polymerization of dithienothiophenes. *Chem. Lett.* 69: 285–286.

148 Fujitsuka, M., Sato, T., Segawa, H., and Shimidzu, T. (1995). Photochemical polymerization of oligothiophene and dithienothiophene. Author links open overlay panel. *Synthetic Met.* 69: 309–310.

149 Piletsky, S.A., Piletska, E.V., Karim, K. et al. (2004). Photochemical polymerization of thiophene derivatives in aqueous solution. *Chem. Commun.* 19: 2222–2223.

150 Crivello, J.V. (1984). Cationic polymerization-Iodonium and sulfonium salt photoinitiators. *Adv. Polym. Sci.* 62: 1–48.

151 Rabek, J.F., Lucki, J., Zuber, M. et al. (1992). Polymerization of pyrrole by cationic photoinitiators. *Polymer.* 33: 4838–4844.

152 Yagci, Y., Yilmaz, F., Kiralp, S., and Toppare, L. (2005). Photoinduced polymerization of thiophene using iodonium salt. *Macromol. Chem. Phys.* 206: 1178–1182.

153 Yagci, Y., Jockusch, S., and Turro, N.J. (2007). Mechanism of photoinduced step polymerization of thiophene by onium salts: reactions of phenyliodinium and diphenylsulfinium radical cations with thiophene. *Macromolecules.* 40: 4481–4485.

154 Yagci, Y., Kornowski, A., and Schnabel, W. (1992). N-Alkoxy-pyridinium and N-alkoxy-quinolinium salts as initiators for cationic photopolymerizations. *Polym. Chem.* 30: 1987–1991.

155 Ohyama, J., Hitomi, Y., Higuci, Y. et al. (2008). One-phase synthesis of small gold nanoparticles coated by a horizontal orphyrin monolayer. *Chem. Commum.* 47: 6300–6302.

156 Jonas, F. and Schrader, L. (1991). Conductive modifications of polymers with polypyrroles and polythiophenes. *Synth. Met.* 41: 831–836.

157 Groenendaal, L.B., Jonas, F., Freitag, D. et al. (2000). Poly(3,4-ethylenedioxythiophene) and its derivatives: past, present, and future. *Adv. Mater.* 12: 481–494.

158 Grpenendaal, L.B., Zotti, G., Aubert, P.H. et al. (2003). Electrochemistry of poly(3,4-alkylenedioxythiophene) derivatives. *Adv. Mater.* 15: 855–879.

159 Meng, H., Perepichka, D.F., Bendikov, M. et al. (2003). Solid-state synthesis of a conducting polythiophene via an unprecedented heterocyclic coupling reaction. *J. Am. Chem. Soc.* 125: 15151–15162.

160 Yamamoto, T., Shiraishi, K., Abla, M. et al. (2002). Neutral poly(3,4-ethylenedioxythiophene-2,5-diyl)s: preparation by organometallic polycodensation and their unique p-doping behavior. *Polymer.* 43: 711–719.

161 Tran-Van, F., Garreau, S., Louarn, G. et al. (2001). A fully undoped oligo(3,4-ethylenndioxythiophene): spectroscopic properties. *Synth. Met.* 119: 381–382.

162 Yamamoto, T. and Abla, M. (1999). Synthesis of non-doped poly(3,4-ethylene-dioxythiophene) and its spectroscopic data. *Synth. Met.* 100: 237–239.

163 Tran-Van, F., Garreau, S., Lousrn, G. et al. (2001). Fully undoped and soluble oligo(3,4-ethylenedioxythiophene)s: spectroscopic study and electrochemical characterization. *J. Mater. Chem.* 11: 1378–1382.

164 Koh, J.K., Kim, J., Kim, B. et al. (2011). Highly efficient, iodine-free dye-sensitized solar cells with solid-state synthesis of conducting polymers. *Adv. Mater.* 23: 1641–1646.

165 Chen, L., Jin, J.P., Shu, X., and Xia, J.B. (2014). Solid state synthesis of poly(3,4-ethylenedioxythiophene) as counter electrode for dye-sensitized solar cell. *Journal of Power Sources.* 248: 1234–1240.

166 Yin, X., Wu, F., Fu, N.Q. et al. (2013). Facile synthesis of poly(3,4-ethylenedioxythiophene) film via solid-state polymerization as high-performance Pt-free counter electrodes for plastic dye-sensitized solar cells. *ACS Appl. Mater. Interfaces.* 5: 8423–8429.

167 Baughman, R.H. (1974). Solid-state synthesis of large polymer single crystals. *J. Polym. Sci.: Polym. Phys. Ed.* 12: 1511–1535.

168 Enkelmann, V., Schleier, G., Wegner, G. et al. (1977). Structure, luminescence and polymerization of a diacetylene crystal: 1,6-di-(*n*-carbazolyl)-2,4-hexadiin (DCH). *Chem. Phys. Lett.* 52: 314–319.

169 Yee, C.K. and Chance, R.R. (1978). Synthesis and properties of a new polydi-acetylene: Poly[1,6-di(*N*-carbazolyl)-2,4-hexadiyne]. *J. Polym. Sci.: Polym. Phys. Ed.* 16: 431–441.

170 Cohen, M.J., Garito, A.F., Heeger, A.J. et al. (1976). Solid state polymerization of sulfur nitride (S_2N_2) to $(SN)_x$. *J. Am. Chem. Soc.* 98: 3844–3848.

171 Walatka, V.V. Jr., Labes, M.M., and Perlstein, J.H. (1973). Polysulfur nitride—a one-dimensional chain with a metallic ground state. *Phys. Rev. Lett.* 31: 1139.

172 Spencer, H.J., Berridge, R., Crouch, D.J. et al. (2003). Further evidence for spontaneous solid-state polymerisation reactions in 2,5-dibromothiophene derivatives. *J. Mater. Chem.* 13: 2075–2077.

173 Patra, A., Wijsboom, Y.H., Zade, S.S. et al. (2008). Poly(3,4-ethylenedioxyselenophene). *J. Am. Chem. Soc.* 130: 6734–6736.

174 Chen, S.A., Xu, J.K., Lu, B.Y. et al. (2011). Solid-state polymerization of 2,5-dibromothiophene derivatives and properties of corresponding conducting polymers. *Advanced Materials Research Vols* 239: 924–929.

175 Chen, S., Lu, B.Y., Duan, X.M., and Xu, J.K. (2012). Systematic study on chemical oxidative and solid-state polymerization of poly (3,4-ethylenedithiathiophene). *J. Polym. Sci. Part A: Polym. Chem.* 50: 1967–1978.

176 Tusy, C., Peng, K., Huang, L.L., and Xia, J.B. (2015). Effect of flexible linker length in 3,4-ethylenedioxythiophene derivatives for solid state polymerization. *RSC Advances* 5: 16292–16301.

177 Tusy, C., Huang, L.L., Peng, K., and Xia, J.B. (2014). Parallel design strategy and rational study of crystal engineering of novel 3,4-ethylenedioxythiophene derivatives for solid state polymerization. *RSC Advances* 4: 29032–29041.

178 Yin, Y.H., Li, Z.X., Jin, J.P. et al. (2013). Facile synthesis of poly (3,4-ethylenedioxythiophene) by acid-assisted polycondensation of 5-bromo-2,3-dihydro-thieno[3,4-*b*][1,4]dioxine. *Synthetic Metals.* 175: 97–102.

179 Bonillo, B. and Swager, T.M. (2012). Chain-growth polymerization of 2-chlorothiophenes promoted by Lewis acids. *J. Am. Chem. Soc.* 134: 18916–18919.

180 Balasubramanian, A., Ku, T.C., Shih, H.P. et al. (2014). Chain-growth cationic polymerization of 2-halogenated thiophenes promoted by Brønsted acids. *Polym. Chem.* 5: 5928–5941.

181 Tsumura, A., Koezuka, H., and Ando, T. (1986). Macromolecular electronic device: field effect transistor with a polythiophene thin film. *Appl. Phys. Lett.* 49: 1210–1212.

182 Bronstein, H., Chen, Z.Y., Ashraf, R.S. et al. (2011). Thieno[3,2-*b*]thiophene-diketopyrrolopyrrole-containing polymers for high-performance organic field-effect transistors and organic photovoltaic devices. *J. Am. Chem. Soc.* 133: 3272–3275.

183 Assadi, A., Svensson, C., Willander, M., and Inganas, O. (1988). Field-effect mobility of poly(3-hexylthiophene). *Appl. Phys. Lett.* 53: 195–197.

184 McCullough, R.D., Tristram-Nagle, S., Williams, S.P. et al. (1993). Self-orienting head-to-tail poly(3-alkvlthiophcnes): new insights on structure-property relationships in conducting polymers. *J. Am. Chem. Soc.* 115: 4910–4911.

185 Bao, Z.N., Dodabalapur, A., and Lovinger, A.J. (1996). Soluble and processable regioregular poly(3-hexylthiophene) for thin film field-effect transistor applications with high mobility. *Appl. Phys. Lett.* 69: 4108–4110.

186 Ong, B.S., Wu, Y.L., Liu, P., and Gardner, S. (2004). High-performance semiconducting polythiophenes for organic thin-film transistors. *J. Am. Chem. Soc.* 126: 3378–3379.

187 Yi, K.Y., Chen, X.S., Jin, Z.P. et al. (2019). A two-dimensional cross-linked polythiophene network. *J. Mater. Chem. C.* 7: 9362–9368.

188 Mehmood, U., Al-Ahmed, A., and Hussein, I.A. (2016). Review on recent advances in polythiophene based photovoltaic devices. *Renew. Sust. Energ. Rev.* 57: 550–561.

189 Thompson, B.C. and Fréchet, J.M.J. (2008). Polymer-fullerene composite solar cells. *Angew. Chem. Int. Ed.* 47: 58–77.

190 Hotta, S., Soga, M., and Sonoda, N. (1988). Novel organosynthetic routes to polythiophene and its derivatives. *Synthetic. Met.* 26: 267–279.

191 Hou, J.H., Chen, T.L., Zhang, S.Q. et al. (2009). An easy and effective method to modulate molecular energy level of poly(3-alkylthiophene) for high-V_{oc} polymer solar cells. *Macromolecules.* 42: 9217–9219.

192 Zhang, M.J., Guo, X., Yang, Y. et al. (2011). Downwards tuning the HOMO level of polythiophene by carboxylate substitution for high open-circuit-voltage polymer solar cells. *Polym. Chem.* 2: 2900–2906.

193 Huo, L.J., Zhou, Y., and Li, Y.F. (2009). Alkylthio-substituted polythiophene: absorption and photovoltaic properties. *Macromol. Rapid Commun.* 30: 925–931.

194 Koppe, B.M., Scharber, M., Brabec, C. et al. (2007). Polyterthiophenes as donors for polymer solar cells. *Adv. Funct. Mater.* 17: 1371–1376.

195 Salehi, A., Fu, X.Y., Shin, D.H., and So, F. (2019). Recent advances in OLED optical design. *Adv. Funct. Mater.* 29: 1808803.

196 Dyer, A.L., Thompson, E.J., and Reynolds, J.R. (2011). Completing the color palette with spray-processable polymer electrochromics. *ACS Appl. Mater. Interfaces* 3: 1787–1795.

197 Wang, F., Wilson, M.S., and Rauh, R.D. (2000). Electrochromic linear and star branched poly(3,4-ethylenedioxythiophene-didodecyloxybenzene) polymers. *Macromolecules* 33: 2083–2091.

198 Argun, A.A., Aubert, P.H., Thompson, B.C. et al. (2004). Multicolored electrochromism in polymers: structures and devices. *Chem. Mater.* 16: 4401–4412.

199 Sapp, S. and Luebben, S. (2006). Work function and implications of doped poly(3,4-ethylenedioxythiophene)-*co*-poly(ethylene glycol). *Appl. Phys. Lett.* 88: 152107.

200 Lin, K., Kumar, R.S., Cheng, L.S., and Jin, C.S. (2009). Performance improvement of carbon-nanotube-incorporated transparent conducting anode film for organic device application. *IEEE. T. Electron. Dev.* 56: 31–37.

201 Vashchenko, A.A., Vitukhnovsky, A.G., Taidakov, I.V. et al. (2014). Organic light emitting devices with multi-shell quantum dots connected with polythiophene derivatives. *Semiconductors* 48: 377–380.

202 McQuade, D.T., Pullen, A.E., and Swager, T.M. (2000). Conjugated polymer-based chemical sensors. *Chem. Rev.* 100: 2537–2574.

203 Heeger, P.S. and Heeger, A.J. (1999). Making sense of polymer-based biosensors. *Proc. Natl. Acad. Sci. U. S. A.* 96: 12219–12221.

204 Zhang, C., Ji, J.K., Shi, X.Y. et al. (2018). Synthesis of structurally defined cationic polythiophenes for DNA binding and gene delivery. *ACS Appl. Mater. Interfaces.* 10: 4519–4529.

205 Faid, K. and Leclerc, M. (1998). Responsive supramolecular polythiophene assemblies. *J. Am. Chem. Soc.* 120: 5274–5278.

206 Nilsson, K.P.R., Herland, A., Hammarstrom, P., and Inganas, O. (2005). Conjugated polyelectrolytes: conformation-sensitive optical probes for detection of amyloid fibril formation. *Biochemistry.* 44: 3718–3724.

207 Ho, H.A., Boissinot, M., Bergeron, M.G. et al. (2002). Colorimetric and fluorometric detection of nucleic acids using cationic polythiophene derivatives. *Angew. Chem. Int. Ed.* 41: 1548–1551.

208 Li, C., Numata, M., Bae, A. et al. (2005). Self-assembly of supramolecular chiral insulated molecular wire. *J. Am. Chem. Soc.* 127: 4548–4549.

209 Nilsson, K.P.R. and Inganas, O. (2004). Optical emission of a conjugated polyelectrolyte: calcium-induced conformational changes in calmodulin and calmodulin-calcineurin interactions. *Macromolecules* 37: 9109–9113.

210 Houston, J.E., Kraft, M., Mooney, I. et al. (2016). Charge-mediated localization of conjugated polythiophenes in Zwitterionic model cell membranes. *Langmuir* 32: 8141–8153.

211 Martin, R.E., Geneste, F., and Holmes, A.B. (2000). Synthesis of conjugated polymers for application in light-emitting diodes (PLEDs). *C. R. Acad. Sci., Ser. IV: Phys., Astrophys.* 1: 447–470.

212 Liu, C.L. and Chen, W.C. (2011). Donor-acceptor polymers for advanced memory device applications. *Polym. Chem.* 2: 2169–2174.

213 Feng, F., Liu, L., Yang, Q., and Wang, S. (2010). Water-soluble conjugated polymers for fluorescent-enzyme assays. *Macromol. Rapid Commun.* 31: 1405–1421.

214 Vashchenko, A.A., Vitukhnovsky, A.G., Taidakov, I.V. et al. (2014). Organic light emitting devices with multi shell quantum dots connected with polythiophene derivatives. *Semiconductors* 48: 392–395.

215 Camurlu, P., Giovanella, U., Bolognesi, A. et al. (2009). Polythiophene-polyoxyethylene copolymer in polyfluorene-based polymer blends for light-emitting devices. *Synthetic. Met.* 159: 41–44.

216 Wang, X.J., Andersson, M.R., Thompson, M.E., and Inganas, O. (2004). Electrophosphorescence from substituted poly(thiophene) doped with iridium or platinum complex. *Thin Solid Films* 468: 226–233.

217 Park, T., Park, C., Kim, B. et al. (2013). Flexible PEDOT electrodes with large thermoelectric power factors to generate electricity by the touch of fingertips. *Energy Environ. Sci.* 6: 788–792.

218 Hiraishi, K., Masuhara, A., Nakanishi, H. et al. (2009). Evaluation of thermoelectric properties of polythiophene films synthesized by electrolytic polymerization. *Jpn. J. Appl. Phys.* 48: 071501.

219 Qian, Y., Wang, T., Qiu, X.Q. et al. (2016). Conductivity enhancement of poly(3,4-ethylenedioxythiophene)/lignosulfonate acid complexes via Pickering emulsion polymerization. *ACS Sustainable Chem. Eng.* 4: 7193–7199.

220 Leon, A.C.C., Pernites, R.B., and Advincula, R.C. (2012). Superhydrophobic colloidally textured polythiophene film as superior anticorrosion coating. *ACS Appl. Mater. Interfaces.* 4: 3169–3176.

221 Gopi, D., Saraswathy, R., Kavitha, L., and Kim, D.K. (2014). Electrochemical synthesis of poly(indole-co-thiophene) on low-nickel stainless steel and its anticorrosive performance in 0.5 molL-1 H_2SO_4. *Int. J. Polym. Sci.* 63: 280–289.

222 Liesa, F., Ocampo, C., Aleman, C. et al. (2006). Application of electrochemically produced and oxidized poly(3,4-ethylenedioxythiophene) as anticorrosive additive for paints: influence of the doping level. *J. Appl. Polym. Sci.* 102: 1592–1599.

9

High Refractive Index Sulfur-Containing Polymers (HRISPs)

Johannes M. Scheiger[1,2] *and Patrick Theato*[1]

[1] *Karlsruhe Institute of Technology (KIT), Institute for Chemical Technology and Polymer Chemistry, Engesser Str. 18, D76131, Karlsruhe, Germany*
[2] *Soft Matter Synthesis Laboratory, Institute for Biological Interfaces III (IBG3), Karlsruhe Institute of Technology (KIT), Hermann-von-Helmholtz-Platz 1, D-76344, Eggenstein-Leopoldshafen, Germany*

9.1 Introduction

Modern technologies rely on the ability to manipulate light. Simple objects such as lenses (e.g. in eyeglasses, telescopes, cameras, lasers), optical fibers (e.g. used for imaging, communication, fiber lasers, sensors), or mirrors (e.g. used in lasers, solar energy harvesting, architecture, analytical instruments) are capable of collimating or dispersing, guiding or reflecting light, respectively (Figure 9.1). More complex optical devices can be constructed from these building blocks to serve applications in science and industry as well as in our daily life. The fabrication of optical devices can be a multi-disciplinary venture and it is up to chemists and material scientists to come up with suitable molecules, compounds, and composites to suit the respective needs of application.

Function follows form is a key principle for optical devices; thus chemists need to keep the aspect of processability in mind when synthesizing new materials for optical applications. Historically, the foundation of optical materials was based on glasses for visible light applications. Later, crystalline materials (mainly alkali halide salts) enabled control over invisible UV and IR light. The demand for cheap, resistive, and lightweight optical materials finally drove the development of polymeric optics [1]. The fundamentals of optics, design strategies, and concepts for polymeric materials used for optical applications are introduced in the following chapters, with a focus on materials and polymers containing sulfur. Further, background information on optical parameters and techniques used for the analysis of polymeric materials is provided.

Sulfur-Containing Polymers: From Synthesis to Functional Materials,
First Edition. Edited by Xing-Hong Zhang and Patrick Theato.

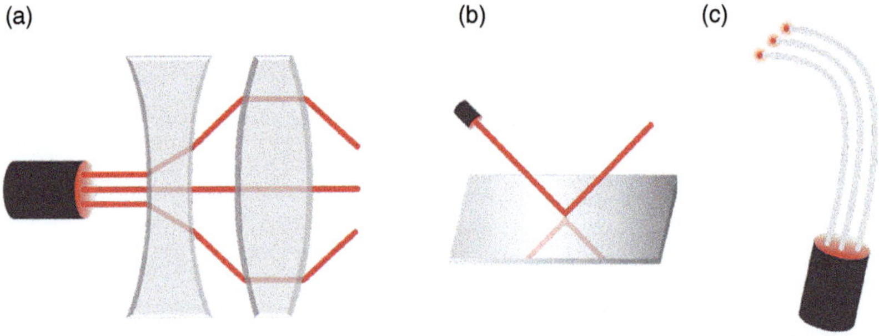

Figure 9.1 Simple optical devices: lenses (a), reflector (b), and waveguides (c) and their influence on light.

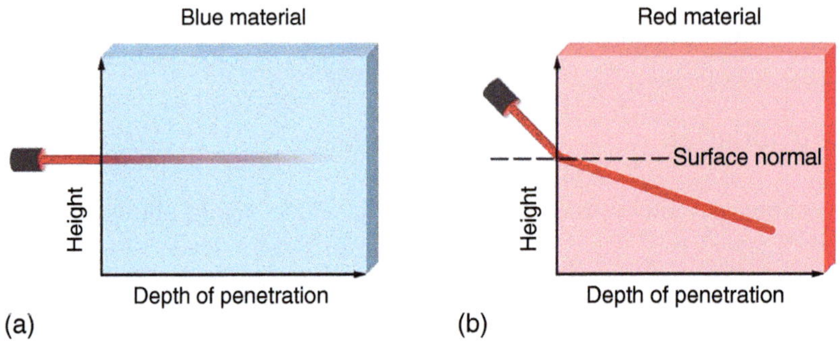

Figure 9.2 Effect of absorption and refraction of red light entering a blue (a) and red (b) material. The blue material is assumed to possess a red-light transmittivity < 100%. The red material is assumed to be 100% transparent for red light and to have a refractive index higher than its surrounding. In (a) the light intensity decreases exponentially with increasing depth of penetration, while in (b) only the direction of the light changes.

9.2 Basics of Optics

9.2.1 Absorption and Refraction

Light can interact with matter in various ways, the most fundamental being absorption and diffraction (Figure 9.2). If the energy of a photon matches with the energy required to trigger a process in a molecule or atom, it can get absorbed. Different photon energies are capable to trigger different processes. Typical examples are transitions of electrons from one orbital to an orbital of higher energy (X-ray, UV–vis) or transitions of vibrational (IR) and rotational (microwaves) states of molecules.

When light passes through an absorbing medium, the intensity of the light beam decreases exponentially according to the Beer law (Eq. (9.1)).

$$I = I_0 e^{-\varepsilon cd} \tag{9.1}$$

Table 9.1 UV cutoff values of selected solvents.

Solvent[a)]	UV cutoff value (nm)
H_2O	190
Acetonitrile	190
Ethanol	210
Ethyl acetate	256
Toluene	284
Cuvette material	**Determination cutoff[b)] (nm)**
Quartz	190
Poly(methyl methacrylate)	320
Polystyrene	340

a) Value given for a cuvette thickness of 1 cm.
b) Values from BrandTech Inc.

In Eq. (9.1) I is the light intensity and I_0 is the initial light intensity in (both in W), ε is the molar extinction coefficient in $\mathrm{l\,mol^{-1}}$, c is the analyte concentration in $\mathrm{mol\,L^{-1}}$, and d is the depth of penetration in cm.

Beer's law relevant in UV–vis, IR, and NIR spectroscopy, where the absorbance of a sample is probed. Dissolved samples are measured in cuvettes against a fully transmitting background. The solvent and the cuvette material must be highly transmissive for light in the spectral range of interest. An indication for the suitability of the cuvette and solvent is given by the so-called (UV-) cutoff value, reflecting the highest wavelength where the intensity of transmitted light is reduced to 10% (Table 9.1). Below 200 nm even air or quartz glass will start to absorb light strongly, rendering this wavelength area unsuitable for convenient use in analysis. For solids (films) a homogenous thickness in the cross section of light transmission must be ensured.

Refraction is a phenomenon occurring when light passes through the interface of two materials with a different refractive index. Light changes its direction towards the surface normal when it is traversing from a medium with a smaller refractive index into a medium with a higher refractive index (Figure 9.1b), thus the incident angle is always smaller than the refracted angle (both relative to the surface normal). This happens when light moves from air into another medium, since the refractive index of air is very small (1.0003) [2]. The process of refraction is described by either geometrical optics or wave optics. Geometrical optics is applicable if the dimensions of objects, e.g. lenses or mirrors, are orders of magnitude bigger than the wavelength. For objects in the size range of the wavelength, refraction effects need to be explained with wave optics [2].

9.2.2 Refractive Index

The refractive index defines the factor by which electromagnetic waves are slowed down in matter relative to the speed of light in vacuum. It is a key parameter in

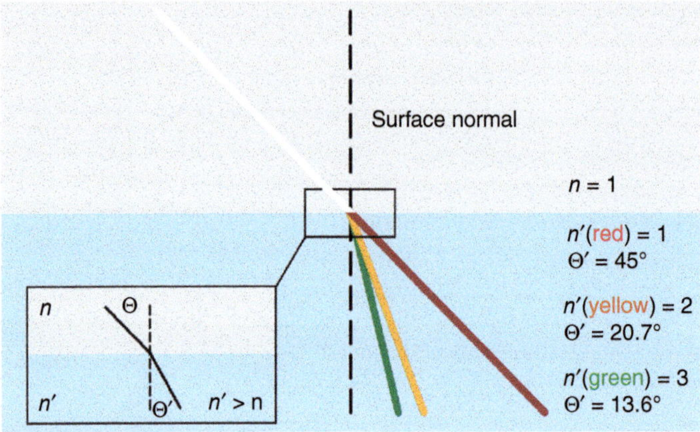

Figure 9.3 Refraction of white light with an incident angle Θ of 45° approaching from vacuum onto a hypothetical material with a refractive index of $n' = 1$ for red light, $n' = 2$ for yellow light, and $n' = 3$ for green light. *Inset*: Definition of incident angle θ and the refraction angle θ'.

optics, since the difference in refractive indices of two materials determines how light is refracted at their interface.

$$n = \frac{c_0}{c} \tag{9.2}$$

In Eq. (9.2), n is the refractive index (dimensionless), c_0 resembles the speed of light in vacuum, and c is the speed of light in a given material. The dependence between the incident angle (θ) and the angle of refracted light (θ') and the refractive index is given by Snell's law [3, 4].

$$\frac{n'}{n} = \frac{\sin(\theta)}{\sin(\theta\prime)} = \text{const.} \tag{9.3}$$

The influence of the refractive index on the refraction behavior is illustrated in Figure 9.3. The higher the refractive index of a material, the stronger light will be refracted, leading to a smaller refraction angle. This is desirable for collecting lenses, since their focal point can then be close to the lense.

The refractive index is dependent on the wavelength, the aggregate state, the temperature, and the pressure. For this reason, the refractive index should be given for a set of defined conditions, but at least a specific wavelength. Often the refractive index is given for the Na-D emission line at 589 nm and is then termed n_D. In fact, there are two emissions lines emitted from sodium vapor lamps, i.e. Na-D_1 (589.592 nm) and Na-D_2 (589.995 nm). Since they are very close to each other, it can be assumed that the refractive index for Na-D_1 and Na-D_2 is equivalent (Table 9.2).

9.2.2.1 Refractive Index Determination

A standard method to determine the refraction index is with an Abbe refractometer, named after its developer Ernst Abbe [5]. There are different variations of refractometers, ranging from simple handheld devices to bench-top instruments

Table 9.2 Refractive index of selected molecules and materials [2].

Material	*n* (589 nm)
Vacuum	1
H_2O (l)	1.333
H_2O (s)	1.310
CO_2 (g)	1.0045
CS_2 (l)	1.628
Quartz glass	1.459
Diamond	2.417

$T = 20\,°C$ and $p = 1013$ mbar.

with autosamplers. The working principle for most devices is the determination of the critical angle, which is the smallest angle of incidence at which total internal reflection occurs. The refracted angle θ' is exactly 90° then, i.e. tangential to the surface and thus cannot leave the material in this plane. This phenomenon can only occur when light travels from a medium with a high refractive index towards a medium with a lower refractive index. For a refraction angle of 90°, Snell's law (Eq. (9.3)) simplifies to Eq. (9.4).

$$\frac{n'}{n} = \sin(\theta) \tag{9.4}$$

A spectator looking at the surface cannot see light with an incident angle equal or bigger than the critical angle. The critical angle is thus observed as the border between an illuminated and a dark part of a measuring prism. As the refractive index of the prism is known, the refractive index of the sample is readily calculated from Eq. (9.4). The light source in a typical Abbe refractometer is a sodium vapor lamp (Na-D line at 589 nm). Modern refractometers can cover a broader range of wavelengths. A drawback is that the refractive index of the measuring prism must be higher than that of the sample, rendering it unsuitable for the determination of very high refractive indices ($n > 1.7$) [6].

For high refractive index polymer (HRIP) films, the prism-coupling method is commonly used [7, 8]. A strength of this method is that the thickness of a film and its refractive index can be determined simultaneously. The polymer film is used as a waveguide for a laser, which is coupled to the polymer film via a prism [9]. Strong laser coupling to the waveguide (sample) occurs only for certain incident angles θ. These angles are then determined by measuring the light intensity as a function of the incident angle. The coupling angles are related to the phase velocity of light in the waveguide (sample) and are thus related to its refractive index, which is then calculated from the dispersion equation of a planar waveguide [8]. Since the prism is in contact with the polymer film, care must be taken not to induce anisotropic optical properties in the film due to mechanical stress.

Another commonly applied method to determine the refractive index is ellipsometry. An advantage in contrast to the prism coupling method is that ellipsometry is a contact free method and films down to the nanometer range can be investigated [10–12]. In ellipsometry, a surface is irradiated with a laser of known polarization and the polarization of the refracted beam is detected. The refractive index can then be calculated from the Airy equation with the experimentally obtained reflection coefficients [13].

Interestingly, the refractive index and the extinction coefficient can be expressed as the real and imaginary parts of the complex refractive index, respectively, and are related by the Kramers–Kronig relation. Thus, the phenomena absorption and reflection are connected by the complex refractive index.

On a molecular level, the refractive index is governed by the polarizability α (cm^3 in cgs units) of a molecule or atom, i.e. its capability to deform its electron charge distribution (electronic structure) in response to an external electromagnetic field (i.e. light) [14]. The link between the microscopic polarizability α and the macroscopic refractive index n is given by the Lorentz–Lorenz equation (Eq. (9.5)) [15]. When expressed in terms of a materials dielectric constant (ε_r) instead of its refractive index, it is termed Clausius–Mossotti relation [16].

$$\frac{n^2 - 1}{n^2 + 2} = \frac{R_M}{V_M} \tag{9.5}$$

In Eq. (9.5), V_M is the molecular volume and R_M is the molar refraction, both in cm^3 (cgs units). The molar refraction R_M contains the mean molecular polarizability α (the average of the polarizability in $x, y,$ and z direction). For a spherical shape (9.6) is valid [17].

$$R_M = \frac{\alpha}{3\varepsilon_r} N_A = \frac{4\pi\alpha}{3} N_A \tag{9.6}$$

In Eq. (9.6) ε_r is the dimensionless dielectric constant (in cgs units) and N_A is the Avogadro constant. Equation (9.5) can be transformed into Eq. (9.7). As evident from Eq. (9.7), the refractive index of a molecule increases with increasing molar refraction R_M and decreasing molar volume V_M. In polymer chemistry, the R_M and V_M values of repeating units can be used to estimate the refractivity of a polymer.

$$n = \sqrt{\frac{1 + 2\left({R_M}/{V_M}\right)}{1 - \left({R_M}/{V_M}\right)}} \tag{9.7}$$

Physicists and chemists in the early nineteenth century tried to break down R_M to the atomic scale by assigning atoms a certain refractivity. The molecular refraction of a molecule is then calculated as the sum of the atomic refractions, which led to drastical errors for many molecules. Soon it was found that atoms like oxygen or sulfur exhibited different R_M values based on their chemical environment (e.g. oxygen in alcohols and carbonyls). Consequently, the molecular refractions of elements in specific functional groups were determined and tabulated (Table 9.3) [23].

The polarizability of an element is proportional to its van der Waals radius ($r_{VdW} \propto \alpha^{1/7}$), which helps to understand and estimate the polarizability of an atom

Table 9.3 Molecular refractivity's of selected atoms and functional groups [18–22].

Atom or group	R_D [a)]
H (CH_2)	1.100
C (CH_2)	2.418
Phenyl	25.463
Naphthyl	43.00
O (alcohol)	1.525
O (ether)	1.643
O (carbonyl)	2.211
F	0.95
Cl	5.967
Br	8.865
I	13.900
S (thiol)	7.69
S (dithia)	8.11
S (thiocarbonyl)	7.97
CS (xanthates)	13.07
SCN (thiocyanates)	13.40
NCS (isothiocyanates)	15.615

a) Measured at the Na *D* line (589 nm).

[24]. In addition, so-called bond refractivities were developed, which correspond to the molar refractivity for a specific bond between two atoms [18, 25]. In terms of accuracy, both the models of atomic refractions and the model of bond refractions perform similarly. However, there are less bond refraction values available. Table 9.3 shows that sulfur and heavy halogen (Br, I) containing bonds or molecules possess a high molar refractivity. This is because of the high polarizabilities α of these elements (e.g. $19.4 \pm 0.1\,cm^3$ for sulfur) [14]. The presence of π-electrons contributes significantly to a high polarizability. The molar refraction of sulfur in thioisocyanates is higher than in thiocyanates, which is due to the higher share of polarizable π-electrons, as expected from the resonance formulas. It is important to mention that while the polarizability of an atom is isotropic, this is mostly not the case for molecules. Usually, the random movement and orientation of molecules average out orientational dependences. For certain crystals and polymers this is not the case, which causes a phenomenon called birefringence (9.2.4).

9.2.3 Dispersion

The refractive index is wavelength dependent. Consequently, some wavelengths are refracted stronger than others, a phenomenon called dispersion. Dispersion is responsible for effects such as rainbows or the "fire" observed for gemstones such as diamonds or zirconia. Dispersive materials are useful to split up light into individual wavelengths, which is required for some spectroscopic techniques. Oppositely, it can cause undesired effects like chromatic aberration in imaging devices such as cameras. This effect is denoted by a difference in focus points for different wavelengths after passing a lens, thus "smearing" the colors along a line instead of a single point. To classify materials according to their dispersity, the Abbe number ν_D was introduced.

$$\nu_\mathrm{D} = \frac{n_\mathrm{D}}{n_\mathrm{F} - n_\mathrm{C}} \tag{9.8}$$

The descriptors D, F, and C relate to the wavelength of the Fraunhofer D (589 nm), F (486 nm), and C (656 nm) line. The Abbe number increases with decreasing dispersion. An evident flaw of the Abbe number is its applicability only for visible light.

There is no generally valid expression for the refractive index n as function of the wavelength λ ($n(\lambda)$). An empirical approach for $n(\lambda)$ is given by the Cauchy (Eq. (9.9)) and later the Sellmeier equation (9.10). The latter can be considered an extension of the Cauchy equation, being more accurate for wavelengths below and above the visible spectrum (UV and NIR). Both expressions do not consider absorption and are thus only valid within the transparent regime of a material.

$$n = B + \sum_{j=1}^{i} \frac{C_j}{\lambda^{2j}} \tag{9.9}$$

The Cauchy equation is typically stopped after the first term. The parameters B (unitless) and C_j (nm^{-2j}) are called Cauchy parameters.

$$n^2(\lambda) = 1 + \sum_{j=1}^{i} \frac{B_j \lambda^2}{\lambda^2 - C_j} \tag{9.10}$$

In Eq. (9.10) the parameters B_j (unitless) and C_j (μm^2) are experimentally determined coefficients. Since C_j is positive, a monotonous increase of the refractive index with decreasing wavelength is observed ($dn/d\lambda < 0$). This behavior is for example observed for polystyrene (Figure 9.4). A generalized Cauchy equation was reported in 2001 [27]. For strongly absorbing materials (metals, semiconductors), other dispersion models are applied [28].

9.2.4 Birefringence

A birefringent material shows two refracted light beams upon refraction. This phenomenon occurs when the polarizability α of molecules or crystals is not isotropic, i.e. when the polarizability in x, y, and z directions are not equal. Consequently, the refractive index depends on the polarization and propagation direction of light. For small molecules in solution, the optical anisotropy is averaged out by their free movement and rotation. Birefringence occurs typically in optically anisotropic crystals and in polymers with restricted orientational isotropy. Crystallinity or the processing history of a polymer can cause molecular orientation and thus induce birefringence.

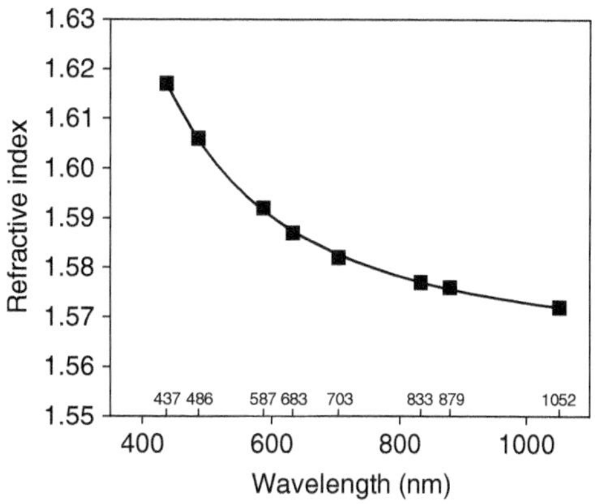

Figure 9.4 Wavelength dependence of the refractive index for atactic polystyrene. Rectangles correspond to experimental values of the refractive index and the line corresponds to a Cauchy fit (IX). The Abbe index ν_D is 30.5. The Cauchy parameters obtained are as follows: $B = 1.5639 \pm 7.3 \times 10^{-4}$, $C_1 = 8.67 \times 10^3 \pm 574$ [λ^{-2}], and $C_2 = 2.86 \times 10^8 \pm 9.2 \times 10^7$ [λ^{-4}]. Source: Based on Ref. [26]. © 2013 Springer Nature.

Polymeric liquid crystals are birefringent, too. Birefringent materials show an ordinary refracted ray (obeying Snell's law) and an extraordinary ray, which is refracted even when the angle of incidence is 0° (disobeying Snell's law). The polarization of the ordinary ray is perpendicular to the polarization of the extraordinary ray. Birefringence can be visualized with polarization microscopes, making it a useful tool to detect mechanically stressed domains in polymeric materials. The birefringence of a material is given as the difference of the ordinary and extraordinary refractive index, n_o and n_e, respectively.

$$\Delta n = n_e - n_o \tag{9.11}$$

The averaged refractive index, n_{av}, is obtained from the refractive indices measured for light of perpendicular polarization (Eq. (9.12)).

$$n_{av} = \left[\frac{(2n_{TE}^2 + n_{TM}^2)}{3}\right]^{\frac{1}{2}} \tag{9.12}$$

In Eq. (9.12) n_{TE} is the in-plane refractive index and n_{TM} is the out-of-plane refractive index. They are determined using linearly polarized laser light parallel (in-plane or transverse electric) and perpendicular (out-of-plane or transverse magnetic) to the polymer surface.

9.3 High Refractive Index Polymers (HRIPs)

Polymers with a refractive index n greater than 1.5 are called HRIPs. Commodity polymers such as polyethylene ($n = 1.476$), polypropylene ($n = 1.474$), or PMMA ($n = 1.489$) at 633 nm, respectively, do not exceed this value due to their lack of polarizable functional groups or atoms. However, polystyrene (1.587) is already an HRIP by this definition [29]. Polymers compete with glasses (e.g. flint glass, crown glass) for applications in visible optics. Glasses are cheap and offer high transparency and high refractive indices ($n \approx 1.45$–1.8 in the visible light spectrum) [30]. Polymers are the preferred materials when integration of multiple features, low weight, flexibility, and high impact resistance are desired. As such, HRIPs are important candidates for the fabrication of optoelectronic devices, substrates for advanced displays [31], optical adhesives or encapsulants for organic light emitting diodes (OLEDs) [32, 33], antireflective coatings [34], immersion lithography [35], or sensors for complementary metal-oxide-semiconductor (CMOS) or charge coupled device (CCD) technology [36].

9.3.1 General Strategies and Applications

Polymer chemists rely mainly on the Lorentz–Lorenz equation to design macromolecules with beneficial optical properties. Usually, a high refractive index, high Abbe number (low dispersion), high transparency, and low birefringence are the desired properties. In order to achieve a high refractive index, a high molar refractivity and low molecular volume are required. The influence of the molar refractivity can be demonstrated for a series of alternating copolymers prepared from equimolar amounts of styrene and an *N*-substituted maleimide (Figure 9.5). The maleimide contains either diphenylether, diphenylthioether, or diphenylselenide substituents [37].

The incorporation of functional groups with high molar refractivity and low molar volume into polymers is thus a key strategy in the synthesis of HRIPs. Frequently, aromatic rings and sulfur or halogen (Cl, Br, I) containing functionalities are utilized. The optical properties of the monomers can already give a hint about the properties of the polymer. An increase in the refractive index is often observed upon polymerization, which is due to an increase in packing density and thus reduction of molar volume. A similar effect can be observed for the influence of the molecular weight on the refractive index. For the selenide compound shown in Figure 9.5, the refractive index increased from 1.648 (633 nm) at 9.300 g mol^{-1} to 1.698 (633 nm) at 27.300 g mol^{-1} [37]. However, it must be noted that except for ideal gases there is no general linear dependence of a materials refractive index and its density [38].

The Kramer–Kronig relation shows that there is an influence of absorption on the refractive index. Absorption bands can thus cause a high fluctuation of the wavelength dependant refractive index from a Cauchy-like behavior, which results in an

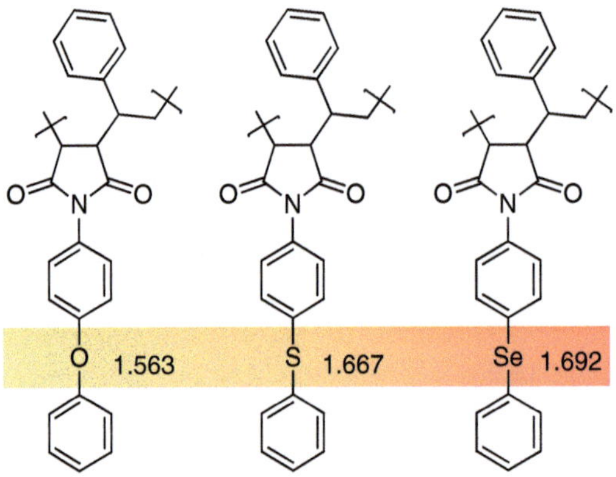

Figure 9.5 Comparison of the refractive indices (633 nm) for diphenylether, diphenylthioether, and diphenyl-selenid N-substituted styrene-*alt*-maleimide (molar ratio 1 : 1) copolymers.

increase of the dispersion. Naturally, functional groups with polarizable functional groups such as aromats or (thio)carbonyls possess conjugated π-systems, whose light absorption can reach from the UV into the Vis region. Thus, many reported HRIPs suffer from high dispersions (low Abbe numbers). To compensate the high dispersion of high refractive index materials, so-called achromats (multi-element lenses) can be employed. As birefringence can be caused by optical anisotropy on the molecular level, it is important that this compensated by a random orientation of molecules in space. Otherwise, as in the case of crystalline polymers or polymers with stress-induced molecular orientation, the materials are birefringent. To tackle birefringence on a synthetical level, chemists can attempt to reduce the optical anisotropy of molecules as well as preventing the formation of crystalline domains. The latter is readily done by the introduction of bulky sidechains or branching, but care must be taken since an increase in molar volume of the polymer chains will reduce the refractive index. Modern modeling tools can assist the synthetical chemist by predicting optical properties of polymers from few input parameters [39].

9.4 Sulfur-Containing HRIPs

9.4.1 Polyimides and Polyamides

The synthesis of sulfur-containing HRIPs based on polyimides was extensively reviewed in 2009 and 2015 by Ueda and coworkers as well as in 2014 by Shaver and coworkers [22, 36, 40]. Sulfur can be incorporated into polyimide backbones as thioether, sulfone, thiophene, thiadiazole, and thianthrene. The preparation procedure involves synthesis of sulfur-containing aromatic diamines and dianhydrides, which are reacted in a two-step polycondensation [41–50]. First, the diamine reacts with the dianhydride at room temperature to form polyamic acid. Heating from 70 to 300 °C then yields the polyimide under elimination of water. Residual water causes premature hydrolysis of the anhydrides and must be removed. To synthesize sulfur-containing aromatic diamines and dianhydrides, aromatic dithiols (e.g. 4,4′-thiobisbenzenethiol) are used. Sulfur-containing diamines were prepared by the reaction of aromatic dithiols with 2 equiv of 4-chloronitrobenzene and subsequent reduction of the nitro groups, while reaction with 2 equiv of 4-bromo-1,8-naphthalic anhydride resulted in the respective anhydride (Figure 9.6).

Refractive indices of sulfur-containing polyimides are mostly above 1.7 in the visible range and tend to rise with increasing sulfur rank (Table 9.4). Consequently, sulfur-rich thiaanthracenes and tetrathiaanthracenes consisting of up to 28.5 wt% sulfur content show the highest refractive indices between ca. 1.76 and 1.77 at 633 nm. Besides the sulfur rank, the type of the thio-functionality influences the refractive index of the polymer. For example, the lower molar refractivitiy and the lower packing density of sulfones compared to thioethers are assumed to be the reason for the lower refractive indices of polyimides containing sulfonyl groups compared to thioether containing polyimides of similar sulfur rank. Due to the flexible nature of thioether and sulfonyl linkages, the naturally high birefringence

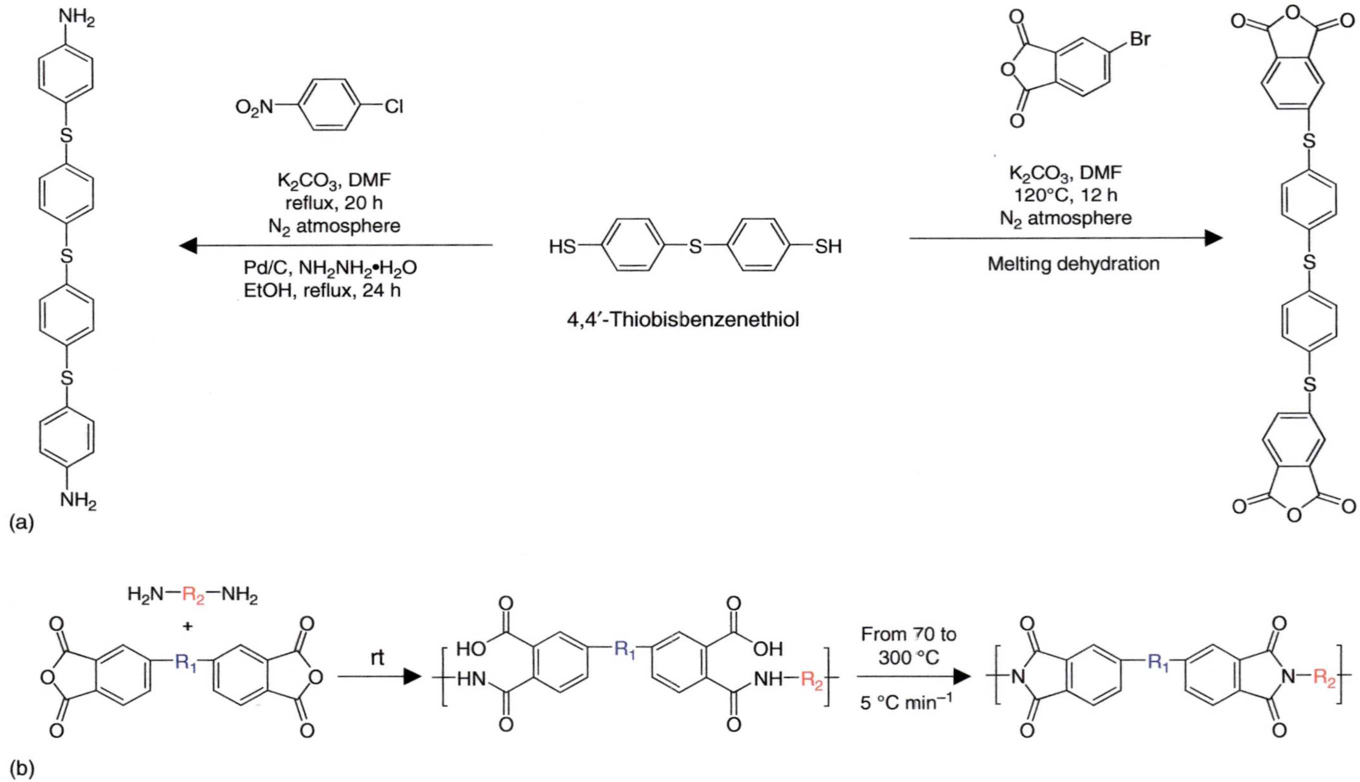

Figure 9.6 Monomer synthesis (a) for the preparation of high refractive index polyimides. Aromatic dithiols are converted to thioether linked diamines and dianhydrides, respectively. Polycondensation (b) of the monomers yields polyamic acids at room temperature and the respective polyimides at elevated temperatures.

Table 9.4 Optical properties of selected sulfur-containing polyimides sorted by descending refractive index.

Polyimide	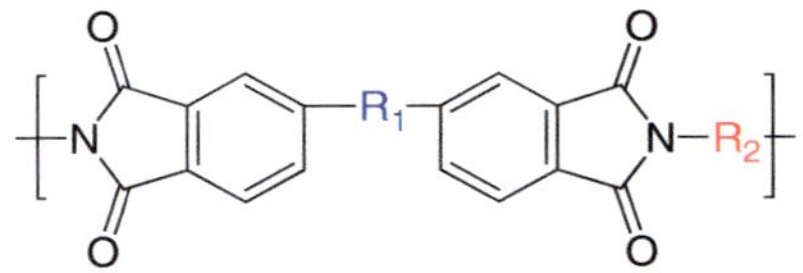			
R_1	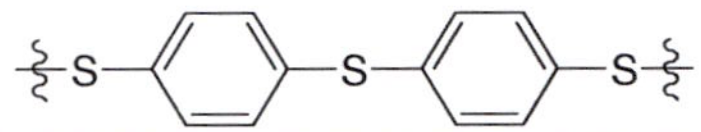			
R_2	**S (wt%)**	***n* (633 nm)**	**Δ*n***	**References**
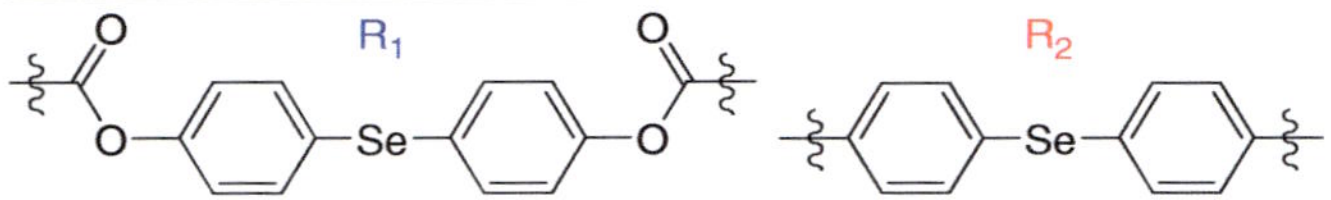	18.7[a)]	1.968[b)]	0.025	[51]
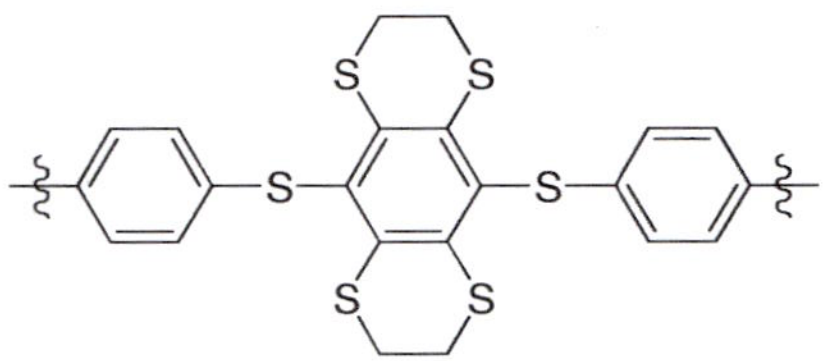	28.5	1.7692	0.0093	[49]
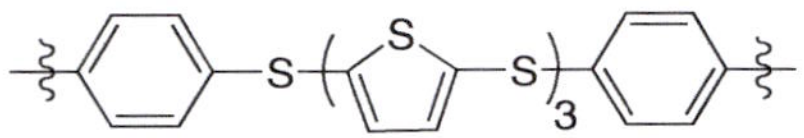	30.1	1.7677	0.0072	[50]

(Continued)

Table 9.4 (Continued)

Polyimide	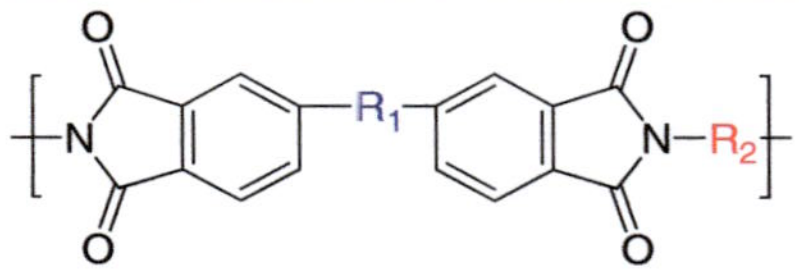			
R_1	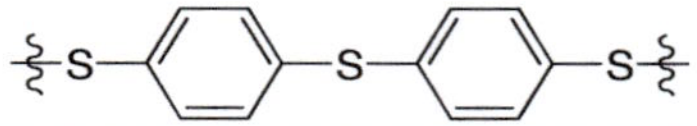			
R_2	S (wt%)	*n* (633 nm)	Δ*n*	References
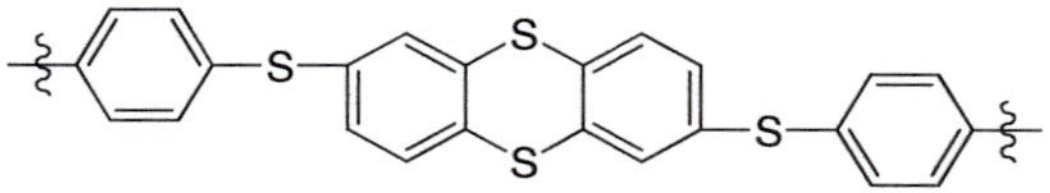	23.2	1.7600	0.0084	[41]
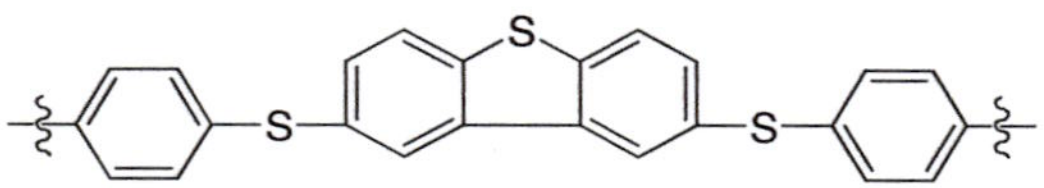	20.5	1.7578	0.0084	[45]
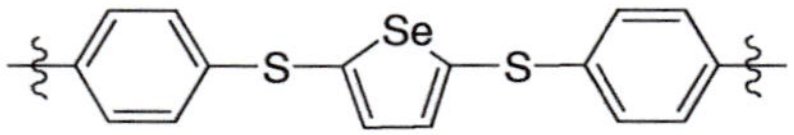 	18.1	1.7594	0.0081	[52]
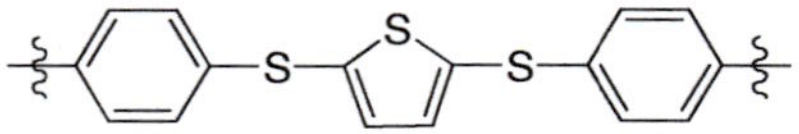	23.0	1.7527	0.0074	[50]

Table 9.4 (Continued)

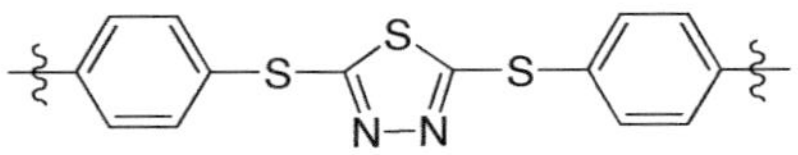	22.9	1.7490	0.0093	[36]
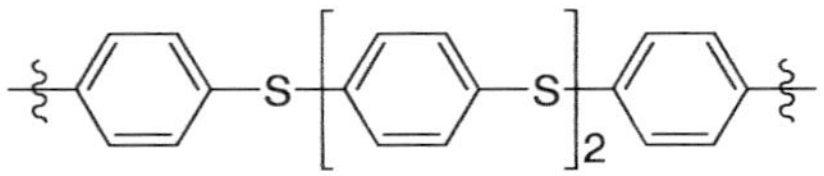	20.5	1.7482	0.0068	[43]
	19.2	1.7461	0.0081	[45]
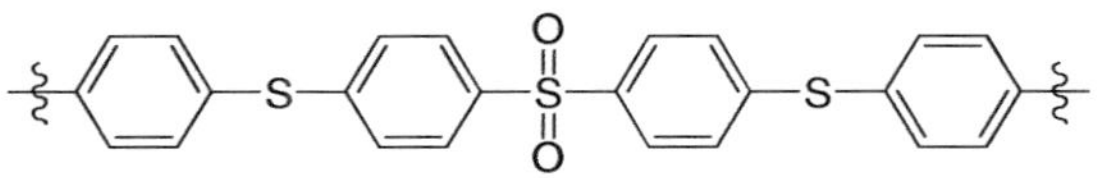	19.8	1.7326	0.0077	[46]
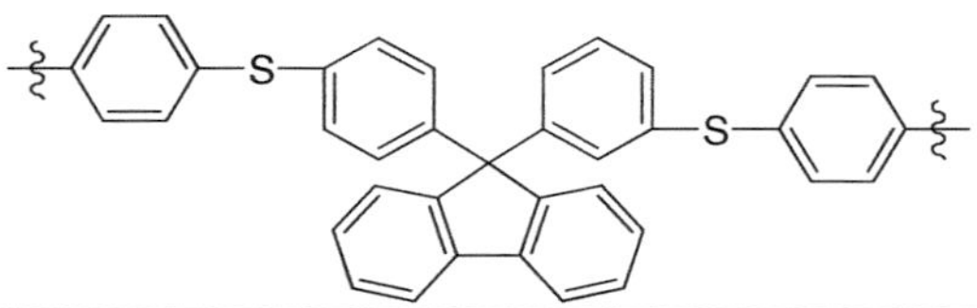	15.0	1.7258	0.0059	[47]

a) Se wt%.
b) Only three decimals reported.

of polyimides can be suppressed. Further, bulky sidechains can be utilized to enhance the flexibility of the polyimide chains. For example, a fluorenyl group was introduced with an orientation perpendicular to the polyimide backbone. The refractive index was only 1.7258 (633 nm) due to the high molecular volume of the fluorenyl group, but the birefringence was the lowest one reported for a sulfur-containing polyimide (Δn = 0.0059). Incorporation of a phosphine oxide into the polyimide backbone along with sulfur did not alter the properties strongly, whereas replacement of sulfur with selenium yielded the polyimide with the highest refractive index reported (1.968 at 633 nm) [51–53]. Refractive indices above 1.75 for sulfur-containing polyimides are only achieved with a high sulfur rank (> 20 wt%) or inorganic fillers (zirconia) [54]. Table 9.4 summarizes sulfur based polyimide HRIPs sorted by their refractive index at 633 nm. Typical visible light transmissions are between 80 and 85 T% (film thickness 10 μm) above 450–500 nm. Their thermomechanical properties are excellent (T_G >200 °C) and the flexible and tough films can withstand temperatures above 400–500 °C (5–10 wt% loss). Nonetheless, there is a clear tradeoff between achieving a high refractive index or a high Abbe number: For sulfur-containing polyimides with a refractive index above 1.7 the Abbe number is typically below 20. Additionally, the lack in transmittance below 450–500 nm (≈ violet–blue light) limits their scope of visible light applications. The use of sulfonyl groups instead of thioethers improves the transparency but lowers the refractive index [41, 46]. A similar effect is observed, when the polyimide backbone is prepared with *meta*- instead of *para*-substituted aromats. The increase in transmission is connected to the decrease in electronic chain–chain interaction, while the refractive index decreases due to the increase in molar volume V_m. To design polyimide HRIPs in the future, computational predetermination of optical properties can be undertaken, leading to more targeted synthesis [55].

Polyamides, the close relatives of polyimides, received less attention as HRIPs. This might be attributed to early findings that some polyamides showed an unusually high birefringence even for fully amorphous polymers [56]. Nonetheless, the reported refractive indices of substituted (e.g. -MeO, -Br, $-CF_3$, $-NO_2$) aromatic polyamides between 1.7 and 1.9 (at 589 nm) seemed promising for applications as HRIPs. The same diamines used for the preparation of polyimides can be used to synthesize sulfur-containing polyamides [57–62]. As carbonyls either carboxylic acid chlorides or aromatic carboxylic acids are used, which can also contain sulfur moieties.

Amongst sulfur-containing polyamides, thioether and thiosulfone containing polyamides have been studied the most, but also disulfide or thiophene containing polyamides were investigated for their optical properties. Polythioamides can be synthesized by direct conversion of diamines with elemental sulfur [63]. Since polyamides are structurally related to polyimides, similar synthetical strategies can be employed to tackle the high birefringence. Introducing bulky (e.g. naphthalene) side groups or flexible (e.g. thioether or sulfanyl) backbones prevents formation of crystalline domains and increases the molecular mobility [64]. As of today, the birefringence of sulfur-containing polyamides can be reduced to ca. 0.004–0.01. The transmission T% at 450 nm typically exceeds 80–90% (thickness 10 μm), with refractive indices above 1.7 and Abbe numbers up to 30 (Table 9.5). In terms of thermal properties, sulfur-containing polyamides and polyimides are similar:The T_G is high (200–250 °C) and decomposition does not occur below 400 °C. An advantage of polyamides over polyimides is the improved solubility in polar aprotic solvents such as *N*-methyl-2-pyrrolidone (NMP), dimethyl sulfoxide (DMSO), *N*,*N*-dimethylacetamide (DMF), and *N*,*N*-dimethylacetamide (DMAC), which is relevant for the casting of polymer films from solution.

9.4.2 Poly(meth)acrylates and Polythioacrylates

Polymethacrylates find widespread applications as windows in construction or aquariums because of their high visible light transparency. The π-electrons of ester bonds require more energy to be excited into the respective π*-orbital than the π-systems of aromats. For this reason, PMMA cuvettes are used instead of PS cuvettes for UV–vis spectroscopy requiring wavelengths below 320 nm. Further, poly(meth)acrylates are attractive for the design of HRIPs due to the variability of functional groups that can be introduced as ester side group. The well-studied thermal and photoinduced radical polymerization of many functional methacrylates as well as the availability of controlled polymerization protocols is beneficial for the fabrication of defined polymers. UV curing can be conveniently used for poly(meth)acrylate synthesis due to the acceptable UV transparency of most (meth)acrylate monomers, allowing efficient photoinitator cleavage [65]. This is particularly beneficial to produce thin films or to spatially control the synthesis of polymers. Different sulfur species have been incorporated into the methacrylate ester group, such as aliphatic or aromatic thioethers, thioacetals, thiourethanes, dithiocarbonates, thianthrenes, and thiophenes (Figure 9.7) [65–71]. For some polymers, sulfur was introduced as a thioacrylate [72, 73].

Sulfur-containing poly(meth)acrylates and polythio(meth)acrylates typically achieve refractive indices between 1.55 and 1.7 in the visible spectrum, which is sufficient to term them HRIPs ($n > 1.5$) [74]. However, the refractive indices are generally lower than those achieved for sulfur-containing polyimides. Yet, high Abbe numbers (around 40) and the resulting low chromatic aberration are a strength of sulfur-containing polymethacrylate HRIPs. A very high Abbe number (48.6) with a refractive index above 1.68 between 486–567 nm was reported for an alicyclic polythioether sulfone [75]. The birefringence and absorption of polymethacrylate HRIPs are also typically lower than for polyimides, which can mainly be attributed to the lower content of aromatic moieties and decreased degree of crystallization (Table 9.6). Another advantage of poly(meth)acrylate HRIPs is that synthesized sulfur-containing (meth)acrylates can readily be copolymerized with other methacrylate monomers to alter the attributes of the final polymeric materials in terms of processability, thermostability, and optical properties [70, 76, 77]. The incorporation of aromatic thioether or thianthrene moieties generates homopolymers with high refractive indexes (Figure 9.7 (v) and (vi), respectively). Data about their dispersion properties is not available, but the birefringence's were found to be low ($\Delta n = 0.0039$ and 0.0069, respectively).

Table 9.5 Sulfur-containing polyamides and polythioamides and their optical properties.

Polymer	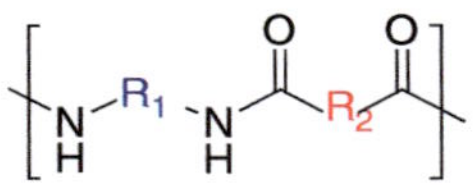	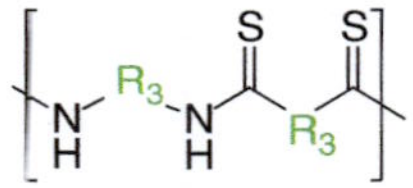				
R₁	**R₂**	**S (wt%)**	***n* (633 nm)**	**Δ*n***	**ν_D**	**References**
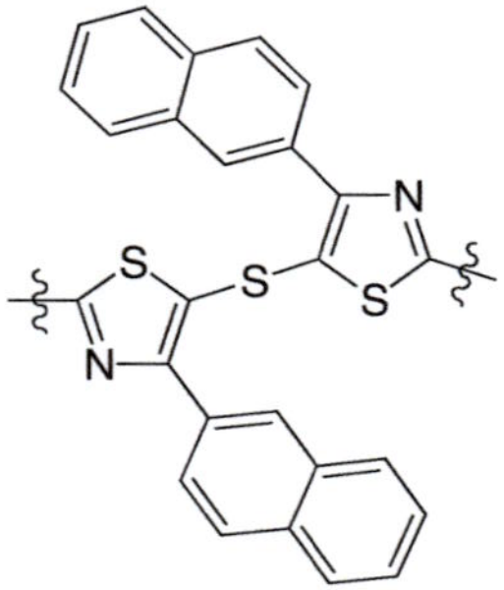	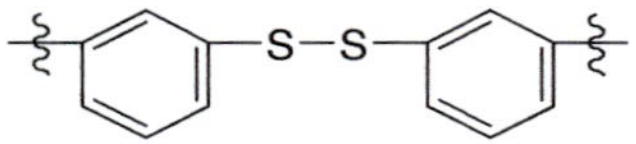	21.3	1.7701	0.005	26.79	[59]
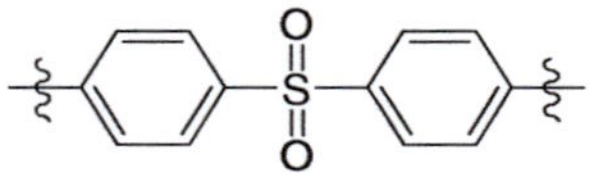	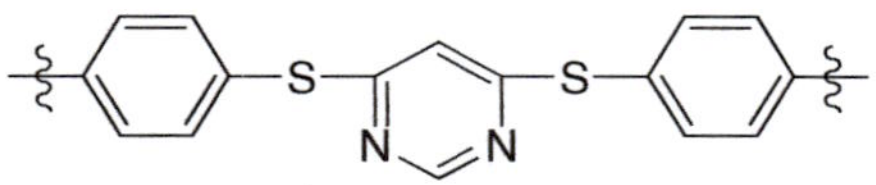	19.7	1.730	0.006	—	[61]
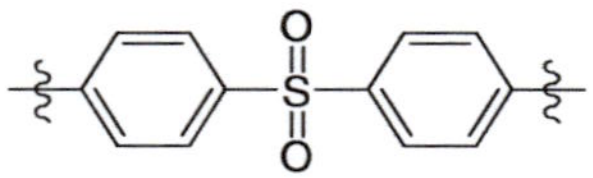	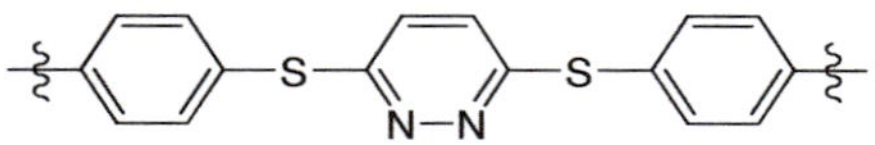	19.7	1.732	0.005	—	[61]
R_1		S (wt%)	*n* (589 nm)	Δ*n*	ν_D	References
		23.7	1.58	—	13.2	[63]
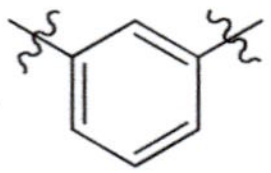		23.7	1.66	—	5.5	[63]

Table 9.6 Sulfur-containing poly(meth)acrylates and polythio(meth)acrylates and their optical properties.

Polymer X = O or S R = org. rest	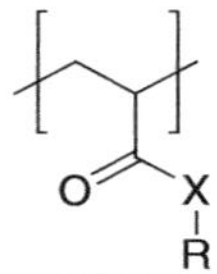						
Polymer	**S (wt%)**	***n* (486 nm)**	***N* (546 nm)**	***n* (589 nm)**	***n* (656 nm)**	ν_D [a)]	**References**
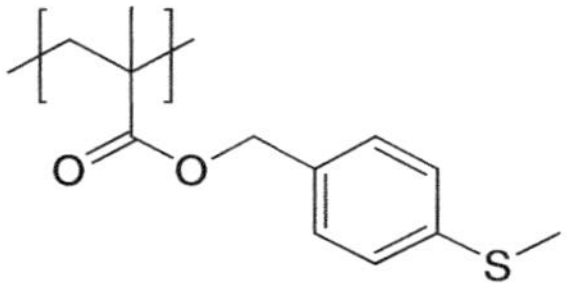	14	1.634	1.628	1.624	1.612	40.9	[67]
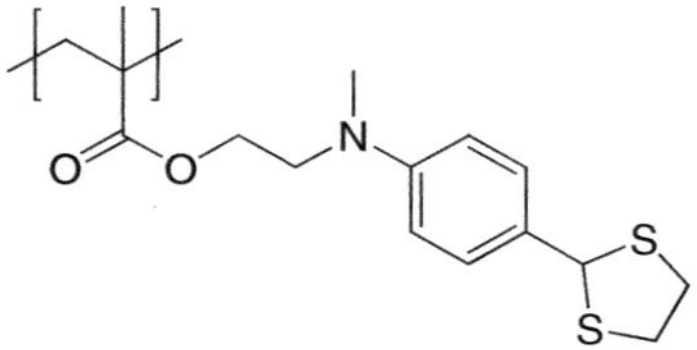	20	1.702	1.692	1.683	1.675	25.4	[67]
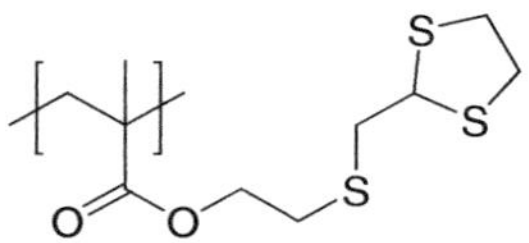	36	1.745	1.731	1.725	1.718	29.0	[67]

(Continued)

Table 9.6 (Continued)

Polymer X = O or S R = org. rest	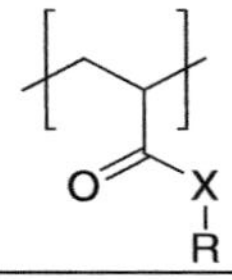						
Polymer	**S (wt%)**	***n* (486 nm)**	***N* (546 nm)**	***n* (589 nm)**	***n* (656 nm)**	ν_D a)	**References**
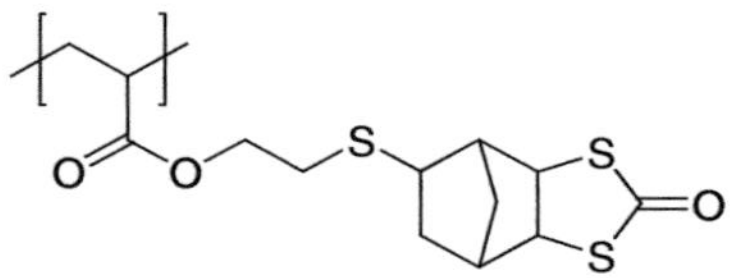	30	1.630	—	1.618	1.615	40.2	[66]
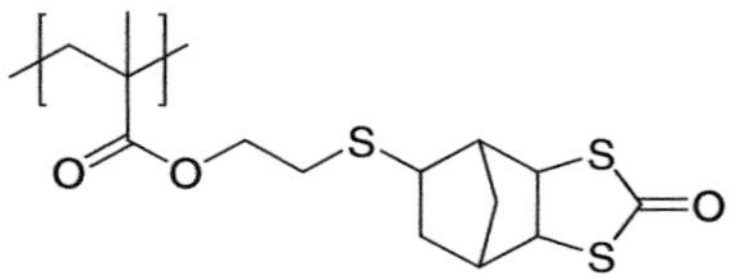	29	1.619	—	1.609	1.606	44.5	[66]
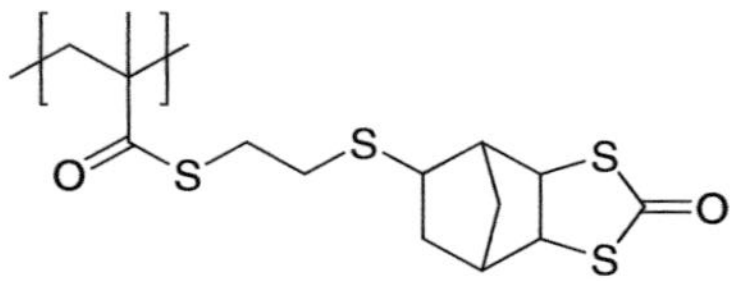	37	1.651	—	1.640	1.636	42.1	[66]

a) Abbe's number (ν_D) was calculated using the refractive indices at 589, 486, and 656 nm.

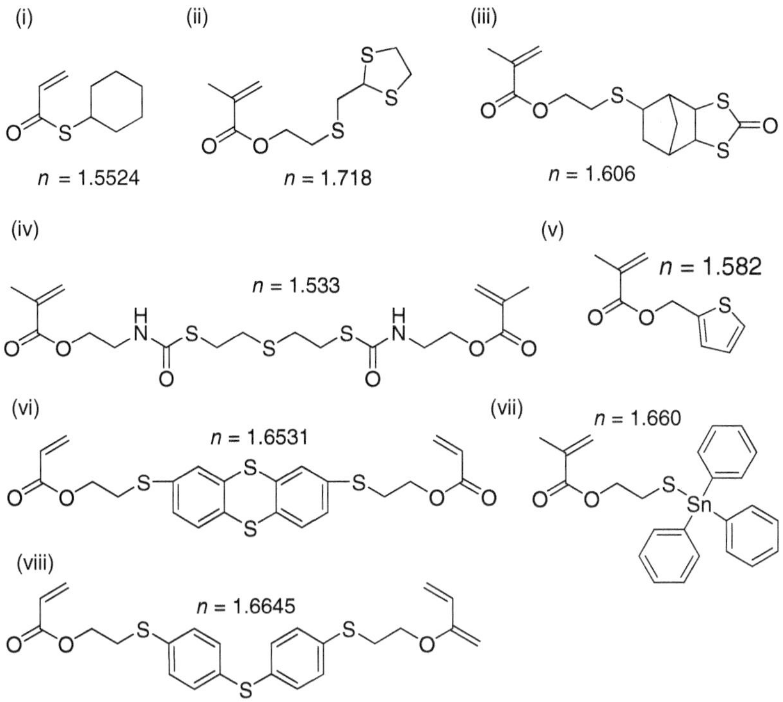

Figure 9.7 Sulfur-containing (meth)acrylates used for the fabrication of HRIPs. Aliphatic thioacrylate (i), thioether and thioacetal (ii), dithiocarbonate (iii), and thiourethane (iv), thiophene (v), thianthrene (vi), thiostannan (vii), and aromatic thioether (viii). Refractive indices are given for homopolymers at 633 nm (vi), 637 nm (i,iv), and 656 nm (ii, iii, vii). For (v) and (viii) no wavelength was specified and for (viii) the value is given for a composite. Source: Refs. [66–71].

9.4.3 Polycarbonates and Polyesters

Since their first systematic description in 1956 by chemists of the Bayer AG, polycarbonates quickly raised interest for optical applications. They possess a high refractive index and transparency and are used for windows or lenses [78]. Due to the higher price of polycarbonates compared with polymethacrylates, they are utilized when a high impact resistance is required (e.g. safety goggles, eyeglasses, and windows). Polycarbonates possess high mechanical, thermal, and chemical stability, however they tend to degrade under UV light and are prone to abrasion, causing turbidity on the surface [79]. To overcome these issues, polycarbonates for outdoors applications are often coated.

Aromatic polycarbonates are synthesized via polycondensation reactions, typically from aromatic diols and phosgene or transesterification of diphenyl carbonate. Their refractive indices are typically between 1.55 and 1.59, a result of their high density and high content of aromatic moieties. Even though polycarbonates are typically amorphous, the birefringence can be high due to the rigidity of the polymer backbone. In the case of the optically anisotropic monomer bisphenol A (2,2-bis(4-hydroxyphenyl)propane), this causes a high birefringence of $\Delta n = 0.106$.

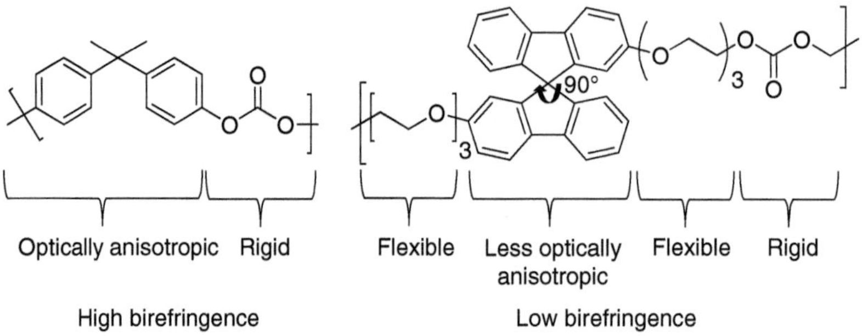

Figure 9.8 Reduction of birefringence by introducing flexible segments and less optically anisotropic diols. Source: Based on Ref. [80].

Table 9.7 Molar refractivity of O–CO–O, O–CO–S, and S–CO–S [74].

Functional group	O–CO–O	O–CO–S	S–CO–S
R_{LL} [a)] ($cm^3\ mol^{-1}$)	7.74	13.63 ± 0.34	24.45 ± 0.61

There are not many polytrithiocarbonates characterized for their optical properties, but they presumably all exhibit a high refractive index.

a) Calculated with the Lorentz–Lorenz equation.

The birefringence can be drastically reduced by increasing the optical isotropy and the amount of flexible bonds in the diol, which was shown for polycarbonates with oligo(ethylene glycole) and 9,9′-spirobifluorene repetition units. Since the bulky fluorene moieties in the spiro-dimer are perpendicular to each other, the optical anisotropy of the repetition unit is reduced and a birefringence of $\Delta n = 0.0001$ could be achieved (Figure 9.8) [80].

Another strategy to decrease the rigidity of the polycarbonate backbone is the use of mono-, di-, or trithiocarbonates [81]. The low polarity of thiocarbonates leads to amorphous polymers [82]. Due to the high polarizability of sulfur, refractive indices higher than 1.6 are readily achieved for polydithiocarbonates throughout the visible spectrum. The contribution of thiocarbonate or dithiocarbonate functional groups to the molar refractivity of the respective poly(di)thiocarbonate were determined for structurally related polymers (Table 9.7).

Aliphatic polytrithiocarbonates are prone to degrade into cyclic trithiocarbonates, which are also a possible side product of polymerization [83]. Another disadvantage is that carbon disulfide has to be used as C=S building block, which is less reactive than phosgene, and thus, a catalyst or activation of the thiol groups is required [84–86]. Typically, polythiocarbonates are formed via interfacial reactions of a dithiol in aqueous NaOH solution and the C_1 building block ($CSCl_2$, $COCl_2$) in an organic phase to yield the respective polydithiocarbonates or polytrithiocarbonates (Figure 9.9a). This is also possible for CS_2 using a phase transfer catalyst [85]. Metal catalyzed reactions of CS_2 with derivatives of ethyleneoxide or -sulfide are performed in bulk or solution (Figure 9.9b,c). In recent years, autoclave reactions

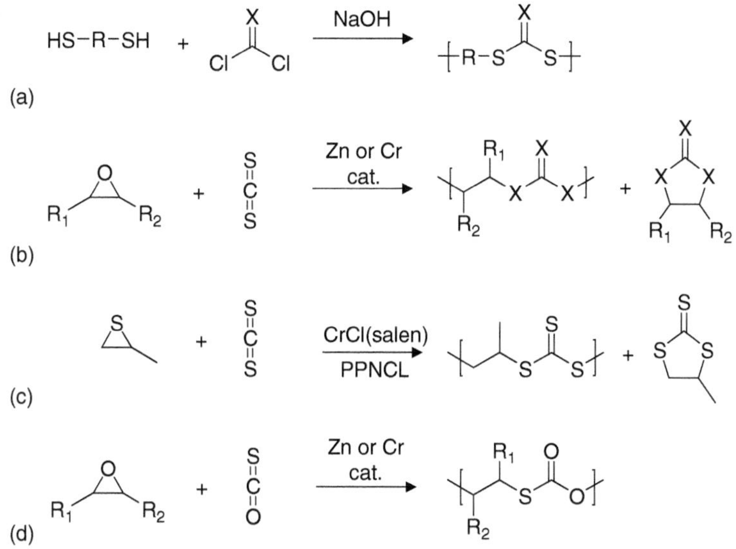

Figure 9.9 Synthesis strategies for polythiocarbonates, introducing sulfur via the nucleophile (a) or via the carbonyl (b–d). In (b) there can be 1–3 sulfur or oxygen atoms in a repeating unit due to O/S exchange reactions. For (d) this is inhibited by using low temperatures. X indicates a sulfur or oxygen atom while R_1 and R_2 denote organic rests. PPNCL (Bis(triphenylphosphoranylidene) ammonium chloride) is a co-catalyst. Source: (d) Refs. [87–90].

of the air pollutant COS as sulfur source together with epoxides produced high refractive index polythiocarbonates (Figure 9.9d) [87–90].

Polyurethanes and polyesters are structurally related to polycarbonates and are relevant in various industrial applications such as foams and bottles. Their sulfur-containing analogues (e.g. polythiourethanes, polythioesters) do not play a role in industry and only few reports about their optical properties exist [91–94]. Polythioesters synthesized with cyclic aliphatic dithiols have an excellent visible light transmission T% of almost 100% above 400 nm and a high Abbe numbers around 30. The optical properties of some polythiocarbonates are shown in Table 9.8.

9.4.4 Thermosets

All HRIPs discussed can in principle be transformed into a thermosetting formulation by adding crosslinkers (radical polymerization) or multifunctional amines, thiols, or alcohols (polycondensation and polyaddition), which can improve the mechanical stability of optical devices [95]. The fabrication process for thermosets is different from solution or melt processable thermoplastics, since they cannot be reshaped after the polymerization. A simple method is cast molding, where a polymer resin is cured in a mold of the intended final shape. Robust reactions in terms of oxygen and moisture sensitivity are required, and the curing conditions should be mild (low temperature). Additional factors, such as the density changes during polymerization, are relevant in order to duplicate the shape of a cast mold. To ensure

Table 9.8 Optical properties of polythiocarbonates.

Polymer	S (wt%)	n_D (589 nm)	ν_D	References
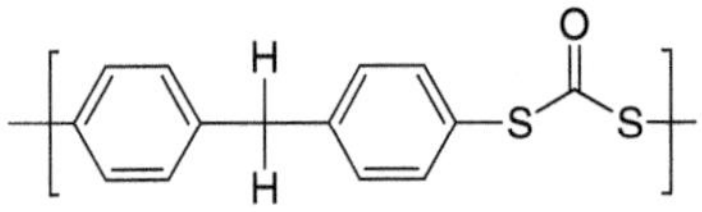	24.8	1.684	—	[74]
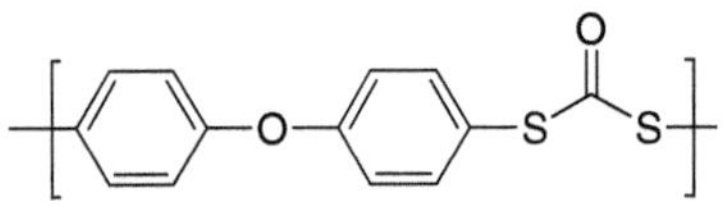	24.7	1.681	—	[74]
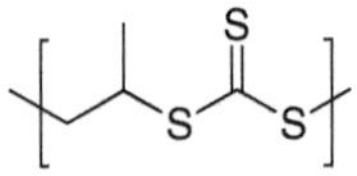	64.0	1.78	—	[83]
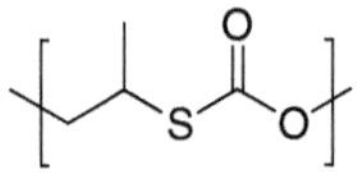	20.3	1.63	15.9	[89]
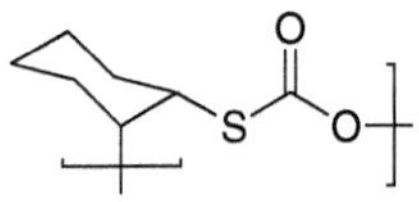	27.1	1.70	32.2	[88]
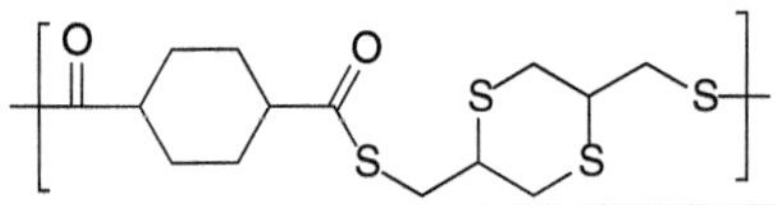	36.6	1.6271	30.0	[94]

that no potentially harmful monomers are left in the final product, the reactions must be fast and quantitative. Reactions to form thermosetting sulfur-containing HRIPs often include thiol chemistry, such as thiol-thiiran ring-opening polymerization, thiol-isocyanate polyaddition, and thiol-ene polyaddition (Figure 9.10) [96–99]. Further, sulfur-containing di(meth)acrylates have been synthesized to produce sulfur containing poly(meth)acrylate thermosets [95].

9.4.5 Inverse Vulcanization

The emergence of the inverse vulcanization process in 2013 initiated a search for potential applications of high sulfur content polymers in which they could be superior to materials such as carbon-based polymers, ceramics, or glasses. Polymers synthesized via the inverse vulcanization process can contain up to 90 wt% of sulfur, which is beneficial to produce materials with a high refractive index. For that reason, polymers synthesized via inverse vulcanization deemed to be suitable candidates for optical applications. A limitation of these materials is their strong absorption of blue light, resulting in a yellow to red color. The light absorption of an optical device should be neglible in its working range; thus, these materials are not ideal for the fabrication of optical devices for the UV or visible light spectrum. The

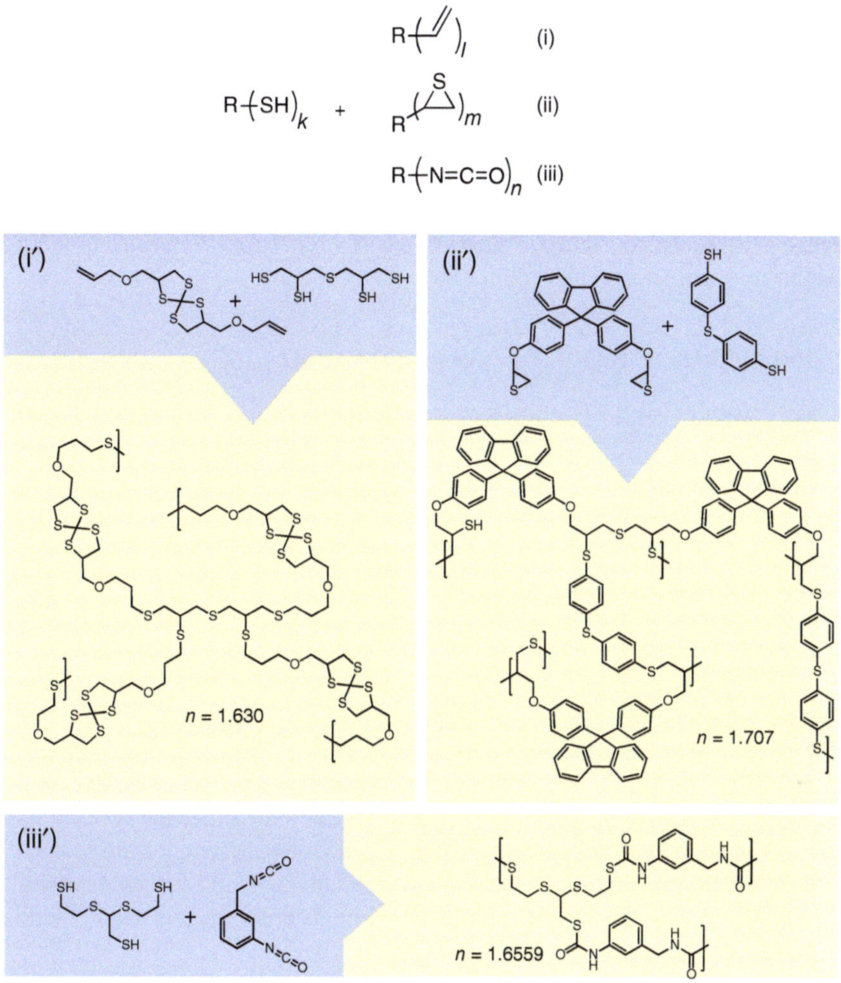

Figure 9.10 Reactions of oligothiols with alkenes (i), thiiranes (ii), and isocyanates (iii) and their respective network structure (i′–iii′). At least two trifunctional (*k* – *n*) and one bifunctional moiety (*k* – *n*) are required to yield a crosslinked polymer via polyaddition. In (ii) it is enough to use two bifunctional moieties ($k = 2$, $m = 2$), since the nucleophilic attack on the thiiran generates a thiol capable of propagating. Source: (i) Refs. [96–99].

research focus for HRIPs prepared via inverse vulcanization thus shifted entirely to the IR spectral range, where they show high transparency in the near-infrared (0.8–2.5 μm or 14 000–4 000 cm^{-1}) and parts of the mid-infrared (3–5 μm) spectrum. These IR applications of polymers prepared via inverse vulcanization have been reviewed recently [100].

When molecules absorb IR light is, they get excited from a vibrational ground state into an excited state. The IR absorbance of a molecule is governed by the change of the dipole moment of a molecule during vibration, whereas the energy required to excite vibrations only depends on the masses and force constants between bound atoms. This is reflected in the selection rules for vibrational transitions and the energy eigenvalues of the harmonic oscillator, a simplified quantum mechanical model for molecular vibration.

$$I \propto R_v^2 \text{ with } R_v = \left(\frac{d\mu}{dz}\right)_{\text{equiv}} \int \Psi_{v\prime}^* z \Psi_v dz \tag{9.13}$$

In Eq. (9.13) the intensity I of a transition from one vibrational state $\Psi_v'^*$ into a higher vibrational state Ψ_v'' is proportional to the square of the transition dipole moment R_v. It is apparent that the dipole moment μ must change during vibration along the bond axis (z-axis) for R_v and I not to become zero.

$$E_v = (v+1)^2 * \hbar\sqrt{\frac{k}{m_{\text{red}}}} \text{ with } m_{\text{red}} = \frac{1}{m_1} + \frac{1}{m_2} \tag{9.14}$$

In Eq. (9.14) the energy of the eigenstates of the harmonic oscillator is given. The energy only depends on the reduced mass of the two vibrating bodies and the force constant between them. In an IR spectrum, S-X (e.g. S-H, S-C) peaks are thus generally not intense and are found at lower wavenumbers than their respective oxygen analogues (e.g. O–H stretching: high intensity at 3550–3200 cm^{-1}, S–H stretching: low intensity at 2600–2550 cm^{-1}). Vibrations between S—S bonds produce peaks with even lower intensity and at very low energies. The S–S vibration in dialkyl compounds is found between 505 and 550 cm^{-1}, values observed for diaryldisulfides such as dibenzyldisulfide (507 cm^{-1}) are also in this range or even lower as in the case of diphenyldisulfide (473 cm^{-1}). For polysulfides (R–S_n–R with $n = 3$–5) a further shift to lower energies is observed (ca. 484–491 cm^{-1}) [101]. Functional groups such as C—H, C—O, C=O, or C≡N produce very intense absorption bands in the mid-infrared range (4000–400 cm^{-1}), which forbids use of hydrocarbon based polymers for IR optics. In contrast, polymers made via inverse vulcanization are promising for infrared optics and strive to compete with commercially available materials (e.g. semimetals such as germanium and silicon, semiconductors such as zinc selenide or other minerals such as sapphire and zinc sulfide).

High sulfur content polymers were shown to possess high transparency in the near-infrared regime from 0.8 to 2.5 μm (14 000–4 000 cm^{-1}) or mid-infrared from 3 to 5 μm (3333–2000 cm^{-1}). The first report about utilizing the optical properties of polymers fabricated via inverse vulcanization was published in 2014. Since then, progress has been made in terms of processability, performance, and characterization. Optical devices fabricated from polymers synthesized via inverse vulcanization were mainly windows or lenses, but also the fabrication of a Bragg mirror, a polarizer or a waveguide were demonstrated. Materials with high ($n > 1.5$) and "ultrahigh" ($n > 1.8$) refractive index (between ca. 630 and 1550 nm) could be fabricated via simple and cheap procedures. Strategies to yield so called "ultrahigh" refractive index materials relied on the incorporation of elements heavier than sulfur, either into the polymer backbone (selenium), or into the organic crosslinker (tin) [102, 103].

Due to the presumably random order in which building blocks are incorporated, polymers made via inverse vulcanization are commonly termed "sulfur-*random*-crosslinker," or short "S-*r*-X" (with X being the acronym of a crosslinker, Figure 9.11). One of the first polymers prepared via inverse

Figure 9.11 Molecular structures of the crosslinkers employed for IR materials as well as the proposed structure of the poly(S-*r*-Se-*r*-DIB) terpolymer.

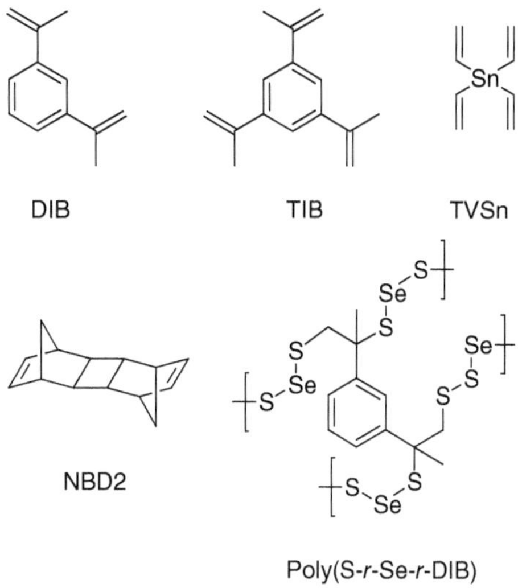

vulcanization, S-*r*-DIB (diisopropylidenebenzene), was also the first polymer to be systemically examined for its application prospects in mid-infrared optics [104]. S_x-*r*-DIB_{100-x} was prepared with x being 50, 60, 70, and 80 wt%. With increasing sulfur rank an increase in the refractive index at 633 nm of 0.11 was observed from 1.75 (50 wt% S) to 1.86 (80 wt% S). These copolymers were melt-processable and capable to form lenses. Films of defined thickness could be casted from solution in 1,2-dichlorobenzene with thicknesses ranging from 1 to 60 μm. Investigation of a 10 μm thick S_{80}-*r*-DIB_{20} film showed a high transparency (>80%) in the range of 650 to 2600 nm as well as a very high transparency (>95%) between 3500 and 5500 nm. High quality imaging in the mid-infrared (3–5 μm) region using a S_{80}-*r*-DIB_{20} lens was demonstrated (Figure 9.12). An investigation of non-linear optical properties of S_{50}-*r*-DIB_{50} and S_{70}-*r*-DIB_{30} revealed their second order refractive indices to be 2.45×10^{-15} cm^2 W^{-1} and 3.06×10^{-15} cm^2 W^{-1}, respectively [105].

Later, scratched poly(S-*r*-DIB) copolymers were shown to be healable via heat curing, which is an advantage over state-of-the-art materials [106]. Other advantages include high impact resistance, low cost, low weight (density), and ease of fabrication and processing (solution and melt). Germanium as an IR transparent material is not permeable for visible light, whereas poly(S-*r*-DIB) is transparent for longer wavelengths of the visible spectrum (i.e. yellow and red), which is an advantage for manually sighted infrared devices. To improve the thermomechanical properties by increasing the crosslinking density, DIB can be replaced with 1,3,5-triisopropenylbenzene (TIB) [107]. In the temperature interval between 90 and 140 °C, the storage modulus for poly(S-*r*-TIB) copolymer (70 wt% S) was roughly ten times higher (~10^6 Pa) compared to the poly(S-*r*-DIB) copolymer (~10^5 Pa) with the same sulfur rank. Moreover, the poly(S-*r*-TIB) copolymer with 50 wt% sulfur was found to have a T_G above 100 °C and maintain a high refractive index of $n > 1.7$ between 635 and 1554 nm. In contrast to that, the T_G of poly(S-*r*-DIB) copolymer was only 37 °C, whereas the refractive index was similar ($n > 1.7$ between 633 and

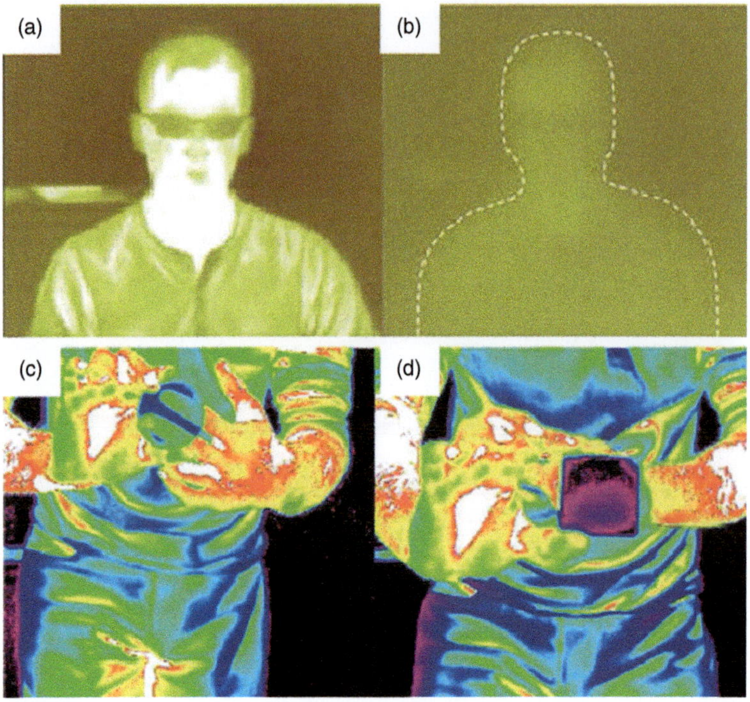

Figure 9.12 Mid-infrared (3–5 μm) imaging. (a) Thermogram of a subject taken through a poly(S-*r*-DIB) copolymer film containing 80 wt% sulfur and (b) thermogram of the same subject through PMMA. Thermogram of the subject's torso holding a poly(S-*r*-DIB) copolymer lense (c) and a glass lense (d) in its hand. The images are colored from white (corresponding to 37 °C) to purple (corresponding to 24 °C) [104].

1554 nm). Due to their higher degree of crosslinking, poly(S-*r*-TIB) copolymers were insoluble in organic solvents, thus lenses were fabricated by melt processing.

To exceed refractive index values of $n > 1.8$, elements with a higher polarizability than sulfur have to be incorporated into sulfur polymers. Selenium is a group 16 element located directly below sulfur. As such, it has many similarities with elemental sulfur like the tendency to polymerize when heated up. The polarizability of selenium (28.9 ± 1.0) is significantly higher than that of sulfur (19.4 ± 0.1) [14]. In 2017, the groups of Pyun and Sanghera independently reported the incorporation of selenium atoms into polymeric chalcogenide polymers to achieve refractive indexes between 1.9 and 2.1 in the range from 633 to 1554 nm, with the latter being the highest refractive index ever reported for a polymer (Figure 9.12).

A key finding is that a molten mixture of sulfur and selenium at 160 °C is capable to undergo an inverse vulcanization reaction to form materials with a high content of both sulfur and selenium, i.e. containing up to ca. 40 wt% of each element in one material. During this time, both the term CHIPs ("Chalcogenide Hybrid Inorganic/Organic Polymers") and Ormochalk ("Organically modified Chalcogenides", in analogy to organosilane-based "Ormosils") were coined. These sulfur-selenium

Table 9.9 High refractive index polymers synthesized via inverse vulcanization.

Polymer composition	*n* (633 nm)	*n* (816 nm)	*n* (984 nm)	*n* (1030 nm)	*n* (1305 nm)	*n* (1554 nm)	References
S_{50}-*r*-DIB_{50}	1.7565[a)]	1.7367[a)]	—	1.7229[a)]	1.723[b)d)]	1.7194[a)]	[104, 106]
S_{60}-*r*-DIB_{40}	1.8387[a)]	1.8136[a)]	—	1.7965[a)]	—	1.7929[a)]	[104, 106]
S_{70}-*r*-DIB_{30} (I)	1.8561[a)]	1.8318[a)]	—	1.8135[a)]	—	1.8097[a)]	[104, 106]
S_{70}-*r*-DIB_{30} (II)	1.834	1.809	—	—	1.791	1.787	[106]
S_{70}-*r*-DIB_{30} (III)	1.840[c)b)]	—	1.805	—	—	1.792[d)]	[109]
S_{80}-*r*-DIB_{20}	1.8696[a)]	1.8480[a)]	—	1.8316[a)]	1.832[b)]	1.83292[a)]	[104, 106]
S_{50}-*r*-TIB_{50}	1.75[a)e)]	1.73[a)]	—	—	1.723[a)]	1.721[a)]	[107]
S_{70}-*r*-TIB_{30}	1.83[a)e)]	1.81[a)]	—	—	1.796[a)]	1.792[a)]	[107]
S_{50}-*r*-Se_{20}-DIB_{30}	1.963	1.945	—	—	1.916	1.912	[110]
S_{42}-*r*-Se_{42}-DIB_{16}	2.140	2.099	—	—	2.087	2.035	[109]
S_{57}-*r*-$TVSn_{43}$[f)]	1.980[c)]	—	1.938	—	—	1.924[d)]	[111]
S_{70}-r-$NBD2_{30}$	1.83[g)]	1.82[g)]	—	—	1.80[g)]	1.80[g)]	[108]
S_{50}-r-$NBD2_{50}$	1.77[g)]	1.75[g)]	—	—	1.74[g)]	1.74[g)]	[108]

a) Averaged values for TE and TM polarizations.
b) Value reported for 1310 nm.
c) Value reported for 636.4 nm.
d) Value reported for 1548.4 nm.
e) Value reported for 632 nm.
f) Mass ratio calculated assuming the same amount of sulfur was used for S-TVSn as for the S-DIB reference.
g) Reported values extracted from figure (not tabulated).

hybrid polymers (e.g. poly(S-*r*-Se-DVB)) significantly improved the image quality of thermograms.

The use of tetravinyl tin (TVSn) as a crosslinker in an inverse vulcanization process led to the development of the first high sulfur content polymer crosslinked with a metal organic crosslinker. The refractive indices were all above $n > 1.9$ between 636 and 1548 nm, whereas the transparency was rather low (30%) for most parts of the IR spectrum between 1.3 and 5.3 μm. Between 5.3 and 6.3 μm the poly(S-*r*-TVSn) copolymer is superior to a poly(S-*r*-DIB) reference of comparable thickness (ca. 1.6 mm). Thus, it showed to be applicable in the far-infrared range (7.5–14 μm), where poly(S-*r*-DIB) cannot be used. A similar result is obtained by using an alicyclic norbornyl-dimer as crosslinker (NBD2). The resulting S_{70}-*r*-$NBD2_{30}$ copolymers exhibit a high transparency between 30–50% (60 μm thickness) in the far-infrared range [108]. Characterized optical materials prepared via inverse vulcanization are listed in Table 9.9.

Apart from rather simple lenses and windows, the fabrication of a bragg mirror was demonstrated. A bragg mirror consists of layers of two materials, one with a high and one with a low refractive index. If the thickness of those layers is one half of the pathlength of the incident light ($\lambda/2$), reflection is maximized. Thus, there are wavelengths (λ_0) for which reflection is maximized, while other wavelengths around this so-called stop-band are not reflected [112]. The breadth of the stop-band increases with increasing difference in refractive index of both layers. As a high-refractive polymer a poly(S_{50}-*r*-Se_{20}-*r*-DIB_{30}) terpolymer and as low refractive index polymer cellulose acetate was used, resulting in a refractive index difference of roughly 0.5 in the range between 663 and 1554 nm. For 11 spin-coated bilayers with layer thicknesses of roughly 260 nm for poly(S-*r*-Se-DIB) and 340 nm for cellulose acetate, the determined IR reflectivities at 2113, 1450, and 998 nm were between 93% and 94%. The fabrication of other optical materials, such as a mid-infrared linear polarizer film prepared via nanoimprint lithography of S_{70}-*r*-DIB_{30} or a mid-infrared waveguide were also demonstrated [113, 114].

9.5 Conclusion and Outlook

Sulfur plays a crucial role in the field of HRIPs due to its high polarizability and chemical versatility. New materials with improved optical properties, e.g. a high refractive index, low dispersion, and low birefringence, could be fabricated from sulfur-containing HRIPs. Sulfur moieties can be incorporated into all polymer classes relevant for optical applications. By introducing sulfur-containing functional groups into polyimides and polyamides, polycarbonates and polyesters, or poly(meth)acrylates, their optical properties could be improved. Sulfur-containing functional groups can be placed in both the polymer backbone and the side chains, enabling various polymer design approaches. The large scope of polymers that can be functionalized with sulfur moieties is helpful to tailor HRIPs for different applications and requirements (i.e. mechanical, chemical, or thermal resistance). Moreover, developments in the field of polymeric sulfur demonstrated to be a driver for innovative new materials for optical applications. For example, the introduction of the inverse vulcanization enabled the use of polymers for IR optics, a field which had been inaccessible to polymers before. The fabrication of optical devices is a multidisciplinary venture. Thus, chemists should pay attention to a careful and thorough characterization of the optical properties of their polymers before passing them on into development. Researchers down the stream in the development of optical devices rightfully complain about the lack of data provided for newly introduced materials. An indication is that not all mentioned reports provide full insight into the optical properties of the synthesized materials. This reduces their attractivity in terms of follow-up investigations and ultimately their real-life applications. As a minimum, the refractive index of synthesized polymers should be given for a reasonable number of different wavelengths and a meaningful expression for the dispersion should be derived (e.g. Cauchy parameters, Abbe number). Absorption bands in between the wavelengths used for the refractive index determination falsify the expressions for the dispersion. Thus, it is helpful to support the characterization of the dispersion with UV–vis spectroscopy. The latter also provides relevant information about the optical loss of materials. Follow-up investigations can then clarify application-specific properties such as long-term UV and oxygen stability or thermal properties. HRIPs as optical materials will continue to compete with glasses or semiconductors and thus need to demonstrate their strengths in terms of (cheap) manufacturing and processing. In that regards, micro- and nanofabrication are areas where HRIPs could excel.

References

1 Sudarsan, V. (2012). *Functional Materials: Preparation, Processing and Applications*, 285. Amsterdam: Elsevier.

2 Hering, E., Martin, R., and Stohrer, M. (2002). *Physik für Ingenieure: Mit 102 Tabellen und 2 Falttafeln*. Berlin: Springer.

3 Iizuka, K. (2019). *Engineering Optics*. Cham, Switzerland: Springer.

4 Ribeiro, P., Andrews, D.L., and Raposo, M. (2019). *Optics, Photonics and Laser Technology 2017*. Cham: Springer International Publishing.

5 Herrmann, P.P. (1980). *Appl. Opt.* 19: 3261.

6 Rawle, A.F. (2017). *Encyclopedia of Spectroscopy and Spectrometry*, 936. Amsterdam, Boston, Heidelberg: Elsevier AP Academic Press.

7 Ay, F., Kocabas, A., Kocabas, C. et al. (2004). *J. Appl. Phys.* 96: 7147.

8 Ulrich, R. and Torge, R. (1973). *Appl. Opt.* 12: 2901.

9 Feldman, A. (1983). *Appl. Opt.* 22: 2380.

10 Vedam, K. (1998). *Thin Solid Films* 313–314: 1.

11 Tompkins, H.G. (ed.) (2010). *Handbook of Ellipsometry*. Norwich, NY: Andrew.

12 Nestler, P. and Helm, C.A. (2017). *Opt. Express* 25: 27077.

13 Yeh, P. (2005). *Optical Waves in Layered Media*. Hoboken: Wiley-Interscience.

14 Schwerdtfeger, P. and Nagle, J.K. (2019). *Mol. Phys.* 117: 1200.

15 Daintith, J. (ed.) (2008). *A Dictionary of Chemistry*. New York: Oxford University Press.

16 Hannay, J.H. (1983). *Eur. J. Phys.* 4: 141.

17 Brocos, P., Piñeiro, Á., Bravo, R., and Amigo, A. (2003). *Phys. Chem. Chem. Phys.* 5: 550.

18 Denbigh, K.G. (1940). *Trans. Faraday Soc.* 36: 936.

19 Gennaro, A.R. and Remington, J.P. (2000). *Remington: The Science and Practice of Pharmacy*, *20*e. Baltimore, MD: Lippincott Williams & Wilkins.

20 Lide, D.R. (ed.) (2005). *CRC Handbook of Chemistry and Physics: A Ready-Reference Book of Chemical and Physical Data*. Boca Raton: CRC Press.

21 Speight, J., Lange, N.A., and Dean, J.A. (2005). *Lange's Handbook of Chemistry*, *16*e. New York, USA: McGraw-Hill Professional Publishing.

22 Higashihara, T. and Ueda, M. (2015). *Macromolecules* 48: 1915.

23 Le Fèvre, R.J.W. (1965). *Molecular Refractivity and Polarizability*, vol. 3, 1. Academic Press Inc. (London) Ltd. Elsevier.

24 Fedorov, D.V., Sadhukhan, M., Stöhr, M., and Tkatchenko, A. (2018). *Phys. Rev. Lett.* 121: 183401.

25 von Steiger, A.L. (1921). *Ber. Dtsch. Chem. Ges. A/B* 54: 1381.

26 Sultanova, N.G., Kasarova, S.N., and Nikolov, I.D. (2013). *Opt. Quant. Electron.* 45: 221.

27 Smith, D.Y., Inokuti, M., and Karstens, W. (2001). *J. Phys. Condens. Matter* 13: 3883.

28 Gao, L., Lemarchand, F., and Lequime, M. (2011). *Thin Solid Films* 520: 501.

29 Sultanova, N., Kasarova, S., and Nikolov, I. (2009). *Acta Phys. Pol. A* 116: 585.

30 Smith, D.Y. and Karstens, W. (2010). *J. Phys. Conf. Ser.* 249: 12034.

31 Choi, M.-C., Kim, Y., and Ha, C.-S. (2008). *Prog. Polym. Sci.* 33: 581.

32 Granstrom, J., Swensen, J.S., Moon, J.S. et al. (2008). *Appl. Phys. Lett.* 93: 193304.

33 Miller, M.S., O'Kane, J.C., Niec, A. et al. (2013). *ACS Appl. Mater. Interfaces* 5: 10165.

34 Li, X., Yu, X., and Han, Y. (2013). *J. Mater. Chem. C* 1: 2266.

35 Blakey, I., Conley, W., George, G.A. et al. (2006). *Advances in Resist Technology and Processing XXIII* (ed. Q. Lin), 61530H. SPIE.

36 Liu, J.-G. and Ueda, M. (2009). *J. Mater. Chem.* 19: 8907.

37 Li, Q., Ng, K.L., Pan, X., and Zhu, J. (2019). *Polym. Chem.* 10: 4279.

38 Burnett, D. (1927). *Math. Proc. Camb. Philos. Soc.* 23: 907.

39 Badur, T., Dams, C., and Hampp, N. (2018). *Macromolecules* 51: 4220.

40 Macdonald, E.K. and Shaver, M.P. (2015). *Polym. Int.* 64: 6.

41 Liu, J.-G., Nakamura, Y., Shibasaki, Y. et al. (2007). *Macromolecules* 40: 4614.

42 Liu, J.-G., Nakamura, Y., Shibasaki, Y. et al. (2007). *Polym. J.* 39: 543.

43 Liu, J.-G., Nakamura, Y., Shibasaki, Y. et al. (2007). *J. Polym. Sci. A Polym. Chem.* 45: 5606.

44 Liu, J.-G., Nakamura, Y., Suzuki, Y. et al. (2007). *Macromolecules* 40: 7902.

45 Liu, J.-G., Nakamura, Y., Terraza, C.A. et al. (2008). *Macromol. Chem. Phys.* 209: 195.

46 Suzuki, Y., Liu, J.-g., Nakamura, Y. et al. (2008). *Polym. J.* 40: 414.

47 Terraza, C.A., Liu, J.-g., Nakamura, Y. et al. (2008). *J. Polym. Sci. A Polym. Chem.* 46: 1510.

48 You, N.-H., Suzuki, Y., Higashihara, T. et al. (2009). *Polymer* 50: 789.

49 You, N.-H., Suzuki, Y., Yorifuji, D. et al. (2008). *Macromolecules* 41: 6361.

50 Fukuzaki, N., Higashihara, T., Ando, S., and Ueda, M. (2010). *Macromolecules* 43: 1836.

51 Li, Q., Zhang, J., Pan, X. et al. (2018). *Polymers (Basel)* 10 (4): 417.

52 You, N.-H., Fukuzaki, N., Suzuki, Y. et al. (2009). *J. Polym. Sci. A Polym. Chem.* 47: 4428.

53 Kim, H., Ku, B.-C., Goh, M. et al. (2018). *Polymer* 136: 143.

54 Nam, K.-H., Lee, A., Lee, S.-K. et al. (2019). *J. Mater. Chem. C* 7: 10574.

55 Afzal, M.A.F., Haghighatlari, M., Ganesh, S.P. et al. (2019). *J. Phys. Chem. C* 123: 14610.

56 Rogers, H.G., Gaudiana, R.A., Hollinsed, W.C. et al. (1985). *Macromolecules* 18: 1058.

57 Shockravi, A., Javadi, A., and Abouzari-Lotf, E. (2011). *Polym. J.* 43: 816.

58 Kiani, H., Nasef, M.M., Javadi, A. et al. (2013). *J. Polym. Res.* 20: 1474.

59 Javadi, A., Najjar, Z., Bahadori, S. et al. (2015). *RSC Adv.* 5: 91670.
60 Javadi, A., Abouzari-Lotf, E., Mehdipour-Ataei, S. et al. (2015). *J. Polym. Sci. A* 53: 2867.
61 Zhang, G., Bai, D.-t., Li, D.-s. et al. (2013). *Polym. Int.* 62: 1358.
62 Zhang, G., Huang, G.-s., Wang, X.-j. et al. (2011). *J. Polym. Res.* 18: 1261.
63 Sun, Z., Huang, H., Li, L. et al. (2017). *Macromolecules* 50: 8505.
64 Shockravi, A., Javadi, A., and Abouzari-Lotf, E. (2013). *RSC Adv.* 3: 6717.
65 You, N.-H., Higashihara, T., Ando, S., and Ueda, M. (2010). *J. Polym. Sci. A* 48: 2604.
66 Okutsu, R., Ando, S., and Ueda, M. (2008). *Chem. Mater.* 20: 4017.
67 Maheswara, M., Oh, S.-H., Ju, J.-J. et al. (2010). *Polym. J.* 42: 249.
68 Matsuda, T., Funae, Y., Yoshida, M., and Takaya, T. (1999). *J. Macromol. Sci. A* 36: 1271.
69 Okubo, T., Kohmoto, S., and Yamamoto, M. (1998). *J. Macromol. Sci. A* 35: 1819.
70 Matsuda, T., Funae, Y., Yoshida, M. et al. (2000). *J. Appl. Polym. Sci.* 76: 45.
71 Nakayama, N. and Hayashi, T. (2008). *Prog. Org. Coat.* 62: 274.
72 Lee, J., Yeo, H., Goh, M. et al. (2015). *Macromol. Res.* 23: 960.
73 Liu, S., Tian, L., Zheng, Z., and Wang, X. (2009). *J. Appl. Polym. Sci.* 113: 3498.
74 Marianucci, E., Berti, C., Pilati, F. et al. (1994). *Polymer* 35: 1564.
75 Okutsu, R., Suzuki, Y., Ando, S., and Ueda, M. (2008). *Macromolecules* 41: 6165.
76 Gao, C., Yang, B., and Shen, J. (2000). *J. Appl. Polym. Sci.* 75: 1474.
77 Kwon, J.Y., Kim, B.G., Do, J.Y. et al. (2007). *Macromol. Res.* 15: 533.
78 Schnell, H. (1956). *Angew. Chem.* 68: 633.
79 Diepens, M. and Gijsman, P. (2007). *Polym. Degrad. Stab.* 92: 397.
80 Seto, R., Sato, T., Kojima, T. et al. (2010). *J. Polym. Sci. A* 48: 3658.
81 Rey-Stolle, M.F., Fernádez-Martín, F., Tagle, L.H., and Hernández-Fuentes, I. (1997). *Eur. Polym. J.* 33: 295.
82 Berti, C., Celli, A., and Marianucci, E. (2002). *Eur. Polym. J.* 38: 1281.
83 Nakano, K., Tatsumi, G., and Nozaki, K. (2007). *J. Am. Chem. Soc.* 129: 15116.
84 Braun, D. and Kiessel, M. (1965). *Monatsh. Chem.* 96: 631.
85 Leung, L.M., Chan, W.H., and Leung, S.K. (1993). *J. Polym. Sci. A Polym. Chem.* 31: 1799.
86 Rokicki, G., Kuran, W., and Kiełkiewicz, J. (1979). *Makromol. Chem.* 180: 2779.
87 Luo, M., Zhang, X.-H., and Darensbourg, D.J. (2016). *Acc. Chem. Res.* 49: 2209.
88 Luo, M., Zhang, X.-H., Du, B.-Y. et al. (2015). *Polym. Chem.* 6: 4978.
89 Luo, M., Zhang, X.-H., Du, B.-Y. et al. (2013). *Macromolecules* 46: 5899.
90 Berti, C., Marianucci, E., and Pilati, F. (1988). *Makromol. Chem.* 189: 1323.
91 Gomez, C.M., Gutierrez, D., Asensio, M. et al. (2017). *J. Elastom. Plast.* 49: 77.
92 Rogulska, M. (2018). *Polym. Bull.* 75: 1211.
93 Kricheldorf, H.R. and Schwarz, G. (2007). *J. Macromol. Sci. A* 44: 625.
94 You, N.-H., Higashihara, T., Yasuo, S. et al. (2010). *Polym. Chem.* 1: 480.
95 Kim, J.-S., Cho, B.-S., Kweon, J.-O., and Noh, S.-T. (2014). *Prog. Org. Coat.* 77: 1695.

96 Aoyagi, N. and Endo, T. (2019). *J. Polym. Sci. A* 57: 1160.
97 Jia, Y., Shi, B., Jin, J., and Li, J. (2019). *Polymer* 180: 121746.
98 Tang, Y., Pina-Hernandez, C., Niu, Q. et al. (2018). *J. Mater. Chem. C* 6: 8823.
99 Lü, C., Cui, Z., Wang, Y. et al. (2003). *J. Appl. Polym. Sci.* 89: 2426.
100 Kleine, T.S., Glass, R.S., Lichtenberger, D.L. et al. (2020). *ACS Macro Lett.* 9: 245.
101 Trofimov, B.A., Sinegovskaya, L.M., and Gusarova, N.K. (2009). *J. Sulf. Chem.* 30: 518.
102 Griebel, J.J., Glass, R.S., Char, K., and Pyun, J. (2016). *Prog. Polym. Sci.* 58: 90.
103 Chalker, J.M., Worthington, M.J.H., Lundquist, N.A., and Esdaile, L.J. (2019). *Top. Curr. Chem.* 377: 16.
104 Griebel, J.J., Namnabat, S., Kim, E.T. et al. (2014). *Adv. Mater.* 26: 3014.
105 Babaeian, M., Diaz, L.R., Namnabat, S. et al. (2018). *Opt. Mater. Express* 8: 2510.
106 Griebel, J.J., Nguyen, N.A., Namnabat, S. et al. (2015). *ACS Macro Lett.* 4: 862.
107 Kleine, T.S., Nguyen, N.A., Anderson, L.E. et al. (2016). *ACS Macro Lett.* 5: 1152.
108 Kleine, T.S., Lee, T., Carothers, K.J. et al. (2019). *Angew. Chem. Int. Ed.* 58: 17656.
109 Anderson, L.E., Kleine, T.S., Zhang, Y. et al. (2017). *ACS Macro Lett.* 6: 500.
110 Boyd, D.A., Baker, C.C., Myers, J.D. et al. (2016). *Chem. Commun.* 53: 259.
111 Boyd, D.A., Nguyen, V.Q., McClain, C.C. et al. (2019). *ACS Macro Lett.* 8: 113.
112 Kleine, T.S., Diaz, L.R., Konopka, K.M. et al. (2018). *ACS Macro Lett.* 7: 875.
113 Nishant, A., Anderson, L.E., Kleine, T.S. et al. (2019). Integrated Optics: Devices, Materials, and Technologies XXIII, SPIE 2019, 14.
114 Berndt, A.J., Hwang, J., Islam, M.D. et al. (2019). *Polymer* 176: 118.

10

Selenium-Containing Dynamic Polymers: From Synthesis to Functions

Jiahao Xia and Huaping Xu

Tsinghua University, Key Laboratory of Organic Optoelectronics and Molecular Engineering, Department of Chemistry, Mengminweikeji Building South Building, Beijing 100084, China

10.1 Introduction

Selenium is a semimetallic element lying in the group XVI of the periodic table. Modern medical research has revealed that selenium is related to a list of human diseases like cardiovascular disease, cancer, diabetes, central nervous system diseases, etc. [1]. As a result, many selenium-rich foods, such as selenium rich rice and eggs, have prevailed during the past few years. A recent discovery reported by Long et al. indicated that there is a very close connection between the three major mass extinction events in history and the severe Se depletion in the oceans [2]. This discovery showed Se is a paramount element not only in human health but also in the evolution of life. Despite the vital role of Se, it did not draw much attention from chemists since it was first discovered by Jöns Jacob Berzelius in 1817, probably due to the unpleasant odor of hydrogen selenide. However, in 1970s selenium was found to be a necessary trace element in mammals [3]. Specifically, Rotruck et al. discovered that the glutathione peroxidase (GPx) in mice blood cell was a kind of selenium-containing enzyme and that selenium played an important role in its catalytic activity [4]. GPx can maintain the balance of redox equilibrium in mammal bodies by catalyzing the decomposing of peroxides. Consequently, many efforts have been dedicated to the research of selenium-containing small molecule for drugs and the organic selenium chemistry was increasingly explored [1, 5–8].

In the chemistry perspective, it is reasonable to deduce that the chemical properties of selenium resemble sulfur in many ways. However, it should be noted that selenium has relatively lower electronegativity and larger atomic radius compared with sulfur. These features endow selenium-containing compounds with higher reactivity and sensitivity. Thus, it is believed that by incorporating selenium into polymers, new features and functions may be created. Although organic selenium chemistry has been developing fast, the successful introducing of selenium into polymer is rather scarce [9]. This is particularly obvious in contrast with its sulfur counterparts. Fortunately, we have seen a drastic rising trend in the area

Sulfur-Containing Polymers: From Synthesis to Functional Materials,
First Edition. Edited by Xing-Hong Zhang and Patrick Theato.

of selenium-containing polymers for the past decade due to the contribution of many research fellows throughout the world [10]. Many preparation methods for selenium-containing polymers have been established. The unique selenium-based dynamic covalent chemistry was developed and further utilized in dynamic materials with distinctive features.

10.2 Synthesis of Selenium-Containing Polymers

Although the synthesis of selenium-containing polymer was initiated in the 1960s, there were two major flaws of the Se-polymer prepared in the early stage. One was their limited solubility and the other was the poor stability. In 2010 Xu, Zhang, and coworkers reported the first example of selenium-containing polymer that was both soluble and stable in common solvents. After that, many other selenium-containing polymers with different structures were synthesized through a series of approaches including step growth polymerization, free radical polymerization, ring-opening polymerization, and so on.

10.2.1 Step Growth Polymerization

The conventional method used to prepare selenium-containing polymers was step growth polymerization. Some preliminary examples include the synthesis of polyselenomethylene reported by Prince and Bremer in 1967. It was simply achieved by reacting sodium selenide with methylene bromide (Figure 10.1a) [11]. Kroll and Bolton reported the synthesis of a series of selenium-containing polymers in 1970 including polyesters, polyureas, and polyurethanes (PU) [12]. Bis(2-hydroxyethyl) selenide or bis(2-aminoethyl) selenide was utilized to react with different types of diisocyanates, terephthaloyl chlorides, anhydrides, and epoxy resins to form selenium-containing polymers. Salzman and coworkers summarized some of the preparation methods for polydiselenides in 1972 [13]. Seçkin and coworkers synthesized a kind of selenium-containing polyimide with increased solubility (Figure 10.1b) in 2005 [14]. Bis (1,3-p-dimethylaminobenzylpyrimidyl-2-ylidene) was used to react with selenium element to obtain selenium carbene-containing (C=Se) monomer. This monomer was further reacted with various dianhydrides to obtain selenium-containing polyimides. For most of these researches mentioned earlier, important features of the polymer like molecular weights were not thoroughly characterized. It should be noted that polyselenophenes have been relatively more investigated as a kind of conductive polymer because of their potential applications as optoelectronic materials [15, 16]. But the development of alkyl-based selenium-containing polymers was rather slow, mainly due to two drawbacks of the obtained selenium-containing polymers. One is that the prepared polymer showed poor solubility in common organic solvents, thus prohibited their further usage in real applications. The other problem is the stability of the polymer since both mono-selenide and diselenide are very sensitive to oxidative agents like oxygen in air.

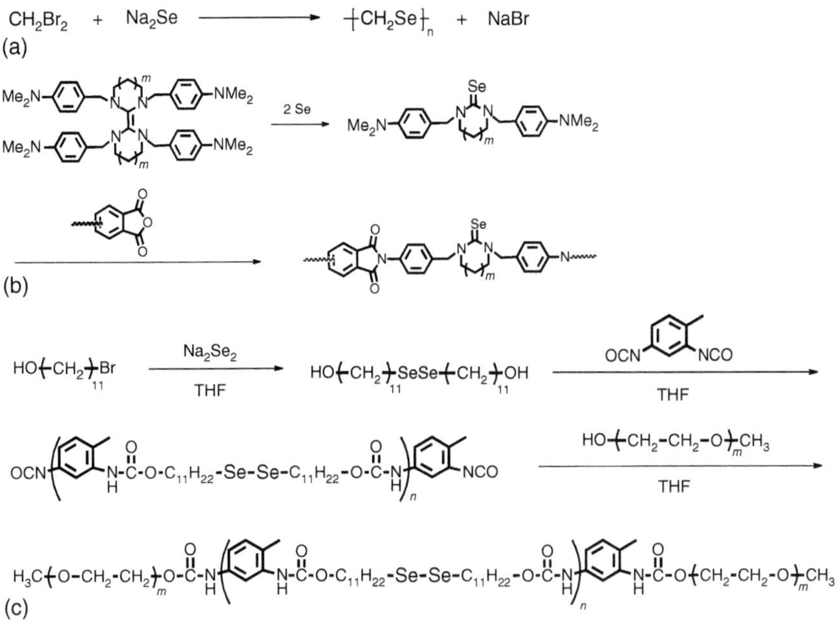

Figure 10.1 Step growth polymerization of selenium-containing polymers. (a) Synthesis of polyselenomethylene. (b) Synthesis of selenium-containing polyimide. (c) Synthesis of PEG-PUSeSe-PEG diselenide-containing polyurethane with good solubility in common organic solvents.

In 2010, Xu and coworkers firstly proposed a diselenide-containing PU with good solubility in common organic solvents like tetrahydrofuran (THF), *N,N*-dimethylformamide (DMF), dichloromethane (DCM), etc. (Figure 10.1c) [17]. A diselenide diol compound was first synthesized through the reaction of disodium diselenide (Na_2Se_2) and 11-bromoundecanol. It was then reacted with 1.1 equivalence of toluene diisocyanate (TDI) to obtain linear diselenide-containing polyurethane (PUSeSe) capped by NCO groups in both ends of the polymer chain. Finally, the NCO groups were terminated by polyethylene glycol (PEG) monomethyl ether to obtain ABA type amphiphilic block polymer denoted as PEG-PUSeSe-PEG. The obtained polymer usually has a molecular weight around 10 000 Da with the PDI close to 2 by gel permeation chromatography (GPC) test. In the meantime, monoselenide-containing PU could be prepared in similar way with disodium diselenide replaced by sodium hydroselenide (NaHSe) in the first step [18]. And the obtained polymer was donated as PEG-PUSe-PEG. The long alkyl chain of the selenide monomer largely facilitated the solubility of the polymer in common organic solvents. Due to the amphiphilic nature of the block polymers, they can self-assemble to form stable aggregates in aqueous solutions. Selenide bonds are hydrophobic and are embedded into the core of the assemblies, thus prevented from oxidation in solution. This synthetic approach represents the first successful example that addresses both the solubility and stability challenges and allows the selenium-containing polymer to be used as a novel stimuli-responsive drug delivery vehicle.

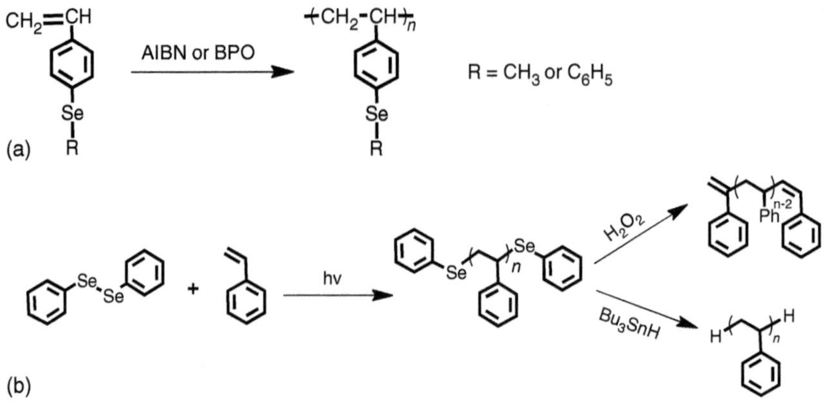

Figure 10.2 Radical polymerization to prepare selenium-containing polymers. (a) Selenium-containing vinyl free radical polymerization. (b) Diselenide compound used as photoiniferter in radical polymerization initiated by UV light irradiation.

10.2.2 Radical Polymerization

In addition to step growth polymerization, selenium-containing polymer could also be prepared by radical polymerization. Kondo and coworkers found that p-methylselenostyrene and p-phenylselenostyrene could be polymerized to afford polymer with molecular weights ranging from 20 000 to 150 000 Da in 1996 (Figure 10.2a) [19]. The reaction was based on free radical polymerization mechanism upon adding radical initiators like 2,2′-azobisisobutyronitrile (AIBN) or benzoyl peroxide (BPO). Later on they discovered that diaryl diselenide itself could be used as a photoiniferter to perform living radical polymerization (Figure 10.2b) [20]. Specifically, diselenide bond could be homolytically dissociated to generate phenylseleno radicals under 100 W high pressure mercury lamp, which accounted for the initiation of the polymerization process. The polymerization exhibited a polydispersity at 1.6–2.1. The prepared polymer has arylseleno groups on both chain ends, which could be transformed to hydrogen or carbon double bond by reductive elimination or oxidation, respectively. In 1999, Yuki and coworkers used this polymerization method to synthesize a grafting polymer with propylene sulfide (PS) as the backbone and poly(methyl methacrylate) (PMMA) as the branches [21]. Later on, they further used this reaction to synthesize a star-shape polystyrene from a tetrafunctional phenyl selenides photoiniferter in 2003 [22]. The molecular weight could reach 153.9 kDa though the polydispersity was high (3.2). In general, using diaryl diselenide as a photoiniferter to initiate the polymerization was the first trial to synthesize selenium-containing polymer by radical polymerization. However, being a photoiniferter means that selenide is both an initiator and an effective chain transfer agent, which inevitably leads to a large polydispersity.

Reversible addition-fragmentation chain transfer polymerization (RAFT) is widely recognized as a versatile technique to perform living radical polymerization [23, 24]. Chain transfer agents like thiocarbonylthio plays a vital role in controlling

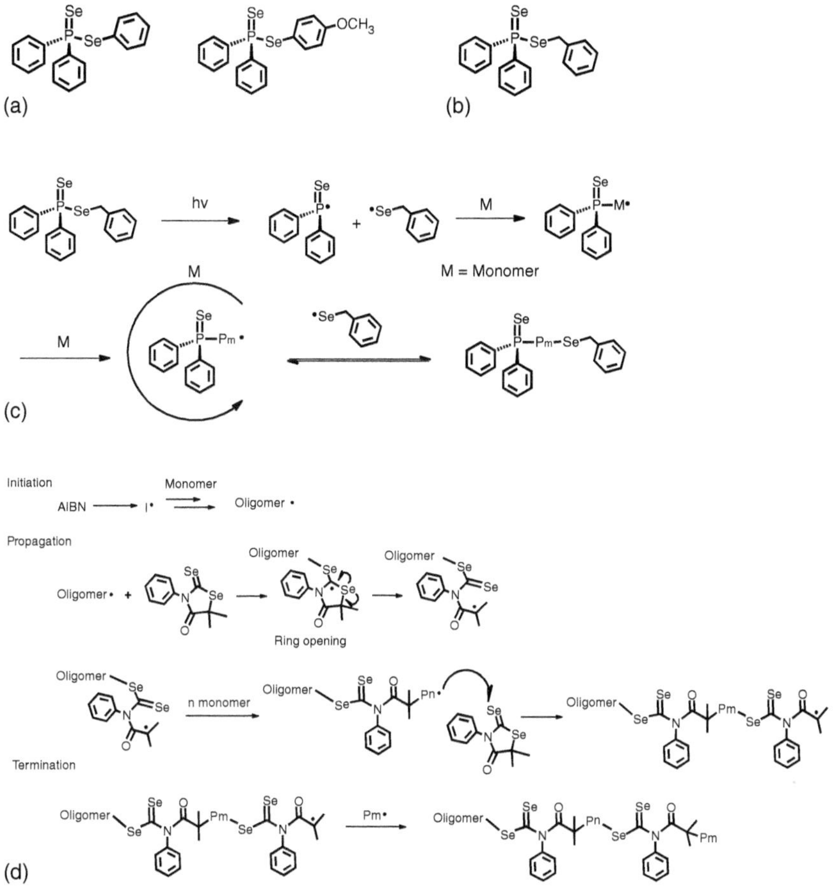

Figure 10.3 Selenium-containing compounds used in controlled polymerization. (a) The structure of phosphinodiselenoic acid esters used in polymerization. (b) A new type of phosphinodiselenoic acid esters (DPPSB) for controlled polymerization. (c) The iniferter polymerization mechanism of DPPSB. (d) RAFT polymerization mechanism by cyclic selenium-containing RAFT agent RAFT-Se.

the polymerization process [25]. Considering the chemical similarity of selenium and sulfur, selenium may also be employed to mediate RAFT polymerization. In 2008, Lee and coworkers introduced a series of phosphinodiselenoic acid esters into the thermally initiated styrene polymerization (Figure 10.3a) [26]. During the polymerization the number-average molecular weight (M_n) increased almost linearly with degree of monomer conversion and time, showing the features of living polymerization. However, the polydispersity increased with the growth of molecular weight, ranging from 1.6 to 2.5. Later on, Zhu and coworkers synthesized a new type of phosphinodiselenoic acid esters (DPPSB) and used it to mediate the polymerization of styrene under UV irradiation (Figure 10.3b) [27]. It was found that a linear polymerization kinetics was observed and the molecular weight increased linearly with the consumption of monomers. The polydispersity of the

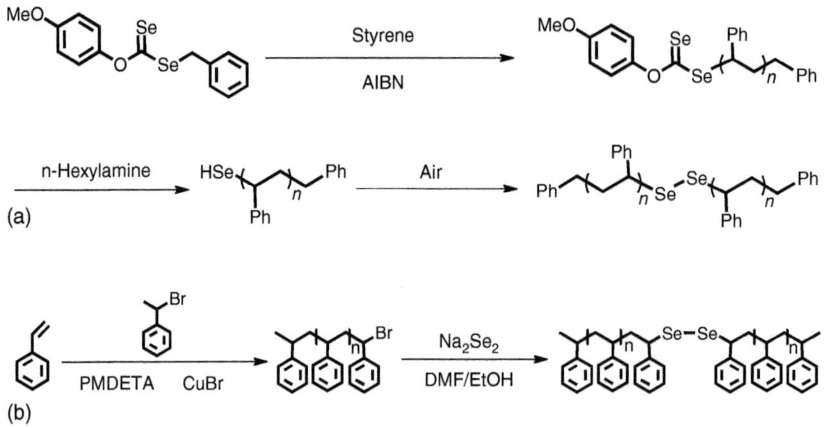

Figure 10.4 Diselenide-containing polymers prepared by (a) aminolysis-oxidation reaction (b) ATRP.

obtained polymer was from 1.5 to 2.0. Further study of the end groups revealed that this polymerization underwent an iniferter mechanism (Figure 10.3c). In 2013, they synthesized a cyclic selenium-containing RAFT agent and compared its function with sulfur counterpart (Figure 10.3d) [28]. RAFT-Se showed better control in terms of molecular weights compared with RAFT-S in the synthesis of polyvinyl acetate. Besides, it was believed that RAFT-Se was not only in the ending groups but also incorporated into the backbone of polyvinyl acetate (PVAc). They have also reported diselenocarbonates as a new type of RAFT agent [29]. The PVAc was obtained with controlled molecular weight. But the polydispersity rose with the increase of molecular weight, from 1.3 to more than 1.7.

In 2015, Pan and coworkers reported a facile transformation from diselenocarbonate-mediated RAFT polymer to diselenide-containing polymer via an efficient aminolysis followed by a spontaneous oxidation coupling reaction (Figure 10.4a) [30, 31]. Specifically, diselenocarbonate could react with amines to produce selenol, which could be oxidized by oxygen in air to obtain diselenide bonds. This reaction could be monitored by UV–Vis, ^{1}H NMR, and GPC.

Atom transfer radical polymerization (ATRP) is another wildly used method to prepare polymers with controlled structures and polydispersity [32–34]. However, the traditional ATRP usually requires cuprous ions as the mediator, which would coordinate with selenium-containing compounds and lost its function. Fortunately, polymers prepared via ATRP usually leave an active halogen group like Br in the end of the chain. This could be efficiently coupled by NaHSe or Na_2Se_2 to form selenium-containing polymers. Xu and coworkers first reported this strategy to synthesize diselenide-containing polystyrene $(PSSe)_2$ in 2014 [35]. In 2015, Zhang and coworkers demonstrated this approach in great details (Figure 10.4b) [36]. Firstly, polystyrene was synthesized by ATRP, then the terminal Br was reacted with Na_2Se_2 through nucleophilic attack to obtain PS-SeSe-PS. The M_n of the polymer was 4000 Da and the polydispersity was only 1.1 from GPC results. The structure of the polymer was testified by GPC, NMR, matrix assisted laser desorption time-of-flight (MALDI-TOF), and oxidation test. From these characterization results it was concluded that almost all of the PS-Br were reacted to generate PS-SeSe-PS. However, when other commonly used monomers like methyl methacrylate were employed, the coupling reaction showed less efficiency. This implied the reaction yield partially depends on the structure of the polymer and further purification may be required to separate the final product.

10.2.3 Ring-Opening Polymerization

Except for step growth polymerization and radical polymerization, recent studies have revealed that ring-opening polymerization is also a promising way to prepare selenium-containing polymers. In 2017, Pan et al. reported the use of selenolactone as the precursor of the polymerization and obtained a series of diselenide-containing polymers with different topologies (Figure 10.5a) [37]. Selenolactone could be easily synthesized via the reaction of sodium hydrogen selenide (NaHSe) and 4-chlorobutanoyl chloride. Selenolactone could be opened by diamine like 4,9-dioxadodecanediamine to form selenol structures, which could be further oxidized to form diselenide bond. The diselenide-containing polymer had a molecular weight M_n of 5.0 kDa and a polydispersity of 1.8 according to SEC measurement. In 2018, Lang and coworkers developed an approach to synthesize well-defined selenium-containing aliphatic polycarbonates via the ring-opening polymerization of a selenic cyclic carbonate dimer monomer (M_{Se}) (Figure 10.5b) [38, 39]. This monomer was prepared by intermolecular cyclization of di(1-hydroxyethylene) selenide and diphenyl carbonate catalyzed by lipase CA. Then the monomer underwent ring-opening polymerization to obtain selenium-containing aliphatic polycarbonates under the same catalyst. The obtained polymer has a M_n of 35.9 kDa and a polydispersity of 1.3. There was a linear relationship between time and $\ln([M]_0/[M_t])$, indicating the living feature of the polymerization. Besides, this approach is expandable as this monomer could copolymerize with other carbonate monomers to prepare polycarbonates with different chemical structures. Recently, Zhang et al. reported alternating and regioselective selenium-containing polyesters synthesized by ring-opening copolymerization of γ-selenobutyrolactone with epoxide derivatives (Figure 10.5c) [40]. 1,4-Benzenedimethanol or benzyl alcohol were used as protic initiators and a phosphazene base was selected as catalyst. The polymerization was performed under room temperature and could reach a 90% yield within 20 minutes. Interestingly, the obtained polyester has an alternating structure with selenide and ester groups coexisting in a single repeating unit. The molecular weight was usually among thousands of Dalton with PDI around 1.0–1.2.

10.2.4 Synthesis of Dendrimer and Hyperbranched Selenium-Containing Polymer

Dendrimer and hyperbranched polymers have been extensively used as catalysts, drug delivery vehicles, self-assembly monolayer, etc. due to their precise structures

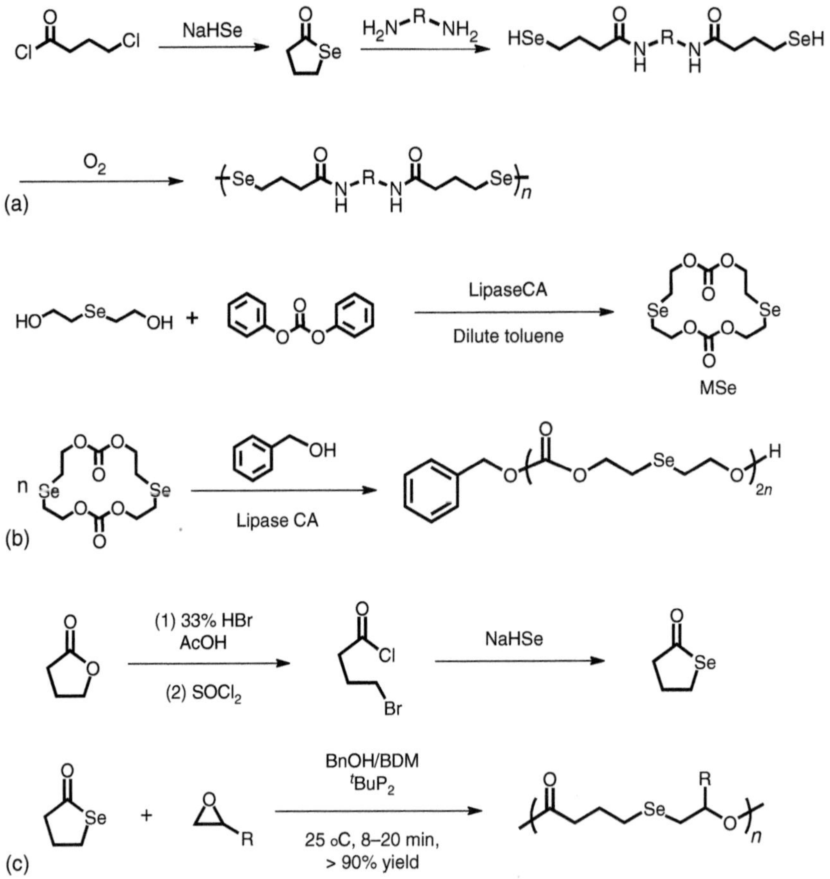

Figure 10.5 Selenium-containing polymers obtained from ring-opening polymerization. (a) Selenolactone used as the precursor to prepare diselenide-containing polymers. (b) Selenic cyclic carbonate dimer used to prepare selenium-containing aliphatic polycarbonates. (c) Using γ-selenobutyrolactone and epoxides to prepare alternating selenium-containing polyesters.

and high functionalities [41]. In 2003, Detty and coworkers synthesized Fréchet-type dendrimers terminated by phenylselenide and phenylseleninic acid groups [42, 43]. This type of dendrimers could be used to catalyze the oxidation of bromide with hydrogen peroxide (Figure 10.6a). In 2004, Takagi and coworkers synthesized a hyperbranched selenium-containing polystyrene via self-condensing vinyl polymerization (Figure 10.6b). Specifically, the monomer p-(phenylselenomethyl) styrene, which has both vinyl and selenide group, could be photopolymerized to generate hyperbranched selenium-containing polystyrene under ultraviolet irradiation [44]. In 2004, Zhang et al. synthesized a three-generation poly(aryl ether) dendrimer with a diselenide core (Figure 10.6c) [45]. In 2006, Smet and coworkers prepared a hyperbranched polyselenides simply by reacting 1,3,5-tris-bromomethyl-2,4,6-trimethyl-benzene with NaHSe (Figure 10.6d) [46]. The dendrimer and hyperbranched polymer could be used as GPx mimics with high

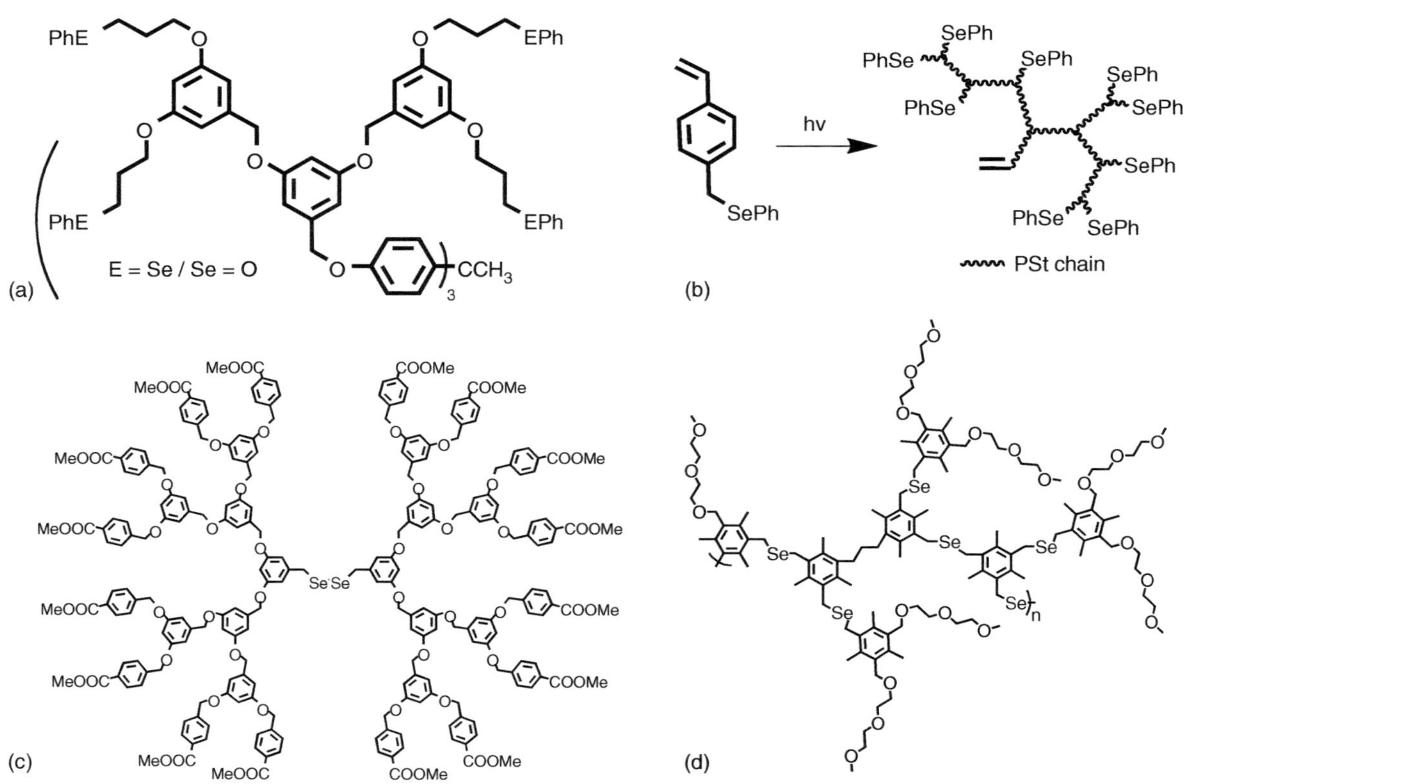

Figure 10.6 Synthesis of dendrimer and hyperbranched selenium-containing polymers. (a) Fréchet-type dendrimers terminating in phenylselenide and phenylseleninic acid groups. (b) Synthesis of a hyperbranched polymer by self-condensing vinyl polymerization. (c) Structure of a three-generation poly (aryl ether) dendrimer with a diselenide core. (d) A hyperbranched polyselenide prepared by 1,3,5-tris-bromomethyl-2,4,6-trimethyl-benzene with NaHSe.

catalytic efficiency. Huang, Yan and coworkers synthesized a hyperbranched poly-diselenide encompassing alternating diselenide groups and phosphate segments in the backbone. Bis-hydroxyl diselenide monomer was first prepared and then reacted with the phosphorus oxychloride ($POCl_3$) in the presence of triethylamine to obtain the polymer. The product was well characterized by NMR, fourier transform infrared spectrometer (FTIR), and differential scanning calorimetry (DSC). This polymer was amphiphilic and can self-assemble into nanomicelles as a broad spectrum anticancer agent [47, 48].

In conclusion, selenium-containing polymer could be prepared by a series of polymerization methods including step growth polymerization, radical polymerization, ring-opening polymerization, etc. Different synthesis routes lead to polymers with various structures and functions. We believe in the future more monomers and polymerization techniques will be designed to facilitate the synthesis of selenium-containing polymers. And the successful preparation of those polymers can provide new building blocks for further applications.

10.3 Selenium-Containing Dynamic Covalent Chemistry

Dynamic covalent bond (DCB) is a kind of chemical bond that can cleave, reform, and exchange under external stimuli [49, 50]. The concept of DCB was first put forward by Professor J. M. Lehn in 1999 and has since then draw much attention due to their dynamic and adaptive features [51]. Many classic dynamic covalent chemistries like Diels–Alder reaction, amide formation/exchange, acylhydrazone formation/exchange, and disulfide exchange have been well studied and applied in a series of research areas [52–56]. Due to the unique features of DCBs, every discovery of a new one would boost research in the field of supramolecular chemistry and smart responsive materials. Different DCBs usually answer to different stimuli. Among those stimuli, light is particularly intriguing since it can provide time and spatial control over the responsive process. In the past few years a series of selenium-containing bonds have been proved to be DCBs and each of them can respond to a different stimuli [57, 58].

10.3.1 Diselenide Bond

Disulfide bond has long been known as a type of DCB and has been applied to many interesting areas [59]. The disulfide exchange reaction is usually trigged by either UV irradiation, base or catalytic reductants. From energy perspective, the bond energy of S—S is 240 kJ mol^{-1} whilst that of Se—Se bond is only 172 kJ mol^{-1} [60]. Single molecule force spectroscopy (SMFS) study also revealed that the rupture force of one S—S bond was ~1450 pN, while that of one Se—Se bond was ~1100 pN [61]. A smaller rupture force means Se—Se bond is more liable to break under external force. Both evidences implied that Se—Se bond may also be a DCB with higher sensitivity than S—S bond. Xu and coworkers first reported that Se—Se bond is indeed a selenium-containing DCB in 2014 [35]. Two symmetric diselenide-containing molecules could undergo exchange reaction to generate an asymmetric diselenide-containing molecule when exposed to visible light without adding catalysts or reactants (Figure 10.7a). The final composition is 25% of each reactant and 50% of exchanged product due to the chemical equilibrium. The metathesis reaction could be easily monitored by ^1H NMR and the reaction kinetics could be investigated via the changes of the integration of the distinct peaks adjacent to selenium atoms based on spectrum (Figure 10.7b). ^{77}Se NMR also offers a unique perspective to illustrate the diselenide metathesis reaction. The new peaks evolved after metathesis clearly indicated the occurrence of the reaction and the asymmetric exchanged product (Figure 10.7c). Apart from visible light stimuli, it was found that the metathesis reaction could also happen by heating over 70 °C. The reaction was proved to be radical mechanism as the radical scavenger like 2,2,6,6-tetramethylpiperidinooxy (TEMPO) could significantly curb the reaction (Figure 10.7d). It is worth noting that diselenide metathesis does not show solvent selectivity and can extend to polymer system as well. Two polystyrenes with different molecular weights could exchange to obtain medium-molecular weight polystyrene though it took longer time to reach equilibrium due to the entanglement of the long polymer chains.

The fact that diselenide metathesis only requires visible light allows it to be a promising candidate in surface/interface chemistry. Hence, it is interesting to exam if diselenide metathesis can occur at the surface/interface [62]. An interface model was designed based on the Pendant Drop test (Figure 10.8a). The oil drop containing hydrophobic dibenzyl diselenide were hanging inside the water phase containing hydrophilic ditetraglycol diselenide. When giving visible light irradiation, a dramatic decrease of the interface tension could be spotted as two diselenide molecules exchanged at the interface to give amphiphilic diselenide product. The latter served as surfactants that helped to stabilize the oil/water interface (Figure 10.8b). It is believed this discovery opened up new avenues for the combination of interface chemistry with diselenide dynamic chemistry. Additionally, how to shift the equilibrium of a dynamic process is a challenging yet essential question in the dynamic covalent chemistry. Applications like dynamic combinatory chemistry require a "lock in" step, where the system could be biased to a favorable state to obtain target products. The mixture of hydrophilic diselenide polymer ($mPEGSe_2$) and hydrophobic diselenide polymer ($PSSe_2$) could generate amphiphilic block polymer (mPEGSeSePS) after exchange reaction, which could self-assemble into micelles in water (Figure 10.8c,d). The unreacted $mPEGSe_2$ could be removed by dialysis and the remaining $PSSe_2$ precipitate could be removed by filtration, though some of the unreacted $PSSe_2$ tend to co-assemble into the hydrophobic core of the micelles. Further experiments proved that the resulting micelles showed light stability even though the composition is not at its stable stage. This means the micelle induced stabilization could serve as an efficient way to shift the dynamic equilibrium of the diselenide metathesis. Recent study also showed that diselenide bond can respond to osmotic pressure [63]. A diselenide bond containing amphiphilic polymer was synthesized and self-assembled to form vesicles in aqueous solution. Adding electrolytes like sodium chloride in water

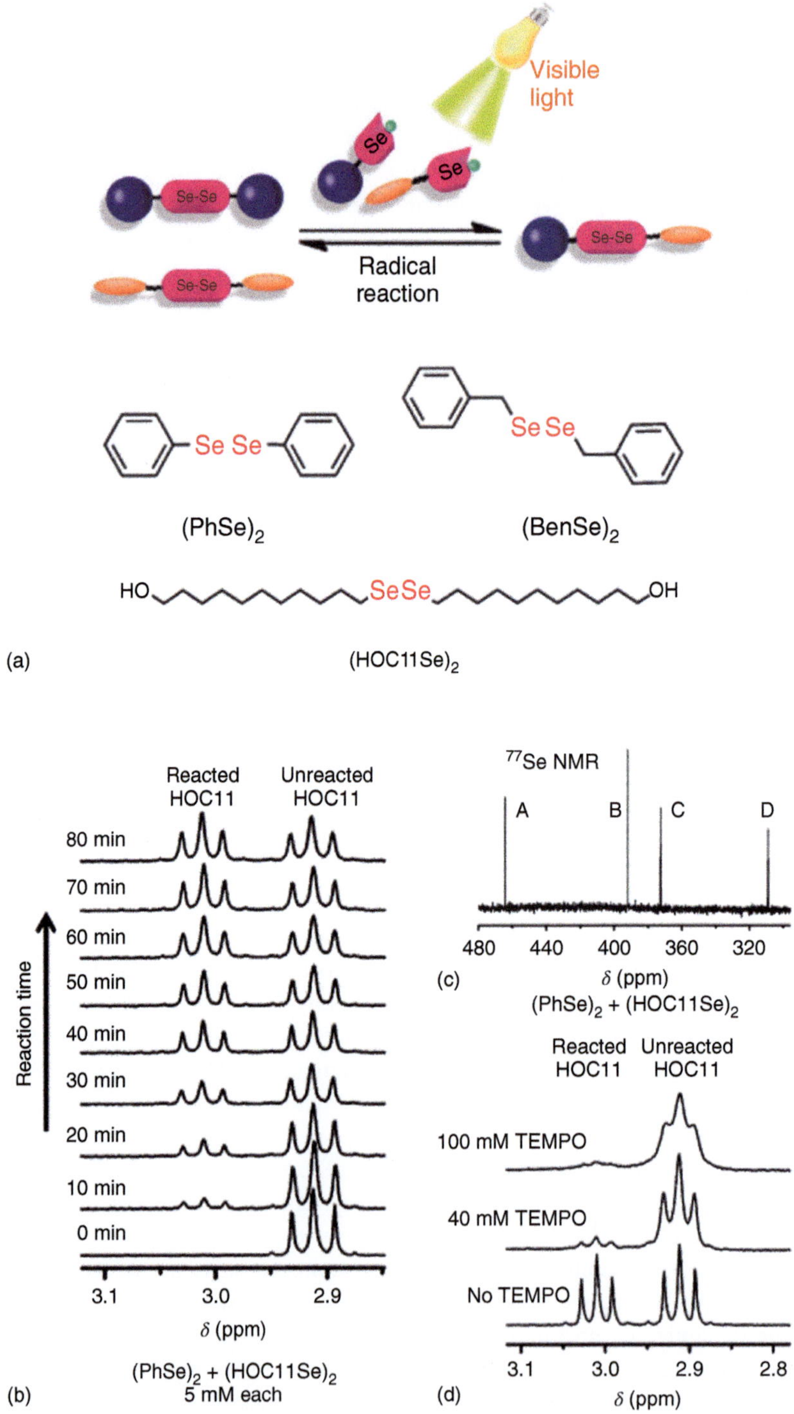

Visible light
Se
Se-Se
Radical reaction
Se Se
(PhSe)$_2$
(BenSe)$_2$
HO SeSe OH
(a)
(HOC11Se)$_2$
Reacted HOC11
Unreacted HOC11
80 min
70 min
60 min
50 min
40 min
30 min
20 min
10 min
0 min
Reaction time
3.1
3.0
2.9
δ (ppm)
(PhSe)$_2$ + (HOC11Se)$_2$
5 mM each
(b)
^{77}Se NMR
A
B
C
D
480
440
400
360
320
δ (ppm)
(c)
(PhSe)$_2$ + (HOC11Se)$_2$
Reacted HOC11
Unreacted HOC11
100 mM TEMPO
40 mM TEMPO
No TEMPO
3.1
3.0
2.9
2.8
δ (ppm)
(d)

Figure 10.7 (a) Diselenide metathesis under visible light irradiation. (b) Peak shift in ^{1}H NMR monitored during metathesis. (c) Two new peaks showed up after diselenide metathesis indicating the appearance of the exchange product. (d) Radical scavenger like TEMPO significantly prohibits the reaction, proving the radical mechanism of the reaction. Source: Reproduced from Ref. [35]. Copyright 2014 John Wiley and Sons.

could apply osmotic force between two sides of the vesicles' shell, which could break the diselenide bond and led to a controlled release manner.

10.3.2 Se–S Bond

From the Pauling equation one can calculate the bonding energy of Se—S bond is 203 kJ mol^{-1} [60], which is in between that of Se—Se and S—S, suggesting Se—S may also be a DCB. In 2018, Xu and coworkers reported that Se—S bond was another light regulated DCB (Figure 10.9a) [64]. Se—S bond could be generated by mixing diselenide and disulfide molecules under 254 nm UV light irradiation (Figure 10.9b). This process was confirmed by both ^{1}H NMR and ^{77}Se NMR (Figure 10.9c,d). The exchange ratio was also 50% due to the chemical equilibrium. When the system was further irradiated by 410 nm light, from NMR one can see Se—S product gradually reduced while Se—Se product increased. This is because the energy of 410 nm light is sufficient to break Se—Se and Se—S bond but not high enough to break S—S bond. Consequently, Se—S would exchange back to Se—Se and S—S but S—S will stay dormant under this circumstance, leading to a reverse reaction. This result showed that the combination of selenium-/sulfur-containing DCBs could build a wavelength-controlled oscillation system since the dynamic exchange of each DCB has a distinct wavelength threshold. Pittelkow and coworkers also established a dynamic combinatorial library (DCL) where diselenides, disulfides, and selenenylsulfides coexist [65]. The DCL was constructed by mixing a bis-diselenide macrocycle and a bis-disulfide cycle together in the presence of 5-mercaptoisophthalic acid acting as the catalyst. Additionally, diselenide was proved to accelerate the formation of disulfide based dynamic combinatorial libraries at physiological pH.

10.3.3 Se–Te Bond

UV light is often criticized for its damage to bio-samples and its limited penetration depth. The current research trend is to develop DCBs with milder responsive behavior. Since the bond energy of Se—Te bond is only 167 kJ mol^{-1} [60], it is reasonable to believe Se—Te bond could be a DCB with super-sensitivity to light. In 2019 it was demonstrated that Se—Se and Te—Te molecules can exchange to form Se—Te bond under visible light with wavelength longer than 600 nm (Figure 10.10a) [66]. The rose of new peak (peak b) proved the occurrence of exchange reaction in Figure 10.10b. This metathesis could be switched on and off in the presence or

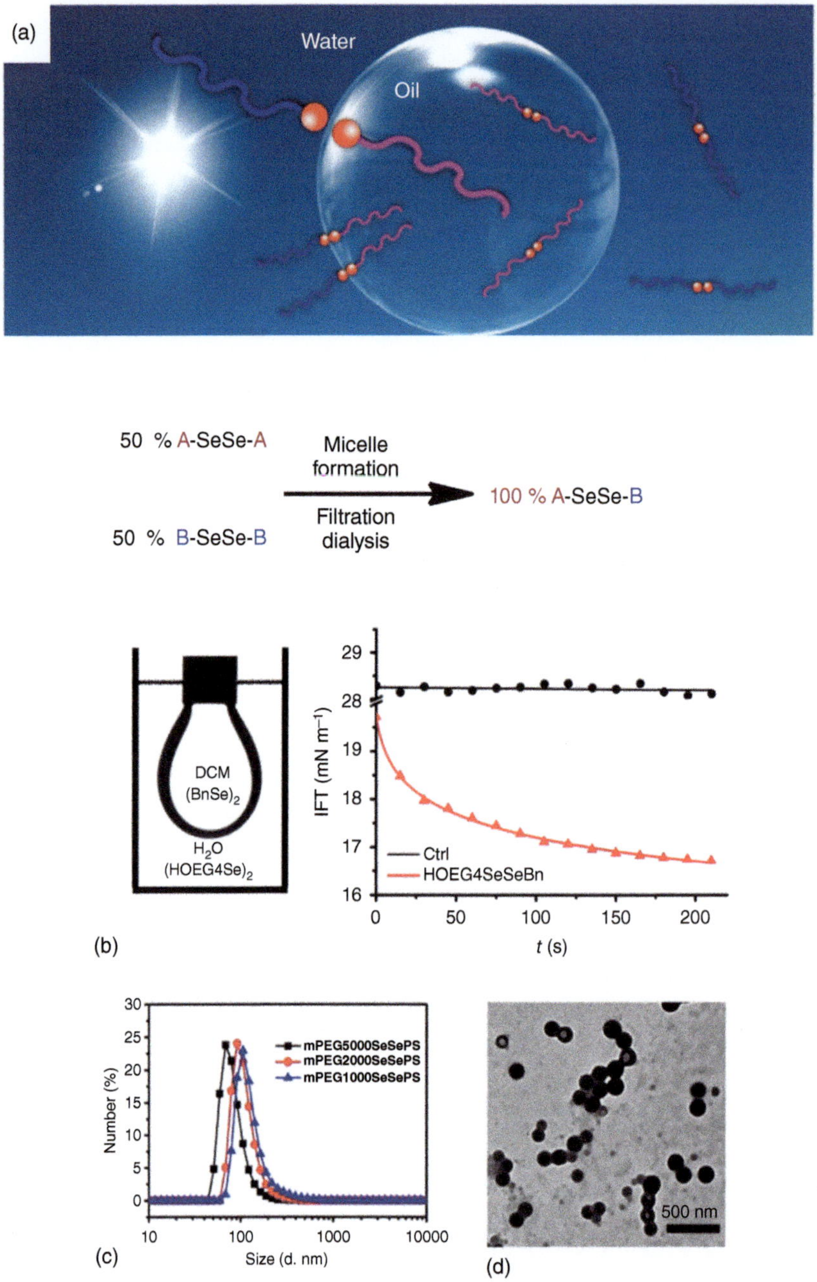

Figure 10.8 (a) Diselenide metathesis at the oil/water interface. (b) Pendant Drop model, the decrease of the surface energy indicated the occurrence of the dynamic diselenide metathesis. (c) DLS results and (d) TEM image of the micelles formed by amphiphilic diselenide-containing polymer. Source: (a, d) Reproduced from Ref. [62] by permission of The Royal Society of Chemistry.

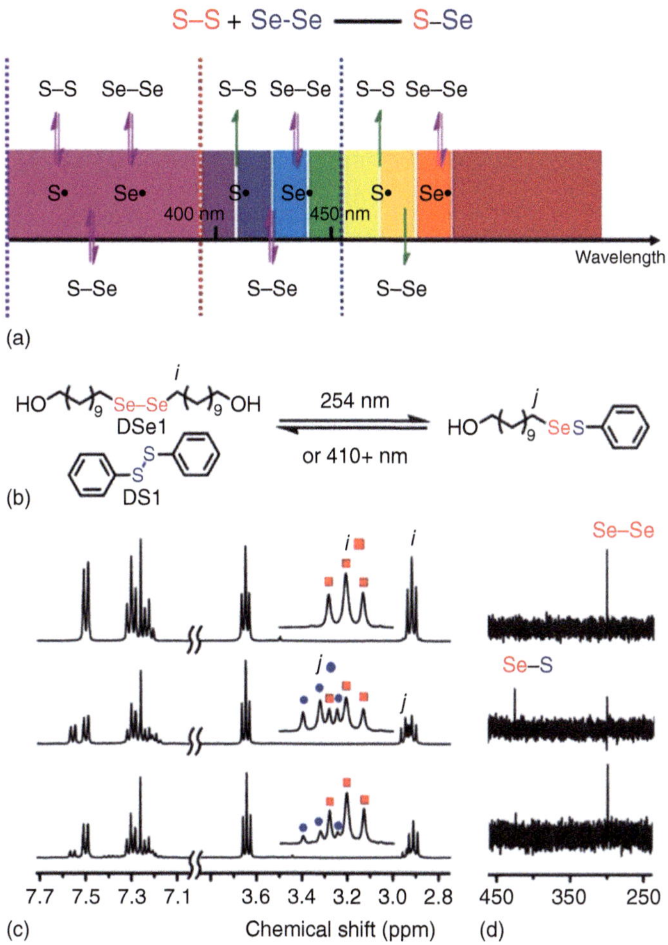

Figure 10.9 (a) Using light with different wavelengths to regulate the exchange reaction between selenium/sulfur-containing DCB. (b) Reversible exchange reaction between Se—Se, S—S, and Se—S bond under light irradiation. (c) ^{1}H NMR of the reversible exchange reaction. (d) ^{77}Se NMR of the reversible exchange reaction. Source: Reproduced from Ref. [64]. Copyright 2018 John Wiley and Sons.

absence of visible light, respectively (Figure 10.10c). EPR spectra further supported the radical mechanism (Figure 10.10d). Moreover, Kinetic investigation was conducted to obtain the insight of this mild response. As shown in Figure 10.10e, this is a first order reaction with an apparent activation energy only at 28.1 kJ mol^{-1}.

10.3.4 Se—N Bond

From the research of GPx, scientists have found that selenium-nitrogen interaction played an important role in their catalytic activities [67]. Xu and coworkers first proposed the discovery of Se—N DCB [68]. The bond energy of Se—N is estimated to be 193 kJ mol^{-1} based on the Pauling equation [60]. It is found that

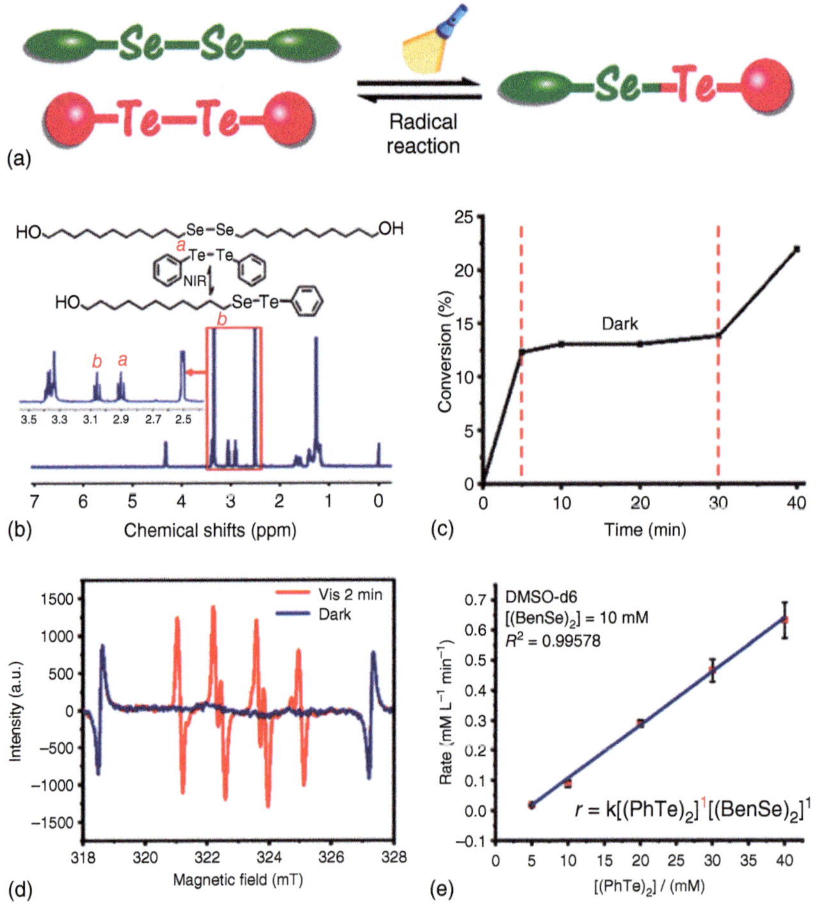

Figure 10.10 (a) Exchange between diselenide and ditelluride bonds to form Se—Te bond under visible light irradiation. (b) ^{1}H NMR of the exchange reaction. (c) Switching the exchange reaction in the presence or absence of visible light. (d) Tellurol radical signal captured by EPR. (e) The initial reaction rate showed first order kinetics with respect to $[(PhTe)_2]$ with a fixed concentration of $(BenSe)_2$. Source: Reproduced from Ref. [66] by permission of The Royal Society of Chemistry.

Se of phenylselenyl halogen species (PhSeBr) and N of pyridine derivative could form Se-N DCB under sonication within one minute. And the formed Se—N DCB could be efficiently cleaved by heating, adding acid, or replacing by stronger electron-donating pyridine derivatives (Figure 10.11a). The formation of the Se-N DCB was achieved by sonication of 1 : 1 mixture of PhSeX (X = Br or Cl) and polystyrene-b-poly(4-vinylpyridine) (PS-b-P4VP) (Figure 10.11b). The shift of ^{77}Se NMR signals from 868.7 to 1180.3 ppm before and after reaction indicated the chemical environment of selenium atoms have changed. Evidences of FTIR, X-ray photoelectron spectroscopy (XPS), transmission electron microscopy (TEM), ^{1}H NMR, and dynamic light scattering (DLS) also confirmed the formation of

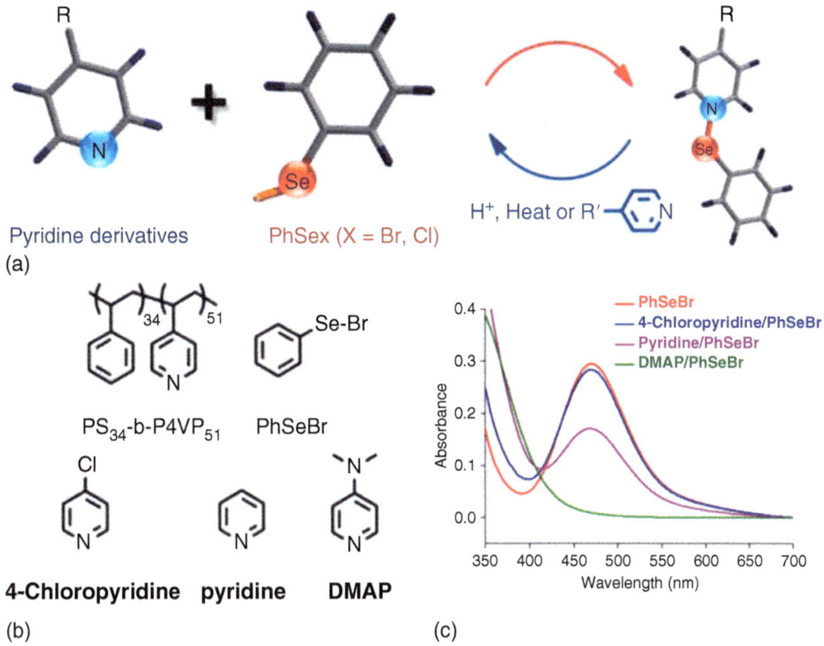

Figure 10.11 (a) Se—N covalent bond formed by pyridine derivatives and PhSeX. (b) Structure of the polymer and pyridine derivatives. (c) UV–Vis spectra of PhSeBr with pyridine derivatives. The signal at 470 nm decreased as electron-donating ability of the N atoms in the model molecules get stronger. Source: Reproduced from Ref. [68]. Copyright 2013 John Wiley and Sons.

Se—N DCB. Then the cleavage of the Se—N DCB was realized by heating and adding 4-dimethylaminopyridine (DMAP) or trifluoroacetic acid. PhSeBr has a unique UV–Vis absorption at 470 nm due to the auxochrome group of selenyl bromide, which was utilized to monitor the concentration of PhSeBr in solution (Figure 10.11c). This absorption changed with the adding of CF_3COOH and pyridine in sequence for several cycles, which proved its reversibility under pH stimuli.

Ebselen is an anti-inflammatory, anti-oxidant drug that contains an Se—N bond in the molecular structure [7, 8]. Xu and coworkers designed and synthesized a compound composed of two ebselen moieties [69]. When DL-dithiothreitol (DTT) was added into the DMSO solution of the monomer, the Se—N bond could be transformed to Se—S bond and formed a novel selenium-containing macrocycle. Adding carboxylate groups as guest template could facilitate the transformation from Se—S bond into Se—Se bond and form a diselenide-containing macrocycle. This work proved that selenium-containing DCBs are able to transform from one to another under certain stimuli.

To summarize, a group of selenium-containing DCBs have been developed in the recent years, among them Se—Se, Se—S, and Se—Te bonds have shown unique light responsiveness. By combining them together, complicated adaptive systems could be achieved. It is believed that selenium-containing DCB family could be further expanded and the transformation rules between them should be established.

10.4 Selenium-Containing Dynamic Materials

Introducing selenium-containing DCBs into materials could endow them with dynamic features. Selenium-containing DCBs have been used to prepare elastomers, surface/interface materials, and nanomaterials and they have shown prospects in many applications. It is worth noting that except for dynamic materials, selenium-containing polymers are also important biomaterials serving as drug delivery vehicles and anti-cancer agents. However, due to the limited space this part will not be covered in this chapter and readers who are interested could find more information in some recent reviews [10, 58, 70–72].

10.4.1 Selenium-Containing Elastomer

With the mild response of the diselenide DCB, it was successfully incorporated into the fabrication of self-healing materials with visible light healing property (Figure 10.12a) [73]. A series of PU with different diselenide bond contents were synthesized by different ratios of TDI, polytetramethylene glycol (PTMG), and di-(1-hydroxyundecyl) diselenide (DiSe). A visual proof of the self-healing property was first demonstrated. PU was cut and attached together, which was followed by 24 hours visible light irradiation from a common table lamp (Figure 10.12b,c). The healed PU was able to bear a weight of 200 g, which was calculated to be 0.2 MPa in terms of tensile strength. Then the mechanical properties of PUs were further tested to quantitatively study the self-healing behavior (Figure 10.12d). According to the strain-stress tests, the original breaking strain and stress of PU was 380% and 1.3 MPa, respectively. After visible light irradiation for 24 and 48 hours, the breaking stress could reach 45% and 72% to its original states. Since it is the dynamic diselenide metathesis that accounts for the healing process, the entanglement of polymer chains is unable to be fully recovered, which explains the breaking stress difference before and after cutting-healing treatment. However, Young's modulus was almost fully recovered, from 2.8 MPa for original to 2.4 MPa after 24 hours healing and 2.8 MPa after 45 hours healing. In order to optimize the self-healing performance, blue laser was employed to replace table lamp. By doing so the healing period was significantly shortened from 48 hours to 30 minutes, meanwhile the healing efficiency was further boosted to 84% (Figure 10.12e). Additionally, using laser as the light source allows potential for remote healing as laser is spatial controllable and light intensive. Sardon and coworkers also reported a kind of self-healing materials based on diselenide DCB [74]. The obtained PU was found to be able to self-heal under room temperature and showed enhanced reprocessing characteristic compared with disulfide counterpart. Furthermore, the commonly used catalyst dibutyltin dilaurate (DBTDL) for PU preparation was found to boost diselenide exchange rate, thus accelerating the self-healing process.

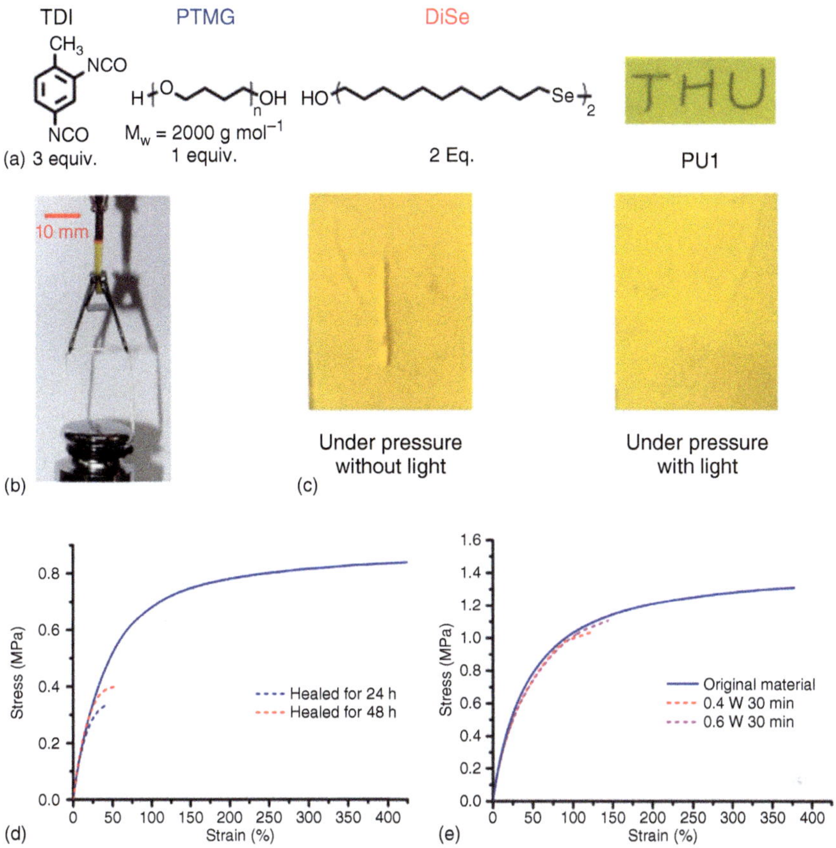

Figure 10.12 (a) Formula of the diselenide-containing polyurethane elastomer. (b) Visual proof of the self-healing property of the polymer, the red part was stained with Nile Red for clarity. (c) Healing behavior under pressure, the crack disappeared after 24 hours light irradiation. (d) Mechanical properties of the polymer before and after self-healing. (e) Self-healing behavior under laser irradiation. Source: Reproduced from Ref. [73]. Copyright 2015 John Wiley and Sons.

Apart from self-healing materials, the diselenide DCB was also employed in shape memory materials with visible light induced plasticity [75]. 4,4′-Diphenylmethane diisocyanate (MDI), PTMG, DiSe, and crosslinker Glycerol was used in different ratios to fabricate thermoset PUs with different diselenide bond contents varying from 0% to 100%. It was found that the shape memory efficiency increased with the rise of the diselenide bond content. Subsequently, the visible light induced plasticity was investigated. As expected, a higher diselenide bond content led to better stress relaxation behavior (Figure 10.13a). The strain fixation was determined to increase with the increase of diselenide bond content which was consistent with the stress relaxation behavior (Figure 10.13b). The mechanism behind is the diselenide metathesis reaction triggered by visible light. For most of the thermoset shape memory materials, the permanent shape was fixed once being fabricated. But the combination of shape memory property and light induced plasticity allowed the

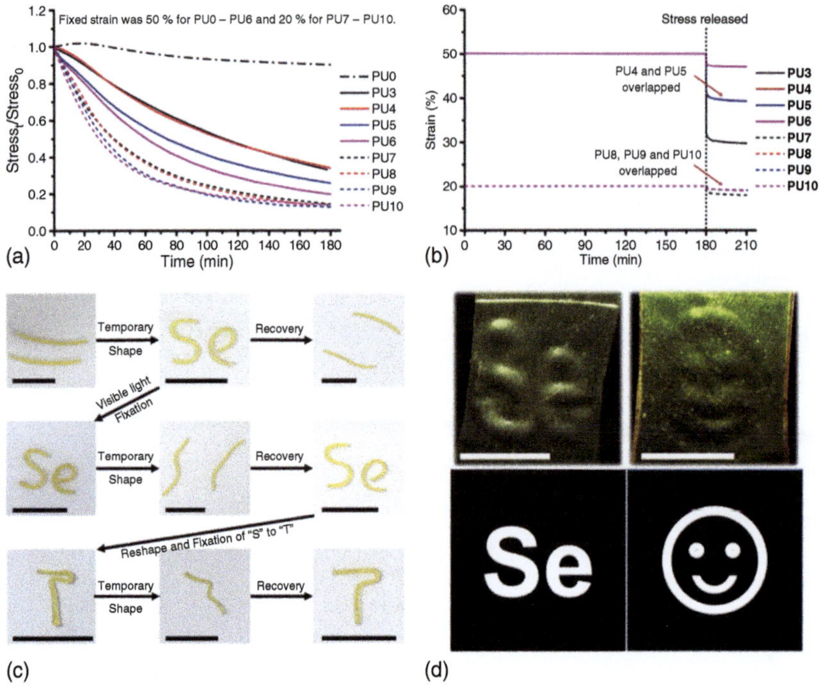

Figure 10.13 (a) Stress relaxation and (b) strain fixation behavior of elastomer with different contents of diselenide bonds. (c) Shape memory cycles and visible light-induced transition of the permanent shape. Scale bar: 2 cm. (d) Visible light patterning based on a commercialized projector. Source: Reproduced from Ref. [75]. Copyright 2017 the American Chemical Society.

transition from a temporary shape to a permanent shape simply by visible light irradiation during the shape memory cycles (Figure 10.13c). As a demonstration, certain images were projected on the pre-stretched materials by a commercialized projector. The projected area underwent stress relaxation while the other parts remained intact. When the stretch was released from the sample, the stress relaxed area bulged to give the pattern of the projected images (Figure 10.13d).

Surface patterning of materials has wide applications in oil water separation, advanced anticounterfeiting, and optical devices. However, the traditional photopatterning could only generate 2D patterns. Xu and coworkers demonstrated that diselenide containing elastomer can achieve 3D patterning by swelling-induced mechanophotopatterning (Figure 10.14a) [76]. A diselenide containing crosslinked PU was synthesized by reacting HDI with PTMG, DiSe, and glycerol. When the elastomer was sunk in good solvents like ethyl acetate, the three-dimensional polymer network would be stretched due to swelling and reach an equilibrium when the swelling force equaled to the elastic contractile force. Upon regional light irradiation under a photomask, the elastic contractile force would be partially relaxed. The polymer network could then be further swelled until a new equilibrium was achieved. Consequently, the difference in swelling extent of light/dark zones

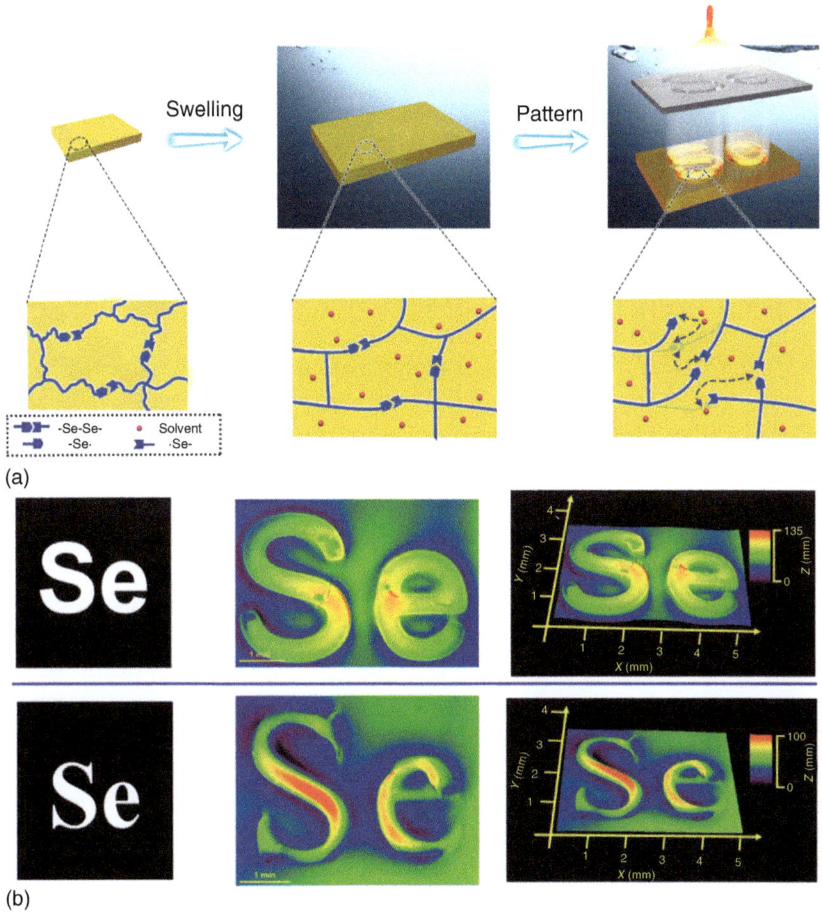

Figure 10.14 (a) Swelling-induced mechanophotopatterning of diselenide containing elastomer. (b) 2D and 3D images of the pattering character "Se" in different fonts. Source: Reproduced from Ref. [76] by permission of The Royal Society of Chemistry.

could cause a 3D patterning. As the demonstration shown in Figure 10.14b, two similar "Se" characters with different fonts were printed on the materials. Although two characters were almost identical in 2D pictures, notable differences could be observed in 3D images, exhibiting their excellence in advanced anticounterfeiting. It is worth noting that this method could be applied in many shapes including thin films, bulk, and cylindrical materials.

Another example of selenium-containing dynamic elastomer was an attachable/cleavable material based on the light modulation of diselenide/disulfide exchange reaction [64]. A diselenide and a disulfide-containing elastomers were synthesized and attached together. Upon UV light (256 nm) irradiation, diselenide and disulfide could exchange at the interface to form Se—S bond that adhered two pieces together. Then a 457 nm blue light was used to replace UV light, rupturing the Se—S bond and detaching two pieces.

10.4.2 Selenium-Containing Surface/Interface Materials

Many selenium-containing DCBs respond to light, making them suitable candidate for surface/interface applications. Xu and coworkers reported a surface modification method based on the diselenide dynamic bond [77]. As shown in Figure 10.15a, diselenide bond was first introduced to substrates like quartz. Then different diselenide compounds could be modified onto surface by diselenide metathesis under visible light irradiation, endowing surface with distinct chemical features. Due to the dynamic property of DCB, fine tuning of surface wettability could be achieved by substituting hydrophilic or hydrophobic molecules on surface (Figure 10.15b). More importantly, one can achieve visible light controlled liquid motion by this method (Figure 10.15c). The inner wall of capillary tube was modified by diselenide bonds and inserted into an aqueous solution of a hydrophilic diselenide molecule $(HOEG4Se)_2$. When giving visible light irradiation, the wettability of the inner surface would increase, leading to the climbing of the liquid surface inside the tube. Besides, this method could also realize photopatterning and bioconjugation. Thanks to the mild diselenide dynamic chemistry, this surface modification approach was mild, fast, and reversible.

Due to the unique oxidation response and visible light response features, the diselenide DCB could be applied to the surface functionalization of two-dimensional materials like graphene oxide (GO) [78]. GO is an important member of two-dimensional materials and has been widely applicable to many research fields. Yet the current existing covalent surface functionalization methods usually requires harsh conditions, long reaction time, and low reactivity, while the non-covalent methods are inevitably unstable with amounts of competitive species in physiological environment. Since GO is oxidative and radical attackable due to its unique chemical structure, it was found that diselenide-containing molecules could react with GO to realize *in situ* covalent surface functionalization in a relatively mild way. The chemical structure of the obtained GOSe was characterized by FT-IR, time of flight secondary ion mass spectrometry (ToF-SIMS), atomic force microscope (AFM), etc. The reaction mechanism was deduced to involve both redox reaction and radical addition reaction. The selenium modified GO material was proved to be capable of modulating the balance of reactive oxygen species *in vitro*. Feng and coworkers designed a field-effect transistors (FET) based on nitrogen-doped graphene (NG) [79]. The FET is a type of transistor which that uses an electric field to control the flow of current. Nitrogen atoms on NG can reversibly react with phenylselenyl bromine (PhSeBr) to form Se-N DCB, altering the electrical properties of EFT in a controlled manner.

10.4.3 Selenium-Containing Nanomaterials

There are also some nanomaterials derived from selenium-containing DCBs. For example, Wang and coworkers used Se—N DCB to adjust the amphiphilicity of the polymer PS-*b*-P4VP and to achieve the one step double emulsion in DCM/H_2O system [80]. The PhSeBr could react with P4VP segment to obtain Se—N cation and Br

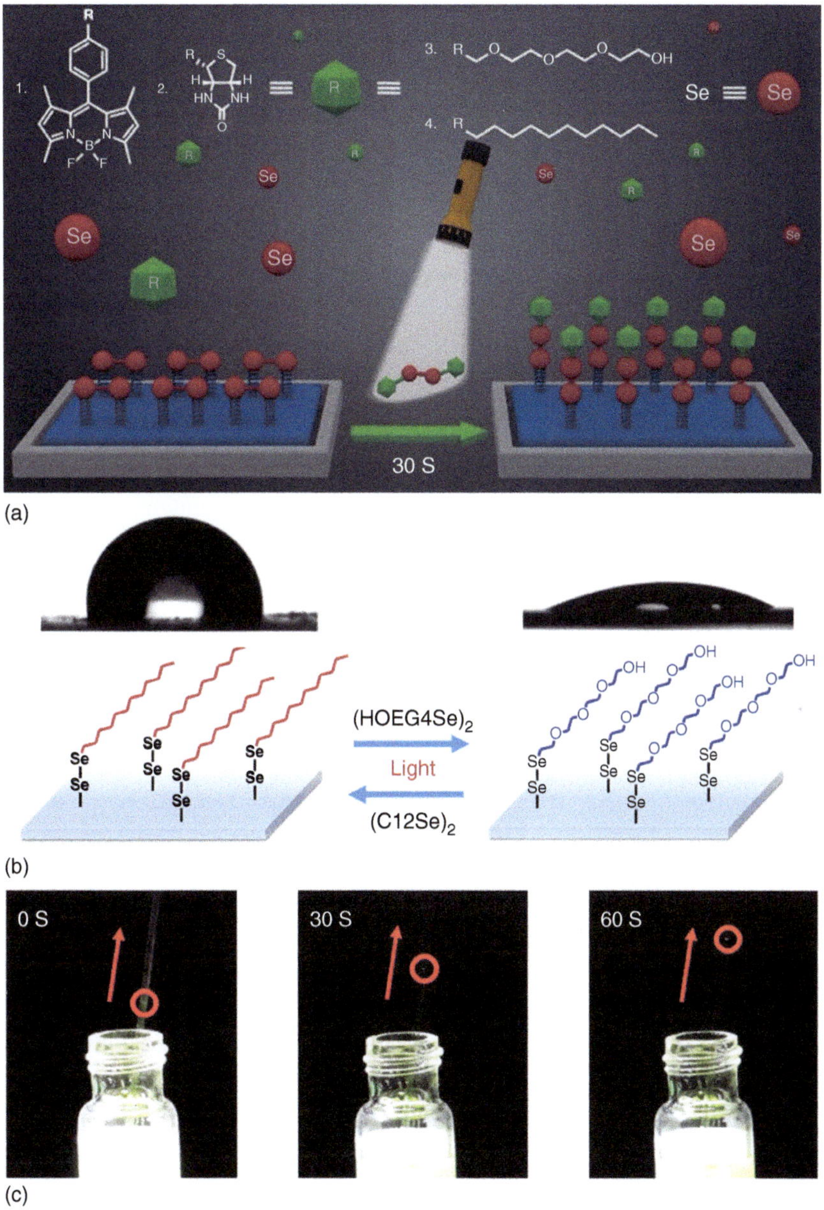

Figure 10.15 (a) Surface modification based on diselenide metathesis under visible light. (b) Switching the wettability by replacing diselenide molecules attached to the substrate. (c) Liquid motion driven by visible light irradiation. Source: Reproduced from Ref. [77]. Copyright 2019 John Wiley and Sons.

counter ion, thus enhance the hydrophilicity to the polymer. When tuning the ratio of pyridine group to PhSeBr, a series of particles with different morphologies could be prepared. Since all of these particles had selenium-containing active centers, they could be used as GPx mimics to catalyze the reduction of hydroperoxides. Li and coworkers developed an initiation-growth strategy that can produce hierarchical nanostructures with well controlled sizes and morphologies based on a diblock copolymer P2VP-*b*-PFMA (poly(2-vinyl pyridine)-*b*-poly(2-(perfluorooctyl)ethyl methacrylate) [81]. PhSeBr can react with copolymer to form complex with Se—N bond. This complex was unstable and would subsequently grow into well-ordered nanostructures driven by liquid crystal ordering effect. Due to the reversible nature of Se—N bond, the obtained nanostructure could be disrupted at high temperature and reformed at low temperature.

In conclusion, selenium-containing DCBs have been extensively used in the field of dynamic materials. Elastomers with diselenide bond can be self-healing materials, shape memory materials with light plasticity, mechanophotopatterning substrates, and attachable/cleavable materials. Selenium-containing DCBs can also be developed into a versatile, fast, reversible surface modification method and to modify 2D materials with tunable features. Furthermore, dynamic properties of selenium-containing DCBs can be used in nanomaterials and brought up novel applications.

10.5 Conclusion and Outlook

The unique chemical properties of selenium element have endowed selenium-containing polymer a series of fascinating features. The methods to synthesize selenium-containing polymer have been widely expanded in recent years. Step growth polymerization allows an efficient way to obtain polymers with a relatively high content of selenium because selenium was incorporated into each repeating unites. However, the shortcoming of this method is obvious, the structure and the polydispersity cannot be precisely controlled. As for radical polymerization, where selenium-containing compounds were used as initiator or RAFT agents, there are only limited amount of selenium-containing residual groups left in the polymer chain, which may limit their application in certain areas. Additionally, selenium-containing monomers are usually incompatible with living radical polymerization like ATRP and RAFT since selenium-containing groups may react with initiator, metal catalysts (ATRP), and sulfur-containing RAFT agents via radical reactions or coordination reactions. Ring-opening polymerization seems to offer a balance between the selenium content and controllable polymer structures. Yet there are only limited monomers to be selected. We expect that the challenge of controlled radical polymerization of selenium-containing monomer could be overcome in the future, which may provide a universal preparation method that can offer structure controllable polymers with high selenium contents.

Selenium-containing DCB is a new area full of opportunities. We have introduced a series of selenium-containing DCBs like Se—Se, Se—S, Se—Te, and Se—N. From the chemistry perspective, more selenium-containing DCBs should be discovered and their internal transformation rules should be established. In the meantime, the thermodynamic and kinetic process of the dynamic covalent reactions should be thoroughly investigated. From the application perspective, firstly scientists in different fields can work together to seek for more practical applications for the existing selenium-containing DCBs. For instance, since many enzymes and proteins contain S—S bond, the Se—S DCB could offer opportunities in the investigation and utilization of protein-based materials. Secondly, it has been proved that a combination of two selenium-containing DCBs in one system can generate an interesting oscillation system. It can be expected that more intelligent and adaptive materials may be achieved by incorporating more than one type of DCB. Although we are still in the primitive stage of developing adaptive materials, it is anticipated that the mild, sensitive, and unique selenium-containing DCB family will play an important role in it.

Acknowledgments

This work was financially supported by the National Science Foundation for Distinguished Young Scholars (Grant 21425416), the National Natural Science Foundation of China (Grant 21734006), and the National Basic Research Plan of China (2018YFA0208900).

References

1 Cai, X., Wang, C., Yu, W. et al. (2016). *Sci. Rep.* 6: 19213.
2 Long, J.A., Large, R.R., Lee, M.S.Y. et al. (2016). *Gondw. Res.* 36: 209.
3 Boyd, R. (2011). *Nat. Chem.* 3: 570.
4 Rotruck, J.T., Pope, A.L., Ganther, H.E. et al. (1973). *Science* 179: 588.
5 Mugesh, G. and Singh, H.B. (2000). *Chem. Soc. Rev.* 29: 347.
6 Mugesh, G., du Mont, W.W., and Sies, H. (2001). *Chem. Rev.* 101: 2125.
7 Muller, A., Cadenas, E., Graf, P., and Sies, H. (1984). *Biochem. Pharmacol.* 33: 3235.
8 Wendel, A., Fausel, M., Safayhi, H. et al. (1984). *Biochem. Pharmacol.* 33: 3241.
9 Kishore, K. and Ganesh, K. (1995). *Polym. Synth. Polym. Eng.* 121: 81.
10 Xu, H., Cao, W., and Zhang, X. (2013). *Acc. Chem. Res.* 46: 1647.
11 Prince, M. and Bremer, B. (1967). *J. Polym. Sci. B Polym. Lett.* 5: 843.
12 Kroll, H. and Bolton, E.F. (1970). *J. Appl. Polym. Sci.* 14: 2319.
13 Gunther, W.H.H. and Salzman, M.N. (1972). *Ann. N. Y. Acad. Sci.* 192: 25.
14 Koytepe, S., Seçkin, T., Cetinkaya, E., and Gok, Y. (2005). *J. Inorg. Organomet. Polym. Mater.* 15: 269.
15 Glenis, S., Ginley, D.S., and Frank, A.J. (1987). *J. Appl. Phys.* 62: 190.
16 Yoshino, K., Kohno, Y., Shiraishi, T. et al. (1985). *Synth. Met.* 10: 319.
17 Ma, N., Li, Y., Xu, H. et al. (2010). *J. Am. Chem. Soc.* 132: 442.

18 Ma, N., Li, Y., Ren, H. et al. (2010). *Polym. Chem.* 1: 1609.
19 Ando, T., Kwon, T.S., Kitagawa, A. et al. (1996). *Macromol. Chem. Phys.* 197: 2803.
20 Kwon, T.S., Kumazawa, S., Yokoi, T. et al. (1997). *J. Macromol. Sci. Part A: Pure Appl. Chem.* 34: 1553.
21 Kwon, T.S., Kondo, S., Takagi, K. et al. (1999). *Polym. J.* 31: 483.
22 Kwon, T.S., Takagi, K., Kunisada, H., and Yuki, Y. (2003). *Eur. Polym. J.* 39: 1437.
23 Chiefari, J., Chong, Y.K., Ercole, F. et al. (1998). *Macromolecules* 31: 5559.
24 Moad, G., Rizzardo, E., and Thang, S.H. (2008). *Polymer* 49: 1079.
25 Keddie, D.J. (2014). *Chem. Soc. Rev.* 43: 496.
26 Moon, J., Nam, H., Kim, S. et al. (2008). *Tetrahedron Lett.* 49: 5137.
27 Zeng, J., Zhu, J., Zhang, Z. et al. (2012). *J. Polym. Sci. Part A: Polym. Chem.* 50: 2211.
28 Zhan, Y., Zhang, Z., Pan, X. et al. (2013). *J. Polym. Sci. Part A: Polym. Chem.* 51: 1656.
29 Ma, F., Zhu, J., Zhang, Z. et al. (2013). *J. Polym. Sci. Part A: Polym. Chem.* 51: 3159.
30 Gao, F., Pan, X., Zhu, J. et al. (2015). *Polym. Chem.* 6: 1367.
31 Cai, Z.X., Lu, W.H., Gao, F. et al. (2016). *Macromol. Rapid Commun.* 37: 865.
32 Wang, J.S. and Matyjaszewski, K. (1995). *Macromolecules* 28: 7901.
33 Wang, J.S. and Matyjaszewski, K. (1995). *J. Am. Chem. Soc.* 117: 5614.
34 Matyjaszewski, K. (2012). *Macromolecules* 45: 4015.
35 Ji, S., Cao, W., Yu, Y., and Xu, H. (2014). *Angew. Chem. Int. Ed.* 53: 6781.
36 Fan, C.W., Ding, C.L., Pan, X.Q. et al. (2015). *Macromol. Rapid Commun.* 36: 903.
37 Pan, X.Q., Driessen, F., Zhu, X.L., and Du Prez, F.E. (2017). *ACS Macro Lett.* 6: 89.
38 Wei, C., Zhang, Y., Yan, B. et al. (2018). *Chem. A Eur. J.* 24: 789.
39 Wei, C., Xu, Y., Yan, B. et al. (2018). *ACS Macro Lett.* 7: 336.
40 Zhang, C.-J., Cao, X.-H., and Zhang, X.-H. (2020). *Macromolecules* 53: 203–211.
41 Bosman, A.W., Janssen, H.M., and Meijer, E.W. (1999). *Chem. Rev.* 99: 1665.
42 Drake, M.D., Bright, F.V., and Detty, M.R. (2003). *J. Am. Chem. Soc.* 125: 12558.
43 Francavilla, C., Drake, M.D., Bright, F.V., and Detty, M.R. (2001). *J. Am. Chem. Soc.* 123: 57.
44 Takagi, K., Nishikawa, Y., Kwon, T.S. et al. (2000). *Polym. J.* 32: 970.
45 Zhang, X., Xu, H., Dong, Z.Y. et al. (2004). *J. Am. Chem. Soc.* 126: 10556.
46 Xu, H., Gao, J., Wang, Y.P. et al. (2006). *Chem. Commun.* 0: 796.
47 Liu, J., Pang, Y., Chen, J. et al. (2012). *Biomaterials* 33: 7765.
48 Li, C.T., Huang, W., Zhou, L.Z. et al. (2015). *Polym. Chem.* 6: 6498.
49 Rowan, S.J., Cantrill, S.J., Cousins, G.R.L. et al. (2002). *Angew. Chem. Int. Ed.* 41: 898.
50 Corbett, P.T., Leclaire, J., Vial, L. et al. (2006). *Chem. Rev.* 106: 3652.
51 Lehn, J.M. (1999). *Chem. A Eur. J.* 5: 2455.
52 von Delius, M., Geertsema, E.M., and Leigh, D.A. (2010). *Nat. Chem.* 2: 96.

53 Wilson, A., Gasparini, G., and Matile, S. (2014). *Chem. Soc. Rev.* 43: 1948.
54 Hou, H., Yin, J., and Jiang, X. (2016). *Adv. Mater.* 28: 9126.
55 Herrmann, A. (2014). *Chem. Soc. Rev.* 43: 1899.
56 Moulin, E., Cormos, G., and Giuseppone, N. (2012). *Chem. Soc. Rev.* 41: 1031.
57 Ji, S., Xia, J., and Xu, H. (2016). *ACS Macro Lett.* 5: 9.
58 Xia, J., Li, T., Lu, C., and Xu, H. (2018). *Macromolecules* 51: 7435.
59 Black, S.P., Sanders, J.K.M., and Stefankiewicz, A.R. (2014). *Chem. Soc. Rev.* 43: 1861.
60 Kildahl, N.K. (1995). *J. Chem. Educ.* 72: 423.
61 Xia, J., Li, H., and Xu, H. *Acta Chim. Sin.* https://doi.org/10.11777/j.issn1000-3304.2019.19166.
62 Xia, J., Ji, S., and Xu, H. (2016). *Polym. Chem.* 7: 6708.
63 Xia, J., Zhao, P., Pan, S., and Xu, H. (2019). *ACS Macro Lett.* 8: 629.
64 Fan, F., Ji, S., Sun, C. et al. (2018). *Angew. Chem. Int. Ed.* 57: 16426–16430.
65 Rasmussen, B., Sorensen, A., Gotfredsen, H., and Pittelkow, M. (2014). *Chem. Commun.* 50: 3716.
66 Liu, C., Xia, J., Ji, S. et al. (2019). *Chem. Commun.* 55: 2813.
67 Yi, Y., Fa, S., Cao, W. et al. (2012). *Chem. Commun.* 48: 7495.
68 Yi, Y., Xu, H., Wang, L. et al. (2013). *Chem. A Eur. J.* 19: 9506.
69 Ji, S., El Mard, H., Smet, M. et al. (2017). *Sci. China Chem.* 60: 1191.
70 Cao, W., Wang, L., and Xu, H. (2015). *Nano Today* 10: 717.
71 Cao, W. and Xu, H. (2019). *Mater. Chem. Front.* 3: 2010.
72 Fan, Z. and Xu, H. (2019). *Polym. Rev.* https://doi.org/10.1080/15583724.2019.1641515.
73 Ji, S., Cao, W., Yu, Y., and Xu, H. (2015). *Adv. Mater.* 27: 7740.
74 An, X.W., Aguirresarobe, R.H., Irusta, L. et al. (2017). *Polym. Chem.* 8: 3641.
75 Ji, S., Fan, F., Sun, C. et al. (2017). *ACS Appl. Mater. Interfaces* 9: 33169.
76 Fan, F., Liu, C., Wang, S. et al. (2019). *J. Mater. Chem. C* 7: 10777.
77 Xia, J., Zhao, P., Zheng, K. et al. (2019). *Angew. Chem. Int. Ed.* 58: 542.
78 Xia, J., Li, F., Ji, S., and Xu, H. (2017). *ACS Appl. Mater. Interfaces* 9: 21413.
79 Zheng, N., Feng, Y., Zhang, Y. et al. (2019). *ACS Appl. Mater. Interfaces* 11: 24360.
80 Huang, X., Fang, R., Wang, D. et al. (2015). *Small* 11: 1537.
81 Jin, B., Sano, K., Aya, S. et al. (2019). *Nat. Commun.* 10: 2397.

11

Poly(disulfide)s

Raju Bej and Suhrit Ghosh

School of Applied and Interdisciplinary Sciences, Indian Association for the Cultivation of Science, 2A and 2B Raja S. C. Mullick Road, Kolkata, 700032 India

11.1 Introduction

Sulfur-containing materials have been utilized in wide-range applications spanning from electrochemical to biomedical field [1]. Likewise polymers with sulfur-containing functional groups such as poly(thioether) [2–4], poly(thioester) [5, 6], poly(thiocarbonate) [7, 8], poly(thioamide) [9, 10], poly(thiourea) [11, 12], poly(thiourethane) [13], poly(thiophene) [14], poly(sulfide) [15], and others have been studied for different purpose [16]. Poly(disulfide)s (PDS) belong to a distinct class of macromolecule where the backbone is constructed by the disulfide bond (Figure 11.1). In comparison with carbon—carbon (C—C) bond, disulfide (S—S) bond has longer bond length, lower bond dissociation energy, and rotational barrier along the S—S axis, which makes it dynamic in nature [17–19].

Disulfide functionality is also ubiquitous in biological systems. For example, the dynamic nature of the disulfide bond plays an important role in the folding, structure, and stability of many proteins. One such example is hair that contains a disulfide containing protein called keratin and the softness of the hair depends on the disulfide content. Disulfide bond can be easily formed from free thiol (sulfhydryl) group upon oxidation with various oxidizing agents such as oxygen (from air), dimethyl sulfoxide (DMSO), hydrogen peroxide (H_2O_2), and iodine. Disulfide bond can be reversibly reduced to free thiol by treatment with suitable reducing agents such as dithiothreitol (DTT), 2-mercaptoethanol (β-ME), tris (2-carboxyethyl) phosphine (TCEP), and glutathione reduced (GSH). GSH, a tripeptide, acts as a reducing agent in intra-cellular environment [20]. GSH cleaves a disulfide bond to free thiols and in turn gets oxidized to its dimer (GSSG), which is stabilized through H-bonding in a β sheet-like manner (Figure 11.2a). In cells, concentration of GSH is maintained (Figure 11.2b) by the glutathione redox couple (GSSG/GSH), [22] which is controlled by the glutathione enzymatic system [21].

Sulfur-Containing Polymers: From Synthesis to Functional Materials,
First Edition. Edited by Xing-Hong Zhang and Patrick Theato.

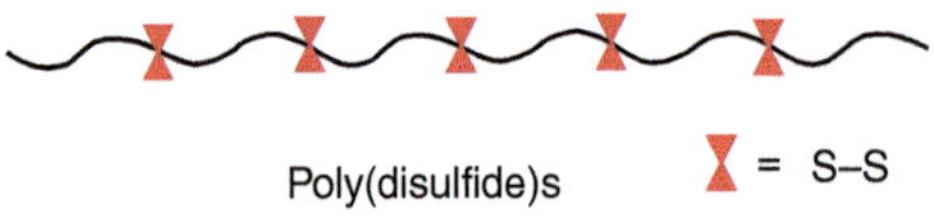

Figure 11.1 General representation of a poly(disulfide)s chain.

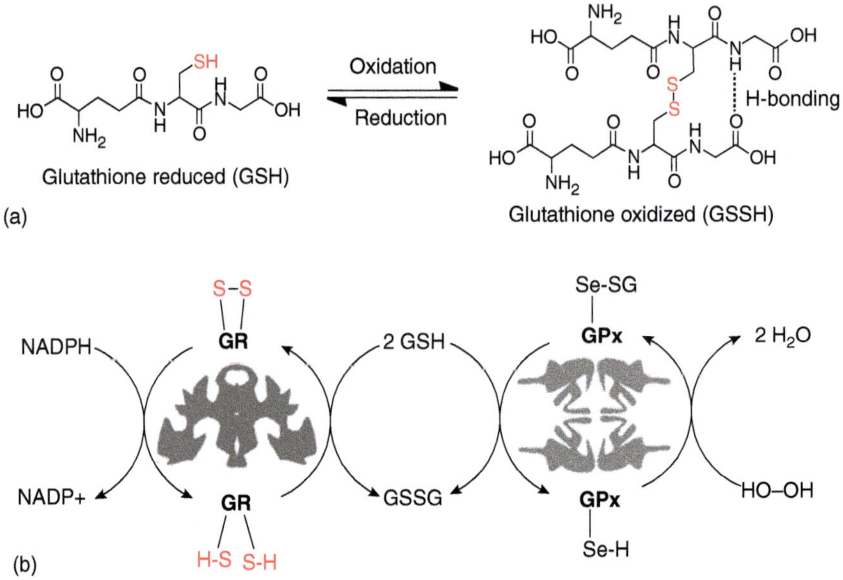

Figure 11.2 (a) Oxidation and reduction of glutathione (GSH). (b) Diagram of glutathione enzymatic system. Source: Reproduced with permission from Ref. [21]. Copyright © 2012, IUPAC.

The constituents of glutathione enzymatic system (Figure 11.2b) are glutathione reductase enzyme (GR), glutathione (GSH or GSSG), selenoprotein glutathione peroxidase (GPx), nicotinamide adenine dinucleotide phosphate (NADPH), and hydrogen peroxide. Upon oxidation of NADPH to $NADP^+$, it supplies two protons that eventually reduce the GR. Now returning back to its oxidized state, GR transfers two protons to oxidized glutathione to generate reduced glutathione. Then GSH reduces the GPx and gets oxidized to GSSG. Oxidized GPx returning to its reduced form oxidizes hydrogen peroxide to form water. By virtue of this redox balance, GSH can perform the role of an important intracellular radical scavenger and protect the cellular components from harmful reactive oxygen species and also transfers the signal in cellular environment.

Redox potential values of GSSG/GSH couple [23] can be estimated by the Nernst equation (11.1)

$$E_{\mathrm{GSH}} = E^{0'}_{\mathrm{GSH}} - \frac{\mathrm{RT}}{2F} \ln \frac{[\mathrm{GSH}]^2}{[\mathrm{GSSG}]} \tag{11.1}$$

where E_{GSH} is half-cell reduction potential, $E^{0'}_{\mathrm{GSH}}$ is the standard half-cell reduction potential ($E^{0'}_{\mathrm{GSH}} = -240\,\mathrm{mV}$), $R = 8.315\,\mathrm{J\,K^{-1}\,mol^{-1}}$ is the gas constant, $T = 298\,\mathrm{K}$ the temperature, and $F = 96\,485\,\mathrm{C\,mol^{-1}}$ the Faraday constant. From Eq. (11.1) it is evident that the reduction potential is related to the ratio of GSH/GSSG and concentration of GSH. In cellular environment, thiol and disulfide exchange is a continuous process but at any given time, the oxidized and reduced forms of glutathione maintain a balance so that >90% glutathione remains in the reduced form and only ~10% is present in the oxidized form (GSSG). In the human body, intracellular GSH concentration is much higher (1–10 mM) compared with extracellular fluids (20–40 μM) [24, 25] due to the rapid catabolism of GSH in the extracellular medium. It is also well known that the concentration of GSH is substantially higher in cancer cells than that of normal cells [26, 27]. Inspired by such concentration gradients of glutathione, in recent years, a large number of polymeric systems containing disulfide functionality at the backbone or pendant have been reported [28–32]. Few representative examples of disulfide-containing polymers are shown in Scheme 11.1. In **P1** [33], the disulfide group is located in the pendant while in **P2** [34], it is a part of the crosslinker. On the other hand, in **P3** [35], **P4** [36], and **P5** [37], the disulfide group is a part of the backbone repeating unit. In block copolymers **P6** [38], **P7** [39], and **P8** [40], the disulfide group acts as the linker between the two different blocks.

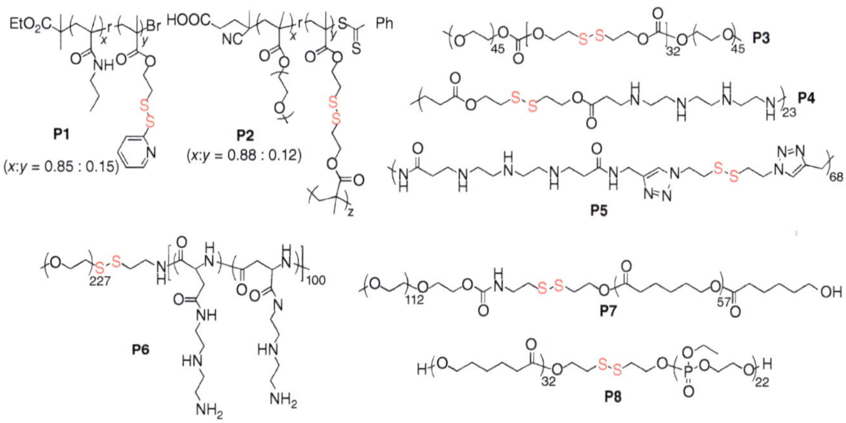

Scheme 11.1 Representative examples of disulfide containing polymers reported in literature [33–40].

However, none of them are polydisulfides in true sense as the polycondensation reaction does not generate the disulfide group in the backbone. Instead they consist of other polymeric backbones in which the disulfide group was embedded by suitable derivatization of the monomeric unit(s) before polymerization. In this chapter all those examples on disulfide containing polymers have been excluded; instead we have focused on development of various synthetic methodologies for polydisulfides and their implications in biological and other applications.

11.2 Synthesis of Poly(disulfide)s

Poly(sulfide) where repeated units contain multiple adjacent sulfur atoms was first reported by Patrick and Mnookin in 1926 while doing a reaction between

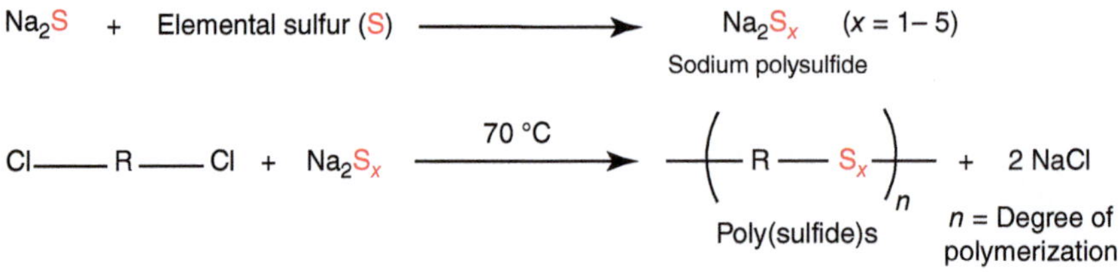

Figure 11.3 Synthesis of poly(sulfide)s by thiokol method.

ethylene dichloride and sodium polysulfide aiming to get an inexpensive antifreeze compound. But unexpectedly they got a gum type product with terrible odor, which was insoluble in any solvent. That is considered as the first synthetic rubber in America and later on this gummy compound was known as thiokol [41–43]. In their methodology known as Thiokol method, firstly sodium poly(sulfide) (Na_2S_x) was prepared by reacting aqueous sodium sulfide with elemental sulfur, then α,ω-dihalogenated linear organic compounds were reacted with them in aqueous medium at 70 °C to get desired poly(sulfide)s (Figure 11.3). Though they showed control over the rank of sulfur (number of consecutive sulfur atom in polysulfide) in synthesized poly(sulfide)s by varying the reaction condition, but exclusively PDS was not obtained by this methodology.

In next three decades it drew attention by industry due to its extensive commercial uses. The synthesis and characterization of industrially important poly(sulfide)s (mainly on disulfide or tetrasulfide polymers) are demonstrated in a review by Ganesh and coworkers [15]. The first academic report [44] about PDS was by Olson and coworkers in 1957. After that, researchers have expanded this field by introducing various methodologies for synthesis of PDS scaffolds as described as follows.

11.2.1 Oxidative Polymerization of Dithiols

PDS formation by the oxidative polymerization of dithiol in air was first reported by Olson and coworkers in 1957 [44]. In their methodology, various dithiol monomers were suspended in emulsion formed by lauric acid and potassium hydroxide and then the mixture was treated with compressed air, bubbled for one to four days, to get PDS. In presence of selenious acid as a catalyst, polymerization rate was accelerated. Later in 1960 Banerjee and coworkers [45] reported the air oxidation polymerization in soap emulsion system using functionalized siloxane as dithiol monomers. The effect of temperature on molecular weight of the polymer was studied and it was found that molecular weight of polymer increased with increasing the reaction temperature up to 80 °C. In 1967, Hoffman and coworkers have studied [46] oxidative polymerization of sugar thiol by using air oxygen as the oxidizing agent and selenium oxide as a catalyst.

Using air as the oxidizing agent, the molecular weight of the synthesized PDS was very low. Endo and coworkers [47] reported (Figure 11.4a) oxidative polymerization of bisphenol dithiol monomers by oxygen in presence of triethyl amine as a base catalyst (Figure 11.4b). The polymerization was carried out at ambient temperature

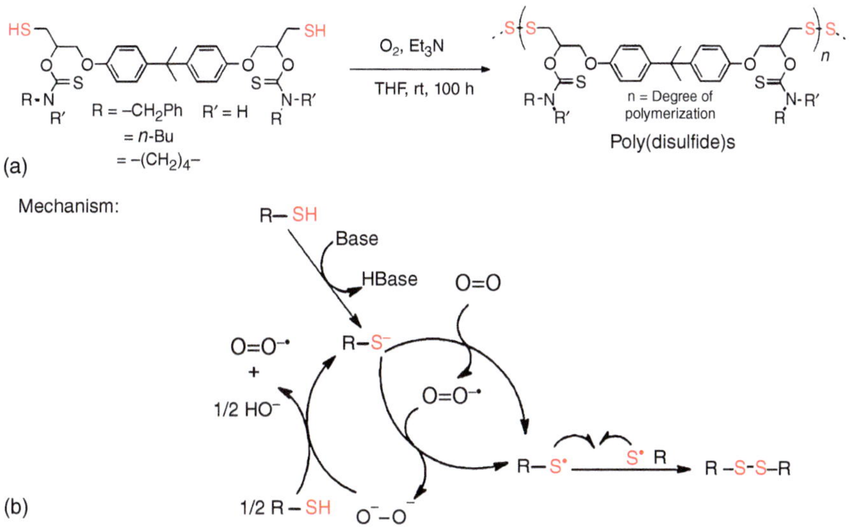

Figure 11.4 (a) Synthesis of poly(disulfide)s by base catalyzed oxidation of dithiol by oxygen. (b) Mechanism for base catalyzed disulfide bond formation. Source: Adapted with permission from Ref. [48]. Copyright © 2011, American Chemical Society.

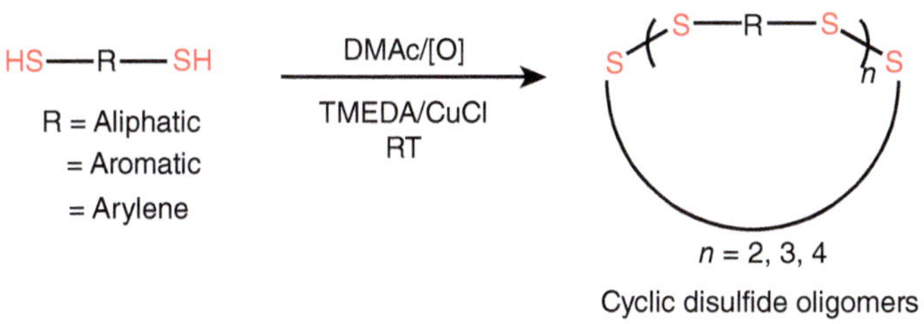

Figure 11.5 Synthesis of cyclic disulfide oligomers by air oxidation in presence of copper catalyst and base. Source: Adapted with permission from Ref. [49]. Copyright © 1996, American Chemical Society.

for 100 hours and reasonably high molecular weight PDS was obtained. In absence of base low molecular weight PDS was obtained.

Cyclic disulfide oligomers have been synthesized (Figure 11.5) using catalytic oxidation of an aliphatic [50], aromatic [49], and arylene [51] dithiols with oxygen in presence of copper amine catalyst. Aromatic macrocycles have also been synthesized [52] from their linear version via cyclo-depolymerization upon heating or in presence of catalytic amount of dithiol.

In 1963, Karabinos and coworkers [53] have used DMSO as an oxidizing agent for selective oxidation of thiol to disulfide without affecting the other oxidizable groups such as aldehyde and amine. In this reaction DMSO played the dual role of oxidizing agent as well as solvent. They have isolated the disulfide with excellent purity and also characterized the byproduct dimethyl sulfide by infrared spectroscopy. Inspired by the oxidizing power of DMSO, in 1968 Goethals et al.

(a)

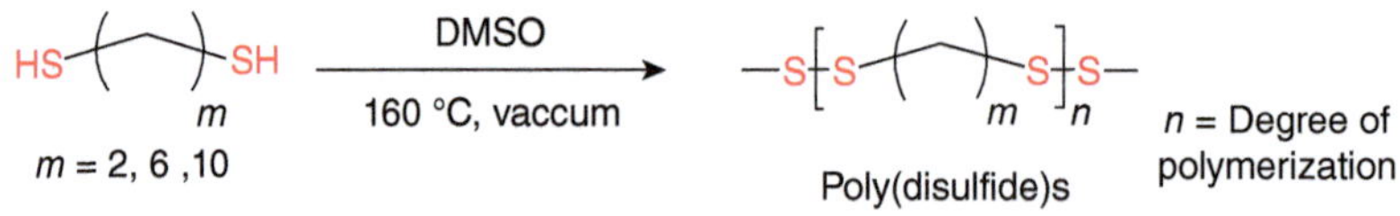

(b)

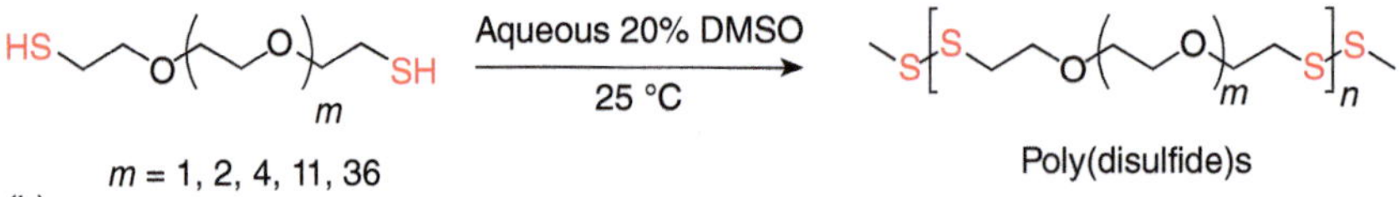

Figure 11.6 Synthesis of poly(disulfide)s by oxidation of (a) aliphatic dithiols. (b) oligoethylene dithiols in presence of DMSO. Source: (b) Adapted with permission from Ref. [55]. Copyright © 2005, American Chemical Society.

utilized it as an oxidizing agent for synthesis of PDS [54] from a series of alkyl dithiol with various chain length (Figure 11.6a). In a typical polymerization process, both DMSO and dithiol were taken in a heavy glass vial, and polymerization was conducted under sealed condition at 160 °C for one to eight hours. In this method they have reported the synthesis of PDS with maximum molecular weight (M_n) = 10 500 g mol^{-1}. Park and coworkers have reported [55] the synthesis of PDS by oxidation of oligoethylenedithiol (Figure 11.6b). During the optimization of polymerization condition, they have played with DMSO concentration in aqueous solution as well as reaction temperature. Eventually with 20% aqueous DMSO, at 25 °C, a series of oligoethylene dithiols were polymerized for 48 hours to produce corresponding PDS.

In 2005, Meng et al. reported [52] the synthesis of aromatic PDS by oxidizing aromatic dithiol, namely, 4,4′-oxybis(benzenethiol) at 80 °C. As the synthesized polymer was insoluble, they could not measure the molecular weight of these polymers.

Hydrogen peroxide (H_2O_2) has also been used as the oxidizing agent for synthesis of PDS. However, in presence of H_2O_2, free dithiols get over oxidized to sulfonic acid instead of disulfide. Therefore, nascent H_2O_2 can't be used for PDS preparation. The kinetics of oxidation of thiol to disulfide by using aqueous hydrogen peroxide as an oxidizing agent has been studied by Tarbell and coworker and the effect of metal ion on reaction rate was measured [56]. Puskas and coworkers have reported [48] the synthesis of PDS by oxidative polymerization using dilute hydrogen peroxide as an oxidizing agent. The polymerization was initiated by the abstraction of protons of thiol groups by triethylamine to form thiolate anions which upon oxidation by H_2O_2 formed diradicals and propagated the polymerization (Figure 11.7). In this polymerization, large disulfide rings also exist along with linear PDS.

Oligomers with terminal thiol groups can undergo oxidative polymerization to produce PDS. Oupický and coworkers have reported [57] thiocarbonate terminated telechelic polymer, which upon aminolysis produced free thiol groups at the chain end and could be subjected to oxidative polymerization with common oxidizing agents such as iodine (I_2) or DMSO or oxygen in water to produce PDS (Figure 11.8). Whittaker et al. have also studied [58] such polymerization and found

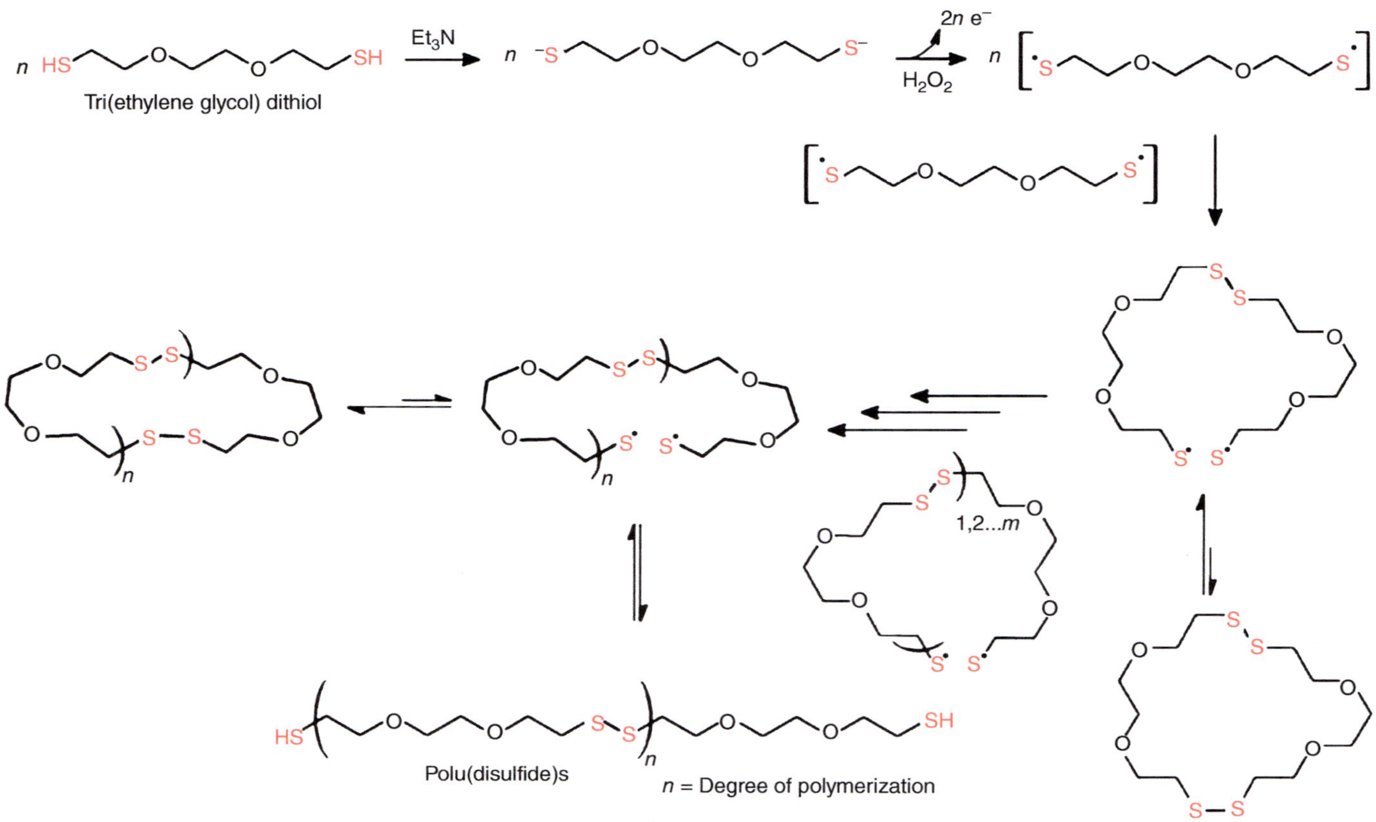

Figure 11.7 Synthesis of poly(disulfide)s in presence of 20% aqueous H_2O_2 as oxidizing agent. Source: Scheme adapted with permission from Ref. [48]. Copyright © 2011, American Chemical Society.

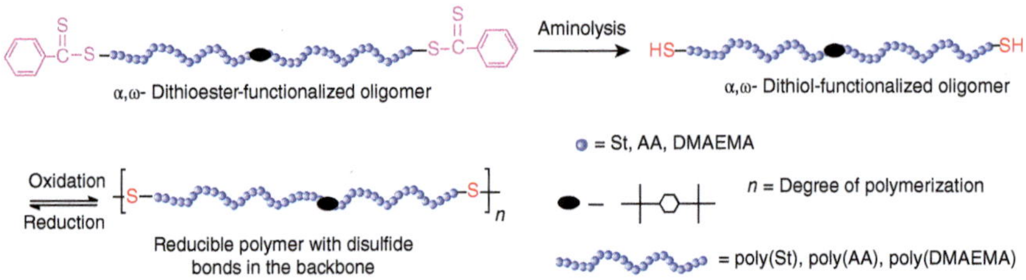

Figure 11.8 Synthesis of reducible polymers containing multiple disulfide bonds in the backbone via oxidation of telechelic polymers prepared by RAFT polymerization. Source: Adapted with permission from Ref. [57]. Copyright © 2007, American Chemical Society.

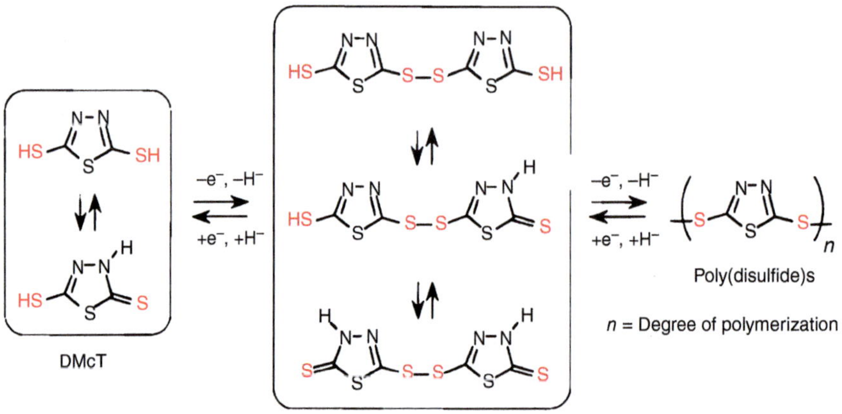

Figure 11.9 Synthesis of poly(disulfide)s by electrochemical oxidation. Source: Scheme adapted with permission from Ref. [59]. Copyright © 1997, American Chemical Society.

that depending on the concentration of dithiol macromonomers either they form cyclic disulfide or liner PDS.

Oyama and coworkers have studied electrochemical oxidation of 2,5-dimercapto-1,3,4-thiadiazole (DMcT) to PDS at a carbon electrode and an indium-tin oxide (ITO) electrode in 0.1 M $LiClO_4$-acetonitrile [59]. They have also discussed the electrochemical depolymerization of PDS (Figure 11.9). The reduction potential of polymerization-depolymerization of DMcT in presence of base and monomer (DMcT) has been studied. Electrochemical polymerization of alkane dithiols to PDS was also reported by Endo and Bu [60, 61] with a series of aliphatic dithiols. They reported the polymerization activity to be dependent on the chain length of monomers and molecular weight of the synthesized polymer was closely related with the applied current.

11.2.2 Ring-Opening Polymerization (ROP) of Cyclic Disulfide

Ring-opening polymerization (ROP) of cyclic oligo-disulfide upon heating at a temperature ranging from 180 to 250 °C without any catalysts or initiators leads to formation of PDS [49, 62–64]. Meng and coworkers have performed [62] the

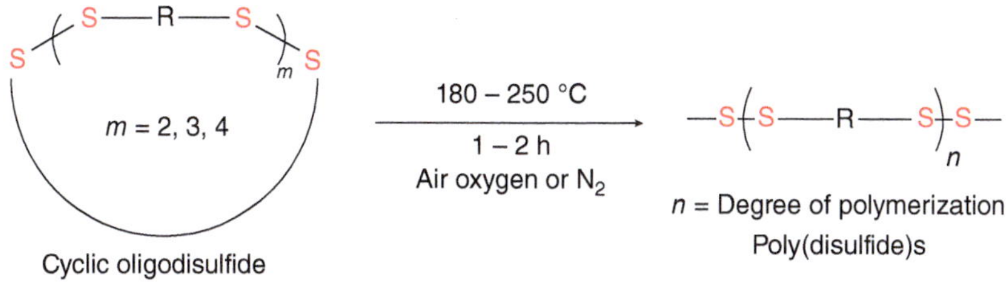

Figure 11.10 Ring opening of cyclic disulfide oligomers to produce poly(disulfide)s.

polymerization both in air and nitrogen atmosphere (Figure 11.10). They found the presence of thiosulfonates (oxidation of one sulfur atom of disulfide) within the polymer. The amount of thiosulfonate in a polymer chain was dependent on the reaction temperature and environment of the reaction. The reaction with higher temperature and in presence of air atmosphere leads to the highest percentage of thiosulfonate in the polymer.

Endo and coworkers have studied [65] the ROP of lipoic acid (LPA) (Figure 11.11a). In bulk polymerization they have heated the LPA monomer above its melting point and the polymer was isolated in high purity with cyclic interlocked structure. Similarly, they have also studied [64, 66] the bulk polymerization of six-membered cyclic disulfide namely 1,2-dithiane (DT) and observed cyclic interlocked structure of the resulting polymers (Figure 11.11b). They have also reported the copolymerization of 1,2 dithiane and LPA in bulk to produce DT-co-LPA copolymer with high molecular weight [67]. The copolymer appeared as polycatenane structures consisting of the entanglements of cyclic polymer.

Schlaad and coworkers have reported [68] entropy-driven disulfide metathesis polymerization of unsaturated macrocycles to produce PDS with hetero-functionality (Figure 11.11c). They have done the polymerization in polar aprotic solvents such as DMA or DMSO and reported very fast polymerization, reaching equilibrium within five minutes. They found that the synthesized polymers were stable in bulk state or in solution when dissolved in weakly polar solvent, but rapidly depolymerized in dilute polar solvent. To overcome the instability in dilute polar solvent, the thiol terminals of the synthesized polymers were end capped with iodoacetonitrile.

11.2.3 Photo-Induced Disulfide Metathesis

Dynamic exchange among disulfide bonds upon photoirradiation [69] has been explored for synthesis of PDS. Otsuka et al. have reported [70] photo-responsive disulfide polymer (DSPES) which upon photoirradiation undergoes disulfide metathesis in bulk state, where reorganization of disulfide bonds takes place (exchanged DSPES) (Figure 11.12).

11.2.4 Fragmentation Polymerization

In 1973, Fujisawa and coworkers have introduced [71–73] a methodology known as fragmentation polymerization for the synthesis of PDS under mild condition.

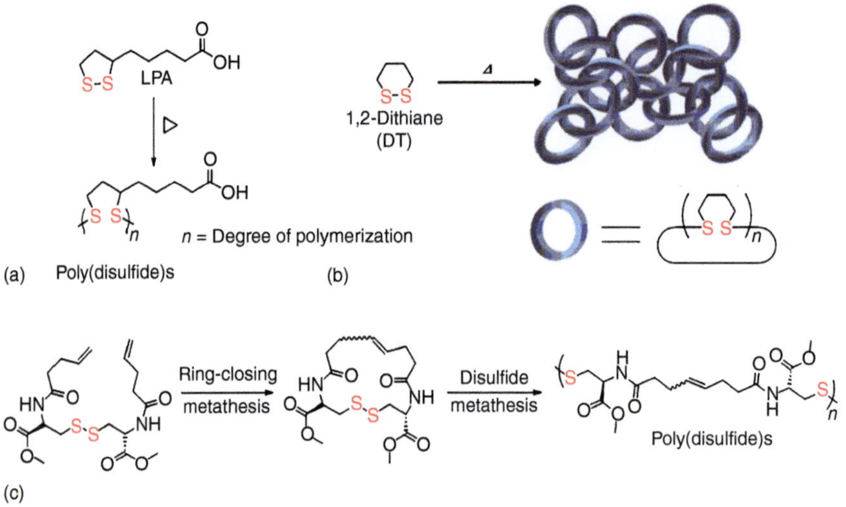

Figure 11.11 Ring-opening polymerization of (a) cyclic five-membered monomer lipoic acid (LPA). (b) 1,2-Dithiane (DT) for poly(disulfide) with interlocked structure. Source: Adapted with permission from Ref. [66]. Copyright © 2008, Springer, Nature. (c) Entropy driven ring opening of L-cystine-based macrocycle for the synthesis of high-molar mass hetero-functional poly(disulfide)s.

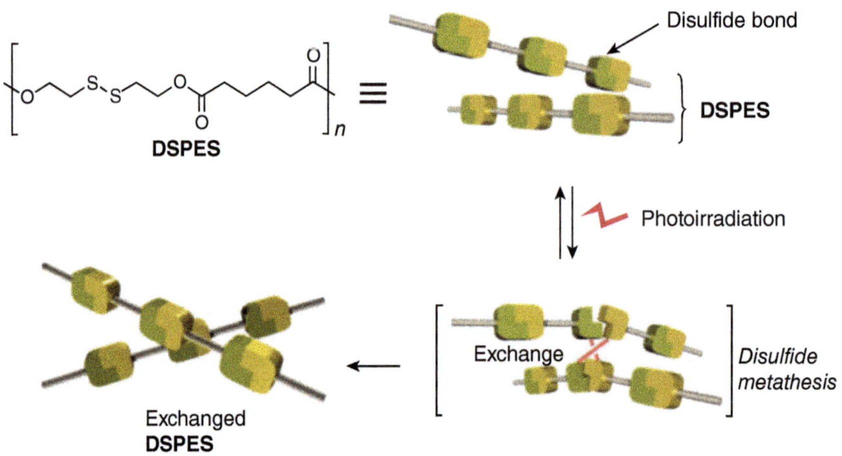

Figure 11.12 Macromolecular crossover reaction of disulfide-containing polyester (DSPES) driven by disulfide metathesis under photoirradiation. Source: Reproduced with permission from Ref. [70]. Copyright © 2010, Royal Society of Chemistry.

In this methodology, bis(sulfenyl thiocarbonates) was reacted with dithiol in the presence of triethylamine (Figure 11.13a). The fragmentation of sulfenyl thiocarbonate induced by thiol was the key reaction for each step of chain extension. By using this methodology they have also synthesized alternate copoly(disulfide)s [73] (Figure 11.13b) using various alkyl spacers between two sulfur atom of the disulfide bond of bis(sulfenyl thiocarbonates). They have also reported the formation of some

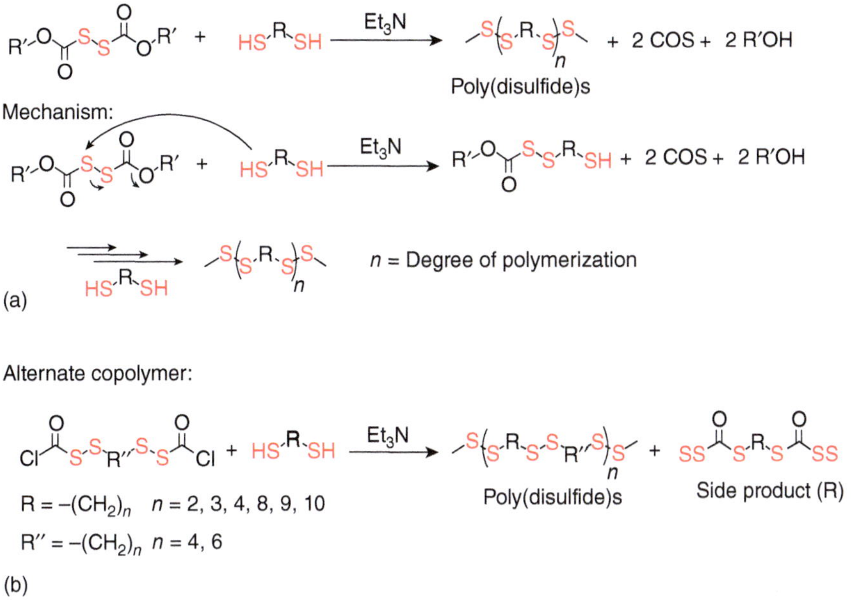

Figure 11.13 (a) Synthesis of poly(disulfide)s by fragmentation polymerization along with mechanism of polymerization. (b) Synthesis of alternate block poly(disulfide)s.

cyclic disulfides and a significant amount of recurring unit (R) during polymerization as the side product (Figure 11.13b).

11.2.5 Self-Organizing Surface-Initiated Polymerization (SOSIP)

Matile and coworkers have introduced [74–79] the concept of self-organizing surface-initiated polymerization (SOSIP) for the synthesis of PDS using transparent oxide as initiating surface. They have used disulfide containing functional chromophore as initiator, DTT as an activator, and strained disulfide containing chromophore as propagator (Figure 11.14a). Firstly, initiator was bound to the indium tin oxide (ITO) surface and activated by DTT to generate free thiols, which then propagated the polymerization with the ring-opening of the strained disulfide of asparagusic acid via thiol-disulfide exchange reaction. The directionality of the polymerization was monitored by supramolecular interaction such as π-stacking of the aromatic core and hydrogen bonding among amide linkage of adjacent chromophores. During polymerization, depending on the nature of propagator used, one could generate alternate, statistical, or self-sorting organization of naphthalenediimide (NDI) chromophores, termed as co-SOSIP [76]. They have also shown the stack exchange of synthesized SOSIP to build covalent double channel photosystems [77]. In order to show the versatility of the methodology, they have used other chromophores apart from NDI, such that perylenediimides (PDI) [78] or oligothiophene [79] to generate complex well-defined architectures. Later on, they have expanded their methodology by moving from ITO surface as initiating site to substrate-initiated synthesis of PDS [80] in solution phase.

Here they have used covalently attached substrate (probes, drug, etc.) as initiator, guanidinium containing strained disulfide as propagator and iodoacetamide as the terminator (Figure 11.14b). They have synthesized a series of functional PDS by this methodology and studied the cellular internalization of the synthesized PDS scaffolds. To overcome the limitation of using iodoacetamides as terminator, they further introduced ethynyl benziodoxolones [81] where click reaction could be performed to introduce functional groups.

11.2.6 Thiol-Disulfide Exchange Reaction

While several possible methods for synthesis of PDS have been discussed in the forgone section, none appears to be suitable for synthesis of structurally diverse polydisulfides with functional group tolerance and difference in connectivity. A particular thiol-disulfide exchange reaction between pyridyl-disulfide (Py-Ds) and thiol appeared promising to us for synthesis of polydisulfide by condensation polymerization approach.

Py-Ds group has been extensively used for synthesis of disulfide containing polymers and also for bioconjugation [28–32]. This reaction is known to be extremely fast and produces a water soluble byproduct (2-pyridothione), which can be easily detected by its yellow color. We have developed a methodology for synthesis of PDS under mild condition with tunable degree of polymerization using this thiol-disulfide reation [82]. In a typical polymerization process, commercially available dithiol (1,6-hexane dithiol) and dipyridyl disulfide (aldrithiol-2) were mixed together in dichloromethane and polymerization was carried out in presence of catalytic acetic acid at room temperature (rt). Progress of the polymerization reaction could be monitored by UV–Vis spectroscopy by following the production of the 2-pyridothione, which suggested the polymerization was over within three hours producing **PDS-1** (Figure 11.15a).

By using stoichiometric excess of the aldrithiol-2, both ends of the synthesized PDS chains could be capped with pyridyl disulfide which was confirmed by MALDI-TOF-MS (Figure 11.15b) analysis. For example, in the MALDI-TOF-MS of **PDS-1**, the peak at m/z 2616.8 corresponds to (16 × mass of the repeating unit) + two pyridyl sulfides at the terminals [(16 × 148.5 (mass of repeating unit) + 2 × 110.01 (mass of pyridyl sulfide) + 22.99 (mass of sodium)]. Therefore, this polymerization technique produced the linear PDS with desired functional group at the chain end. The terminal pyridyl disulfide groups could be selectively replaced with 2-mercapto ethanol without affecting the poly(disulfide) backbone to generate hydroxyl functional group terminated PDS (**PDS-4**). MALDI-TOF-MS further confirmed the presence of terminal hydroxyl groups at the both chain terminal (Figure 11.15c). For example, the peak at m/z 2846.6 corresponds to (18 × mass of the repeating unit) + two 2-mercaptoethanol at the terminals [(18 × 148.5 (mass of repeating unit) + 2 × 77.13 (mass of 2-mercaptoethanol) + 22.99 (mass of sodium)] and also successful conversion. The hydroxyl terminal functional groups were further utilized for ROP of cyclic lactide to produce ABA type triblock copolymer (**PDS-5**).

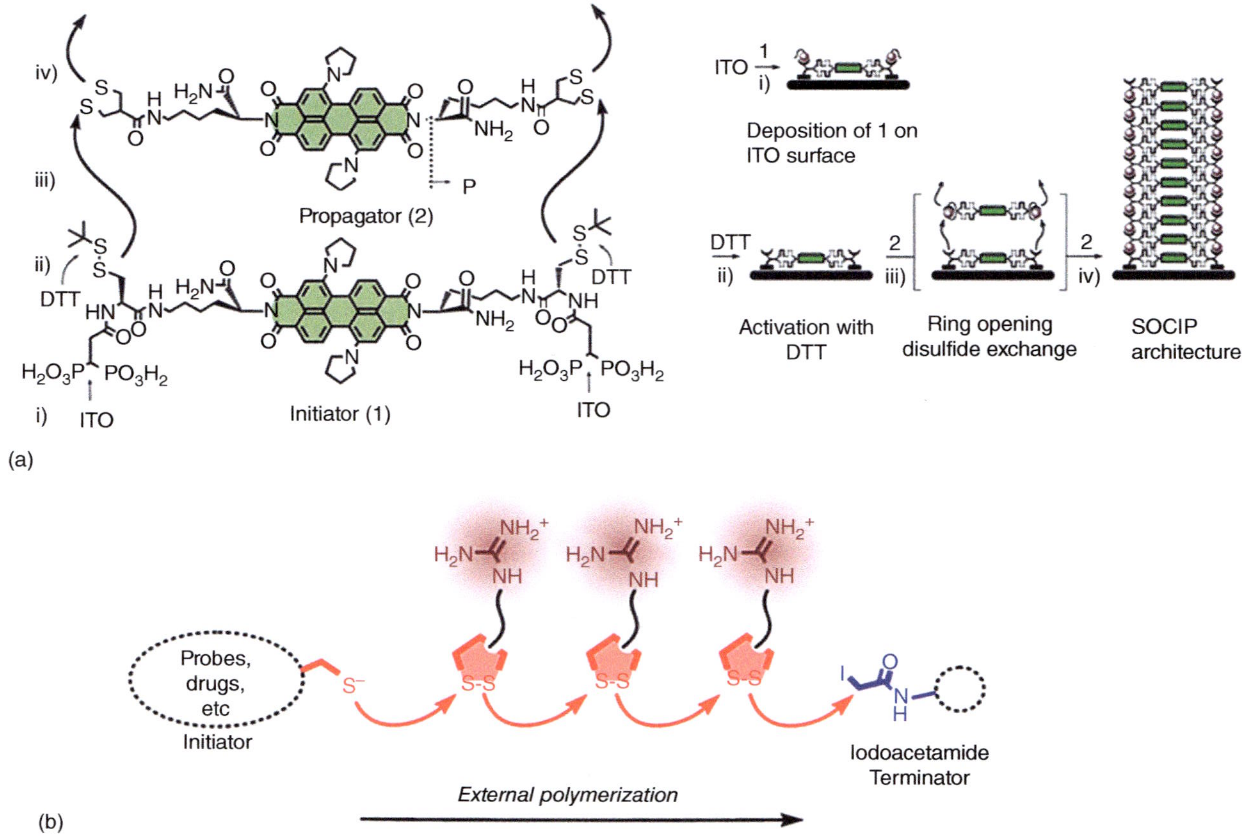

Figure 11.14 Synthesis of poly(disulfide)s by SOSIP from (a) ITO surface. Source: Reproduced with permission from Ref. [78]. Copyright © 2012, Royal Society of Chemistry. (b) Substrate initiated synthesis of poly(disulfide)s. Source: Reproduced with permission from Ref. [80]. Copyright © 2013, American Chemical Society.

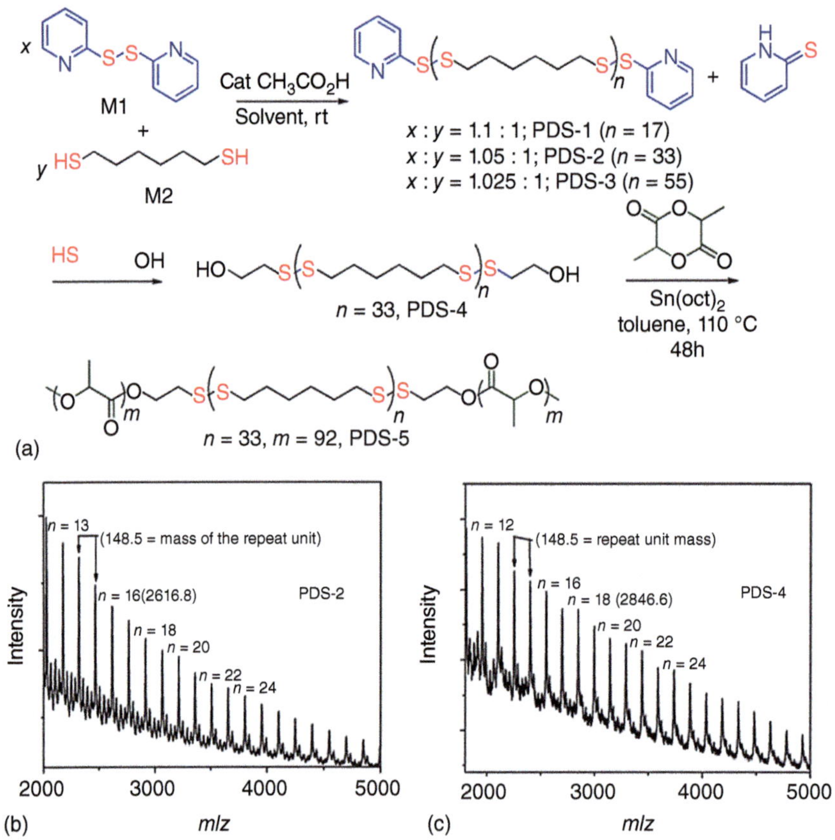

Figure 11.15 (a) Synthesis of telechelic poly(disulfide)s. MALDI-TOF-MS of (b) **PDS-1** and c) **PDS-4**. Source: Adapted with permission from Ref. [82]. Copyright © 2014, John Wiley and Sons.

We have further improved the methodology by introducing monofunctional impurity technique to obtain telechelic PDS in one step with desired functional groups at the chain ends [83]. Using various monofunctional impurity (**MFI-X**), we could show predictable degree of polymerization and placement of desired functional group at the chain ends (Figure 11.16).

By extending this newly developed strategy, we have recently reported [84] a generally applicable methodology for synthesis of hyperbranched polydisulfide (HBPDS) by A2 + B3 condensation approach (Figure 11.17). The B3 monomer contains three Py-Ds groups while a di-thiol compound serves as the A2 monomer. Polycondensation reaction under a very mild condition enables synthesis of reasonably high molecular weight HBPDS with degree of branching (DB) of 0.8. The reactive Py-Ds groups can be selectively (over the backbone disulfide groups) and fully replaced by post-polymerization functionalization using a hydrophilic thiol of choice resulting in bio-reducible and biocompatible amphiphilic HBPDS.

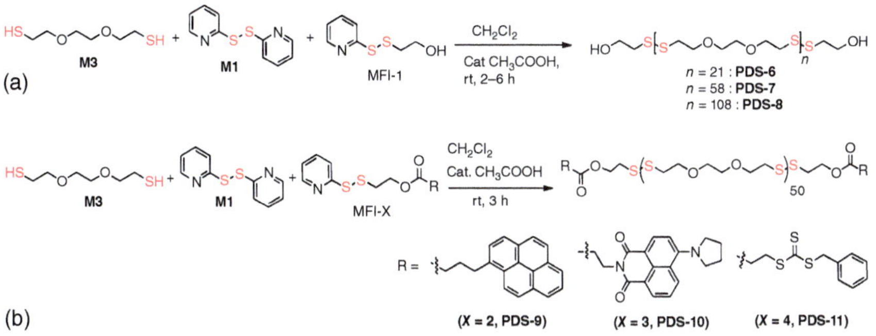

Figure 11.16 Synthesis of poly(disulfide)s with (a) control degree of polymerization and (b) functional end groups.

11.3 Amphiphilic PDS and Drug Delivery Application

Considering the suitability of this approach for synthesis of functional polymers, we utilized this methodology for synthesis of different types of amphiphilic PDS and examined their utility in guest encapsulation and drug delivery application. Using the methodology discussed above, statistical poly(disulfide) copolymer (**PDS-12**) containing pendant hydroxyl groups was synthesized [83]. These pendant hydroxyl groups were reacted with succinic anhydride to produce carboxylic acid which upon treatment with potassium carbonate (pH ~ 8.0) produced carboxylate anion (Figure 11.18a) and therefore makes **PDS-13** as an amphiphilic polymer with hydrophobic backbone and hydrophilic pendant group.

It produced micellar aggregates in aqueous medium (Figure 11.18b) which could encapsulate hydrophobic dye Nile Red (NR) and showed glutathione (GSH) induced backbone degradation and sustained release of the encapsulated cargo (Figure 11.18c).

Amphiphilic block copolymers [85, 86] have been extensively studied for drug delivery applications due to its attractive features such as low critical aggregation concentration, kinetic stability, and guest encapsulation ability. We have synthesized amphiphilic PDS by growing hydrophilic polymer chains from the hydrophobic telechelic PDS. Hydroxyl terminated PDS (**PDS-14** and **PDS-15**) with differing hydrophobicity of the backbone were synthesized by the methodology described in Figure 11.15. They were made as bifunctional macroinitiators by reaction with the α-bromoisobutyryl bromide and utilized to polymerize the hydrophilic monomer (M4) from both chains terminals (Figure 11.19) using ATRP method. This produced two amphiphilic triblock copolymers (**PDS-18** and **PDS-19**) in which the PDS backbone served as the hydrophobic block while terminal PTEGMA chains served as hydrophilic bloks [87]. In aqueous medium both of them formed micellar aggregates with average diameter of ~15 nm and showed ability to sequester hydrophobic guest Nile red. The impact of hydrophobicity on GSH triggered micellar disassembly and dye release kinetics was tested, which revealed slower backbone degradation and release of encapsulated NR from the micellar aggregates with relatively more

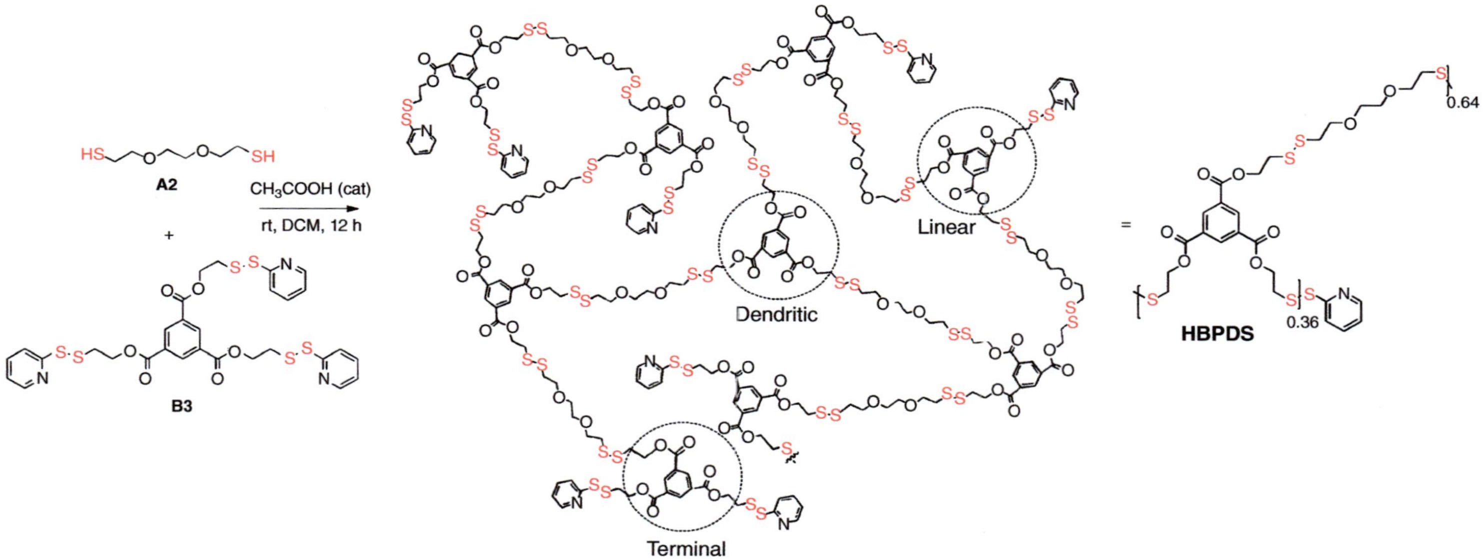

Figure 11.17 Synthesis of hyperbranched polydisulfide. Source: Adapted with permission from Ref. [84]. Copyright © 2020, Royal Society of Chemistry.

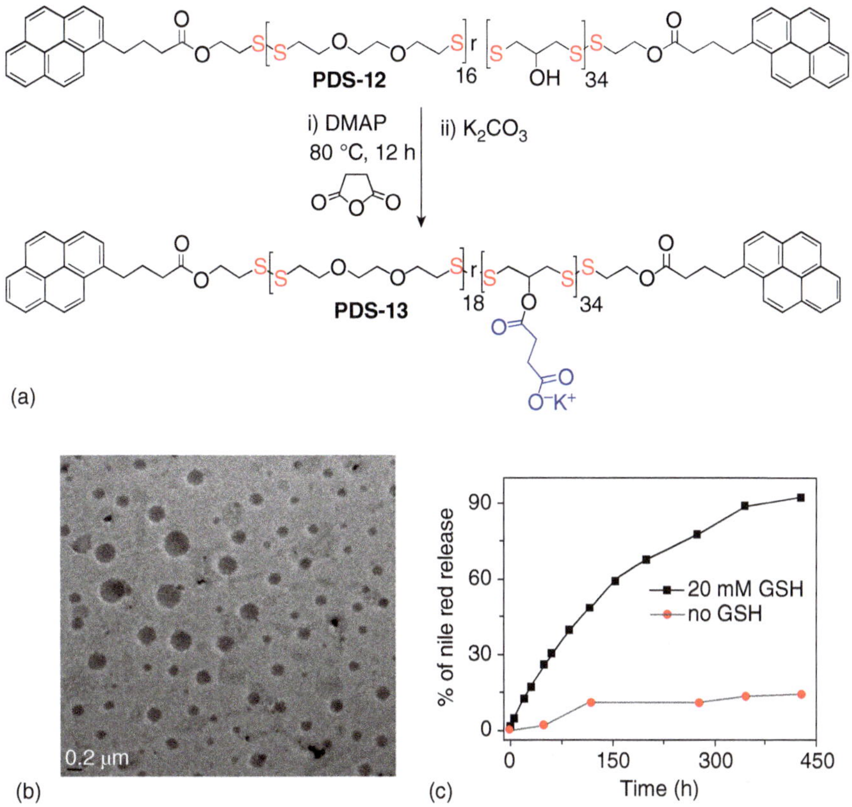

Figure 11.18 (a) Synthesis of amphiphilic poly(disulfide)s **PDS-13**. (b) HRTEM image of aqueous dispersion of **PDS-13** (C = 100 μM). (c) Release profile of encapsulated Nile red in **PDS-13** in the absence (0.0 mM) and presence of 20 mM GSH. Source: Adapted with permission from Ref. [83]. Copyright © 2018, John Wiley and Sons.

hydrophobic core (**PDS-18**) compared with micelles with relatively less hydrophobic core (**PDS-19**). This could be due to relatively faster diffusion of polar GSH towards the less hydrophobic micellar core.

Ni and coworkers [88] have introduced amphiphilicity in the PDS backbone by ROP of 2-ethoxy-2-oxo-1,3,2-dioxaphospholane (EOP) with the terminal hydroxyl groups of **PDS-20** (Figure 11.20a). Biodegradable polyphosphoester-based triblock copolymer **PDS-21** produced spherical nanoparticles (Figure 11.20b) in aqueous medium. **PDS-21** appeared as biocompatible both in L929 and HeLa cell lines (Figure 11.20c). Using these degradable and biocompatible nanocarriers, they have shown anticancer drug doxorubicin delivery in HeLa and HepG2 cell lines.

We have synthesized chain transfer agent (CTA) terminated hydrophobic PDS (**PDS-11**) (Figure 11.16) for further chain extension [89] with hydrophilic monomers such as N-isopropylacrylamide (NIPAM) and 2-(2-(2-methoxyethoxy)ethoxy)ethyl methacrylate (TEGMA) to produce amphiphilic triblock co-polymers PNIPAM-*b*-PDS-*b*-PNIPAM (**PDS-22**) and PTEGMA-*b*-PDS-*b*-PTEGMA (**PDS-23**), respectively

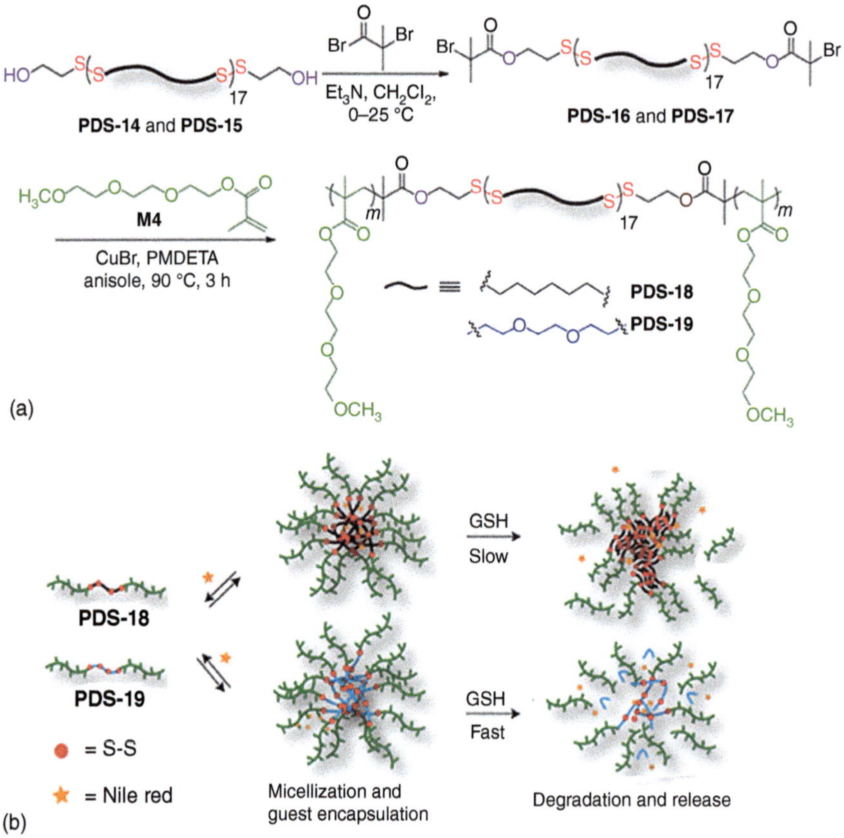

Figure 11.19 (a) Synthesis of amphiphilic poly(disulfide)s **PDS-18** and **PDS-19**. (b) Aqueous aggregation and glutathione (GSH) induced backbone degradation and release of encapsulated Nile red in the micellar core of **PDS-18** and **PDS-19**. Source: Adapted with permission from Ref. [87]. Copyright © 2015, Royal Society of Chemistry.

(Figure 11.21a). The block copolymer with PNIPAM as hydrophilic group (**PDS-22**) formed polymersome structure, while the other (**PDS-23**) formed micelle owing to the difference in their packing parameters (Figure 11.21b). While both aggregates sequestered a hydrophobic guest such as Nile red or an anticancer drug Dox, GSH triggered release was much faster from the polymersome than micelle owing to the complete degradation of the PDS segment in polymersome. Both **PDS-22** and **PDS-23** were found to be highly biocompatible as studied by 3-(4,5-dimethylthiazol-2-yl)-2,5-diphenyl tetrazolium bromide (MTT) and hemolysis assay.

Dox loaded polymersome killed the cancer cells (HeLa) more efficiently than micelle. Fluorescence microscopy studies (Figure 11.21c) indicated comparable cellular uptake for both polymersome and micelle, but the polymersome entrapped drug escaped fast from the cargo and reached the nucleus, while the drug-loaded micelle remained trapped in the perinuclear zone explaining the

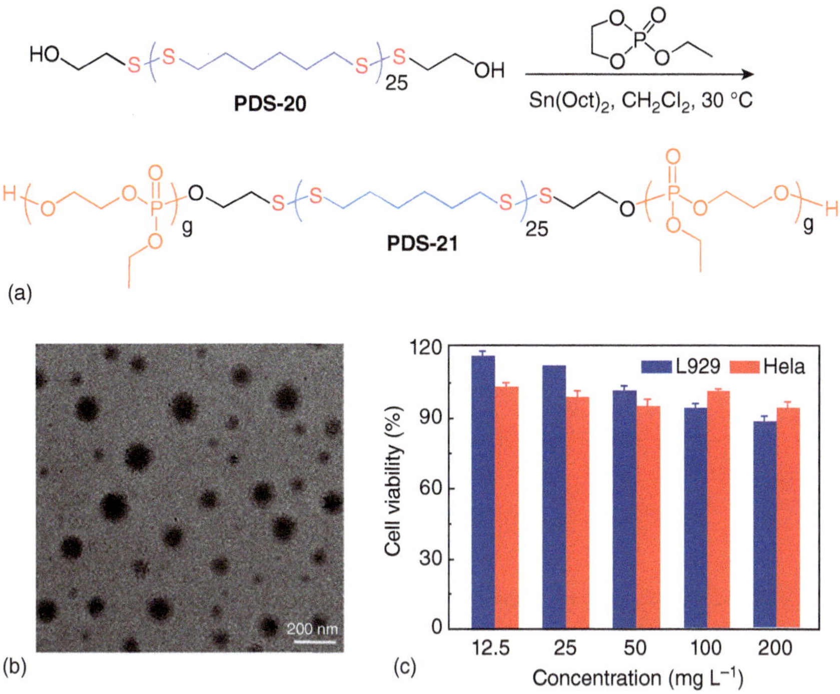

Figure 11.20 (a) Synthesis of reduction-responsive triblock copolymer **PDS-21.** (b) HRTEM images of aqueous aggregates of **PDS-21** ($C = 100\ \mu g\ mL^{-1}$). (c) Cell viability assay of **PDS-21** in L929 and HeLa cell lines for 48h of incubation (data are presented as the average ± standard deviation) ($n = 4$). Source: Adapted with permission from Ref. [88]. Copyright © 2018, Royal Society of Chemistry.

significant difference in the drug delivery performance of polymersome and micelle. Dox-loaded polymers (**PDS-22** and **PDS-23**) however exhibited minimum toxicity to normal cells (C2C12) indicating possibility of achieving targeted drug delivery with such amphiphilic PDS.

In the forgone discussion we have shown non-covalent drug encapsulation and GSH triggered intra-cellular drug release from aggregates of amphiphilic polymers based on bio-reducible PDS scaffold. However, the noncovalent encapsulation approach suffers from the undesired possibilities of premature release and fast clearance from the bloodstream or unpredictable fate of the delivery vehicle, which limit their *in vivo* application. Covalent polymer–drug conjugates are believed to address these issues more successfully [90–92]. We have demonstrated synthesis of hydrophobic PDS with covalently conjugated cancer drug and made them dispersible in water by PEGylation [93], which is known for improving the biocompatibility, blood circulation time, and cellular internalization. We have synthesized [94] pendant hydroxyl group containing poly(disulfide) **PDS-24** (Figure 11.22a) and conjugated it with anticancer drug Camptothecin (CPT) via carbonate linkage to get drug conjugated PDS (**PDS-25**). Finally, the terminal pyridyl sulfide groups were replaced with poly(ethylene)glycol thiol (PEG-thiol) to generate amphiphilic drug

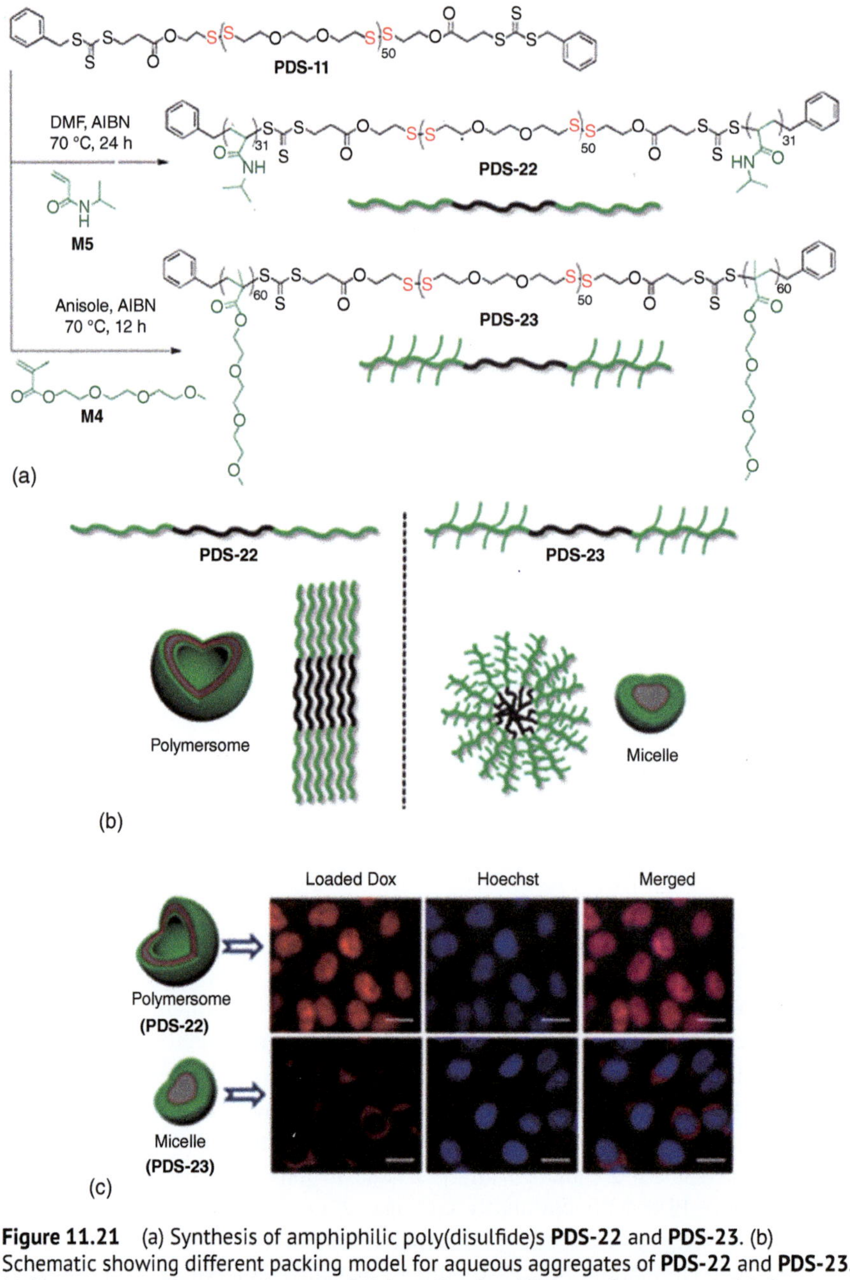

Figure 11.21 (a) Synthesis of amphiphilic poly(disulfide)s **PDS-22** and **PDS-23**. (b) Schematic showing different packing model for aqueous aggregates of **PDS-22** and **PDS-23**. (c) Fluorescence microscopy images of HeLa cells after 48 hours of treatment with Dox loaded polymersome (**PDS-22**) and micelle (**PDS-23**). (Scale bar = 20.0 μm in all images). Source: Adapted with permission from Ref. [89]. Copyright © 2018, John Wiley and Sons.

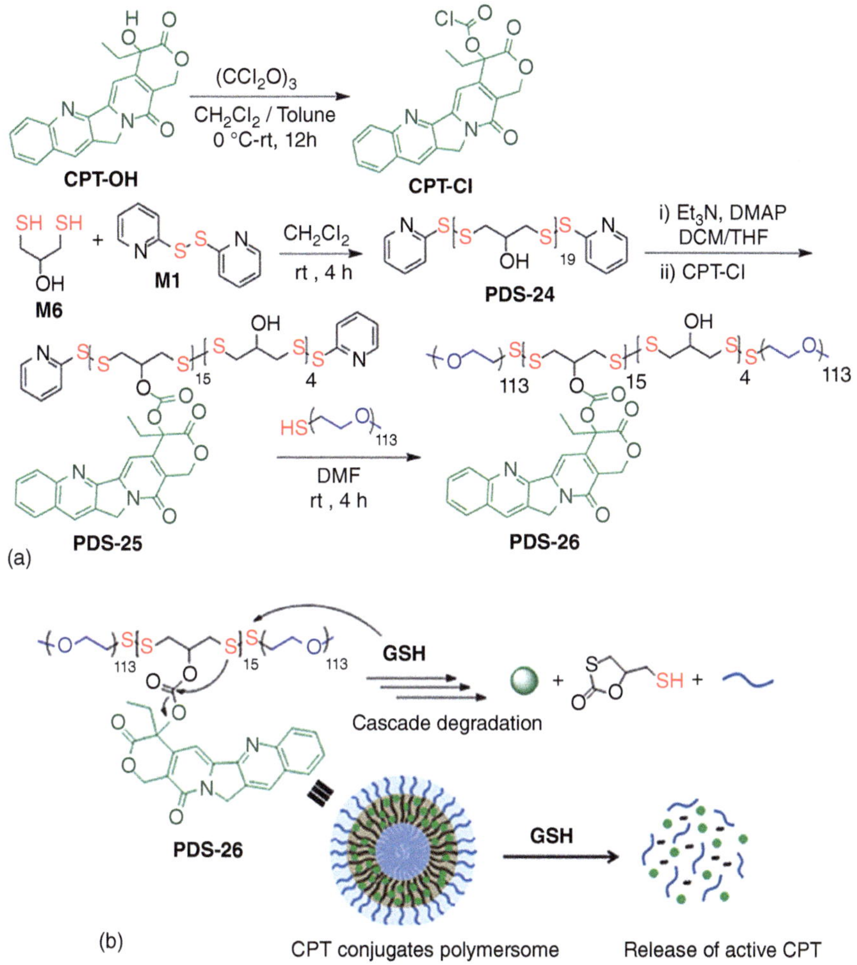

Figure 11.22 (a) Synthesis of CPT conjugated amphiphilic poly(disulfide)s **PDS-26**. (b) Self-assembly and GSH triggered cascade degradation of CPT conjugated amphiphilic poly(disulfide)s **PDS-26**. Source: Reproduced with permission from Ref. [94]. Copyright © 2019, American Chemical Society.

conjugated PDS (**PDS-26**) (Figure 11.22a). It showed redox-responsive fluorescent polymersome assembly in water, which could be disassembled in the presence of GSH. GSH cleaved the backbone disulfide linkage to generate free thiol that further involved in intrachain nucleophilic attack to the adjacent carbonate that triggered a cascade degradation of the backbone and consequently released the appended CPT (Figure 11.22b). The polymeric pro-drug (**PDS-26**) entered into HeLa cells and reached to the nucleus within four hours and showed efficient killing, selectively to cancer cells, with a remarkably low IC_{50} value of 3.1 $\mu g\,mL^{-1}$. The drug appended polymersome showed negligible toxicity to a normal cell, confirming its selectivity towards cancer cells. The non-drug-conjugated PDS analogue of **PDS-26** appeared

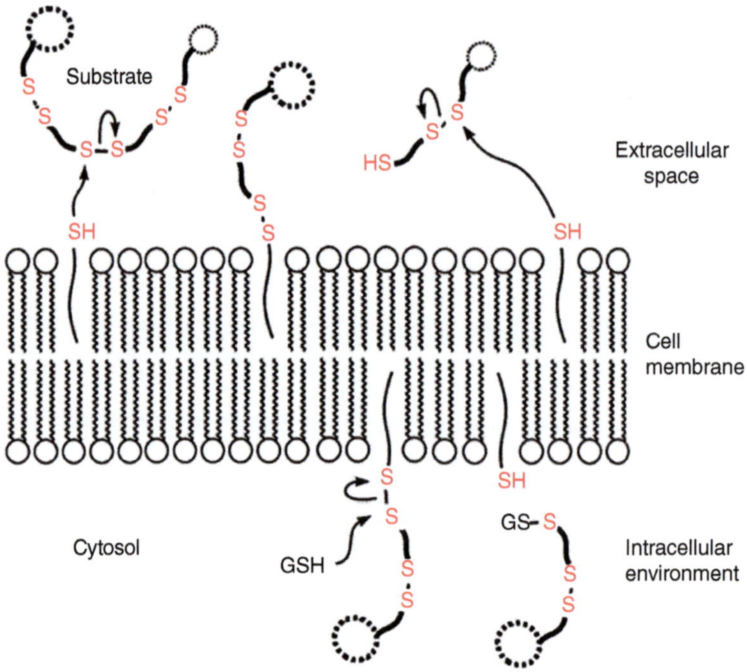

Figure 11.23 Representation of thiol-disulfide exchange reaction on cell surface and internalization of poly(disulfide)s into cell. Source: Adapted with permission from Ref. [95]. Copyright © 2014, American Chemical Society.

to be biocompatible, confirming the nontoxic nature of the polymer backbone and also the degradation products.

11.4 Cell-Penetrating Poly(disulfide)s

Cell-penetrating poly(disulfide)s (CPPDS) have been reported in the literature in the recent years as emerging new materials with promising application. It is understood that the disulfide bonds of CPPDS interacts with cell-surface thiols by thiol-disulfide exchange reaction [95], which facilitates their cellular internalization (Figure 11.23). They enter into cell via energy independent direct translocation (with the aid of disulfide bonds) pathways. Matile and coworkers have synthesized a series of such cell-penetrating materials with the SOSIP technique (Figure 11.14b) [80, 81]. Furthermore, by utilizing the synthesized CPPDS materials, the authors have shown delivery of proteins like streptavidin and quantum dots in the cytosols [96, 97]. Yao and coworkers have also used similar strategy to deliver antibodies and silica nanoparticles in the cancer cells via endocytosis independent pathways [98, 99]. Recently, Abe and coworkers have introduced disulfide groups at the terminus of oligo nucleotide strands and showed efficient delivery of them in the cytosol by energy independent endocytosis pathways [100].

11.5 Summary and Outlook

In this chapter we have collated examples of various synthetic methodologies for PDS and shown their utility in biological applications. From the discussion it is evident that the dynamic nature of the disulfide bond offers attractive options for intra-cellular delivery of drugs or other materials specifically in cancer cells. The advantages, limitation, and mechanism of various synthetic methodologies for PDS have been discussed in details. In one hand they are highly susceptible to reducing environment while they are normally stable under ambient condition as well as physiological environment and thus it is possible to design target specific delivery vehicles using PDS derivatives or conjugates. Although PDS scaffolds have been known for long time, lack of generally applicable synthetic methodologies limited their functional utility until recently when we demonstrated a condensation polymerization approach for synthesis of functional PDS. We have shown the methodology allows to synthesize amphiphilic block copolymers, HBPDSs, and polymer–drug conjugates and demonstrated their utility in drug delivery by *in vitro* experiments. Further exploration of this methodology for synthesis of alternating copolymers, star polymers, and functional polymers are possible and may enable realization of true potential of this bio-reducible polymeric scaffold in diverse biological applications. CPPDS are certainly encouraging and could make a big impact in the cell biology and biomedicine field in the future. However, still the missing link is *in vivo* experiments, which will determine whether this promising class of polymers will be actually useful for practical applications.

References

1 Mutlu, H., Ceper, E.B., Li, X. et al. (2019). *Macromol. Rapid Commun.* 40: 1800650.
2 Vo, C.D., Kilcher, G., and Tirelli, N. (2009). *Macromol. Rapid Commun.* 30: 299.
3 Imai, Y., Kato, A., Ii, M., and Ueda, M. (1979). *J. Polym. Sci. Part C: Polym. Lett.* 17: 579.
4 Macallum, A.D. (1948). *J. Org. Chem.* 13: 154.
5 Kricheldorf, H.R. and Schwarz, G. (2007). *J. Macromol. Sci. Part A: Pure Appl. Chem.* 44: 625.
6 Aksakal, S., Aksakal, R., and Becer, C.R. (2018). *Polym. Chem.* 9: 4507.
7 Tagle, L.H., Diaz, F.R., Vega, J.C., and Alquinta, P.F. (1985). *Makromol. Chem.* 186: 915.
8 Tagle, L.H., Diaz, F.R., and Riveros, P. (1986). *Polym. J.* 18: 501.
9 Gressier, J.C. and Levesque, G. (1980). *Eur. Polym. J.* 16: 1175.
10 Sanui, K., Kishimoto, Y., and Ogata, N. (1971). *Polym. J.* 2: 422.
11 Yamazaki, N., Higashi, F., and Iguchi, T. (1974). *Polym. Lett. Ed.* 12: 517.
12 Tian, T., Hu, R., and Tang, B.Z. (2018). *J. Am. Chem. Soc.* 140: 6156.
13 Arnold, R.G., Nelson, J.A., and Verbanc, J.J. (1957). *Chem. Rev.* 57: 47.

14 Kaloni, T.P., Giesbrecht, P.K., Schreckenbach, G., and Freund, M.S. (2017). *Chem. Mater.* 29: 10248.
15 Kishore, K. and Ganesh, K. (1995). *Adv. Polym. Sci.* 121: 83.
16 Lowe, A.B. and Bowman, C.N. (2013). *Thiol-X Chemistries in Polymer and Materials Science*. UK: RSC.
17 Leesand, W.J. and Whitesides, G.M. (1993). *J. Org. Chem.* 58: 642.
18 Ulrich, S. (2019). *Acc. Chem. Res.* 52: 510.
19 Chakma, P. and Konkolewicz, D. (2019). *Angew. Chem. Int. Ed.* 58: 9682.
20 Hopkins, F.G. and Morgan, E.J. (1936). *Biochem. J.* 30: 1446.
21 Rosenthal-Kim, E.Q. and Puskas, J.E. (2012). *Pure Appl. Chem.* 84: 2121.
22 Go, Y.-M. and Jones, D.P. (2008). *Biochim. Biophys. Acta* 11: 1273.
23 Schafer, F.Q. and Buettner, G.R. (2001). *Free Radic. Biol. Med.* 30: 1191.
24 Halprin, K.M. and Ohkawara, A. (1967). *J. Investig. Dermatol.* 48: 149.
25 Jones, D.P., Carlson, J.L., Samiec, P.S. et al. (1998). *Clin. Chim. Acta* 275: 175.
26 Kuppusamy, P., Li, H., Ilangovan, G. et al. (2002). *Cancer Res.* 62: 307.
27 Russo, A., DeGraff, W., Friedman, N., and Mitchell, J.B. (1986). *Cancer Res.* 46: 2845.
28 Kumar, P., Liu, B., and Behl, G. (2019). *Macromol. Biosci.* 19: 1900071.
29 Sui, B., Cheng, C., and Xu, P. (2019). *Adv. Therap.* 2: 1900062.
30 Quinn, J.F., Whittaker, M.R., and Davis, T.P. (2017). *Polym. Chem.* 8: 97.
31 Oh, J.K. (2019). *Polym. Chem.* 10: 1554.
32 Son, S., Namgung, R., Kim, J. et al. (2012). *Acc. Chem. Res.* 45: 1100.
33 Ghosh, S., Basu, S., and Thayumanavan, S. (2006). *Macromolecules* 39: 5595.
34 Jiang, Z., Cui, W., Prasad, P. et al. (2019). *Biomacromolecules* 20: 435.
35 Xu, Z., Liu, S., Kang, Y., and Wang, M. (2015). *ACS Biomater Sci. Eng.* 1: 585.
36 Yin, Q., Gao, Y., Zhang, Z. et al. (2011). *J. Control. Release* 151: 35.
37 Wang, Y., Zhang, R., Xu, N. et al. (2011). *Biomacromolecules* 12: 66.
38 Takae, S., Miyata, K., Oba, M. et al. (2008). *J. Am. Chem. Soc.* 130: 6001.
39 Sun, H., Guo, B., Cheng, R. et al. (2009). *Biomaterials* 30: 6358.
40 Wang, Y.-C., Wang, F., Sun, T.-M., and Wang, J. (2011). *Bioconjug. Chem.* 22: 1939.
41 Patrick, J.C. (1932) Method of making plastic substances and products obtained thereby. US Patent, 1 890 191, Issued 6 December 1932.
42 Patrick, J.C. (1936). *Trans. Faraday Soc.* 32: 347.
43 Patrick, J.C. and Mnookin, N.M. (1935) Making Plastic substances and product obtained thereby. US Patent, 1 996 486, Issued 2 April 1935.
44 Marvel, C.S. and Olson, L.E. (1957). *J. Am. Chem. Soc.* 79: 3089.
45 Bonsignore, P.V., Marvel, C.S., and Banerjee, S. (1960). *J. Org. Chem.* 25: 237.
46 Whistler, R.L. and Hoffman, D.J. (1967). *J. Polym. Sci. A* 5: 2111.
47 Choi, W., Sanda, F., Kihara, N., and Endo, T. (1998). *J. Polym. Sci. Part A: Polym. Chem.* 36: 79.
48 Rosenthal, E.Q., Puskas, J.E., and Wesdemiotis, C. (2012). *Biomacromolecules* 13: 154.
49 Ding, Y. and Hay, A.S. (1996). *Macromolecules* 29: 6386.
50 Meng, Y.Z., Hllil, A.R., and Hay, A.S. (2004). *Polym. Adv. Technol.* 15: 564.

51 Meng, Y.Z., Tjong, S.C., and Hay, A.S. (2001). *Polymer* 42: 5215.
52 Meng, Y.Z., Liang, Z.A., Lu, Y.X., and Hay, A.S. (2005). *Polymer* 46: 11117.
53 Yiannios, C.N. and Karabinos, J.V. (1963). *J. Org. Chem.* 28: 3246.
54 Goethals, E.J. and Sillis, C. (1968). *Die Makromolekulare Chemie* 119: 249.
55 Lee, Y., Koo, H., Jin, G. et al. (2005). *Biomacromolecules* 6: 24.
56 Pascal, I. and Tarbell, D.S. (1957). *J. Am. Chem. Soc.* 79: 6015.
57 You, Y.-Z., Manickam, D.S., Zhou, Q.-H., and Oupický, D. (2007). *Biomacromolecules* 8: 2038.
58 Whittaker, M.R., Goh, Y.-K., Gemici, H. et al. (2006). *Macromolecules* 39: 9028.
59 Tatsuma, T., Yokoyama, Y., Buttry, D.A., and Oyama, N. (1997). *J. Phys. Chem. B* 101: 7556.
60 Endo, K. and Bu, H.-B. (2001). *J. Electroanal. Chem.* 506: 155.
61 Endo, K. and Bu, H.-B. (2001). *Polymer* 42: 3915.
62 Song, L.N., Xiao, M., Shu, S. et al. (2007). *J. Mater. Sci.* 42: 1156.
63 Thomas, R.C. and Reed, L.J. (1956). *J. Am. Chem. Soc.* 78: 6148.
64 Endo, K., Shiroi, T., Murata, N. et al. (2004). *Macromolecules* 37: 3143.
65 Kisanuki, A., Kimpara, Y., Oikado, Y. et al. (2010). *J. Polym. Sci. Part A: Polym. Chem.* 48: 5247.
66 Ishida, H., Kisanuki, A., and Endo, K. (2009). *Polym. J.* 41: 110.
67 Endo, K. and Yamanaka, T. (2006). *Macromolecules* 39: 4038.
68 Behrendt, F.N. and Schlaad, H. (2018). *Macromol. Rapid Commun.* 39: 1700735.
69 Li, J., Carnall, J.M.A., Stuart, M.C.A., and Otto, S. (2011). *Angew. Chem.Int. Ed.* 50: 8384.
70 Otsuka, H., Nagano, S., Kobashi, Y. et al. (2010). *Chem. Commun.* 46: 1150.
71 Kobayashi, N., Osawa, A., and Fujisawa, T. (1973). *J. Polym. Sci. Polym. Lett. Ed.* 11: 225.
72 Kobayashi, N. and Fujisawa, T. (1972). *J. Polym. Sci. Part A-1* 10: 2591.
73 Kobayashi, N., Osawa, A., and Fujisawa, T. (1973). *J. Polym. Sci. Polym. Chem. Ed.* 11: 679.
74 Sakai, N., Lista, M., Kel, O. et al. (2011). *J. Am. Chem. Soc.* 133: 15224.
75 Orentas, E., Lista, M., Lin, N.-T. et al. (2012). *Nat. Chem.* 4: 746.
76 Lista, M., Areephong, J., Sakai, N., and Matile, S. (2011). *J. Am. Chem. Soc.* 133: 15228.
77 Sakai, N. and Matile, S. (2011). *J. Am. Chem. Soc.* 133: 18542.
78 Charbonnaz, P., Sakai, N., and Matile, S. (2012). *Chem. Sci.* 3: 1492.
79 Areephong, J., Orentas, E., Sakai, N., and Matile, S. (2012). *Chem. Commun.* 48: 10618.
80 Bang, E.-K., Gasparini, G., Molinard, G. et al. (2013). *J. Am. Chem. Soc.* 135: 2088.
81 Morelli, P., Benlloch, X.M., Tessier, R. et al. (2016). *Polym. Chem.* 7: 3465.
82 Basak, D., Kumar, R., and Ghosh, S. (2014). *Macromol. Rapid Commun.* 35: 1340.
83 Bej, R., Sarkar, J., and Ghosh, S. (2018). *J. Polym. Sci. Part A: Polym. Chem.* 56: 194.
84 Bej, R., Rajdev, P., Barman, R., and Ghosh, S. (2020). *Polym. Chem.* 11: 990.

85 Roy, D., Cambre, J.N., and Sumerlin, B.S. (2010). *Prog. Polym. Sci.* 35: 278.
86 Elsabahy, M. and Wooley, K.L. (2012). *Chem. Soc. Rev.* 41: 2545.
87 Basak, D., Bej, R., and Ghosh, S. (2015). *Polym. Chem.* 6: 6465.
88 Ju, P., Hu, J., Li, F. et al. (2018). *J. Mater. Chem. B* 6: 7263.
89 Bej, R., Sarkar, J., Ray, D. et al. (2018). *Macromol. Biosci.* 18: 1800057.
90 Callmann, C.E., Barback, C.V., Thompson, M.P. et al. (2015). *Adv. Mater.* 27: 4611.
91 Louage, B., Nuhn, L., Risseeuw, M.D.P. et al. (2016). *Angew. Chem. Int. Ed.* 55: 11791.
92 Bae, Y., Fukushima, S., Harada, A., and Kataoka, K. (2003). *Angew. Chem. Int. Ed.* 42: 4640.
93 Zhang, Y., Kohler, N., and Zhang, M. (2002). *Biomaterials* 23: 1553.
94 Bej, R. and Ghosh, S. (2019). *Bioconjug. Chem.* 30: 101.
95 Gasparini, G., Bang, E.-K., Molinard, G. et al. (2014). *J. Am. Chem. Soc.* 136: 6069.
96 Derivery, E., Bartolami, E., Matile, S., and Gonzalez-Gaitan, M. (2017). *J. Am. Chem. Soc.* 139: 10172.
97 Gasparini, G. and Matile, S. (2015). *Chem. Commun.* 51: 17160.
98 Fu, J., Yu, C., Li, L., and Yao, S.Q. (2015). *J. Am. Chem. Soc.* 137: 12153.
99 Yu, C., Qian, L., Ge, J. et al. (2016). *Angew. Chem. Int. Ed.* 55: 9272.
100 Shu, Z., Tanaka, I., Ota, A. et al. (2019). *Angew. Chem. Int. Ed.* 58: 6611.

12

Reduction-Responsive Disulfide-Containing Polymers for Biomedical Applications

Xing Wang and Decheng Wu

Chinese Academy of Sciences, Institute of Chemistry, Beijing National Laboratory for Molecular Sciences, State Key Laboratory of Polymer Physics & Chemistry, Zhongguancun North First Street 2, Beijing 100190, China

12.1 Introduction

The human body consists of compartmentalized regions of differing redox potential. Specifically, sulfur is a unique atom that exists in multiple oxidation states; accordingly, sulfur-based materials have been developed to respond to oxidation triggers in various applications (Figure 12.1) [1–3]. In general, the reductive agent of glutathione (GSH) is easily found at a concentration of 2–3 orders of magnitude larger inside the cells than outside [4]. In contrast, oxidizing agents that include hydrogen peroxide are associated with tissue inflammation and injury [5]. These obvious differences in reduction potential between a local tissue/cellular environment and the surroundings present an unique opportunity to create responsive biomaterials that are triggered via oxidation or reduction within the body. More conspicuously, the past few decades has witnessed a booming development of reduction-responsive polymers for active intracellular drug delivery, wherein the GSH tripeptide is the most abundant biological reductive agent with millimolar concentrations within the cytosol and cell nucleus ranges from 2 to 10 mM, which is about 100–1000 times higher than that in body fluids and extracellular matrices (2–10 μM). Moreover, tumor tissues are highly reductive with at least four times higher concentration of GSH compared with normal tissues [6]. Consequently, reduction is a unique and ubiquitous signal for disulfide-containing polymers that can be used as a fascinating trigger to accomplish rapid and efficient drug release in both cancer cells and tumor tissues. Therefore, disulfide-containing materials that respond to biological cues (biochemical differences between the extra-and intracellular environments) are the subject of intense research interest for preparing nanoparticles as the smart drug delivery vehicles, since the disulfide bonds can remain stable under physiological conditions during circulation as well as in the extracellular tissues on account of a low concentration of GSH, but quickly be cleaved upon exposure to a highly reductive environment within the cells [7].

Sulfur-Containing Polymers: From Synthesis to Functional Materials,
First Edition. Edited by Xing-Hong Zhang and Patrick Theato.

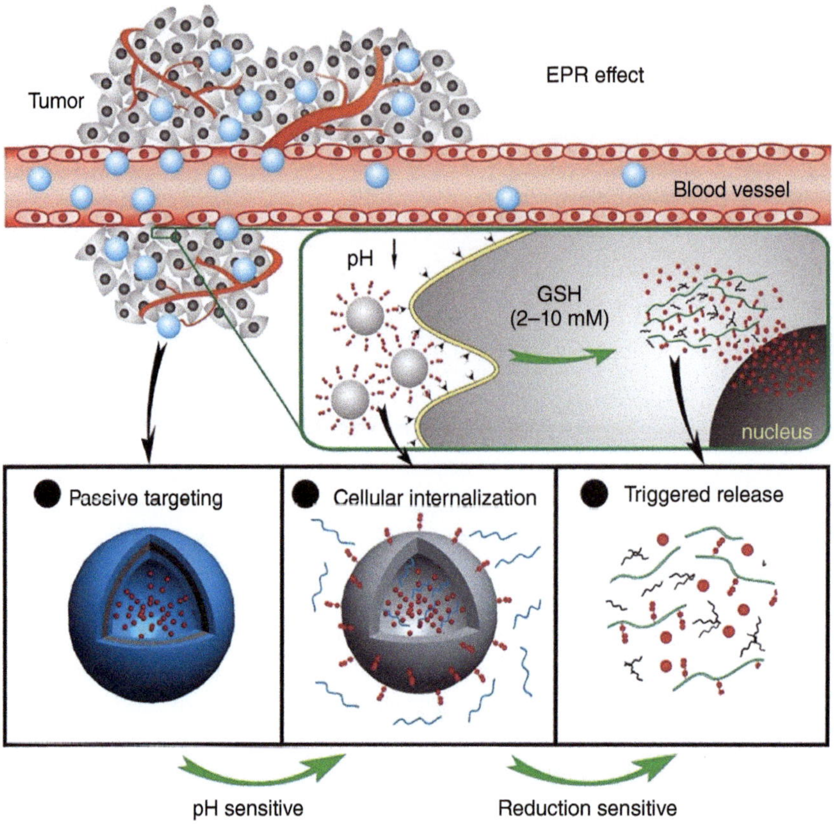

Figure 12.1 Localized regions throughout diseased tissue can be exploited for selective uptake of polymer nanoparticles and triggers for drug delivery. Source: Adapted with permission from Ref. [1]. Copyright 2017, Springer Nature.

Thiol-disulfide exchange is a rapid and readily reversible reaction, which plays an significant role in maintaining proper biological functions of living cells, including the stabilization of protein structures, enzymatic activity, and redox cycles [8]. This quick-response chemical degradation behavior is different from those traditional hydrolytically degradable polymers such as aliphatic polyesters and polycarbonates, in which the ester and carbonate bonds have good capacities to exhibit the gradual degradation kinetics inside the body with degradation times from days to weeks then to months [9, 10]. Therefore, in views of these features of reduction-cleavable disulfide chemistry that disulfide linkage can be cleaved through the thiol-disulfide exchange reactions with reductive molecules of GSH, dithiothreitol (DTT) or others, a number of reduction-responsive polymers and solution nano-assemblies have been synthesized, evaluated, and developed for biomedical applications.

This review summarizes numerous strategies and developments of reduction-responsive topological polymers that have been proposed to prepare nano-assemblies in various bio-applications of targeted drug release, gene transfer, bioimaging, self-healing materials, and others. In consideration of location site of disulfide linkages and architectural topologies, a variety of reduction-responsive disulfide-containing polymers including linear polymers, dendritic polymers, polypeptides and proteins, nanoparticles, and hydrogels have been well designed and developed in recent years. The advanced strategies retaining reductive responsiveness shall allow for the synthesis of disulfide-containing polymers and their self-assembled nanoparticles for further extensive utilization in many biomaterial fields. Notably, these well-organized nano-assemblies are stable under physiological conditions while they can be induced in a controlled fashion as cellular GSH is provided in tumor tissues, which offers unexpected properties and unlimited possibilities for greatly promising applications. We will give several recent examples to highlight the synthetic strategy and application research of these reduction-responsive topological polymers.

12.2 Disulfide-Containing Topological Polymers

There are some general synthetic methods for incorporating the disulfide linkages into various topological architectures by the rapid development of living/controlled polymerization like reversible addition-fragmentation chain transfer polymerization (RAFT), atom transfer radical polymerization (ATRP), and ring-opening polymerization (ROP) techniques, in which disulfide linkages may act as the end-group or be introduced to the end of polymer chains by post-polymerization methods [11, 12]. Wherein, the thiol-disulfide exchange reaction is employed as one of efficient and mild methods to construct multifarious topological polymers.

12.2.1 Systems with the Disulfide Linkages

Disulfide linkages positioned in various locations have been applied broadly in reduction-responsive biomedical systems. Traditional approaches can be classified with the number, position, and location of disulfide linkages (Figure 12.2) [13]. For example, disulfide-containing block copolymers can be designed with disulfide linkages either in the hydrophobic block or at the junction of hydrophilic/hydrophobic blocks. Their self-assembly in aqueous solutions enables the formation of nano-assemblies with disulfide linkages positioned in a single location, as either in hydrophobic cores or at core/corona interfaces. Originating from the environments of disulfide linkages being varied with their numbers and locations, these nano-assemblies respond to a reducing environment in diverse manners as main chain biodegradation and change in hydrophilic/hydrophobic balance. Through these reduction-responsive mechanisms, disulfide-containing nano-assemblies are destabilized and disintegrated, leading to the enhanced release of encapsulated drug molecules.

12.2.2 Disulfide-Containing Linear Polymers

Living polymerization and click chemistry techniques have brought about great progress on preparation of disulfide-containing linear polymers, wherein a survey of

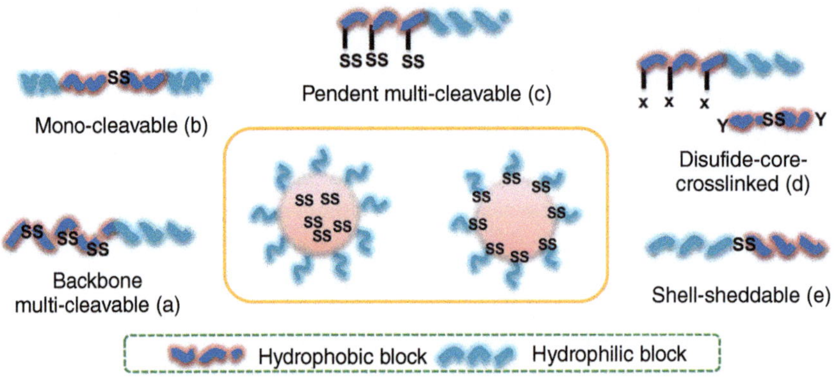

Figure 12.2 General approaches to synthesize reduction-responsive degradable polymers and nano-assemblies with disulfide linkages at various locations and micellar core or core/corona interface. Source: Adapted with permission from Ref. [13]. Copyright 2019, Royal Society of Chemistry.

the literatures demonstrated that reduction-responsive disulfide linkages could be incorporated either in the backbone or at the side chain in the polymeric structures by various and high-efficient synthetic approaches.

12.2.2.1 Linear Polymers with Cleavable Backbones

In the past several decades, incorporation of disulfide bonds into the junction of hydrophobic and hydrophilic blocks has been recognized as a powerful and efficient strategy for construction of biodegradable materials. Based on high-effective Michael-type polyaddition, a series of structurally linear bioreducible poly(amido amine)s with multiple disulfide bonds in their main chains were prepared between cystamine bisacrylamide (CBA) and various amine monomers (Figure 12.3) [14]. Xing and coworkers synthesized a novel copolymer based on polyethylene glycol (PEG) and reducible poly(β-amino ester)s (PBAEs) containing disulfide bonds in the backbone by 2,2′-dithiodiethanol diacrylate, 4,4′-trimethylene dipiperidine, and methoxy-PEG-NH_2, which could self-assemble into stable micelles in a physiological environment but quickly destroy in the presence of 5 mM DTT agent for 1.5 hours from the dynamic light scattering (DLS) results, demonstrating great potential for targeted drug delivery [15]. Oh and coworkers designed a new method for synthesis of biodegradable PLA-based amphiphilic copolymers having disulfide linkages at the block junctions. The method employed combined ROP and ATRP, as well as synthesis of a new disulfide-containing double-head initiator having both terminal OH and Br groups (OH-SS-iBuBr). Then, the PLA-SS-iBuBr and PLA-SS-POEGMA-Br polymers with narrow-distributed molecular weight were successfully synthesized, and the aqueous micellization of PLA-SS-POEGMA polymers could result in the core/shell micelles with disulfide linkages positioned at the interface between PLA cores and POEGMA coronas. Most importantly, reduction-responsive linear polymers and corresponding micelles demonstrated that POEGMA coronas were shed from the PLA cores upon cleavage of disulfide linkages in presence of DTT agent, causing precipitation of PLA core originating

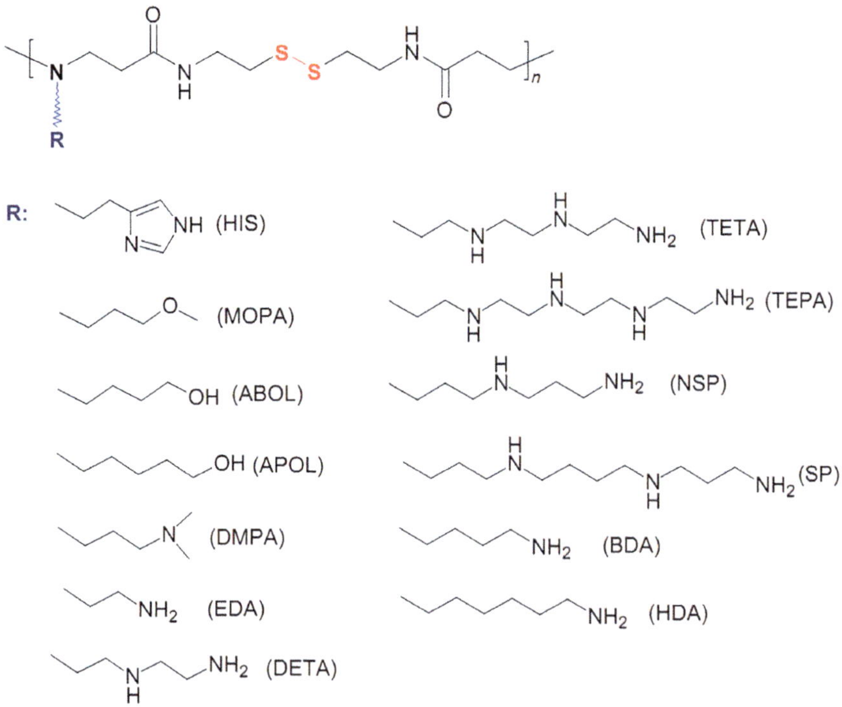

Figure 12.3 Linear bioreducible poly(amido amine)s synthesized by Michael addition polymerization between CBA and various amine monomers. Source: Reference [14] Copyright 2009, Elsevier.

from loss of colloidal stability. These results powerfully pointed out that these biodegradable PLA-based sheddable micelles can be useful as smart drug carriers for tumor-targeted intracellular drug delivery [16].

In addition, the linear polymers with cleavable backbones were sensitive, not only to a single stimulus, but to the simultaneous presence of multiple stimuli. Wang and coworkers reported a facile method to construct pH- and reduction-responsive PBAEs micelles from linear polymer with reducible disulfide linkages in the main chain, in which a large number of hydroxyl groups endowed good hydrophilicity to stabilize the micelles in water, and the abundant disulfide linkages embedded in the core and tertiary amines located in the shell provided multiple responsiveness for the micelles (Figure 12.4). Therefore, these nanoparticles can be rapidly degraded at high levels of reductive agents or acidic environment, presenting the multi-responsive behaviors for targeted drug delivery [7].

12.2.2.2 Linear Polymers with Cleavable Side Chains

Polymers with disulfide bonds on the side chains were also a series of disulfide-containing linear polymers with wide applications. Jiang and coworkers introduced disulfide linkages into the side chains of a thermo- and reduction-responsive P(OEGMA-*co*-Boc-Cyst-MMAm) via copolymerization of OEGMA macromonomer and Boc-Cyst-MMAm monomer via a disulfide linkage, which could self-assemble

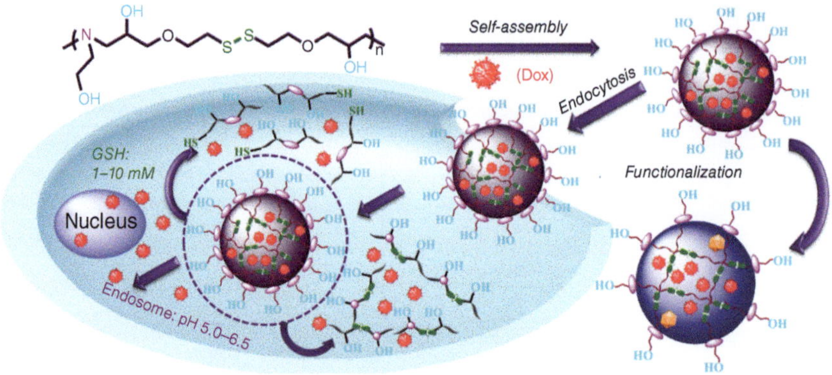

Figure 12.4 Activated intracellular drug release from pH and GSH dual-sensitive biodegradable micelles. Source: Adapted with permission from Ref. [7]. Copyright 2016, American Chemical Society.

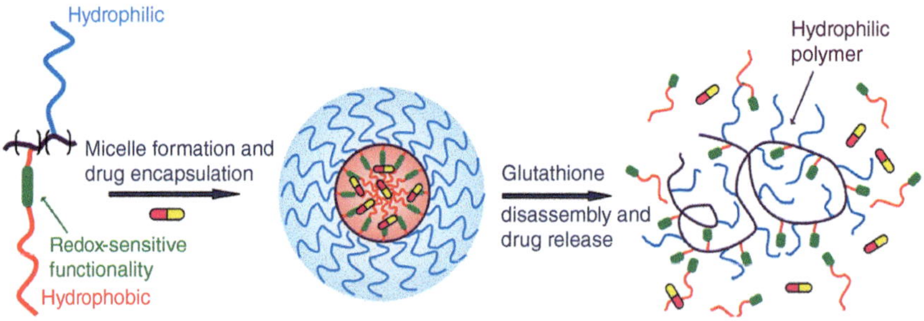

Figure 12.5 Schematic representation of disassembly and consequent drug release of amphiphilic polymeric micelle containing reduction-cleavable hydrophobic chain triggered by GSH. Source: Adapted with permission from Ref. [18]. Copyright 2010, American Chemical Society.

into micelles at room temperature. Once addition of reductive agent DTT, the disulfide linkages in the polymer side chains could be interrupted [17]. Similarly, Thayumanavan and coworkers prepared various amphiphilic polymers through free radical polymerization of tri(ethylene glycol) and an alkyl chain via a disulfide linkage. These amphiphiles could form micelle-type nano-assemblies in water, and further used to encapsulate the hydrophobic drug molecules. However, these micelles disassembled and dissolved in water after treatment with the reduction agents, which ascribed to conversion of amphiphilic polymers into soluble polymers upon cleavage of disulfide linkages and detachment of hydrophobic alkyl chains (Figure 12.5), thus resulting in the rapid release of hydrophobic guest molecules from the self-assembled micelles [18].

12.2.3 Disulfide-Containing Dendritic Polymers

Most of functionalized disulfide-containing nanoparticles were prepared by linear polymers, which however only possessed limited terminal groups that could impair the drug loading ability and further functionalization. Alternatively, disulfide-containing dendritic polymers (hyperbranched polymers and dendrimers), comprising of dendritic, linear, and terminal units with multi-disulfide bonds positioned in the backbones, had been come out and shown great promising in biomedical fields.

12.2.3.1 Disulfide-Containing Hyperbranched Polymers

Compared with conventional linear polymers, hyperbranched polymers are a kind of highly branched macromolecules with a 3D dendritic architecture and a number of terminal groups for further functionalization. In views of their unique topological structures and physical/chemical properties, hyperbranched polymers are more readily accessed to prepare and had many inner cavities that could encapsulate small molecules for construction of smart nanocarriers. Therefore, it could be imagined if the polymeric nanoparticles for anticancer drug delivery were obtained by hyperbranched polymers with disulfide linkages through their whole backbones and/or functionalized terminal groups, an active tumor targeting polymeric drug delivery system could be generated facilely with high dose of drug payload and reductive responsiveness. Yan and coworkers prepared a series of disulfide-containing hyperbranched copolymers of amphiphilic H40-star-PLA-SS-PEP and copolyphosphates (HPHSEP-star-PEPx). In aqueous solutions, these hyperbranched copolymers self-assemble into nano-sized micelles with tunable dimensions reduction-responsive properties. The GSH-mediated intracellular drug delivery was investigated in detail against HeLa cells by pretreating these cells with either glutathione monoester (GSH-OEt) or buthionine sulfoximine (BSO), and found that a faster drug release in GSH-OEt pretreated HeLa cell than that in the non-pretreated cells [19, 20]. Liu and coworkers prepared a model hyperbranched polymer that was synthesized via one-pot two-step Michael addition polymerization of a 1-(2-aminoethyl)piperazine (AEPZ) and a double-molar *N*,*N*′-bis(acryloyl)cystamine (BAC) (Figure 12.6), which could facilitate fine-tuning of the polymeric topology and integration of stimuli-responsive moieties to yield various nanosized motifs in various bio-applications, which proved that such reduction-stimulus nanoparticles were recognized as promising biodegradable drug carriers for improving the efficacy of hydrophobic chemotherapeutic drugs in cancer therapy [21].

Wang and coworkers synthesized a series of aggregation-induced emission (AIE)-active polymer probes for biothiol detection, which was comprised of disulfide-functionalized hyperbranched poly(amido amine) (ssHPA) and hydrophobic AIE-active tetraphenylethene (TPE) fluorogen. Without organic solvents or heavy metal ions as the quencher to eliminate background signals, these AIE polymers were utilized as fluorescence probes for the detection of biothiols because of the relatively rigid dendritic structures. Upon exposure to GSH in an aqueous solution of TPE-ssHPAs, the collapse of dendritic structures could result in the aggregation of TPE units and occurrence of bright fluorescence, offering an effective solution to pre-quench the fluorescence of turn-on type AIE-active sensor [22].

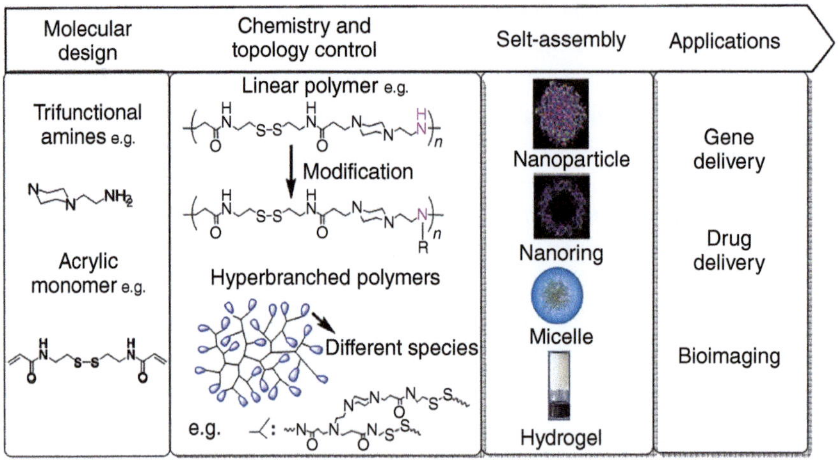

Figure 12.6 Schematic representation of hyperbranched polymer from the Michael addition polymerization of trifunctional amine and acrylic monomer. Source: Adapted with permission from Ref. [21]. Copyright 2016, American Chemical Society.

12.2.3.2 Disulfide-Containing Dendrimers

Dendrimers are highly branched, globular macromolecules with multi-arms emanating from a centered core [23, 24]. The stepwise synthesis of dendrimers affords molecules with a highly regular branching pattern, a unique molecular weight or a low polydispersity index and a precise number of peripheral groups. Therefore, dendrimers with perfect branching resulting from stepwise synthetic processes are distinct from those of hyperbranched polymers with irregular branching by polymerization processes. However, on account of tedious multistep reactions, purification processes, and stringent experimental conditions, the unavailable monomers and prepolymers significantly limited the preparation of disulfide-containing dendrimers, let alone the varieties and scopes of dendritic self-assembly behaviors for biomedical applications.

Fortunately, Malkoch and coworkers designed a family of dendrimers based on 2,2-bismethylol propionic acid (bis-MPA) with large and exact number of dormant and internally queued disulfide bridges. These monomers were attractive building blocks that could participate as key components in materials science via a number of chemical transformation reactions including Michael addition and UV-initiated thiol-ene or thiol-yne click reactions in mild conditions. While these designable and unique dendrimers exhibited a dual function as a consequence of their disassembly capability in a reductive environment and were acted as macromolecular scaffolds that could fully collapse into a novel set of thiol-functional building blocks within 15 minutes once exposure to the DTT agents. Until now, this is the first report of a disulfide-containing dendrimer with rupturing dual capacity for nanomedicine materials (Figure 12.7) [25].

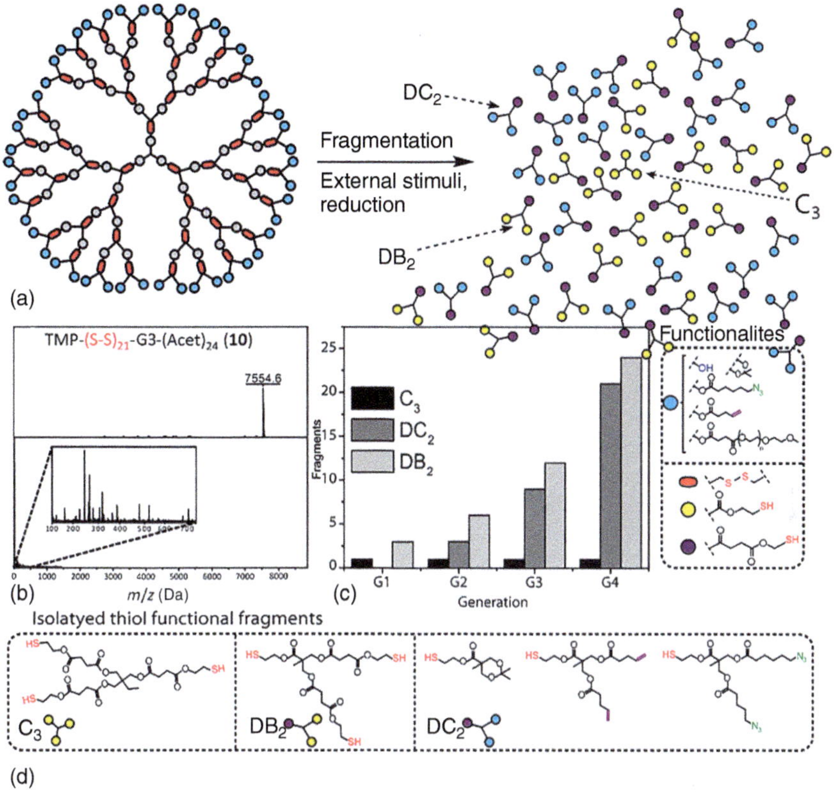

Figure 12.7 (a) Illustration of dendrimer fragmentation with components gained upon fragmentation. (b) MALDI-TOF-MS before and after degradation using DTT for TMP-(S-S)$_{21}$-G3-(Acet)$_{24}$ displaying defined dendrimer peak before and complete disintegration after. (c) Amount of thiols attained in proportion to dendritic generation as well as number of different fragments attained. (d) All of isolated thiol functional fragments. Source: Adapted with permission from Ref. [25]. Copyright 2016, American Chemical Society.

12.2.4 Disulfide-Containing Polypeptides and Proteins

A new class of peptide gene delivery agents had been developed by inserting multiple cysteine residues into short synthetic peptides for the reductively activated peptide gene delivery agents. Rice and coworkers prepared a kind of low molecular weight (LMW) of DNA condensing polypeptide via substitution of one to four lysine residues of Cys-Trp-Lys$_{18}$ with cysteine groups, which could be well oxidized to form interpeptide disulfide bonds after binding to plasmid DNA, generating the small stabilized DNA condensates with a significant enhancement in gene expression [26]. Then, they also found that a minimal peptide sequence of four Lys and two terminal Cys residues could yield reversibly stabilized compact DNA condensates that exhibited high levels of gene expression within the HepG2 cells [27]. Yang and coworkers designed redox-responsive and fluorescent (HA-PEG(SS)-His-Diet-QD) nanocarriers via disulfide-containing poly(ethylene glycol)-histamine-diethylene-triamine covering core/thick-shell CdZnSeS/ZnS QDs for active CD44-targeting

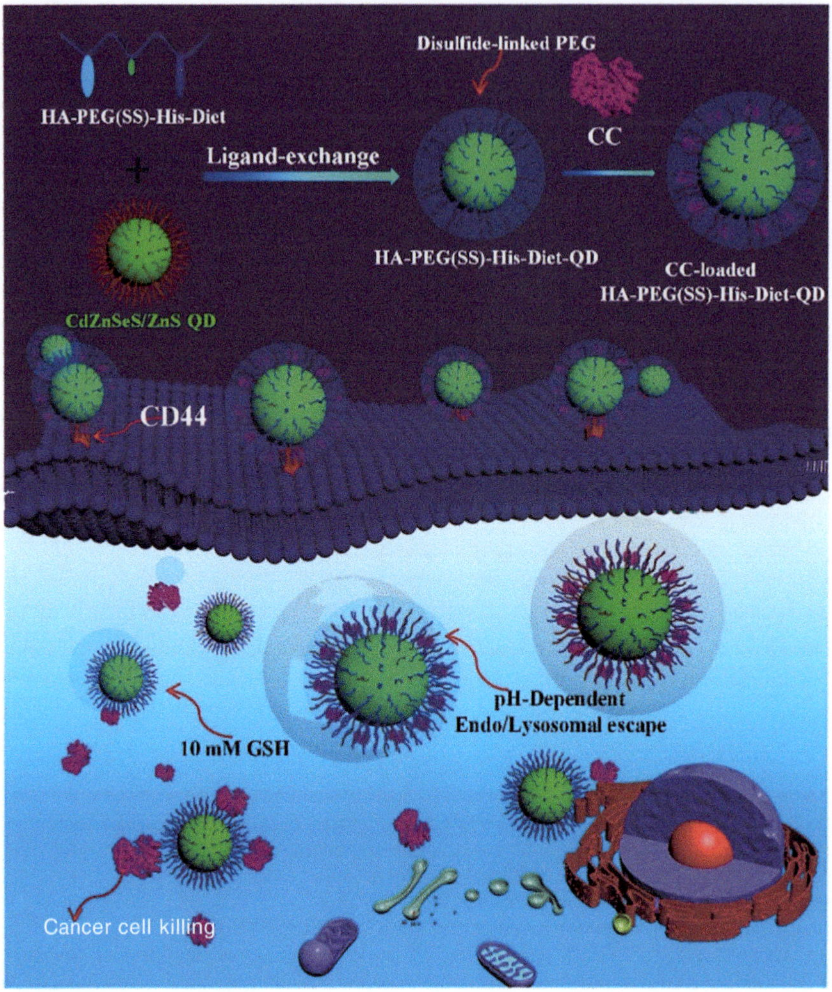

Figure 12.8 Schematic illustration of redox-responsive and fluorescent HA-PEG(SS)-His-Diet-QD nanocarriers for targeted delivery of therapeutic proteins (CC) to CD44-postive cancer cells. Source: Adapted with permission from Ref. [28]. Copyright 2019, Elsevier.

intracellular protein delivery (Figure 12.8). The generating HA-PEG(SS)-His-Diet-QD exhibited excellent stability, green fluorescence property, high QY, and low cytotoxicity. It was also shown that a protein drug of CC could be efficiently loaded and rapidly released in the reductive intracellular environments. Confocal images and MTT assays demonstrated that these HA-PEG(SS)-His-Diet-QD nanoparticles were capable of enhancing endo/lysosomal escape and targeting and releasing proteins to CD44-overexpressed tumors, thus inducing cell apoptosis with high killing effects [28].

Hubbell and Segura designed ABC triblock copolymers of PEG, poly(propylene sulfide) (PPS), and a positively charged peptide (PEG-PPS-peptide), wherein the selected cationic peptides, TAT peptide domain of HIV (RKKRRQRRR), or an oligolysine (Lys_9) were coupled to the PEG-PPS via a disulfide linkage. These ABC triblock copolymers were able to self-assemble with siRNA to form complexes with good biocompatibility compared with the commercially available cationic liposome LF2000. PEG_{45}-PPS_5-K_9 at higher siRNA doses was able to mediate 90% gene expression down-regulation [29]. Uludag and coworkers conjugated bisphosphonates (BPs) that had an exceptional affinity for bone mineral hydroxyapatite to fetuin (a model protein) via a disulfide linkage. The fetuin-BP conjugates (3.3-4.9 BPs/protein) exhibited strong hyaluronic acid (HA) bonding with good stability under aqueous solutions. However, the physiological thiols of L-cysteine, DL-homocysteine, and L-GSH could readily cleave the fetuin-BP conjugates, and therefore releasing the fetuin from the mineral [30].

12.2.5 Disulfide-Containing Polymeric Nanoparticles

12.2.5.1 Disulfide Linker for Amphiphilic Polymers

Reduction-responsive nanoparticles are generally self-assembled from amphiphilic copolymers containing disulfide units that are cleavable under reductive condition while stable under oxidative environment. The disulfide bond has been widely employed as a unique linkage between hydrophilic and hydrophobic segments in block, star, and graft copolymers, which give micelles, vesicles, nanoparticles, or polyplexes that shed off hydrophilic shells in response to intracellular reducing potential [31–39]. In addition, amphiphilic block, graft, and hyperbranched copolymers containing single or multi-disulfide bonds in the hydrophobic regime have been utilized to afford core reductively degradable micelles and nanoparticles for nanomedicines applications [40–48]. With the burgeoning of reduction-responsive and reversibly crosslinked polymeric nanomedicines in the past several years, a series of systems have demonstrated significantly enhanced anti-tumor activity to diverse malignant tumors as compared with their reduction-insensitive counterparts in many excellent reviews [14, 49–52].

Shell-sheddable block copolymers consist of disulfide-containing polypeptides and proteins: disulfide linker for amphiphilic polymers a hydrophobic block covalently conjugated through a disulfide linkage with a hydrophilic block. As a consequence, disulfide bonds positioned at the block junction are located at the interfaces of core/corona in the nanoparticles. In response to reductive reaction, hydrophilic coronas are detached from the hydrophobic cores upon the cleavage of disulfide linkages. This process caused the destabilization of nanoparticles and loose of core structures, thus leading to enhanced drug release. For example, Li et al. synthesized a hyaluronic acid-g-SS-deoxycholic acid (HA-g-SS-DOCA) graft conjugate via a 1-ethyl-3(3-dimethylaminopropyl) carbodiimide (EDC)-mediated coupling reaction. In aqueous media, HA-g-SS-DOCA conjugates self-assembled into nanosized micelles, which rapidly disassembled in response to 20 mM GSH. By using paclitaxel (PTX) as a model drug molecule, the obtained PTX-encapsulated micelles were taken up by human breast adeno-carcinoma cells (MDA-MB-231) via HA receptor-mediated endocytosis and exhibited significantly higher antitumor

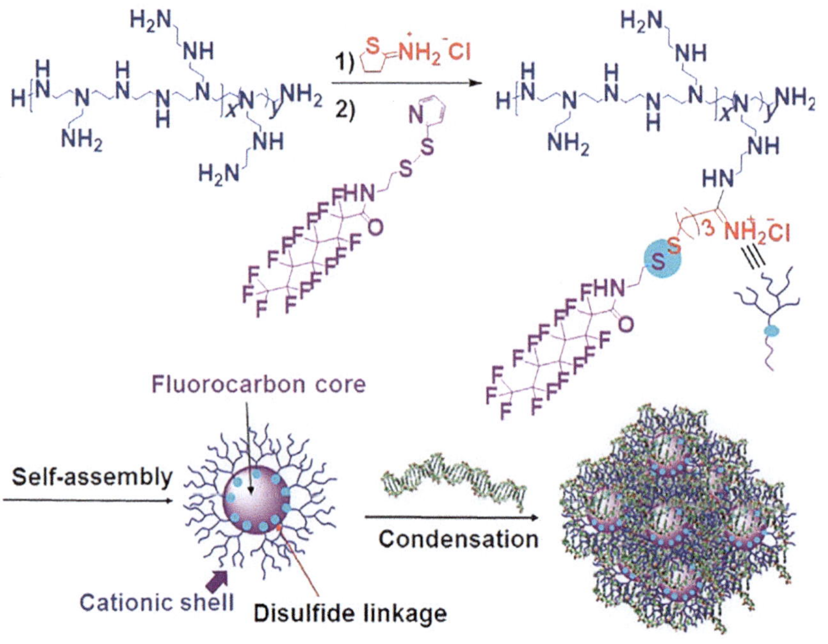

Figure 12.9 Scheme outlining the linking of fluorocarbon chains to PEI, nanoparticle formation, and the complexation of nanoparticles with DNA. Source: Adapted with permission from Ref. [55]. Copyright 2015, John Wiley and Sons.

activity [53]. In addition, Wei et al. fabricated a block copolymer containing a hydrophilic oligoethyleneglycol and a tetraethylenepentamine-grafted polycation through a disulfide linkage via the RAFT and ATRP techniques, which could effectively condense plasmid DNA into salt-stable particles that were de-shield under reducing conditions. On account of the combined properties of targeting and shielding, these particles could retain high transfection efficiency compared to analogous homopolycation vectors for targeted gene delivery [54]. You and coworkers reported a novel class of bioreducible cationic nanoparticles using disulfide bonds to connect the cationic shell to fluorocarbon core. These bioreducible nanoparticles formed strong interactions with DNA and completely condensed DNA at an N/P ratio of 1. Notably, the polyplexes formed by the condensation of these nanoparticles with DNA were very stable under physiological conditions, but were disrupted by intracellular GSH, which led to the release of the entrapped DNA and high levels of gene-transfection efficacy (95%), exhibiting high biocompatibility and effective carrier for gene delivery system (Figure 12.9) [55].

12.2.5.2 Disulfide Linker for Nano-Assemblies

The structural stability and integrity of the nanostructures is a crucial issue in practical applications of traditional micelles due to the inevitable dissociation of non-crosslinked micelles into individual polymeric chains under the shear forces, large volume dilution by gastric, blood or other body fluids, salinity fluctuations, and interactions with cells and biomolecules available in the blood after intravenous administration. Such destabilization may lead to premature release of the loaded drugs and inferior tumor targeting ability, further resulting in severe side effects *in vivo* [56]. In the past decades, the strategies of covalent stabilization via crosslinking of the micellar core or shell were explored extensively to overcome the instable defect [57]. Although crosslinked micelles with improved stability render the possible *in vivo* application, it is worth pointing out that permanently crosslinked micelles were unfavorable for real applications because the renal clearance of crosslinked micelles with relatively large dimension is impossible after drug release. Therefore, using reduction-responsive crosslinkers is capable of responding to the reduction condition in the intracellular environment, which is an elegant approach to address the dilemma of the spontaneous dissociation of non-crosslinked micelles in circulatory systems and impossible renal clearance of permanent crosslinked micelles after drug release at the sites of action.

Apart from the direct disulfide linkage formation reaction, one of the most common strategies was crosslinking by disulfide-containing crosslinkers. Various disulfide containing small molecules like dithiodipropionic acid, bis(2,2′-hydroxyethyl) disulfide, cystamine, and their derivatives have been used to crosslink the polymeric systems, leading to high circulation stability while fast response to GSH [58–62]. Wooley and coworkers raised the concept of "shell-crosslinked knedel-like" (SCK) particles, and latter Armes' and McCormick's groups applied this method to prepare some stimuli-responsive polymers [63–65]. However, crosslinking may slow down the response speed. Zhang and Zhao reported a surface-crosslinked micelle with extremely rapid release of the encapsulated pyrene (Figure 12.10). The pyrene emission showed no change over at least a period of six months, but once adding the HIO_4 with gentle vortexing for less than 1 min, the release signal of pyrene was changed quickly and amazingly. The reason for the fast release profile was associated

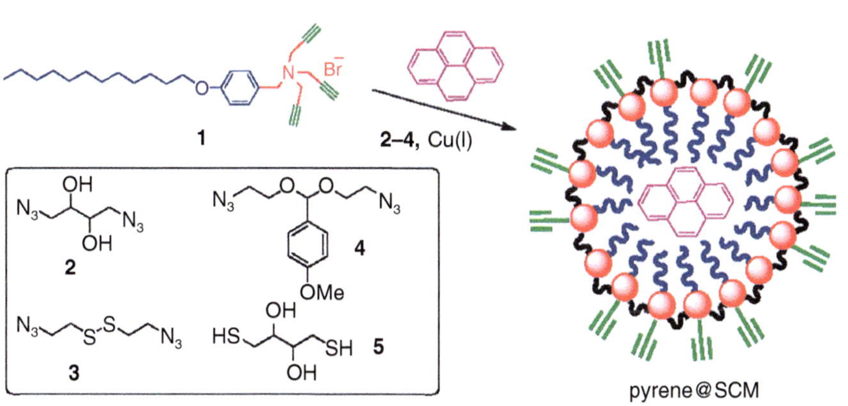

Figure 12.10 Schematic representation of surface crosslinked micelles. Source: Adapted with permission from Ref. [66]. Copyright 2010, American Chemical Society.

with the electrostatic repulsion among the headgroups. Once the covalent crosslinkers were broken by the stimuli, the electrostatic repulsion would be predominant and the micelle exploded rapidly [66].

12.2.5.3 Controlled Fabrication of Solution Nano-Assemblies

Recent advances in the preparation of shape-shifting nano-assemblies are hot topics in development of biomaterial sciences, because many intelligent biofunctions are always relied on their shape and dimension. Wang and coworkers reported a dynamic manipulation of sequential self-assembled evolution via a hierarchical assembly strategy on a tunable thiol-disulfide exchange reaction (Figure 12.11) [67]. By adjusting the external stimuli and reaction time, the solution-assemblies perform a series of attractively architectural and dimensional evolution *in situ* with an instant "on/off" function. In the process of self-assembled evolution, once the thiol-disulfide exchange was interrupted at any moment via neutralizing the system pH or removing UV lamp, the active sites of assemblies were deactivated, and thus no further morphology evolution was conducted any more. However, these stable inert sites could be reactivated whenever needed by re-tailoring the external stimuli; in this case, the desired morphology can be gained and kept stable for a long time, which provided a universal strategy on fabrication of smart carriers with desirable morphologies and dimensions for advanced applications.

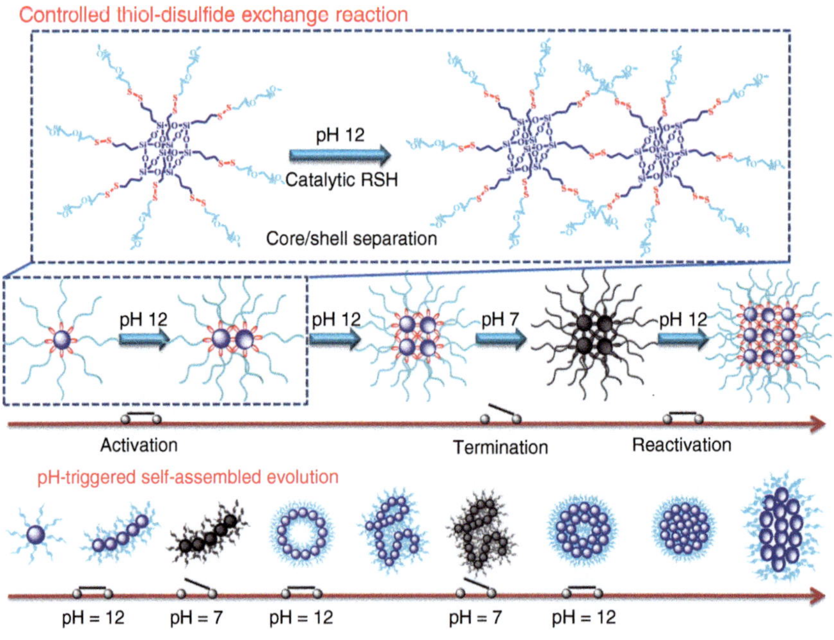

Figure 12.11 Schematic representation of a controlled thiol-disulfide exchange reaction and the pH-triggered self-assembled evolution with a pH-switched "on/off" function. Source: Adapted with permission from Ref. [67]. Copyright 2018, Springer Nature. Licensed under CC BY 4.0.

12.2.6 Disulfide-Containing Crosslinking Hydrogels

12.2.6.1 Disulfide-Crosslinked Micro-/Nanogels

The formation of micro/nanogel particles is attributed to the crosslinking of disulfide linkages; therefore, the hydrogel particles are biodegradable since disulfide bonds can be cleaved by reduction agents. Qian and coworkers prepared a TME-responsive P(CPT-MAA) prodrug nanogel by distillation-precipitation polymerization of methacrylic acid (MAA), CPT monomer (CPTM), and *N*,*N*′-methylenebisacrylamide via the disulfide linkages. By tailoring the reduction-responsive disulfide linkages and pH-responsive PMAA, the CPT release could be accelerated from prodrug nanogels, realizing the "on-demand" release of chemotherapeutic agents in tumor tissue microenvironment (Figure 12.12) [68]. Siegwart and coworkers synthesized a disulfide-crosslinked nanogel by oxidative radical polymerization of 2,2′-(ethylenedioxy) diethanethiol as a monomer with different crosslinkers. After adjustment of the ratio of monomer to crosslinker, the crosslinking density and swelling ratios could be tailored, providing a way to control crosslinking density with constant nanogel size but tunable drug release kinetics [69].

Wu and coworkers reported a controlled crosslinking strategy to prepare the tailor-made hydrogels particles [70]. Combining with an inverse emulsion technique, various micro micro-/nanogels in a confined space were controlled obtained. More importantly, the controlled formation of micro/nanogels is *in situ* fabricated with flexibly designable structures and properties only by manipulating the solution pH at a predetermined time (Figure 12.13), which was attributed to the crosslinking disulfide bonds guided by the thiol-disulfide exchange reactions in aqueous solutions.

12.2.6.2 Disulfide-Crosslinked Macroscopic Hydrogels

The disulfide-crosslinked macroscopic hydrogels are polymeric networks with adaptable linkages that can be quickly broken and reversibly reformed, possessing the unique properties of softness, elasticity, absorption ability, and degradability, which played vital roles in the biomedical fields [71–77]. In recent years, more and more researchers have tried to develop facile and efficient hydrogel synthesis strategies through various functionalities, but the compactness of crosslinked networks was difficult to be further adjusted because of the uncontrollable crosslinking degree and network structure once the disulfide reshuffling reaction was activated, although a few of methods by regulation of pre-set number of crosslinking agents in the initial system [78–80]. To overcome this difficulty, Wu et al. reported the controlled crosslinking strategy to tailor the gelling process with customized crosslinking degree and property from rational design of disulfide-containing hydrogel precursors via controlling thiol-disulfide exchange under various pH conditions [81, 82].

For example, a hyperbranched (poly(*N*,*N*′-bis(acryloyl)cystamine-1-(2-aminoethy) piperazine)-poly(ethylene glycol), BAP) polymer was synthesized via a "one-pot" Michael addition polymerization. In this system, BAP could dissolve in

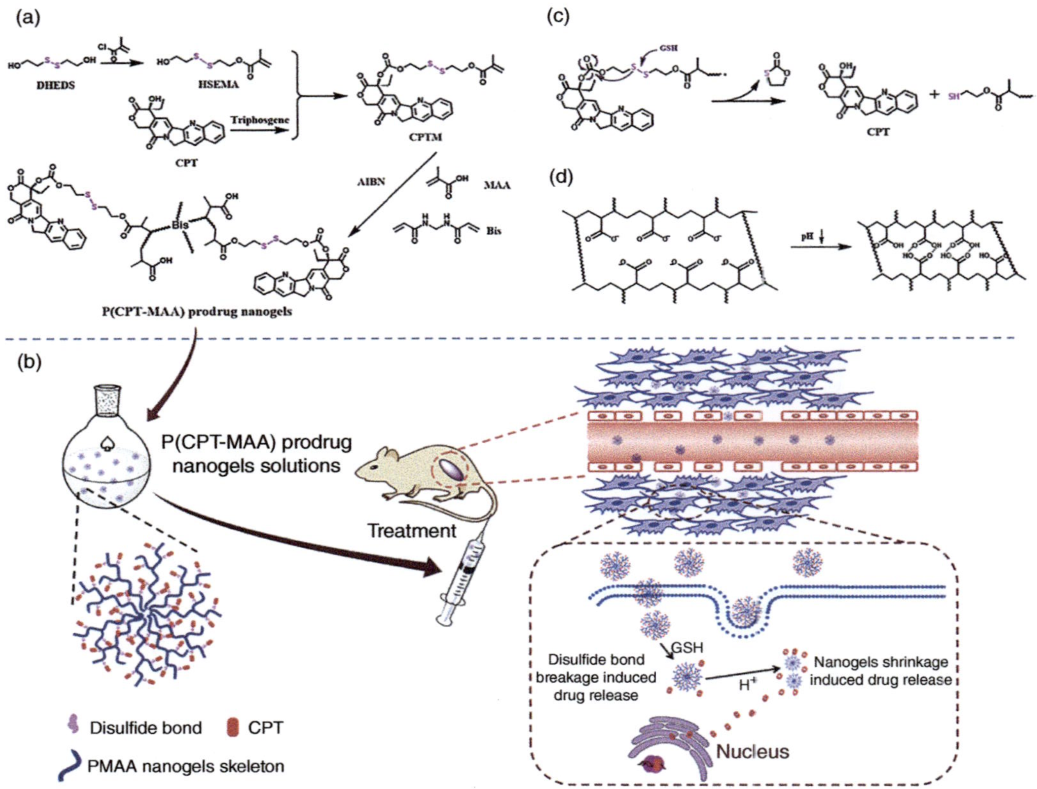

Figure 12.12 (a) Synthesis of the P(CPT-MAA) prodrug nanogels. (b) Schematic illustration of the pH and reduction dual responsive drug release of P(CPT-MAA) prodrug nanogels. (c) The reductive-responsive drug release mechanism. (d) The pH-responsive mechanism. Source: Adapted with permission from Ref. [68]. Copyright 2019, Elsevier.

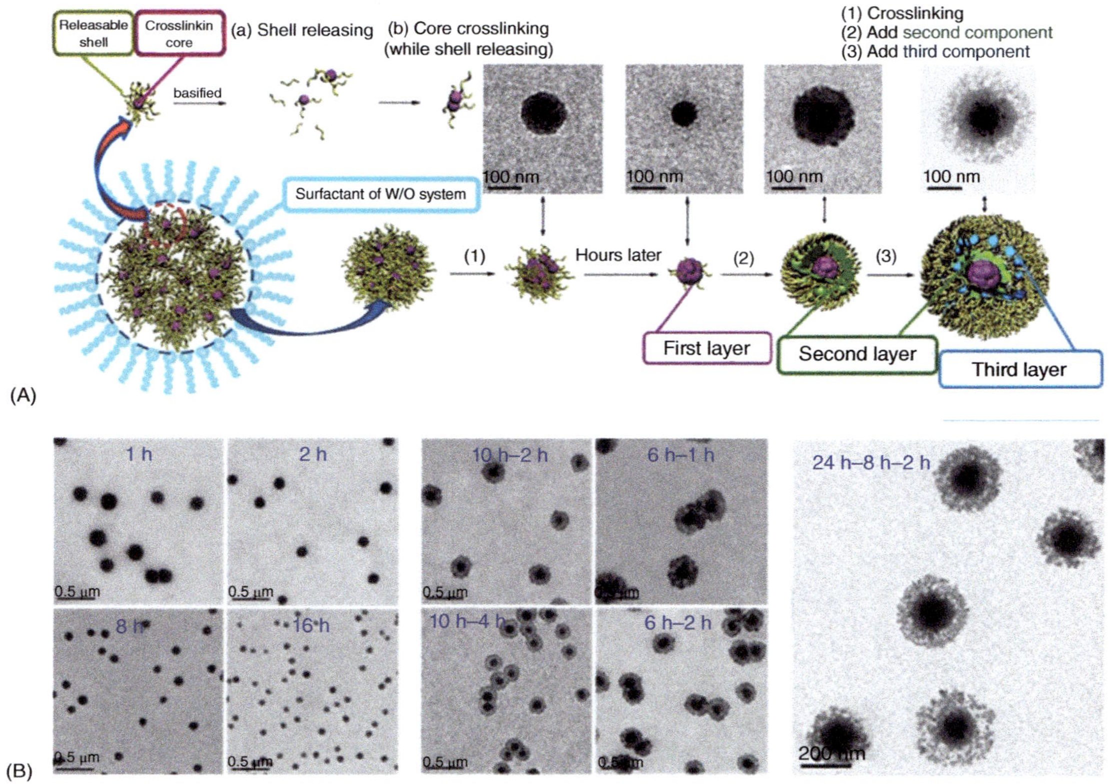

Figure 12.13 (A) Schematic illustration of the core/shell separation process-dissociation of the shells and crosslinking of the cores and depiction of the synthetic approach to controlled formation of (multilayered) hydrogel particles. (B) TEM images of mono-/double-/triple-layered nanogels obtained by crosslinking for various time. Source: Adapted with permission from Ref. [70]. Copyright 2012, American Chemical Society.

water with good stability under low pH solutions but quickly be crosslinked to form the "living" hydrogel as results of the trigger of thiol-disulfide exchange reaction under alkaline conditions. Prolongation of the crosslinking reaction and time, the "living" hydrogel network evolved from the loose to compact with a smaller pore size and denser network (Figure 12.14). Notably, the crosslinking process could be interrupted at any time by neutralizing the solution pH in the system, and the "living" hydrogel was converted to "stable" hydrogel that could be inert for a long time in a pH 7.0 solution. More importantly, this "stable" hydrogel could quickly revert to the "living" hydrogel state by re-basification of system, generating the following size-shrinkage evolution. On account of the disulfide-linked hydrogel structures, these hydrogels were quickly disintegrated after addition of 10 mM DTT, presenting a reduction-responsive degradation behavior. In addition, these hydrogels could be used for *in situ* encapsulation of two active molecular drugs of DOX and PTX, which indicated the advantages of this synthesis strategy on tailor-made hydrogels that would allow the fine modulation of smart drug delivery system to manipulate drug release behaviors in aqueous solutions. The controlled strategy is also suitable for other materials with functional disulfide-linked core/shell structures. Wang et al. produced an organic–inorganic hybrid hydrogel from a precursor of POSS-$(SS\text{-}PEG)_8$ polymer, which could yield a kind of hybrid hydrogels and nanoparticle aggregates with the customized structures and high mechanics with a pH-responsive "on/off" reaction [83].

Compared with the anion-mediated disulfide exchange mechanism, thiol-disulfide exchange reactions could also be controlled via a radical-mediated mechanism under UV irradiation condition [84]. In consideration of the requirement of alkali environment to induce the controlled gelling process, many functional and intelligent hydrogels are not stable in the strong alkali conditions and thus not conducive to biomedical applications. Wu and coworkers developed a photochemical approach to obtain the controlled hydrogels using a comb-like P(EMA-SS-PEG) polymer (poly(ethyl methacrylate)-dithiol-poly(ethylene glycol)) by the UV-triggered thiol-disulfide exchange mechanism, which possessed unique advantages of real-time manipulation and tunably spatiotemporal precision that enabled to realize "turn-off" control over the photochemical process. By varying the irradiation time, the crosslinking network of hydrogels was easily adjusted, acquiring the concomitant transitions in the conformation, mechanical, and swelling properties [85].

12.2.6.3 Disulfide-Crosslinked Natural Hydrogels

Recently, disulfide-crosslinked HA hydrogels have turned to be versatile materials platforms from static to smart and stimuli-responsive systems. For example, Zhang and coworkers prepared the crosslinking HA hydrogels for cell culture. After exposition to air of different thiolated HA derivatives, the hydrogels displayed excellent biocompatibility, biodegradation behavior, and good mechanics that had a great potential for tissue engineering [86]. Yang and coworkers developed a disulfide shuffling strategy to prepare keratin hydrogels via transferring intramolecular disulfide bonds into intermolecular disulfide bonds. In views of reductive

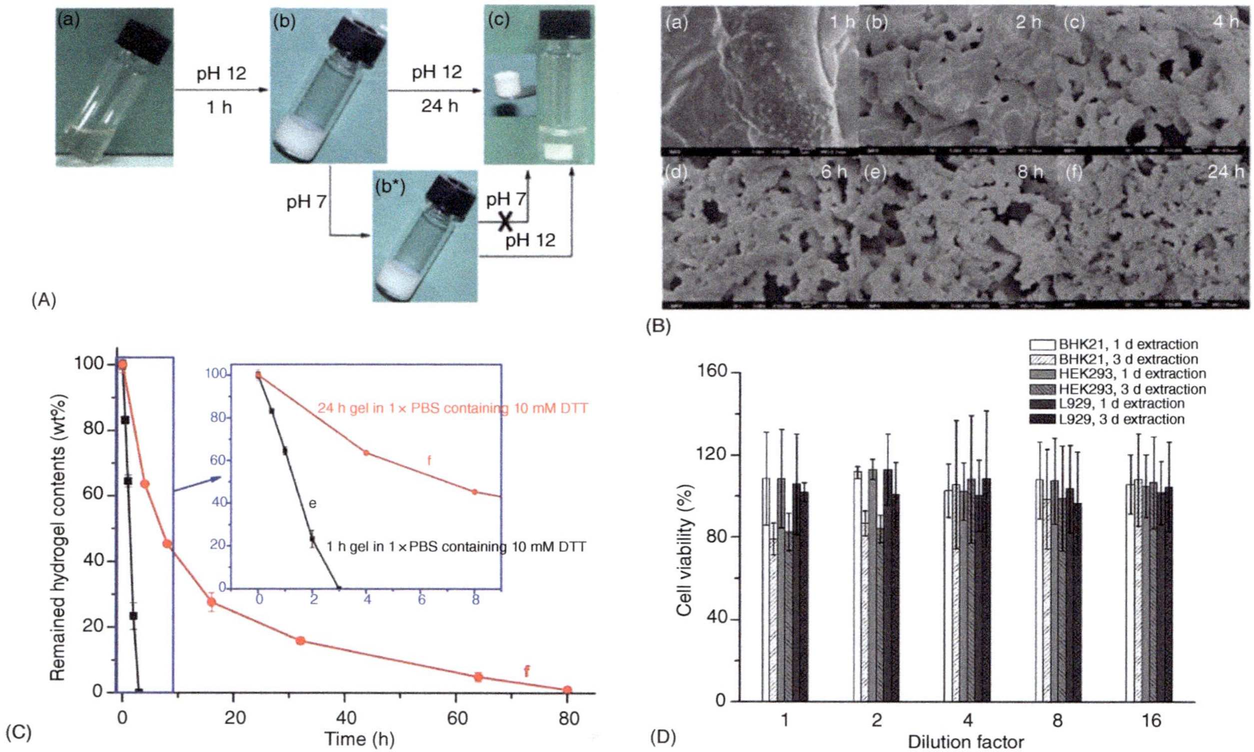

Figure 12.14 (A) Schematic representation of basification of BAP solution and derived methodology for producing loose and compact hydrogels. (a) 10 wt % BAP solution; (b) loose hydrogel after basification for one hour; (b*) inert neutralized loose hydrogel after one hour of crosslinking; and (c) compact hydrogel after basification for 24 hours. (B) SEM images of the hydrogels obtained after basification for various time. (C) Remained amounts of the hydrogels with different gelling time in 10 mM of DTT, at 37 °C as a function of time ($n = 3$). (D) Cytotoxicity of the loose hydrogel after four hours of crosslinking in BHK21, HEK293 and L929 cells after incubation for one and three days ($n = 4$). Source: (a, b) Adapted with permission from Ref. [83]. Copyright 2010, American Chemical Society.

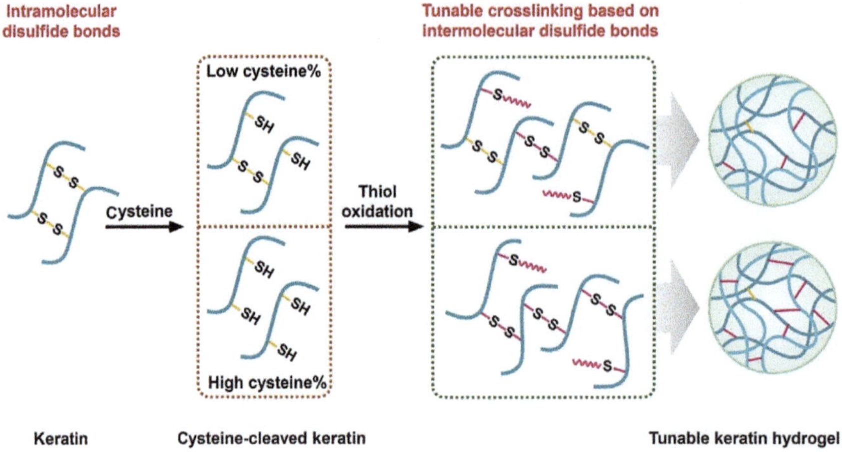

Figure 12.15 Schematic illustration of gelation mechanism by the disulfide shuffling strategy. The intermolecular disulfide crosslinking density can be tuned by the cysteine levels, thus providing altered microstructure and properties including mechanical strength, hydrogel degradation, and drug release. Source: Adapted with permission from Ref. [87]. Copyright 2019, Elsevier.

reagent to cleave the intramolecular disulfide bonds, the liberate free thiol groups were oxidized to form the intermolecular disulfide bonds, generating altered disulfide-containing crosslinking hydrogels: disulfide-crosslinked natural hydrogels network structures and drug release behaviors, providing the extensive applications in tissue engineering (Figure 12.15) [87].

12.3 Disulfide-Containing Polymers for Biomedical Applications

Disulfide bridges were selected as the most promising dormant species mainly on the basis of (i) selective cleavage through the established reductive-responsive mechanism between dormant oxidized disulfide and reactive reduced thiol group and (ii) important biological and responsible functions for scission of disulfides within living organisms. Therefore, it had been exploited extensively for applications of targeted drug and gene carriers, bioimaging, drug conjugates, self-healing materials, etc.

12.3.1 Targeted Drug Delivery

The disulfide bonds were cleaved by high intracellular GSH agents to rapid release of payloads for effective cancer therapy (Figure 12.16), which had been confirmed to be effective in a variety of nano-formulations for biomedical applications [88]. As previously mentioned, it has been well-established that disulfides were degradable in presence of reducing agents. The basic principle of programmable redox-responsive drug delivery systems was to utilize the distinct differences in

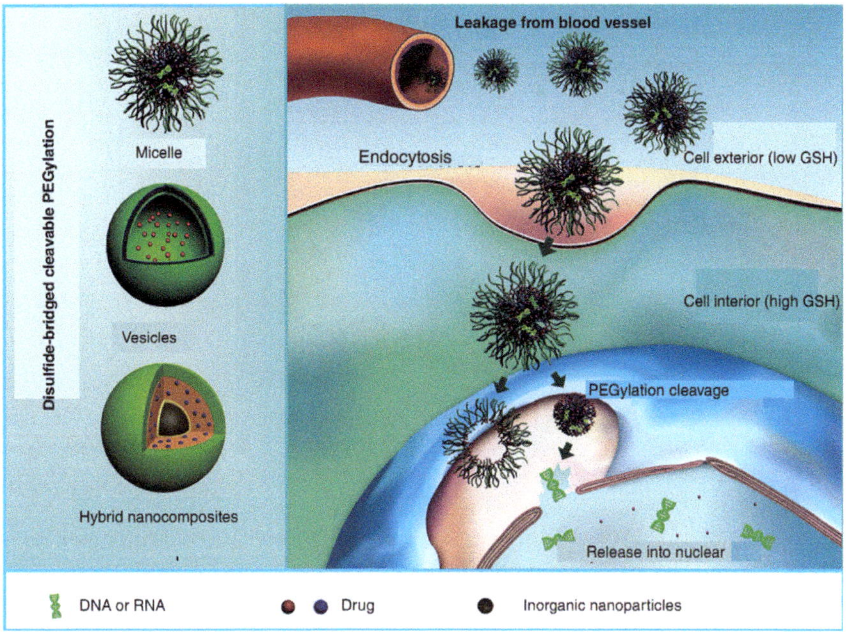

Figure 12.16 Nano-formulations engineered with disulfide-bridged cleavable PEGylation (left) and their function pathway from blood vessel leakage, and cell endocytosis to drug/gene release inside the cell. The release behavior can be regulated by selective intracellular PEGylation cleavage. Source: Adapted with permission from Ref. [88]. Copyright 2015, Future Medicine.

reductive potentials between tumors and normal tissues. Many literatures had reported that GSH/glutathione disulfide (GSSG) was the most abundant reductive molecules couple within animal cells [49, 89, 90]. Moreover, reactive oxygen species (ROS) were believed to be implicated with some serious diseases like arteriosclerosis, heart injury and cancer [91]. All of these pathological signals can be exploited as significant guidance for design of the reductive-responsive drug delivery systems.

12.3.1.1 Disulfide-Containing Nano-Assemblies for Targeted Drug Delivery

Redox-responsive nano-assemblies have own unique potential to facilitate drug release in tumors, which is originated from varied concentrations of reduction species within intracellular and extracellular compartments. McCormick and coworkers have prepared a thermosensitive triblock copolymer PEO-*b*-P(DMA-*stat*-NAS)-*b*-PNIPAM using a RAFT approach, which molecularly dissolved in water at room temperature but transformed into spherical micelles with hydrodynamic diameters of 38 nm at 45 °C. The micellar shell contained reactive *N*,*N*′-acryl-oxysuccinimide units that yielded reversibly shell-crosslinked (SCL) micelles after crosslinking with cystamine. The SCL micelles did not dissociate as the solution temperature was lowered again to room temperature, indicating the stabilization of micellar structures. Furthermore, these micelles were destabilized using DTT or tris (2-carboxyethyl) phosphine (TCEP) and re-crosslinked

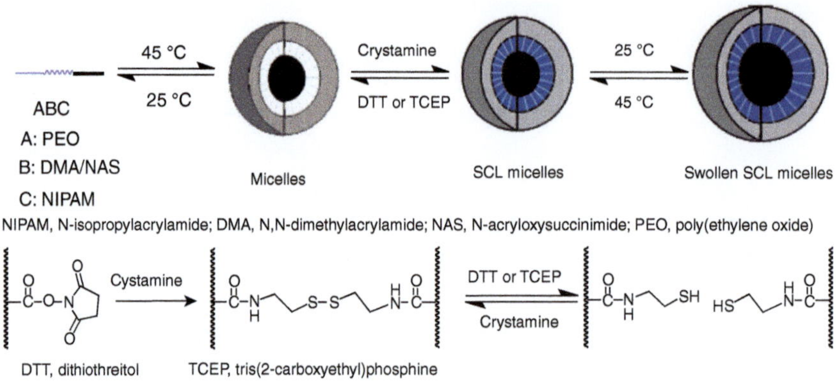

Figure 12.17 Schematic illustration of the formation of reversible shell crosslinked (SCL) micelles from PEO-*b*-(DMA-*s*-NAS)-*b*-NIPAM triblock copolymers. Source: Adapted with permission from Ref. [92]. Copyright 2006, American Chemical Society.

using cystamine, which were applied for sustained release of dipyridamole, wherein the rate of drug release could be readily controlled by the DTT concentration and solution temperature (Figure 12.17) [92]. In a similar way, Liu and coworkers synthesized a thermo-responsive hydrophilic block copolymer of PEO-*b*-P(NIPAM-*co*-NAS), which also molecularly dissolved at 25 °C but self-assembled into micelles at an elevated temperature. The P(NIPAM-*co*-NAS) core was crosslinked by cystamine, generating the reversible core-crosslinked (CCL) micelles with good degradability and targeted drug release behaviors [93]. Ravoo and coworkers designed and prepared a family of amphiphilic cyclodextrin derivatives by grafting the hydrophobic substituents to the primary side of cyclodextrin molecules via a disulfide bond. These amphiphilic cyclodextrins could self-assemble into the vesicles or nanoparticles in aqueous solutions, which released an encapsulated hydrophobic guest from the nanoparticles by adding DTT agents [94].

12.3.1.2 Disulfide-Containing Nanogels for Drug Delivery

As the nanogel particles were formed by the disulfide crosslinkers, these biodegradable nanogels can be used as appropriate carriers for drug delivery. Qi and coworkers designed a temperature-activated drug locking and GSH-triggered drug unlocking nanogel from the connection of HA and poly(*N*-isopropyl-acrylamide) (PNIPAAm) via a disulfide bond. These synthetic HA-SS-PNIPAAm copolymers endure temperature-motivated "lock" behaviors to form nanogels because of the thermosensitive lipophilic transformation of PNIPAAm followed by locking the DOX into cavities of the nanogels, which exhibited GSH-triggered "opening" behavior to unlock drugs for tumor therapy. Compared with free DOX, HA-SS-PNIPAAm@DOX nanogels showed a more significant destructive effect on tumors. HA-SS-PNIPAAm nanogels could not only significantly enhance antitumor efficacy in 4T1 tumor-bearing mice, but also reduce the side effects of DOX *in vivo* (Figure 12.18) [95]. He and coworkers fabricated a new biocompatible and

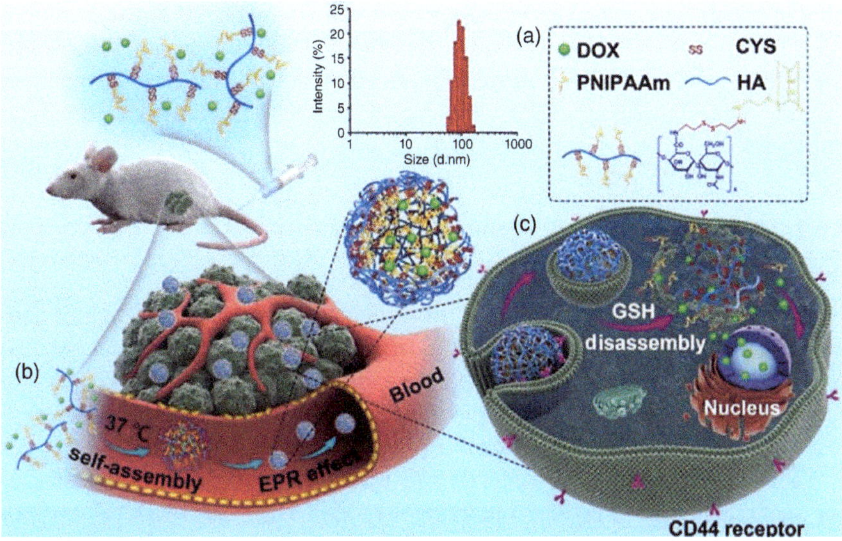

Figure 12.18 (a) Chemical structure of HA-SS-PNIPAAm polymer. (b) Schematic diagram of the self-assembled H-SS-P@DOX nanogels. (c) The intracellular uptake mechanism of nanogels through CD44 receptor-mediated endocytosis and intracellular drug release mechanism triggered by GSH. Source: Reference [95]. Copyright 2019, Royal Society of Chemistry.

self-assembled acylated rapeseed protein isolate (ARPI)-based nanogel by chemical acylation and heat-induced protein denaturation. Compared with the native rapeseed protein isolates and ARPI without the heat treatment, ARPI-based nanogels exhibited significantly altered spatial secondary and tertiary structures, improved surface hydrophobicity, and reduced free sulfhydryl contents of the protein. The self-aggregating ability of ARPI caused the hydrophobic core with formations of covalent disulfide bonds and the hydrophilic shell with succinyl moieties exposed to the water; in this case, ARPI nanogels exhibited high-efficient encapsulation of hydrophobic curcumin (CUR), thus improving the anticancer activity against multiple cancer cell lines [96].

12.3.1.3 Disulfide-Containing Hydrogels for Drug Delivery

Reductive-responsive hydrogels, which undergo reversible volume phase-transition or sol-gel phase-transition in response to reduction potential, had great potential in drug delivery [97–106]. Zhong and coworkers prepared a 10-hydroxy camptothecin (HCPT)-peptide (phenylalanine-phenylalanine-glutamicacid-SS-glutamicacid-glutamicacid (FFEssEE)) hydrogel system via the disulfide linkages. This system possessed recovery, shear-shinning properties, and high drug loading capacity. Moreover, this reductive-responsive hydrogel system was innoxious in hydrogel concentration under normal physiological conditions; however, it possessed a favorable anticancer effect when incubated with cancer cells on account of the drug release by the disrupt of disulfide linkages in reductive environments [107]. Dziubla and coworkers synthesized a PBAE hydrogel formed via reduction-responsive disulfide crosslinkers, which served as a supplement to the environmental cellular antioxidant defense. The results showed that these disulfide-containing hydrogels were cleaved under reducing environments, and therefore the thiolated degradation products varied human umbilical vein endothelial cell (HUVEC) viability IC_{50} values across an order of magnitude [108].

12.3.2 Gene Delivery

Polyethylenimine (PEI) has been widely used nonviral gene vector and presents advantages over other polycations for its strong DNA condensation ability and intrinsic endosomolytic activity [109, 110]. However, high molecular weight of PEI can induce serious cytotoxicity, and strong packing of DNA in PEI/DNA complex. It therefore became a critical hurdle to the release of DNA inside the cytoplasm. Cleavable PEGylation was introduced to PEI for improved biostability, prolonged *in vivo* circulation time, and reduced the toxicity. For example, the chitosan oligosaccharide-based disulfide-containing PEI derivative, PEG-SS-COS-SS-PEI, was found to effectively condense DNA into small particles with improved buffering capacity (44%) compared with PEI1.8k (20%). *In vitro* study showed much lower cytotoxicity of PEGylated reduction-responsive copolymer, but high transfection efficiency as compared with the control branch of 25 KDa PEI [111]. Poly-L-lysine (PLL) was chosen as another regular cationic polypeptide for nonviral gene vector to incorporate disulfide-containing PEGylation to obtain the PEG-SS-PLL via a facile ROP. Due to the cleavable PEGylation, its transfection activity is serum resistant after gene complexation because of the PEG-shielding effect. However, the transfection activities of PLL50 in luciferase expression were significantly suppressed in the presence of 10% serum for both 293T and HeLa cell lines, which was presumably associated with the poor stability of PLL complexes in serum with positive surface charge [112].

Balakirev et al. developed cationic disulfide-containing amphiphiles based on the lipoic acid, which could condense the free DNA into homogeneous spherical particles in the oxidized state while swelled into DNA toroids with subsequent release of free DNA upon reduction conditions. *In vitro* transfection studies showed a several fold increase of transgene expressions compared with 1,2-dioleoyl-3-trimethylammonium-propane (DOTAP), which could be enhanced by attaching nucleus-targeting peptide to the amphiphile [113]. Yang et al. prepared a reducible star-shaped gene vector (POSS-(SS-PDMAE-MA)$_8$) with uniform structure and relatively narrow molecular weight distribution via ATRP of (2-dimethylamino) ethyl methacrylate (DMAEMA) from a POSS macroinitiator, which can effectively bind pDNA into uniform complexes with suitable particle size and zeta potential. Incorporation of disulfide bonds provided the POSS-(SS-PDMAEMA)$_8$ polymers with facile bioreducibility and high transfection efficiency (Figure 12.19) [115].

Park and coworkers developed disulfide-crosslinked HA nanogels via physically encapsulating green fluorescence protein (GFP) for the tumor-targeted intracellular delivery of siRNA [116]. These HA nanogels, possessing a size distribution from

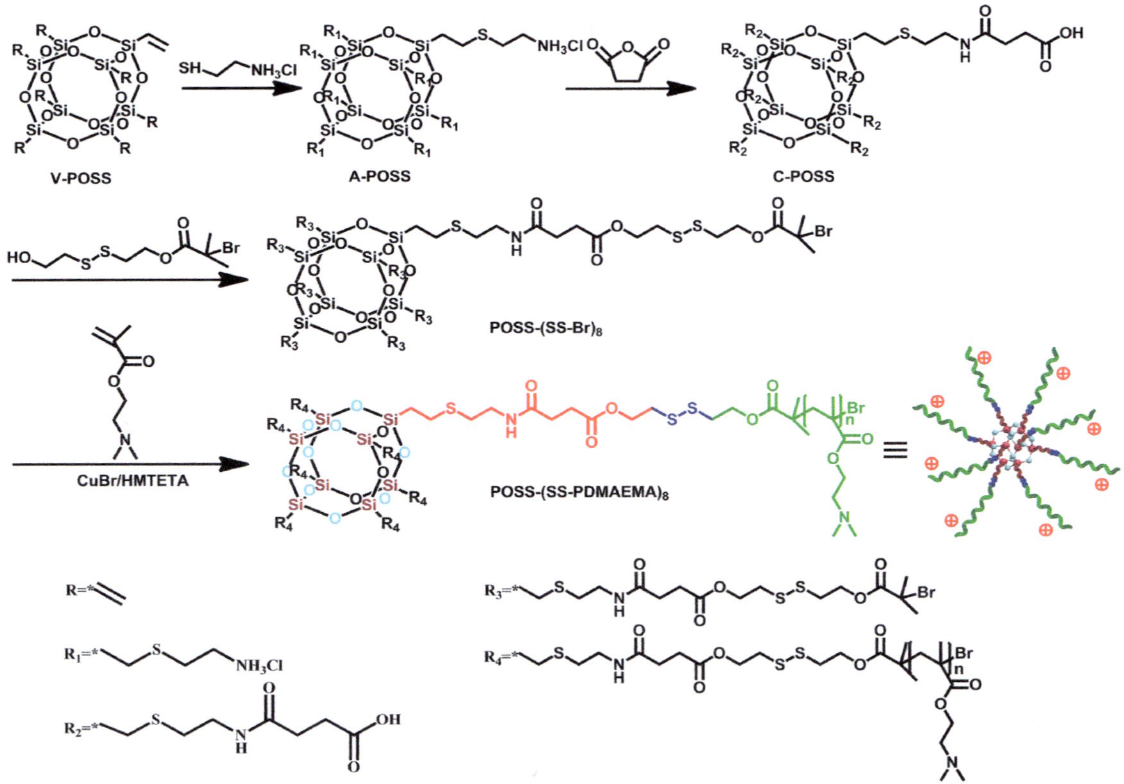

Figure 12.19 Synthetic processes of bioreducible POSS-based gene vectors via ATRP. Source: Adapted with permission from Ref. [114]. Copyright 2014, American Chemical Society.

200 to 500 nm and siRNA loading efficiency of approximately 50%, were prepared from thiol-derivatized HA via an inverse water-in-oil emulsion method, which followed by the self-crosslinking through the formation of disulfide linkages. HA nanogels effectively protected the encapsulated siRNA from enzymatic degradation. HA/siRNA nanogels were readily endocytosed by HCT-116 cells, which possess HA-specific CD44 receptors on their surface but not by CD44 deficient NIH-3T3 cells. Gene silencing studies showed that HA nanogels encapsulating GFP siRNA exerted a significant level of GFP gene inhibition in HCT-116 cells in both 10% serum and non-serum conditions. Park and coworkers conjugated vascular endothelial growth factor (VEGF) siRNA to PEG via a disulfide linkage, which formed polyioncomplex (PIC) micelles by condensing with PEI or KALA (acationic fusogenic peptide). The PIC micelles showed greater stability than naked VEGF siRNA against enzymatic degradation. The intact form of VEGF siRNA was released under a reductive condition mimicking the cytosolic environment by cleavage of the disulfide linkage. Remarkably, VEGF siRNA-PEG/PEI PIC micelles effectively silenced VEGF gene expression in prostate carcinoma cells for up to 96.5% under an optimized formulation condition. Importantly, it should be noted that high level of VEGF gene silencing was also achieved in the presence of 10% FBS [117, 118]. In addition, Bernkop-Schnurch and coworkers prepared chitosan-thiobutylamidine by treating the chitosan with 2-iminothiolane and forming nano-sized coacervates with pDNA, which were prone to disaggregate in a reducing environment, thus quickly releasing pDNA [119].

12.3.3 Biomedical Imaging

Intelligent polymeric nanoparticles provide great potential for accurate cancer theranostics, and therefore the drug-loaded NPs are promising candidates for improving the efficiency of cancer diagnosis and therapy. Wang and coworkers designed a multi-functional polymeric micelle based on a folate (FA)-modified triblock copolymer P (TPF-co-ibuprofen (Ibup))-PAEMA-PPEG as a smart drug delivery system. This copolymer possessed the unique properties of anti-tumor and anti-inflammatory abilities, charge conversion; FA-mediated enhanced cellular uptake and AIE active two-photon bioimaging. After high-efficient encapsulation of anticancer drug CUR, these CUR-loaded micelles exhibited good stability in a physiological environment. The acidic tumor environment would rapidly trigger charge conversion while the GSH in the tumor cell cytoplasm could cleavage the disulfide linkages, leading to rapid release of Ibup and further triggered CUR release. In addition, *in vitro* and *in vivo* experiments found that deep tissue imaging of these AIE micelles were observed with the aid of the two-photon cell. Therefore, these disulfide-containing multi-functional micelles showed great potential for simultaneous cancer therapy, inflammation restraint, and AIE active two-photon bioimaging [120]. In addition, they also prepared the gemcitabine (GEM)-conjugated reductive-responsive prodrug micelles based on a PMPC-*b*-P (DEMA-*co*-SS-GEM-*co*-TPMA) copolymer via GSH-cleavable disulfide linkages, which possessed the advantages of pH-responsive, charge-conventional, and a

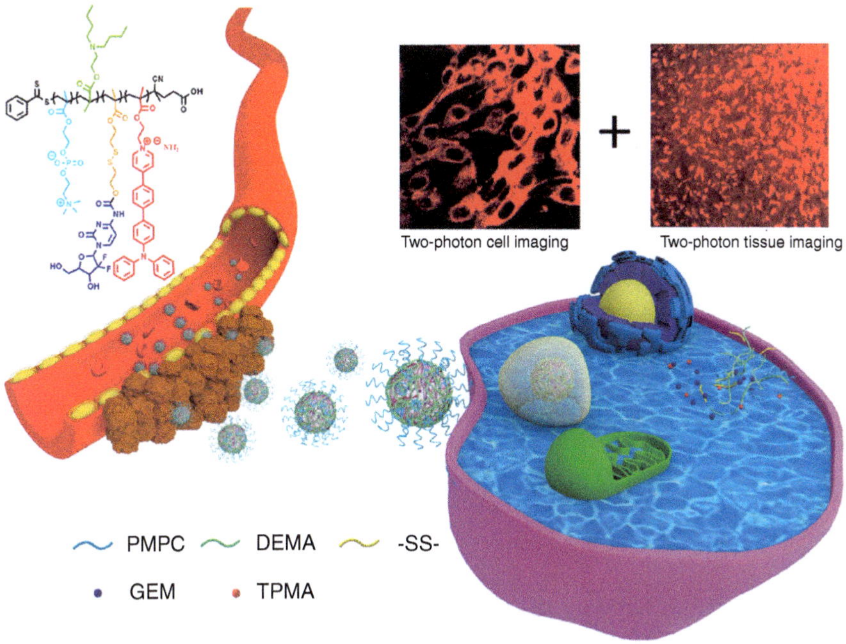

Figure 12.20 Schematic diagram of GEM-conjugated micelles based on PMPC-*b*-P (DEMA-*co*-SS-GEM-*co*-TPMA) copolymer for pH- and GSH-triggered drug delivery and AIE-active two-photon bioimaging. Source: Adapted with permission from Ref. [121]. Copyright 2019, American Chemical Society.

two-photon absorbing AIE fluorescence probe capacities for the achievement of lysosome-targeted drug release and bioimaging. Especially, the high-quality AIE-active two-photon imaging was well confirmed by cell and deep-tissue imaging, and tracing the distribution of drug carriers within the tumors, which would be a potential candidate for cancer diagnosis and therapy (Figure 12.20) [121].

12.3.4 Drug Conjugates

To convert the therapeutic agent into an inactive but more stable prodrug that reverts back to the pharmacologically active moiety triggered by unique biological stimuli, there are various prodrug designs and syntheses that have been developed with the objective to augment aqueous solubility, increased target selectively and alter unfavorable pharmacokinetic properties. Recent reports demonstrate that combination of anticancer small molecule drugs into disulfide-containing linear or branched macromolecules can effectively improve the tumor therapy because of longer circulation time, increased retention at various tumor sites, and controlled drug delivery. For example, Li et al. designed a prodrug polymer of CPT-SS-PEG-SS-CPT by reacting NH_2-SS-PEG-SS-NH_2 with succinate-activated CPT, in which the prodrug spontaneously arranged into the reduction-responsive

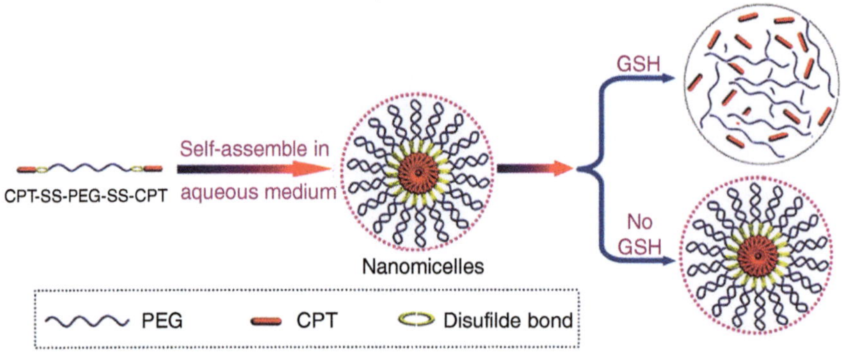

Figure 12.21 Schematic outline of the predicted self-assembly behavior of CPT-SS-PEG-SS-CPT under aqueous conditions and its drug release behavior. Source: Adapted with permission from Ref. [122]. Copyright 2011, Royal Society of Chemistry.

micelles in an aqueous solution. Under tumor-relevant reductive conditions, the cleavage of disulfide bonds initiated the micellar rearrangement associated with rapid release of therapeutic payload. Cell proliferation assays demonstrated pharmacological efficacy of CPT released from drug conjugates under high GSH concentration for effective therapy of HepG2 cells (Figure 12.21) [122]. Feng and coworkers developed a reduction-responsive polymer prodrug, polyethylene glycol-poly(beta-benzyl-L-aspartate) (PEG-PBLA)-SS-paclitaxel (PPSP). The PPSP micelles could significantly release the PTX compared with the mPEG-PBLA-CC-PTX (PPCP) micelles in the tumor physiological environment, displaying the excellent reduction-responsive behavior and improved anti-tumor activity on HepG2, MCF-7 and HL-7702 cell proliferations [123].

12.3.5 Self-healing/Self-repairing

The property of a material to recover from any kind of damage automatically without any peripheral intercession is recognized as "self-healing (self/autonomic repairing) property." [124]. Recently, more and more researchers had developed the various man-made and natural occurrences of self-healing materials. Among different self-healing mechanisms, the dynamic chemistry based on reversible covalent bonds, including Diels–Alders reaction, disulfides, acylhydrazone, and boronic ester bonds, had been employed to design self-healing materials [125–134]. Compared with other dynamic bonds, disulfide linkages were extremely versatile to fabricate self-healing polymers because they featured various self-healing mechanisms like thiol-disulfide exchange, reduction–oxidation, and radical-mediated disulfide fragmentation. Therefore, polymers with intrinsic self-healing abilities were currently of great interest owing to their broad applications as protective coatings, sensors, and healthcare [135–141]. For example, Yang and coworkers developed the self-healing coatings via surface-initiated thiol-ene photo-polymerization of different monomers including

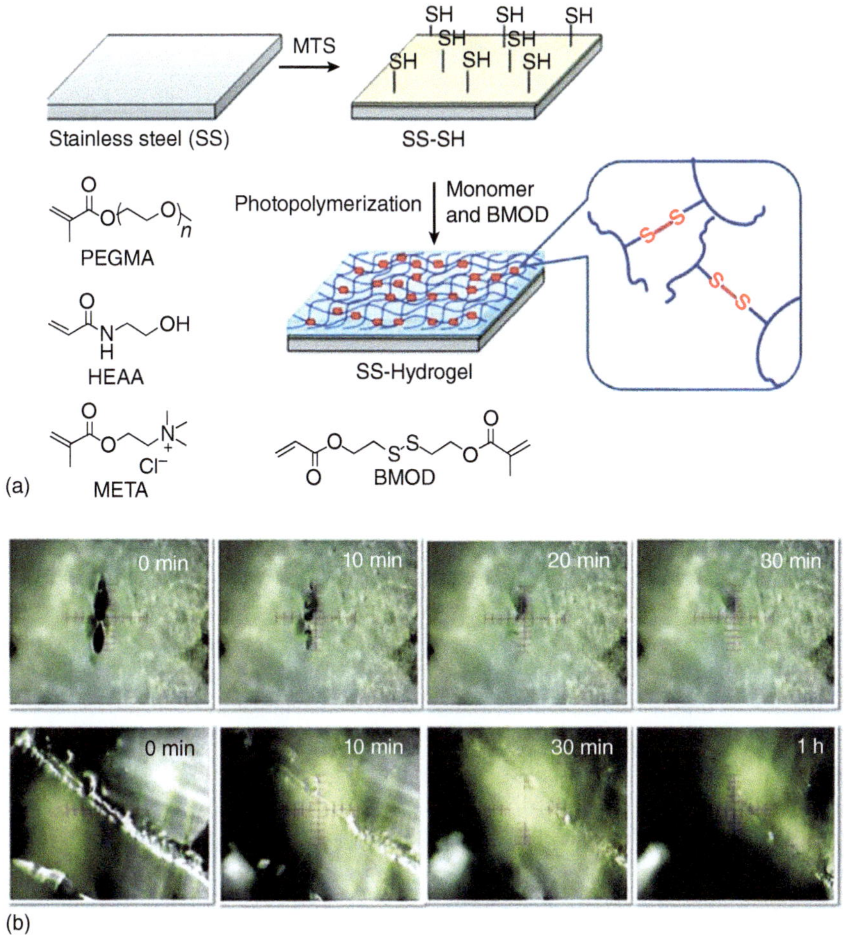

Figure 12.22 (a) Schematic illustration of the preparation of hydrogel-functionalized SS surfaces via surface-initiated thiol-ene photopolymerization. (b) Self-healing behavior of the hydrogel coatings under an optical microscope of a step profiler, including poly(PEGMA) hydrogel coating (top row) and poly(HEAA-*co*-META) hydrogel coating (bottom row). Source: Adapted with permission from Ref. [130]. Copyright 2015, The Royal Society of Chemistry.

poly(ethylene glycol) methyl ether methacrylate (PEGMA), *N*-hydroxyethyl acrylamide (HEAA), and 2-(methacryloyloxy) ethyl trimethylammonium chloride (META) with the aid of crosslinker of bis(2-methacryloyl) oxyethyl disulfide (BMOD). The self-healing behavior of the disulfide-containing functional surfaces was clearly observed with optical microscopy. A small cut of disulfide-containing poly(PEGMA) hydrogel could be well healed within 0.5 hour, while poly(HEAA-co-META) hydrogel coating with about 50 μm could be fully healed within one hour, presenting the excellent self-healing properties (Figure 12.22) [130].

12.4 Summary and Perspectives

In this chapter, we have attempted to provide an overview of recent advances of disulfide-containing polymeric fabrication, materials science, and biological application. Critically, we tried to underline the attractiveness of reduction-responsive behaviors in great biomedicine applications due to the broad variety on the disulfide-containing materials like linear polymers, dendritic polymers, polypeptides/proteins, and crosslinking polymers. Therefore, we intended to give a glimpse of the important physical–chemical strategies on the intelligent fabrication of reduction-responsive nanoparticles like micelles, vesicles, DNA/siRNA-containing nanoparticles, micro/nanogels, natural polysaccharide/protein hydrogels, etc. The literatures discussed in this chapter are by no means complete. We only selected and highlighted some typical examples to raise the reader's interest and awareness about the reduction-responsive disulfide-containing polymers for biomedical applications.

The past decade has witnessed the remarkable advancement of diverse disulfide-containing biomaterials for efficient treatment of various cancers and tumors, especially for triggering active intracellular delivery of hydrophilic and/or hydrophobic drugs and varying their biotherapeutics of proteins, DNA, and siRNA. These reduction-responsive degradable materials meet the conflicting requirements of an ideal delivery system, i.e. high stability in circulation while rapid degradation in tumor tissues and/or inside targeted cells, achieving extraordinary therapeutic effects and minimal carriers-related *in vitro* and *in vivo* cytotoxicity for remarkedly expanding their biomedical applications.

Even though, it should be noted that reduction-responsive delivery systems are still at its infancy. For example, although the disulfide-containing polymers possessed good stability in the non-reductive environment, many structurally asymmetric disulfide monomers and macromolecules (e.g. A-SS-B) were extremely unstable to quickly reform into A-SS-A and B-SS-B monomers even in mild conditions, such as weak alkaline, light irradiation, room temperature, polar solvents, etc., which had be significantly limited in fabrication of polymeric architectures and corresponding applications. Whereas some other disulfide-containing unsymmetrical polymers or natural protein/amino acids that could maintain their structural integrity and stability even in harsh environments. Unfortunately, so far there is no relatively authoritative explanation. With this understanding, we believe that the topologies of the polymers and the location of disulfide bond in the polymer played roles in the rate of responsiveness, which may be due to the exposure time and concentration of reductive agents to the disulfide bonds. For example, compared with the linear polymers, the disulfide-containing hyperbranched polymers with disulfide bonds outside the backbone were more accessible to the reductive agents (e. g. DTT). These DTT molecules could easily contact and break up the disulfide bonds of dendritic polymers; in this case, the hyperbranched topology with more outside disulfide bonds could be quickly responded and easily destroyed than that of liner polymers with interior disulfides. In addition, on account of the instability of disulfide bonds in the asymmetric monomers or macromolecules, the hyperbranched or irregular polymers exhibited the lower stability even in the mild conditions, let alone the exposure of alkaline solutions or UV irradiation. From this perspective, it was also concluded that the responsive rate of disulfide was significantly relative to the polymeric topologies and location of disulfide bond within the polymers. In addition, we know that in order to achieve the clinical applications, multifunctional carriers are desired to simultaneously combine reductive responsivity, good biocompatibility, long circulation period, effective cellular, and/or intracellular targetability. Thus, it is necessary to systemically, carefully, and extensively evaluate the preclinical requirements before going clinics with these systems, such as the pharmacokinetics, efficacy, and toxicity of drug-loaded nanoparticles after repeated administration. For example, it has been known that tumors have very complicated microenvironments that should be considered and well addressed, including special biodistribution, multidrug resistance, and enhanced drug release in target cells. Also, mechanistic studies are also required to deeply investigate where and how the reduction exactly takes place inside the cells, since it is still not clear how the microenvironment of influences the reductive degradation kinetics of disulfide-containing materials. In addition, the contradictory between the multiple functions with simple structures and less cost with specific architectures are awaited to be urgently solved for further rational design of more advanced delivery systems.

It should be also further noted that polymer scientists have now developed a variety of potential structures possessing excellent properties *in vitro*, but in fact we are quite far from the effective clinical applications *in vivo* for a long period of time. For example, although there is enough *in vitro* evidence for the utility and functionality of disulfide-containing biomaterials in drug delivery, the *in vivo* proof is less well-formed until now. There are fewer investigations of the usage of disulfide-containing hybrid materials in animal models, and in many cases *in vivo* testing is a minor element that be given more concerned with material design and synthesis, because there is still a lack of proof to provide a clear and unambiguous evidence for the advantages of incorporating a disulfide linkage. As such, while GSH-responsive polymers have been clearly demonstrated *in vitro*, but the jury remains out on whether this can translate to clear clinical benefits when the same materials are applied to animals or humans. More importantly, how GSH-responsive materials might impact on the extracellular GSH concentrations is also worthy of detailed investigations. It is one thing to say that nanoparticles are stable in the bloodstream, but are they impacting the blood chemistry in possibly deleterious ways? These viewpoints are generally neglected for many years by researchers in studies where the focus in firmly on delivery as opposed to the potential biochemical impact of the materials. Consequently, it is probably time for polymer chemists to collaborate more knowledges and activities with scientists of other fields to evaluate the performance of these disulfide-containing polymers *in vivo* to acquire important and direct feedback about the structure-function relationships, and further be applied to improve the current biomaterials, as well as to synthesize the tailor-made nanocarriers for the drug and gene delivery.

This exciting goal will hopefully be achieved by the scientific community with the lessons learnt from the literature through this chapter. Therefore, we are strongly convinced that with the help of continuous developments of rational polymeric design and exquisite adjustment of their physicochemical properties, more advanced multi-responsive nanoparticles with optimized characteristics will eventually be generated to achieve the greater control, and higher efficacy for various biomedical applications. While newly multi-responsive polymer compositions are hopefully to be continually developed and the ability to obtain macromolecules with topological complexity is excitingly expanding in the next generation of smart biomaterials.

References

1 Lu, Y., Aimetti, A.A., Langer, R., and Gu, Z. (2017). *Nat. Rev. Mater.* 1: 16075.
2 Chung, W.J., Griebel, J.J., Kim, E.T. et al. (2013). *Nat. Chem.* 5: 518.
3 Singh, V.V., Kaufmann, K., de Avila, B.E.F. et al. (2016). *Adv. Funct. Mater.* 26: 6270.
4 Wu, G., Fang, Y.Z., Yang, S. et al. (2004). *J. Nutr.* 134: 489.
5 Napoli, A., Valentini, M., Tirelli, N. et al. (2004). *Nat. Mater.* 3: 183.
6 Smith, C.V., Jones, D.P., Guenthner, T.M. et al. (1996). *Toxicol. Appl. Pharmacol.* 140: 1.
7 Li, D.W., Bu, Y.Z., Zhang, L.N. et al. (2016). *Biomacromolecules* 17: 291.
8 Christina, L., Jelkmann, M., and Bernkop-Schnurch, A. (2019). *Adv. Drug Deliv. Rev.* https://doi.org/10.1016/j.addr.2019.04.007.
9 Mahmood, A. and Bernkop-Schnurch, A. (2019). *Adv. Drug Deliv. Rev.* 142: 91.
10 Ikada, Y. and Tsuji, H. (2000). *Macromol. Rapid Commun.* 21: 117.
11 Siegwart, D.J., Oh, J.K., and Matyjaszewski, K. (2012). *Prog. Polym. Sci.* 37: 18.
12 Boyer, C., Bulmus, V., Davis, T.P. et al. (2009). *Chem. Rev.* 109: 5402.
13 Oh, J.K. (2019). *Polym. Chem.* 10: 1554.
14 Meng, F.H., Hennink, W.E., and Zhong, Z.Y. (2009). *Biomaterials* 30: 2180.
15 Chen, J., Qiu, X.Z., Ouyang, J. et al. (2011). *Biomacromolecules* 12: 3601.
16 Sourkohi, B.K., Cunningham, A., Zhang, Q., and Oh, J.K. (2011). *Biomacromolecules* 12: 3819.
17 Li, L.H., Jiang, X.L., and Zhuo, R.X. (2009). *J. Polym. Sci. Part A: Polym. Chem.* 47: 5989.
18 Ryu, J.H., Roy, R., Ventura, J., and Thayumanavan, S. (2010). *Langmuir* 26: 7086.
19 Liu, J.Y., Pang, Y., Huang, W. et al. (2011). *Biomacromolecules* 12: 1567.
20 Liu, J.Y., Pang, Y., Huang, W. et al. (2011). *Biomacromolecules* 12: 2407.
21 Cheng, W.R., Wu, D.C., and Liu, Y. (2016). *Biomacromolecules* 17: 3115.
22 Wang, B., Li, C., Yang, L.J. et al. (2019). *J. Mater. Chem. B* 7: 3846.
23 Sun, H.J., Zhang, S.D., and Percec, V. (2015). *Chem. Soc. Rev.* 44: 3900.
24 Wang, X., Yang, Y.Y., Gao, P.Y. et al. (2014). *Chem. Commun.* 50: 6126.

25 Andren, O.C.J., Fernandes, A.P., and Malkoch, M. (2017). *J. Am. Chem. Soc.* 139: 17660.

26 McKenzie, D.L., Kwok, K.Y., and Rice, K.G. (2000). *J. Biol. Chem.* 275: 9970.

27 McKenzie, D.L., Smiley, E., Kwok, K.Y., and Rice, K.G. (2000). *Bioconjug. Chem.* 11: 901.

28 Fu, Y., Jangc, M.S., Wu, T.P. et al. (2019). *Carbohydr. Polym.* 224: 115174.

29 Segura, T. and Hubbell, J.A. (2007). *Bioconjug. Chem.* 18: 736.

30 Zhang, S.F., Wright, J.E.I., Bansal, G. et al. (2005). *Biomacromolecules* 6: 2800.

31 Chen, W., Zhong, P., Meng, F.H. et al. (2013). *J. Control. Release* 169: 171.

32 Wang, X.Y., Sun, H.L., Meng, F.H. et al. (2013). *Biomacromolecules* 14: 2873.

33 Li, S.X., Liu, L., Zhang, L.J. et al. (2016). *Polym. Chem.* 7: 5113.

34 Shi, C., Guo, X., Qu, Q. et al. (2014). *Biomaterials* 35: 8711.

35 Tang, Z., Zhang, L., Wang, Y. et al. (2016). *Acta Biomater.* 42: 232.

36 Hu, K., Zhou, H., Liu, Y. et al. (2015). *Nanoscale* 7: 8607.

37 Chandrashekaran, R.B.K.C.V., Cheng, B., Chen, H. et al. (2014, 1897). *Mol. Pharm.* 11.

38 Chen, J., Shi, M., Liu, P. et al. (2014). *Biomaterials* 35: 1240.

39 Chen, W., Zou, Y., Jia, J. et al. (2013). *Macromolecules* 46: 699.

40 Zhang, W.Y., Zheng, W.Y., Mao, M.W., and Yang, Y. (2014). *Biotechnol. Bioeng.* 111: 2520.

41 Huang, L.J., Chiang, C.W., Chen, S.L. et al. (2019). *J. Food Drug Anal.* 27: 531.

42 Chan, N., An, S.Y., and Oh, J.K. (2014). *Polym. Chem.* 5: 1637.

43 Lu, W., Wang, X., Cheng, R. et al. (2015). *Polym. Chem.* 6: 6001.

44 Sun, H.L., Cheng, R., Deng, C. et al. (2015). *Biomacromolecules* 16: 597.

45 Wang, X., Zhang, J., Cheng, R. et al. (2016). *Biomacromolecules* 17: 882.

46 Zhang, P., Zhang, H.Y., He, W.X. et al. (2016). *Biomacromolecules* 17: 1621.

47 Sun, Y., Zou, W., Bian, S. et al. (2013). *Biomaterials* 34: 6818.

48 Zhuang, Y., Deng, H., Su, Y. et al. (2016). *Biomacromolecules* 17: 2050.

49 Cheng, R., Feng, F., Meng, F. et al. (2011). *J. Control. Release* 152: 2.

50 Sun, H.L., Meng, F.H., Cheng, R. et al. (2013). *Expert Opin. Drug Del.* 10: 1109.

51 Guo, X.S., Cheng, Y., Zhao, X.T. et al. (2018). *J. Nanobiotechnol.* 16: 74.

52 Quinn, J.F., Whittaker, M.R., and Davis, T.P. (2017). *Polym. Chem.* 8: 97.

53 Li, J., Huo, M., Wang, J. et al. (2012). *Biomaterials* 33: 2310.

54 Wei, H., Schellinger, J.G., Chu, D.S., and Pun, S.H. (2012). *J. Am. Chem. Soc.* 134: 16554.

55 Wang, L.H., Wu, D.C., Xu, H.X., and You, Y.Z. (2016). *Angew. Chem. Int. Ed.* 55: 755.

56 Wei, H., Quan, C.Y., Chang, C. et al. (2010). *J. Phys. Chem. B* 114: 5309.

57 O'Reilly, R.K., Hawker, C.J., and Wooley, K.L. (2006). *Chem. Soc. Rev.* 35: 1068.

58 Chen, W., Zheng, M., Meng, F.H. et al. (2013). *Biomacromolecules* 14: 1214.

59 Hu, X., Tian, J., Liu, T. et al. (2013). *Macromolecules* 46: 6243.

60 Jin, S., Wan, J., Meng, L. et al. (2015). *ACS Appl. Mater. Interfaces* 7: 19843.

61 Zhan, Y., Goncalves, M., Yi, P. et al. (2015). *J. Mater. Chem. B* 3: 4221.

62 Du, A.W., Lu, H., and Stenzel, M. (2016). *Mol. Pharm.* 13: 3648.

63 Thurmond, K.B., Kowalewski, T., and Wooley, K.L. (1996). *J. Am. Chem. Soc.* 118: 7239.
64 Liu, S. and Armes, S.P. (2001). *J. Am. Chem. Soc.* 123: 9910.
65 Li, Y., Lokitz, B.S., and McCormick, C.L. (2005). *Macromolecules* 39: 81.
66 Zhang, S. and Zhao, Y. (2010). *J. Am. Chem. Soc.* 132: 10642.
67 Wang, X., Gao, P.Y., Yang, Y.Y. et al. (2018). *Nat. Commun.* 9: 2772.
68 Qu, Y., Chu, B.Y., Wei, X.W. et al. (2019). *J. Control. Release* 296: 93.
69 Elkassih, S.A., Kos, P., Xiong, H., and Siegwart, D.J. (2019). *Biomater. Sci.* 7: 607.
70 Zhang, J., Yang, F., Shen, H., and Wu, D.C. (2012). *ACS Macro Lett.* 1: 1295.
71 Jiang, Y., Chen, J., Deng, C. et al. (2014). *Biomaterials* 35: 4969.
72 Yang, Y.Y., Wang, X., Yang, F. et al. (2018). *Adv. Mater.* 30: 1707071.
73 Gajendiran, M., Rhee, J.S., and Kim, K. (2018). *Tissue Eng. Part. B Rev.* 24: 66.
74 Boyd, D.A. (2016). *Angew. Chem. Int. Ed.* 55: 15486.
75 Machado, T.O., Sayer, C., and Araujo, P.H.H. (2017). *Eur. Polym. J.* 86: 200.
76 Azcune, I. and Odriozola, I. (2016). *Eur. Polym. J.* 84: 147.
77 Wang, H. and Heilshorn, S.C. (2015). *Adv. Mater.* 27: 3717.
78 Cao, Q.Q., Wang, X., and Wu, D.C. (2018). *Chin. J. Polym. Sci.* 36: 8.
79 Cohen, Y., Ramon, O., Kopelman, I.J., and Mizrahi, S. (1992). *J. Polym. Sci. B Polym. Phys.* 30: 1055.
80 Hsu, T., Ma, D.S., and Cohen, C. (1983). *Polymer* 24: 1273.
81 Wu, D.C., Loh, X.J., Wu, Y.L. et al. (2010). *J. Am. Chem. Soc.* 132: 15140.
82 You, Y.Z., Yu, Z.Q., Cui, M.M., and Hong, C.Y. (2010). *Angew. Chem. Int. Ed.* 49: 1099.
83 Wang, X., Li, D., Yang, F. et al. (2013). *Polym. Chem.* 4: 4596.
84 Li, L.Y., Song, C.F., Jennings, M., and Thayumanavan, S. (2015). *Chem. Commun.* 51: 1425.
85 Wang, L.L., Li, L., Wang, X. et al. (2016). *Polym. Chem.* 7: 1429.
86 Bian, S., He, M., Sui, J. et al. (2016). *Colloids Surf. B: Biointerfaces* 140: 392.
87 Cao, Y., Yao, Y.Q., Li, Y. et al. (2019). *J. Colloid Interface Sci.* 544: 121.
88 Dong, H.Q., Tang, M., Li, Y. et al. (2015). *Nanomedicine* 10: 1941.
89 Su, Y., Hu, Y., Du, Y. et al. (2015). *Mol. Pharm.* 12: 1193.
90 Kuppusamy, P., Li, H., Ilangovan, G. et al. (2002). *Cancer Res.* 62: 307.
91 Lee, S.H., Gupta, M.K., Bang, J.B. et al. (2013). *Adv. Healthcare Mater.* 2: 908.
92 Li, Y.T., Lokitz, B.S., Armes, S.P., and McCormick, C.L. (2006). *Macromolecules* 39: 2726.
93 Zhang, J.Y., Jiang, X., Zhang, Y.F. et al. (2007). *Macromolecules* 40: 9125.
94 Nolan, D., Darcy, R., and Ravoo, B.J. (2003). *Langmuir* 19: 4469.
95 Chen, J.J., Wu, M., Veroniaina, H. et al. (2019). *Polym. Chem.* 10: 4031.
96 Wang, Z.G., Zhang, R.X., Zhang, C. et al. (2019). *J. Agric. Food Chem.* 67: 887.
97 Hoffman, A.S. (2002). *Adv. Drug Deliv. Rev.* 54: 3.
98 Ahadian, S., Sadeghian, R.B., Salehi, S. et al. (2015). *Bioconjug. Chem.* 26: 198.
99 Peppas, N.A., Hilt, J.Z., Khademhosseini, A., and Langer, R. (2006). *Adv. Mater.* 18: 1345.

100 Caccavo, D., Cascone, S., Lamberti, G., and Barba, A.A. (2015). *Mol. Pharm.* 12: 474.

101 Tomatsu, I., Hashidzume, A., and Harada, A. (2006). *Macromol. Rapid Commun.* 27: 238.

102 Altinbasak, I., Sanyal, R., and Sanyal, A. (2016). *RSC Adv.* 6: 74757.

103 Dutta, S., Samanta, P., and Dhara, D. (2016). *Int. J. Biol. Macromol.* 87: 92.

104 Zhang, R., Li, X., He, K. et al. (2017). *Polym. Adv. Technol.* 28: 1759.

105 Shen, Y., Li, X., Huang, Y. et al. (2016). *Macromol. Res.* 24: 602.

106 Wang, X., Wang, J., Yang, Y.Y. et al. (2017). *Polym. Chem.* 8: 3901.

107 Wu, C., Li, R., Yin, Y. et al. (2017). *Mater. Sci. Eng. C Mater. Biol. Appl.* 76: 196.

108 Lakes, A.L., Jordan, C.T., Gupta, P. et al. (2018). *Acta Biomater.* 68: 178.

109 Cai, X., Zhu, H., Dong, H. et al. (2014). *Adv. Healthcare. Mater.* 3: 1818.

110 Zhao, F., Yin, H., Zhang, Z.X., and Li, J. (2013). *Biomacromolecules* 14: 476.

111 Jia, L.J., Li, Z.Y., Zhang, D.R. et al. (2013). *Polym. Chem.* 4: 156.

112 Cai, X.J., Dong, H.Q., Xia, W.J. et al. (2011). *J. Mater. Chem.* 21: 14639.

113 Balakirev, M., Schoehn, G., and Chroboczek, J. (2000). *Chem. Biol.* 7: 813.

114 Shi, J., Luo, Y., Wei, H. et al. (2014). Modified two-step deposition method for high-efficiency $TiO_2/CH_3NH_3PbI_3$ heterojunction solar cells. *ACS Appl. Mater. Interfaces* 6 (12): 9711–9718.

115 Yang, Y.Y., Wang, X., Hu, Y. et al. (2014). *ACS Appl. Mater. Interfaces* 6: 1044.

116 Lee, H., Mok, H., Lee, S. et al. (2007). *J. Control. Release* 119: 245.

117 Kim, S.H., Jeong, J.H., Lee, S.H. et al. (2006). *J. Control. Release* 116: 123.

118 Lee, S.H., Kim, S.H., and Park, T.G. (2007). *Biochem. Biophys. Res. Commun.* 357: 511.

119 Schmitz, T., Bravo-Osuna, I., Vauthier, C. et al. (2007). *Biomaterials* 28: 524.

120 Hu, J., Zhuang, W.H., Ma, B.X. et al. (2019). *J. Mater. Chem. B* 7: 3894.

121 Yu, T., Zhuang, W.H., Su, X. et al. (2019). *Bioconjug. Chem.* 30: 2075.

122 Li, X.Q., Wen, H.Y., Dong, H.Q. et al. (2011). *Chem. Commun.* 47: 8647.

123 Chang, S., Wang, Y.F., Zhang, T.Y. et al. (2019). *Front. Oncol.* 9: 823.

124 Bekas, D.G., Tsirka, K., Baltzis, D., and Paipetis, A.S. (2016). *Compos. Part B-Eng.* 87: 92.

125 Gandini, A., Silvestre, A., and Coelho, D. (2013). *Polym. Chem.* 4: 1364.

126 Liu, Y.L. and Chuo, T.W. (2013). *Polym. Chem.* 4: 2194.

127 Zhang, J., Niu, Y., Huang, C. et al. (2012). *Polym. Chem.* 3: 1390.

128 Canadell, J., Goossens, H., and Klumperman, B. (2011). *Macromolecules* 44: 2536.

129 Pepels, M., Filot, I., Klumperman, B., and Goossens, H. (2013). *Polym. Chem.* 4: 4955.

130 Yang, W.J., Tao, X., Zhao, T. et al. (2015). *Polym. Chem.* 6: 7027.

131 Liu, F., Li, F., Deng, G. et al. (2012). *Macromolecules* 45: 1636.

132 Deng, G., Tang, C., Li, F. et al. (2010). *Macromolecules* 43: 1191.

133 He, L., Szopinski, D., Wu, Y. et al. (2015). *ACS Macro Lett.* 4: 673.

134 Cromwell, O.R., Chung, J., and Guan, Z. (2015). *J. Am. Chem. Soc.* 137: 6492.

135 Luo, X. and Mather, P.T. (2013). *ACS Macro Lett.* 2: 152.

136 Li, Y., Li, L., and Sun, J.Q. (2010). *Angew. Chem. Int. Ed.* 49: 6129.

137 Bode, S., Zedler, L., Schacher, F.H. et al. (2013). *Adv. Mater.* 25: 1634.
138 Tee, B.C., Wang, C., Allen, R., and Bao, Z. (2012). *Nat. Nanotechnol.* 7: 825.
139 Huynh, T.P. and Haick, H. (2016). *Adv. Mater.* 28: 138.
140 Cao, J., Lu, C.H., Zhuang, J. et al. (2017). *Angew. Chem. Int. Ed.* 56: 8795.
141 Kang, J., Son, D., Wang, G.J.N. et al. (2018). *Adv. Mater.* 30: 1706846.

Index

Sulfur-Containing Polymers: From Synthesis to Functional Materials,
First Edition. Edited by Xing-Hong Zhang and Patrick Theato.

h